Quantum Entanglement in High Energy Physics

Edited by Oliver K. Baker

Published in London, United Kingdom

Quantum Entanglement in High Energy Physics
http://dx.doi.org/10.5772/intechopen.111219
Edited by Oliver K. Baker

Contributors
A. Pedro Aguiar, Alexander Machavariani, Ali Saif M. Hassan, Christopher G. Timpson, Felix A. Buot, Ismail Zahed, Karin Marie Fierke, Kisalaya Chakrabarti, Maciej A. Nowak, Mohamed A. Shukri, Mohammed Boukidi, Mohammed Nadir, Nahid Binandeh Dehaghani, Nancy Lynn Bowen, Oliver K. Baker, Rachid Benbrik, Rafal Wisniewski, Shrobona Bagchi, Waleed S. A. Hasan, Yizhuang Liu, Yoram Kirsh

First published in London, United Kingdom, 2024 by IntechOpen
IntechOpen is the global imprint of INTECHOPEN LIMITED, registered in England and Wales, registration number: 11086078, 167-169 Great Portland Street, London, W1W 5PF, United Kingdom

British Library Cataloguing-in-Publication Data
A catalogue record for this book is available from the British Library

Additional hard and PDF copies can be obtained from orders@intechopen.com

Quantum Entanglement in High Energy Physics
Edited by Oliver K. Baker
p. cm.
Print ISBN 978-0-85466-083-4
Online ISBN 978-0-85466-082-7
eBook (PDF) ISBN 978-0-85466-084-1

For EU product safety concerns:
IN TECH d.o.o., Prolaz Marije Krucifikse Kozulić 3, 51000 Rijeka, Croatia,
info@intechopen.com or visit our website at intechopen.com.

Meet the editor

Oliver K. Baker, the D. Allan Bromley Physics Professor at Yale University, conducts research in experimental particle physics. His leadership role in the construction of the ATLAS Transition Radiation Tracker (TRT), and use of machine learning were essential in the Higgs boson discovery. His current research in quantum entanglement, quantum computing, entanglement entropy, and searches for phenomena that are beyond the Standard Model of Particle Physics uses Higgs production and decay. Baker earned a Ph.D. in Physics, an MS in Physics and Mathematics at Stanford University, and a BS in Physics from the Massachusetts Institute of Technology. He is a member of the American Academy of Arts and Sciences and the American Physical Society. He was the 2006 US Presidential appointee to the board of the National Medal of Science.

Access an insightful interview with the editor by scanning this QR code.

Contents

Preface

Research in the field of quantum information science can include several subfields in High Energy Physics (HEP), including quantum entanglement, entanglement entropy, Bell's inequality violation studies, and quantum computing along with quantum algorithm performance, quantum communications and quantum information and processing, quantum cryptography, and quantum sensors. There may be, overall, additional subfields that exist currently, or that may surface in the near future as the field grows in person-power and technology. Given that the research topics will now include quantum information science at high energies – elementary particle masses of order tens of GeV/c2 and kinetic energies or momenta that may be measured in units of sub-TeV/c – a supplementary challenge that includes determination of relativistic effects in experimental measurements and theoretical definitions of these somewhat new concepts may ensue. It is also clear from the accentuation given in the Introductory Chapter, "Quantum Entanglement at High Energies – Experiment, Theory, and Interdisciplinarity", that this study of physical phenomena in this way, at this level, requires that the theoretical descriptions, calculations, and results must be correct and complete, and the physical reality that is acquired and measured experimentally at high energies and masses must have a match in the theoretical culmination. Interdisciplinarity between theorists and experimentalists in physics, as well as with colleagues in other fields (mathematics, computing, social sciences, etc.) may very likely be essential to success.

The book commences with an introductory chapter that provides readers with an overview of the subject matter.

In Chapter 2, Prof. Shrobona Bagchi of the Korea Institute of Science and Technology develops and then summarizes how fundamentals in the understanding of natural phenomena with robust mathematical construct and subsequent reproducible experimental verifications have been some of the strongest pillars of physics to date.

In Chapter 3, Prof. Felix A. Buot of the University of San Carlos, explains the inverter-chain link viewpoint, the entanglement of two bare qubits, entangled qubits versus emergent qubits, the difference between mixed and pure states showing calculations, and much more.

In Chapter 4, Prof. Yoram Kirsh of the Open University of Israel describes an analysis that gives rise to an additional possibility concerning the tension between quantum entanglement in Quantum Mechanics and information transfer in Special Relativity with regards to cause and effect, are very deserving of such a test.

In Chapter 5, Prof. Rachid Benbrik and Prof. Mohammed Boukidi of Cadi Ayyad University, Marrakech, Morocco, present an intriguing argument supporting experimental vector-like quark searches that if discovered (at the Large Hadron Collider) would be a major breakthrough in particle physics.

In Chapter 6, Prof. Nancy Lynn Bowen of Colorado Mountain College explains that previous nuclear theories have ignored or lessened the electromagnetic force, considering only Coulomb forces between protons, whereas such assumptions are not valid. This leads to potential new rules in quantum mechanics inside the nucleus and nucleon.

In Chapter 7, Professors Nahid Binandeh Dehaghani, A. Pedro Aguiar from the University of Porto, and Rafal Wisniewski from Aalborg University provide an in-depth analysis of properties and phenomena associated with neutral K-mesons, and decoherence effects on entangled kaons.

In Chapter 8, Prof. Christopher G. Timpson of the University of Oxford provides an intriguing review with new insights into the meaning and significance of Bell inequalities violation as tests of whether the world is not locally causal, or the extent to which the criteria for providing these tests would be satisfied in high-energy experiments, employing a range of different philosophical approaches.

In Chapter 9, Prof. Karin Marie Fierke of the University of St. Andrews, educates readers about implications related to quantum social theory for the social sciences, many of which ask how particular socio-political phenomena look different when lifted from the assumptions of classical physics and placed in an uncertain quantum world.

In Chapter 10, Prof. Mohammed Nadir of Tampere University provides an overview (with references) of quantum entanglement, showing that if intertwined with quantum field theory and the AdS/CFT correspondence, quantum gravity, and much more at the high-energy physics frontier, may lead to an incredible, deeper understanding of the universe.

In Chapter 11, Prof. Kisalaya Chakrabarti of the Haldia Institute of Technology considers the case where a particle with spin zero splits into two separate particles, showing the spin magnitudes and direction of these two decay particles before and after the spins are measured. The author also explains why this happens.

In Chapter 12, Prof. Alexander Machavariani of the High Energy Physics Institute of Tbilisi State University uses experimental and relativistic field-theoretical amplitudes of vector meson resonance decays to check the consistency of the approach with a formulation based on Einstein-Podolsky-Rosen concepts. The author also studies the entanglement that can occur in high-energy collisions.

In Chapter 13, Professors Ali Saif M. Hassan of the University of Amran, Waleed S.A. Hasan, and Mohamed A. Shukri of the University of Sana'a and the University of Amran show that using velocity phase matching to produce X waves having orbital angular momentum can be used to determine the length of a dispersive media and more.

In Chapter 14, Professors Yizhuang Liu and Maciej A. Nowak from the Institute of Theoretical Physics at Jagiellonian University, and Ismail Zahed from the Center for Nuclear Theory at Stony Brook University show that in the context of high-energy particle physics, the Koba-Nielsen-Olesen scaling (KNO scaling), formulated many years ago, is of tremendous importance in analysis of many high-energy hadronic multiplicities, and yet usually challenging to derive from first principles in Quantum Chromodynamics (QCD). Results are interpreted from the point of view of quantum entanglement between slow and fast degrees of freedom in QCD.

Oliver K. Baker
Wright Laboratory,
Department of Physics,
Yale University,
New Haven, CT, USA

Section 1

Introduction

Chapter 1

Introductory Chapter: Quantum Entanglement at High Energies – Experiment, Theory, and Interdisciplinarity

Oliver K. Baker

1. Introduction

One of the most recent publications that includes descriptions of quantum entanglement in the context of quantum mechanics is found in the textbook [1]. The book contains passages that describe the development of quantum mechanics, in general, and quantum entanglement. The authors of [1] point out that the divergence of opinions about the theory is trying to inform us about our possible lack of understanding of the nature of reality and that physicists' attention is turned to the frenetic development of different technologies that make use of an understanding of the question, "... what is the interpretation of quantum mechanics?" which in this author's opinion can be expanded to the inquiry "... why are there so many interpretations of quantum entanglement, especially in high energy physics?"

More specifically, proposed interpretations appear to be more concerned with the original form of the theory, seemingly ignoring it's most advanced form. The purpose of the interpretations in [1] is to encourage taking a step back and then attempting to retrace the development of the theory by investigating original sources for the original published papers and letters of the participants.

In a publication on the subject of physical reality, correctness, and completeness [2], questions are asked by the authors and explained concerning physical reality being considered complete when using a quantum mechanical description of that physical reality. As pointed out in that publication, the first question to ask is if the theoretical completion is correct and complete, that is, in this particular case, do the quantum mechanical results line up correctly with the experimental results (the data) that are acquired and measured, and to what degree is there agreement?

The second question asked by the authors is whether the theoretical description is complete in explaining the theory, again the quantum mechanical description in this particular case. The second question was given the priority in [2], that each element of the physical reality acquired and measured experimentally must have a match, a companion in the theoretical completion.

This is especially true and important when questions, especially the second question concerning high energy research results, are at the measurement frontier: (i) The Energy Frontier that makes use of the world's largest high energy facility which is in

Geneva, Switzerland, the Large Hadron Collider (LHC) at CERN [3]; (ii) The Precision Frontier of High Energy Physics where the facilities at Fermi National Accelerator Laboratory (FNAL) in Evanston, Illinois, near Chicago, comprise its epicenter; and (iii) The Cosmic Frontier where space-based and ground-based sensors and detectors, along with their data acquisition systems, are being constructed and tested currently, in several countries worldwide.

2. Quantum entanglement, Higgs bosons, and physical reality

An example of this importance can be understood in the experimental discovery of the theoretically proposed, and now confirmed, Higgs boson. Evidence of this particle's observation was made by both the ATLAS [4] and CMS [5] experimental collaborations at CERN's LHC in 2012. The Nobel Prize for the discovery of this, the first and only Standard Model fundamental boson with zero intrinsic spin angular momentum, was awarded in 2013 for theory contributions that showed how fundamental Standard Model particles acquire mass from the Higgs boson. The Higgs boson's existence (via Spontaneous Symmetry Breaking in the early universe) was considered by many theorists to be the most likely of several other, different, proposals.

Figure 1 shows the Feynman Diagram of a Higgs boson decay to two vector bosons (Z-bosons) and the vector boson decays to leptons in the "Golden Channel" (fewer background events) $H \rightarrow Z^0Z^{0*} \rightarrow$ 4leptons. The Z^0 particle shown in the figure is one of those vector bosons that have a mass of 91.19 GeV/c^2 at the middle of its resonance peak in the data's spectrum, corresponding to the listing in the Particle Data Group booklet, while the Z^{0*} particle shown is the vector boson that has a mass smaller than the Particle Data Group booklet value. The latter boson that has a mass of between 15 GeV/c^2 and 60 GeV/c^2 is given the label "off-mass-shell" Z^{0*}-boson, while the former bosons are simply Z^0-boson, which is "on-mass-shell". The [g] particle lettering in **Figure 1** corresponds to two gluons that collide and couple to form particles as shown by the lines that are in the shape of a triangle in the figure, so this is a gluon-gluon fusion collision process. Gluons mediate the strong interaction in high energy physics. The Higgs boson couples to mass, so the closed triangle in **Figure 1** represents what are mainly the most massive particles of the Standard Model of Particle Physics - top plus anti-top quarks for example. Since the vector bosons shown here are neutral, they each decay into two same flavor, opposite-charge leptons, as shown in this particular type of decay. An electron-positron pair and a positively charged muon along with a negatively charged muon make up the four leptons.

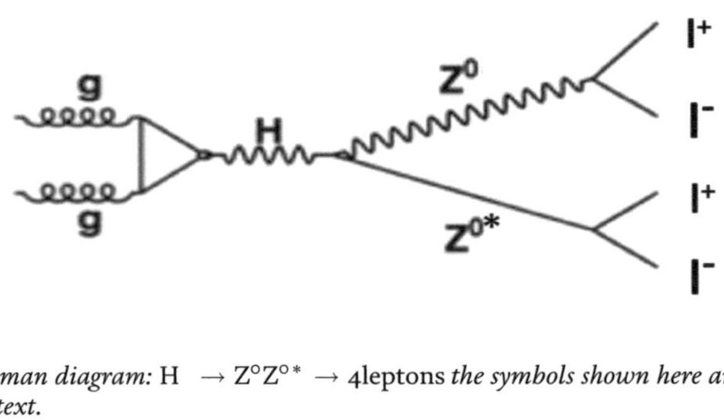

Figure 1.
Higgs boson decay Feynman diagram: $H \rightarrow Z^{\circ}Z^{\circ*} \rightarrow$ 4leptons *the symbols shown here are defined and explained in the main text.*

DOI: http://dx.doi.org/10.5772/intechopen.1004164

Discovery of the Higgs via this process required the coincident measurement of each of these four leptons' position and energy, or momentum, relative to the decayed vector boson linear momentum direction prior to its decay. This **Figure 1** process occurs about one out of every $\mathbf{10}^{n-th}$ proton-proton collisions, where $n \approx 10$ for a 7 TeV proton-proton collision energy, while n is smaller for higher proton-proton collision energies.

The Introduction section of this Introductory Chapter makes use of the Higgs boson production and decay [6] discussion in order to help clarify, in an overview, both quantum entanglement in High Energy Physics as well as the process that leads to the chosen final result. It is based upon research measurements and analysis, followed in order, to correctly search for signals of quantum entanglement, entanglement entropy, Bell's Inequality studies, and quantum computing at collider energies in both particle and nuclear physics. The Higgs decay process discussed and shown above is one of the two decay channel processes that existed in the Higgs boson discovery. The quantum entanglement overview analysis discussed in this chapter makes use of both theoretical and experimental descriptions.

When two or more particles (qubits) are produced and interact with each other so that the final quantum state under analysis will have contributions from both (in a bipartite (qubit A, qubit B) system) or the group (in the system of multiple qubits) of particles, they are both (bipartite) or all (multipartite) entangled with each other. Every particle involved must be described by a quantum state that depends upon the quantum state of the other particle, or particles, involved in the bipartite or multipartite system, respectively. And, there is no dependence on the distance between the states.

The quantum entanglement definition is shown in **Figure 2**. When there is quantum entanglement, there is a sum of terms used in describing the quantum state. It is labeled as superposition in this context.

New discoveries in high energy nuclear and particle physics can give rise to new phenomena that require new interpretations of, or additions to, quantum entanglement [7–9], that is, further experimental and theoretical research may need to be carried out and completed prior to complete physical reality success in discoveries at the energy frontier in order to make fully proper statements about quantum

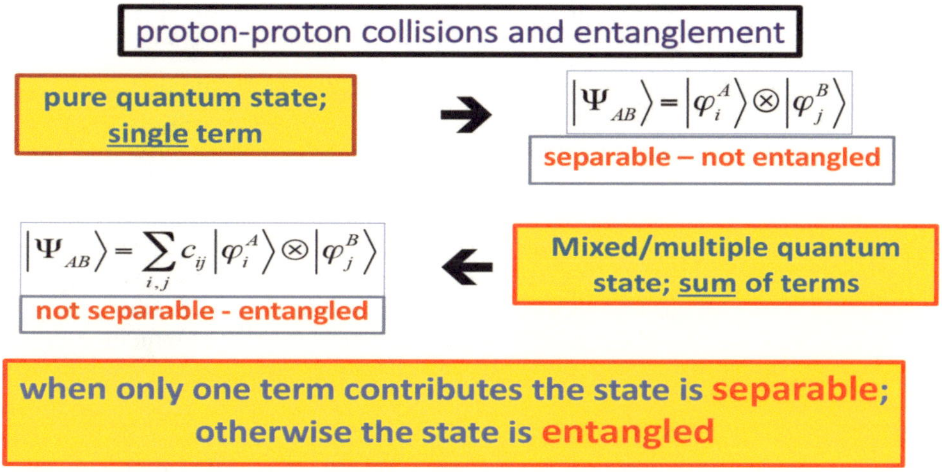

Figure 2.
Quantum entanglement definition. See details in main text.

entanglement in those cases. Quantum entanglement, while shown to exist experimentally and theoretically in quantum mechanics, is so far not observed in classical mechanics.

In production and then decay of the Higgs boson, further measurements and analyses, along with theoretical research, have been carried out [10–15] for CP-violation and further measurements, respectively, as a few examples even after the Higgs boson discovery, in order for them to be considered correct and complete when using a quantum mechanical description, as indicated in the Introduction section of this Introductory chapter.

Experimental searches are currently underway for evidence of quantum entanglement, entropy of entanglement, and Bell's Inequality violation signatures using Higgs bosons' decays to vector bosons, in the ATLAS and CMS experiments at CERN's Large Hadron Collider [16]. Additionally, theory calculations and data analyses are being studied in collaboration with the experimentalists [17–21]. In both cases, Monte Carlo-simulated events are used for analysis at this stage of ongoing research.

3. Conclusions

As underscored in the Introduction of this Introductory chapter [2], each element of physical reality acquired and measured experimentally must have a match, a companion in theoretical completion, that is, both must be correct and complete. And, the elements of each physical reality and theoretical completion sets must match.

We may gain in understanding by reinforcing very similar statements from another giant in science [22] whose quote "The proper method for inquiring after the properties of things is to deduce them from experiments" points to the importance of proper physics experiments at high energies. In addition, a follow-up quote from the same scientist [22], "It is the weight, not numbers of experiments that is to be regarded," emphasizes that there is essentialness and significance in follow-up experiments to a major discovery such as the Standard Model Higgs boson [4, 5, 10–15] as described in the previous section of this Introductory Chapter. Significance is important since it is, in this particular case, the Higgs boson observation along with its production and decay to vector bosons that is used in quantum entanglement measurement significance at these high energies. And finally, the quote [22] "If the experiments I urge be defective, it cannot be difficult to show the defects; but if valid, then by proving the theory, they must render all objections invalid" points to the essence of correct and complete theoretical research (again, in this case, involving the Higgs boson and its decays to vector bosons) for successful measurement and reliability.

The concept, as well as the interdisciplinarity, of quantum entanglement has grown, especially in Condensed Matter Physics, Atomic, Molecular, and Optical Physics (the Nobel Prize in Physics was awarded to Alain Aspect, John Clauser, and Anton Zeilinger in 2022 for research results that tested the purity of quantum mechanics as we know it and use it by detecting evidence for Bell's Inequality violation [23]), quantum communications [24], and most recently, more, new quantum entanglement applications and results in nuclear and particle physics as highlighted in the preceding section (Section 2) of this Introductory Chapter and referenced in [7–9].

It will be interesting to compare the current research topics, their methods, means, and results, listed in this book with those that will show themselves in the next several generations!

DOI: http://dx.doi.org/10.5772/intechopen.1004164

Acknowledgements

The author (OKB) gratefully acknowledges helpful scientific discussions with Professor Steven Lamoreaux for our collaboration in completing this Introductory Chapter of the current book. And sincere thanks to Yale University for providing financial support.

Conflict of interest

The author has no conflict of interest.

Author details

Oliver K. Baker
Wright Laboratory, Department of Physics, Yale University, USA

*Address all correspondence to: oliver.baker@yale.edu

References

[1] Golub R, Lamoreaux S. The Historical and Physical Foundations of Quantum Mechanics. Oxford, United Kingdom: Oxford University Press; 2023

[2] Einstein A, Podolsky B, Rosen N, Institute for Advanced Study, Princeton, New Jersey. Can quantum-mechanical description of physical reality be considered complete? Physical Review. 1935;**47**:777

[3] Peters T, Peterson C. The Higgs boson: An adventure in critical realism. Theology and Science. 2013;**24**: 185-207. DOI: 10.1080/14746700.2013.809948

[4] ATLAS Collaboration. A particle consistent with the Higgs boson observed with the ATLAS detector at the large hadron collider. Science. 2012; **338**(6114):1576-1582. DOI: 10.1126/science.1232005

[5] CMS Collaboration. A new boson with a mass of 125 GeV observed with the CMS experiment at the large hadron collider. Science. 2012;**338**(6114): 1569-1575. DOI: 10.1126/science.1230816

[6] Internet, Constraints on Higgs boson properties.pdf. Available from: https://pure.uva.nl/ws/files/110885883/Constraints_on_Higgs_boson_properties.pdf

[7] Michael Irving. Physicists discover completely new type of quantum entanglement. Available from: https://newatlas.com/physics/new-type-quantum-entanglement-particles/

[8] Department of Energy, Office of Science. New Form of Quantum Entanglement Gives Insight into Nuclei. Washington, DC: Department of Energy, Office of Science; 2023. Available from: https://www.energy.gov/science/articles/new-form-quantum-entanglement-gives-insight-nuclei

[9] Star Collaboration. Tomography of ultrarelativistic nuclei with polarized photon-gluon collisions. Science Advances. 4 Jan 2023;**9**(1):eabq3903

[10] ATLAS Collaboration. Measurement of the CP properties of Higgs boson interactions with tau-leptons with the ATLAS detector. European Physical Journal. 2023;**C83**:563

[11] ATLAS Collaboration. Search for a CP-odd Higgs boson decaying to Zh in pp collisions at $\sqrt{s} = \mathbf{8TeV}$ with the ATLAS detector. PLB. 2015;**744**: 163-183

[12] ATLAS Collaboration. CP properties of Higgs boson interactions with top quarks in the $t\bar{t}$ H and tH processes using $\mathbf{H} \rightarrow \gamma\gamma$ with the ATLAS detector. PRL. 2020;**125**:061802

[13] ATLAS Collaboration. Measurement of the Higgs boson mass with $\mathbf{H} \rightarrow \gamma\gamma$ decays in 140 $\mathbf{fb}^{-1}$ of $\sqrt{s} = 13\,\mathbf{TeV}$ pp collisions with the ATLAS detector. PLB. 2023;**847**:138315

[14] ATLAS Collaboration. Combined measurement of the Higgs boson mass from the $\mathbf{H} \rightarrow \gamma\gamma$ and $\mathbf{H} \rightarrow ZZ^* \rightarrow$ **4leptons** decay channels with the ATLAS detector using $\sqrt{s} = \mathbf{7}, \mathbf{8}$, and 13 TeV pp collision data. Physical Review Letters. 2023;**131**:251802(1-21)

[15] ATLAS Collaboration. Combination of searches for invisible decays of the Higgs boson using 139 $\mathbf{fb}^{-1}$ of proton-proton collision data at $\sqrt{s} = \mathbf{13}$ TeV

collected with the ATLAS experiment. PLB. 2023;**842**:137963

[16] CERN, LHC Experiments; Diverse Experiments at CERN: ATLAS; CMS. Available from: https://home.cern/science/experiments

[17] Fabbrichesi M, Floreanini R, Gabrielli E, Marzola L. Stringent bounds on HWW and HZZ anomalous couplings with quantum tomography at the LHC. Journal of High Energy Physics. 2023; **2023**:1-5. Article number 195. arXiv: 2304.02403 [hep-ph]

[18] Fabbrichesi M, Floreanini R, Gabrielli E, Marzola L. Bell inequality is violated in **B0** $\rightarrow J/\psi K^*(892)^0$ decays. 8 May 2023:1-5. e-Print: 2305.04982 [hep-ph]

[19] Aguilar-Saavedra JA, Bernal A, Casas JA, Moreno JM. Testing entanglement and bell inequalities in **H** $\rightarrow$ **ZZ**. Physical Review D. 20 Jan 2023;**107**:016012-1-016012-12

[20] Barr A. Testing bell inequalities in Higgs boson decays. PLB. 2022;**825**: 136866

[21] Ashby-Pickering R, Barr AJ, Wierzchucka A. Quantum state tomography, entanglement detection and bell violation prospects in weak decays of massive particles. JHEP. 2023; **5**:20

[22] Isaac Newton Quotes. Brainy Quote Publication; 25 Dec 1642-20 Mar 1727. Available from: https://www.brainyquote.com; /authors/isaac-newton-quotes?__cf_chl_tk=ekGfN5xPlEQhhLtPOzFuKns_WJunCP4E1mYVXS8TxJc-1703267060-0-gaNycGzNFDs

[23] Eva Nevelius, MLA style: The Nobel Prize in Physics 2022. Nobel Prize.org. Nobel Prize Outreach AB 2023. Available from: https://www.nobelprize.org/prizes/physics/2022/summary/ [Accessed: December 28, 2023]

[24] Zou N. Quantum entanglement and its application in quantum communication. Journal of Physics: Conference Series. 2021;**1827**:012120. DOI: 10.1088/1742-6596/1827/1/012120

Section 2

Theory

Chapter 2

On the Questions of Spin and Spin Quantum Correlations in Relativistic Quantum Mechanics and Relativistic Quantum Information

Shrobona Bagchi

Abstract

The majority of current understanding of the quantum correlations is in the field of non-relativistic quantum mechanics. To develop quantum information and computation tasks fully, one must inevitably take into account the relativistic effects. In this regard, the spin is one of the central tools. For this purpose, it is of paramount importance to understand and characterize fully the theory of spin in relativistic quantum information theory where the spin states act as qubit. This area is still far from being resolved. As a result, this article will explore the recent studies of the concepts of the spin and spin quantum correlations in inertial frames and some apparent paradoxes regarding this concept. We will mainly focus on the problem of characterizing the spin, reduced spin density matrices and spin quantum correlations in inertial reference frames and the apparent paradoxes involved therein. Another important aspect is the use of tools of quantum field theory to extend several concepts in non-relativistic domain to relativistic one. In this regard, we analyze the development of the theory of relativistic secret sharing and a correlation measure namely the entanglement of purification.

Keywords: relativistic spin quantum correlations, relativistic spin-momentum correlation, relativistic quantum information, spin reduced density matrix in relativistic scenario, entanglement of purification

1. Introduction

Quantum correlations are an important part of modern quantum information theory [1–3]. This area is very well studied, developed, and understood in nonrelativistic quantum information using nonrelativistic quantum mechanics [4]. The study of these correlations has helped in the development and implementation of various quantum information processing protocols such as quantum teleportation, quantum cryptography, and quantum secret sharing. However, this theory is not well understood and developed in relativistic quantum information theory, which uses relativistic quantum mechanics or quantum field theory [5–9]. There are several less-understood areas that require careful analysis and resolution. We handpick a few such areas and explore in detail the fundamental and underlying issues, explore some fundamental quantum information tasks in relativistic quantum information, and end with conclusions and potential future directions. This discussion inevitably falls in the section of quantum information in high-energy physics, since the effect of relativistic velocity or acceleration becomes relevant in high-energy physics. At the end, we also discuss how an important concept of entanglement of purification has been generalized in the area of holography, reflect on the conjecture, and raise a few important issues regarding its operational significance in this area. This analysis shall also be helpful in future space satellite-based quantum communication.

As per our knowledge of quantum mechanics and relativistic quantum mechanics, the 'spin' is usually thought of as an intrinsic 'angular momentum' associated with elementary particles like electrons, protons, or other particles as such. It is understood to be purely a quantum-mechanical property without any counterpart in classical physics. Spin in quantum mechanics is considered an existing 'intrinsic' angular momentum of the particle, and it is not due to the classical rotation of any internal component of the particle. Spin as a relativistic concept is still undergoing revisions to be understood to be in full glory and entirety. It is an essential part in majority of quantum information tasks in modern day quantum information and computation theory and applications. The important details start to emerge as one goes to inertial and noninertial frames and several paradoxes and inconsistencies start to show in this arena in terms of its definitions, conceptualization, and further generalizations such as the reduced spin density matrices in inertial frames. These observations point to the fact this is a partially developed concept, and its development will lead to robust implementation of various quantum information processing tasks in relativistic regimes such as in future satellite-based quantum communication. In the upcoming paragraphs, we visit some of these areas in short, which are explained in detail in the main sections later.

Spin qubit is a well-understood concept in nonrelativistic quantum mechanics, It is understood very well through the implementation of the Stern-Gerlach experiment in the nonrelativistic (low velocity, which is much smaller than the speed of light in vacuum) limit. However, the definition and understanding of the spin in the relativistic regime are still underdeveloped. This has been analyzed in various works spread out around the last decade, and research on this is still ongoing today [10–17]. One of the origins of this difficulty lies in the entangling of the spin degree of freedom with the momentum degree of freedom in the Lorentz-boosted reference frames. The second difficulty arises from the fact that for a quantum particle moving in a superposition of velocities as a quantum-mechanical possibility, it is impossible for it to suddenly transition to its rest frame, wherein its spin is defined and understood properly. Possible remedies for these problems have been proposed by various authors. They eventually propose a solution to this problem and ways of experimentally observing the relativistic features of the spin, which then in turn promises to open up the possibilities of devising quantum information protocols using spin as a qubit in the special-relativistic regime.

Another very important and recurring problem in relativistic quantum information or quantum information in high energy physics is the robust formulation of the reduced spin density matrix in the relativistic regime. Though the reduced density matrix for the spin degree of freedom is well-defined in nonrelativistic quantum mechanics, its definition and formulation run into problems when trying to simply extend this to a relativistic regime. An apparent paradox involving the definition and formalism of spin density matrix in a relativistic regime is given in [11]. It was shown that a model for particle detection wherein a linear application of the Wigner rotations was applied to the state of a massive relativistic particle in a superposition of two counter-propagating momentum states, leads to a paradox. The paradoxical behavior is that the probability of finding the relativistic quantum particle at different positions depends on the reference frame, which is unwanted feature in the theory. A solution to the paradox was given there. According to the proposed solution, the authors argue that we cannot in general linearly apply the Wigner rotations to a quantum state without considering the appropriate physical interpretation.

Again in similar vein, there is another problem similar in nature to the above. An open problem in the field of relativistic quantum information is whether entanglement and the quantitative degree of violation of Bell's inequalities for massive relativistic particles are dependent on the frames of reference or not. At the heart of this question lies the effect that the spin degree of freedom gets entangled with the momentum degree of freedom a relativistic regime. In a more advanced work, the authors here show that the Bell's inequalities for a pair of quantum particles can be maximally violated in a 'special-relativistic regime', even without any postselection of the momentum of the particles, shown via the use of the novel methodology of quantum reference frames. The authors claim that the use of the quantum reference frames allows them to transform the problem to the rest frame of a particle, whose state can be in a superposition of relativistic momenta from the viewpoint of the laboratory frame of reference. In this work, the authors work with several problems of defining the spin density matrix in the relativistic regime and show that when the relative motion of two particles is noncollinear, the appropriate measurements for violating Bell's inequalities in the laboratory frame involve the "coherent Wigner rotations". In this work, the authors also show that the degree of violation of Bell's inequalities is independent of the choice of the newly introduced and defined quantum reference frames, which is a desired feature in theory.

After the full description of the existing line of work spin density matrices in relativistic reference frames, we turn our attention to the use of tools of quantum field theory in implementing quantum information protocols in the high energy regime [18–20]. In this respect, we review a relativistic quantum secret-sharing protocol in a relativistic regime. Here, the authors develop a quantum secret-sharing protocol that relaxes usual assumptions and considers the effects due to the accelerating motion.

Another research area is the extensions and further development of various quantum information correlation measures via quantum field theory and conformal field theory [21, 22]. One such definition is the entanglement of purification, which defines the total correlation measure for a quantum particle in an operational way [23]. Entanglement wedge cross-section has been developed to account for the counterpart of entanglement of purification in quantum field theoretic and conformal field theoretic terms. It has been termed the Holographic Entanglement of Purification by Tadashi Takayanagi and Koji Umemoto [23–25]. They suggest that it is a holographic counterpart of the entanglement of purification, which measures a bipartite correlation in a given mixed-state quantum state defined on an operational basis. We point out in our coming sections that a similar operational footing in the quantum field theoretic terms will be a promising area of research in the future.

2. On the questions of spin and spin quantum correlations in relativistic quantum mechanics and relativistic quantum information

In this section, we briefly review a few key concepts needed to understand the basic analysis of spin in quantum mechanics and the problems associated with it for its formulation in the relativistic regime. These concepts include those of reference frames as defined in the theory of special relativity. Any physical system is defined using a set of coordinates that completely specifies its reference frame. From the principles of special relativity, we know that all physical laws hold the same way in all reference frames. The laws of physics transform covariantly in between the different reference frames. However, there is typically a reference frame that is the most convenient to use, where the system rests. This reference frame is called the rest frame of the physical system. These concepts of the classically defined reference frames, as in classical physics, are very well understood in the absence of intricacies of certain quantum-mechanical phenomena. However, when we start to analyze some quantum properties in detail using the traditionally defined reference frames, we face some problems. In this respect, it was shown that when the external degrees of freedom for example the momentum of the physical system, are in a quantum superposition with respect to the laboratory frame of reference, no classical reference frame transformation, as prescribed by the special theory of relativity, can map the description of physics from the laboratory to the rest frame [12]. This area of physics is not understood well enough. Subsequently, the concept of a quantum reference frame was introduced and leveraged to give an operational footing to the concept of spin in relativistic quantum mechanics [12]. It was claimed in [12] that such a formulation is able to solve some of the paradoxical features related to the transformation of spin in relativistic quantum mechanics. There are several claims in this direction by various authors. As a result, proper experimental verifications are needed to settle the correct theory for spin in relativistic quantum mechanics and relativistic quantum information theory. These concepts are explained later in the later paragraphs.

One of the most important concepts is spin in non-relativistic quantum mechanics, defined operationally via the Stern-Gerlach experiment. The spin of a Dirac spin-1/2 particle is defined by the 2×2 Pauli matrices $\sigma_i (i = 1,2,3)$. The Pauli matrices and the unit matrix generate an irreducible representation of the $SU(2)$ group. It is well known that the spin operator of a nonrelativistic spin-1/2 particle in quantum mechanics is very well understood. Also, we know that there is a clear correspondence between quantum-mechanical operators and classical variables in nonrelativistic quantum mechanics. This correspondence exists for all operators such as the position, momentum, and angular momentum etc. However, in contrast to this observation, the connection between the quantum-mechanical operators and classical variables in relativistic quantum mechanics is much more subtle and complex.

In this respect, we review the concept of spin in relativistic quantum mechanics from different perspectives provided by various authors until now. They involve the analysis of an apparent paradox caused by linear Wigner rotation and the quantum reference frames. These are presented below.

2.1 Apparent paradox of Wigner rotations for relativistic effects of spin of quantum particle

The intricacies of the conceptual and analytical foundation of spin of relativistic quantum particles were presented in [11]. In their analysis, the authors presented an apparent paradox involving the definition and formalism of spin density matrix in a super-relativistic regime. It was shown there that a method of particle detection in combination with linear Wigner rotations, which then corresponds to momentum-dependent changes of the particle spin owing to the fact that spin and momentum degrees of freedom get entangled in the relativistic scenario under Lorentz transformations, applied to the state of a massive relativistic quantum particle in a superposition of two different momentum states leads to a paradox. The paradoxical feature is that the probability of finding the relativistic quantum particle at different positions depends on the reference frame, which should not be the case. As a solution to the paradox, which is also simple, the authors suggested that one cannot in general linearly apply the linear Wigner rotations to a quantum state without considering the appropriate physical interpretation of it. We sketch out the main steps of their analysis here. The initial state taken in this case is of the following form

$$|\Psi\rangle = \frac{1}{\sqrt{2}}[|p\hat{y}, Z\rangle - |p\hat{y}, -Z\rangle], \tag{1}$$

in the frame S_0. Here, $|p, \pm Z\rangle$ represents a state for the particle with 4-momentum $\left(p^0, \vec{p}\right)$, and spin state pointing in the $\pm z$ direction, being the eigenvector of the Pauli matrix σ_z with eigenvalue ± 1 with reference to S_0. Here, the authors have used Wigner's definition of spin, and they have taken $c = 1$. Therefore, from the perspective of the rest frame, the quantum particle is in a superposition of momentum in opposite directions. Using proper algebra, it was found out by the authors that the probability density of finding the relativistic quantum particle around position y obeys the following expression

$$p_0(y) \propto \sin^2\left(\frac{py}{\hbar}\right) \tag{2}$$

If one then makes a change of reference frame to a frame that moves with velocity βz in relation to $S(0)$, each momentum component of the state $|\Psi\rangle$ undergoes a different spin transformation. The spin and the momentum degrees of freedom get entangled when linear Wigner rotations are applied in a nontrivial way. See [11] for details. In the reference frame S_1, the momentum state of the particle changes from before, though the y component remains the same. In this work, the authors then analyze the expressions about the y dependence of the particle wavefunction and do not analyze for z, x direction for simplification without any loss of generality. With some reasonable approximation, the probability expressions found with respect to the reference frame S_1 is given by the following

$$p_1(y) \propto \cos^2\left(\frac{\phi}{2}\right) \sin^2\left(\frac{py}{\hbar}\right) + \sin^2\left(\frac{\phi}{2}\right) \cos^2\left(\frac{py}{\hbar}\right), \tag{3}$$

where the angle ϕ is related to the boost parameter for the reference frame S_1 [11]. Therefore, this new expression for probability points to a paradox that has crept into the calculation in between, done traditionally. The authors in [11] argue that this is a paradox since the probability of finding the particle around some position should depend on the reference frame. As a result, this paradox points toward a deficiency in the current state of the art of the theory of spin in quantum mechanics. Thereafter, the authors in [11] test this observation differently via the use of quantum measurements using detectors. They consider measurements of the particle position using a detector that, by construction, responds only to the charge or the mass of the particle but does not in any case depend on its spin. Using this formalism and again placing a small number of reasonable approximations that are relevant in the experimental setup, they again show the discrepancy in the expressions for probability expressions as calculated concerning different reference frames.

To summarize, they have shown that the application of the momentum-dependent linear Wigner rotations to the quantum state of a massive relativistic particle in a superposition of counter-propagating momentum states along with a model for particle detection leads to a paradox, since the probability of finding the particle at different positions would depend on the reference frame. Considering the physical implementation of the quantum state, they have discussed that the Wigner rotation depends on the preparation method, such that, with a change of the reference frame, the spin transformation of a quantum state in a superposition of different momenta is not exactly equivalent to the linear application of the momentum-dependent Wigner rotation to each momentum component of the state. This, they say, solves the apparent paradox. Their work and a few previous works on the subject show that relativistic quantum transformations cannot generally be computed only by following a mathematical procedure as in the traditional literature of relativistic quantum mechanics. The authors argue that the physical meaning of the transformations must always take precedence before their application in that physical scenario.

Though they have proposed the above solution to the apparent paradox, they also stress that their solution may not be the only viable solution to this apparent paradox. It may be possible that the paradox could be solved by modeling the particle detection by some more complicated scheme, keeping the linearity property of the Wigner rotations. Consequently, different solutions were proposed by different authors to this problem. The next sections discuss one of the main contenders in the name of quantum reference frames [12]. However, it is important to note that settling and reaching a proper solution to this apparent paradox needs experimental verifications under different conditions and approximations.

2.2 Relativistic Stern-Gerlach experiment and quantum reference frames

To provide a consistent description of the relativistic effects on the spin of a quantum particle, more theories were proposed, one of which is based on the relatively recent proposals of quantum reference frames. Along this line, the relativistic treatment of the Stern-Gerlach experiment was proposed and was termed as the relativistic Stern-Gerlach experiment [12]. The theory of the quantum reference frames was evoked there to give an operational interpretation to the spin in relativistic quantum mechanics. It was noted in [12] that when the particle has relativistic velocities, the spin degree of freedom transforms in a momentum-dependent way, as was noted by previous authors. Then, suppose a standard Stern-Gerlach measurement is performed on a particle in a pure quantum state moving in a superposition of relativistic velocities. In that case, the operational identification of the spin fails because it was shown in [12] that no orientation of the Stern-Gerlach apparatus returns an outcome with unit probability. The question that then arises is whether it is possible to find 'covariant measurements' of the spin and possibly momentum, which predict invariant probabilities in different Lorentzian reference frames for the case of a quantum relativistic particle moving in a superposition of velocities [12]. This is therefore an alternate solution to the solution proposed in [11] as described in the previous section. If such measurements are possible to construct, then it would be possible to map the description of spin in the rest frame of the particle to the frame of the laboratory in an unambiguous way. This would enable one to derive the corresponding observables to be measured in the laboratory frame to verify the correct theory of the spin of a relativistic quantum particle.

The trial for finding such covariant measurements that preserve probability values in different reference frames is motivated by the potential applications in which the spin degree of freedom is used as a qubit to encode and transmit quantum information in the relativistic regime. As such, earlier protocols are no longer valid in a relativistic context. These severely constraints and undervalue the wide range of applicability of techniques involving spin as a quantum information carrier in the relativistic regime. It is then important to explore possible methods which can overcome this limitation. In the context of relativistic quantum information, this question has been extensively discussed in relation to Wigner rotations. It has been related to the problem of identifying a covariant spin operator [11–15]. A variety of relativistic spin operators have been proposed to date. Some of them are called the Frenkel, the Pauli-Lubanski, the Pryce, the Foldy-Wouthuysen, the Czachor, the Fleming, the Chakrabarti, and the Fradkin-Good spin operators.

To remedy the above problems, the authors in [13], use the concept of "superposition of Lorentz boosts" which allows them to make the relativistic quantum particle "jump" into the rest frame even if the particle is not in a momentum eigenstate but in a quantum state with a superposition of momentum in general. It is well known that in the rest frame, the spin observables satisfy the $SU(2)$ algebra and are operationally defined through the famous Stern-Gerlach experiment. The authors in [13] aim to make this same concept work in inertial reference frames. In the work [13], the authors transform the set of spin observables in the rest frame to an isomorphic set of observables in the laboratory frame. The transformed observables are generally entangled in the spin and momentum degrees of freedom as expected. The new set fulfills the $SU(2)$ algebra again and is operationally defined through an experiment that the authors label as the "relativistic Stern-Gerlach experiment." In this experiment, the authors construct the interaction term and the measurement term between the spin-momentum degrees of freedom and the electromagnetic field in the laboratory frame, which gives the same probabilities as the Stern-Gerlach experiment in the rest frame, as desired, and stated earlier in the paragraph. This set of observables in the laboratory frame allows the authors to partition the total Hilbert space into two subspaces corresponding to the two outcomes, which can be termed as "spin up" and "spin down." Hence, with techniques of the quantum reference frames, the relativistic spin states can effectively be used to construct a qubit state in an operationally well-defined way, as claimed by the authors in [13]. Thus, the quantum reference frames and the relativistic Stern–Gerlach experiment promise to be robust candidates representing the theory of intrinsic spin of relativistic quantum particles and its transformations between different reference frames. However, the correctness of the corners of this theory is still open to experimental demonstrations to be established fruitfully.

With the above background in mind, we now describe the relativistic Stern–Gerlach experiment, as discussed in [12]. One considers an experiment performed in the laboratory reference frame referred to as C. One allows the particle to have any quantum state and, in particular, to move in a superposition of momenta. This condition implies a nonclassical relationship between the two reference frames. This means that a standard boost transformation does not relate to the rest of frame A and the laboratory frame C as in classical special relativity. The authors in this work [12] implement a method to generalize the boost transformation to this case of the relativistic quantum particle. The coordinates for the mathematical analysis z are used to describe the external degrees of freedom of particle A and the intrinsic spin degrees of freedom as $\tilde{A}$ of the relativistic quantum particle. The state of the particle time $t = 0$ is taken to be the following

$$|\Psi\rangle = \cos\theta|\psi^{+}\rangle + \sin\theta|\psi^{-}\rangle, \quad (4)$$

where again we have the following definitions

$$|\psi^{\pm}\rangle = |\psi_z\rangle_A|\phi_x^{\pm}\rangle_{A\tilde{A}}, \quad (5)$$

is a division of the total wave function into components in x and z directions in rest frame and lab frames as per the notation. It is assumed that the motion along z direction is nonrelativistic, without any loss of generality. Writing these wave functions in terms of the superposition of momentum states, we have the following expressions

$$|\psi_z\rangle_A = \int dp_z \psi_z |p_z\rangle_A, \quad (6)$$

with ψ_z denoting Gaussian wave functions in the momentum variable p_z centered around $p_z = 0$ and standard deviation s_z. The other component is denoted as the following

$$|\phi_x^{\pm}\rangle_{A\tilde{A}} = \int d\mu(p_x)\phi_x|p_x, \Sigma_{p_x}^{\pm}\rangle_{A\tilde{A}}, \quad (7)$$

where $\phi_x^{\pm}$ is a general wavepacket expression and $|\Sigma_{p_x}^{\pm}\rangle_{A\tilde{A}}$ are the eigenvectors of the operator obtained via Lorentz boost and Pauli–Lubanski operator as defined in [12], with eigenvalues ± 1. In the laboratory frame, it is possible to define the observables corresponding to the spin operators in the rest frame by transforming the spin, as defined in the rest frame, with a quantum reference frame which then correspond to the transformation. They are called the manifestly covariant Pauli-Lubanski spin operator. Now, after this, the authors engineer a Hamiltonian with the following interaction term Hamiltonian in the laboratory frame as follows

$$H_{int} = \mu B_z \xi, \quad (8)$$

where $B_z = B_z^0 - \alpha z$. ξ is the term containing the components that are obtained using the components of the manifestly covariant Pauli–Lubanski spin operator modified with parameters dependent on the boost parameters, as defined in [12]. Let us now see how the operators ξ came into the picture. The authors in [12] note that in laboratory frame C, when the particle A is in a superposition of momentum states, no spin measurement in a standard Stern-Gerlach experiment gives a result with probability one because of the following two reasons: the spin and momentum are entangled, and the relation between the laboratory and the rest frame is not a classical special-relativistic reference frame transformation. To devise such measurement that will give consistent probability values in all reference frames as in the rest frame, the authors in [12] note that in the laboratory frame, it is possible to define the observables corresponding to the spin operators in the rest frame by transforming the spin, as defined in the rest frame, with a quantum reference frame transformation, the expression of which is then derived as ξ, the details of which can be found in [12].

Now, let us look at the dynamics of the relativistic quantum particle due to the Hamiltonian as described above. The Hamiltonian term is an appropriate interaction Hamiltonian containing a magnetic field in the z direction of the laboratory frame. The state is then evolved with the action of this Hamiltonian, and its form is written down appropriately in the interaction picture as a function of time. It was shown that under the interaction with the magnetic field, the Gaussian wavepacket z gets split into two wavepackets, moving in opposite directions according to the state of the spin. After this, appropriate projection operators are applied to the wave packet, and probabilities of obtaining spin "up" or spin "down" as value is obtained denoted by $p^{\pm}$. It was shown via this calculation in [12] that for a time when the two wavepackets become distinguishable, the probabilities for obtaining up and down spins are found out to be $\cos^2\theta$ and $\sin^2\theta$, when irrevalent terms are neglected under the appropriate limit and subsequent approximation. The authors claim that in this way, one can solve the problem of ambiguity of finding the correct expressions for probabilities in rest frames in the relativistic Stern-Gerlach experiment.

In this work, the authors claimed to have provided a correct operational description of the spin of a special-relativistic quantum particle, which has been elusive for a while. Such operational description was initially difficult to obtain with standard traditional treatments due to the combined effect of special relativity and quantum-mechanical properties, which makes the spin and momentum entangled and an impossibility of jumping to the rest frame with traditional tools. To remedy this problem, the authors have introduced the concept and mathematical characterization of the "superposition of Lorentz boosts" transformation to the rest frame of a quantum particle, moving in a superposition of relativistic velocities from the point of view of the laboratory reference frame. As a result of their analysis based on the quantum reference frames, probabilities obtained in the relativistic Stern–Gerlach experiment are shown to remain the same in the rest frame and in the laboratory frame, which was a challenging task to accomplish before. This approach is relatively new to some earlier approaches and proposed theoretical remedies. However, it should be emphasized that the theoretical treatment offered in this work has yet to undergo several experimental checks in different limits, experimental conditions, and relevant approximations to be accepted and established as a correct theory for relativistic effects in spin of a quantum particle.

2.3 Other effects related to relativistic treatment of spin of a quantum particle

Several other effects are associated with the correct description of spin in relativistic quantum mechanics. An open question in relativistic quantum information is the invariance of a measure of entanglement and/or the quantitative degree of violation of Bell's inequalities for massive relativistic particles in different frames of reference. Such questions can be extended similarly to other quantum information theoretic correlation measures. Likewise, as before, at the core of this dilemma is the effect that spin gets entangled with the momentum degree of freedom at relativistic velocities. In [13], the authors claim to show that Bell's inequalities for a pair of particles can be maximally violated in a special-relativistic regime, even without any postselection of the momentum of the particles, again via the use of the concept of quantum reference frames. They specifically show that when the relative motion of two particles is noncollinear, the optimal measurements for violating Bell's inequalities in the laboratory frame involve "coherent Wigner rotations" [13]. Thus, they also touch upon a debated concept of the appropriate application of Wigner rotations in this physical setup. In this formalism, the authors also show that the degree of violation of Bell's inequalities is independent of the choice of the quantum reference frame, which is a desired effect. As a result, this work attempts to settle some important open questions involving the fundamental concepts of spin and relativity. However, for it to be established to be a correct theory or otherwise, several experimental checks must be performed consistently and in a reproducible way. As a result, experimental proposals to test out these theories in the laboratory under different conditions pose the next important steps in the full development of this area of research and enquiry.

2.4 Experimental efforts to test relativistic theory of spin of quantum particle

Experimentally, many efforts have recently been made to measure the quantum spin correlations of elementary particles. One of those proposed experiments involves studying relativistic electron pairs' quantum spin correlations to test the nonlocality of relativistic quantum mechanics. Finding the right expression and formalism for spin and spin quantum correlations in relativistic quantum mechanics is an important direction of research both from the perspective of space communications and testing the fundamentals of quantum physics and quantum gravity. Here, we report on two attempts at experiments to measure the spin quantum correlations in relativistic scenarios.

An experiment investigating the quantum spin correlations of relativistic electrons is reported here [16]. The project presented in [16] tries to make the first measurement of the quantum spin correlation function for a pair of massive relativistic particles. This measurement is claimed to be the first attempt to verify the predictions of relativistic quantum mechanics in the domain of spin correlations. This is an interesting research direction since it can settle competing theories of spin quantum correlations experimentally or even point out unknown deficiencies in the foundations of relativistic quantum mechanics. Per their description of the proposed experiment, the measurement is carried out on a pair of electrons in the final state of Moller scattering. The measurement attempts to measure correlations of spin projections on chosen directions for the final state pair after the complete evolution via the chosen dynamics. The detector consists of two Mott polarimeters, in which the spins of both Moller electrons are measured simultaneously. However, the results have not yet been linked to the theoretical predictions of the quantum reference frames or other competing theories of spin in relativistic quantum mechanics. This remains a promising future direction of research.

Another research direction has been the study of quantum spin correlations of relativistic electron pairs to test the nonlocality of relativistic quantum mechanics. The theory developed in this direction has been discussed in the previous sections. Therefore, an experiment that tests the predictions offered by the theory, for example, as offered by the quantum reference frames, will be extremely helpful in settling the correct theory of relativistic effects of quantum spin among many competing theories. This will help advance the understanding of fundamental theory in nature. This project is supposed to be a Polish–German project QUEST that will study relativistic quantum spin correlations of the Einstein–Podolsky–Rosen–Bohm type through appropriate measurements and the corresponding probabilities for relativistic electron pairs. This experiment will also use the Moller scattering method and Mott polarimetry technique.

3. Quantum information with quantum field theory: relativistic quantum secret sharing

In the previous section, we discussed the fundamental question of spin in relativistic quantum mechanics. We discussed how spin is an interesting topic and how its understanding and unraveling of exact nature and experimental verification will help develop intricate quantum technologies that use the spin as a qubit in quantum information processing tasks. However, though it is important to understand the concept of spin in relativistic quantum mechanics, one can also use the concept of quantum information processing tasks in relativistic quantum information using tools from quantum field theory. One of these approaches has been to use cavity dynamics for the case of noninertial motion in general. In this section, we report on the quantum secret-sharing protocol in the relativistic setting. Here, we focus more on the theory of a specific quantum information task called relativistic quantum secret sharing using tools from quantum field theory.

In (2,3)-threshold quantum secret sharing, the “dealer,” one of the parties taking part in the protocol, encodes the quantum secret in three quantum shares in a localized manner. The authors in [12] use the framework of accelerating cavities for this purpose, as it is a suitable choice to study the effect of nonuniform motion on localized quantum fields. Accelerating cavities are popular to study the relativistic effects on quantum information protocols. However, the authors in [12] develop a different approach from the approach of accelerating cavities. They formulate the evolution of the quantum field inside an accelerating cavity as a bosonic quantum Gaussian channel, which they then use to include the effects of the nonuniform motion of the quantum shares.

The authors in [20] focus on a relativistic variant of a (2,3)-threshold quantum secret sharing protocol. In the relativistic protocol presented in [20], similar to the nonrelativistic case, a dealer encodes the quantum secret into several quantum shares and distributes them to all the players. In this setup, the players are all located at different regions in the Minkowski spacetime, and the dealer and the players are all stationary. Under such circumstances, during the dealer’s distribution, the quantum shares experience nonuniform motion (noninertial) as they are transmitted to spacetime points in the future light cone of the dealer. Then, a subset of players within the access structure collaborate to retrieve the quantum secret by sharing their individual shares. However, to reach the same spacetime point, the shares go through phases of accelerating and decelerating motion while being transmitted, rendering the dynamics noninertial in general, in contrast to the special-relativistic regime described in the previous sections. The authors investigate how the noninertial motion of the shares affects the fidelity of the quantum secret-sharing protocol. The authors in this work claim to have solved continuous–variable quantum secret sharing wherein the quantum shares move nonuniformly in Minkowski spacetime. The tools used in this approach mainly comprise the tools developed in continuous-variable quantum information such as the formalism of the Gaussian quantum channels, dynamics of quantum field inside the cavity [20]. The authors do not use spin as the qubit in this relativistic scenario and yet are able to implement the relativistic quantum secret sharing protocol mainly via using the formalism of quantum field inside cavity, which may themselves be in noninertial motion [20]. The authors in [12] specify that they use the framework of Gaussian quantum information to write the evolution and dynamics of the quantum field inside the cavity central to their implementation of the protocol of quantum secret sharing, as a Gaussian quantum channel, in the noninertial regime. They use this channel to study the effect of noninertial motion of the shares on the fidelity of the quantum secret sharing. As a result, this work shows that various methods can be utilized to study the effects of noninertial motion without directly referring to the transformation of spin in a super-relativistic regime.

4. Entanglement of purification and entanglement wedge cross-section

In this section, we discuss the relativistic quantum information from another angle. We discuss the quantification and characterization of correlation measures used in nonrelativistic quantum information theory using quantum field theory in relativistic regime. The quantity that is usually used is the entanglement entropy. The entanglement entropy is used to quantify the entanglement of the pure quantum states. Using the AdS/CFT correspondence, the entanglement entropy has a holographic counterpart given by the area of minimal surface [25]. This characterization, therefore, provides a relationship between spacetime geometry and quantum entanglement.

The entanglement is a pure quantum correlation. However, the mixed quantum states contain both the classical and quantum correlation together contained in the total correlation of the quantum state. Now, mutual information is a usual measure of the total correlation of a mixed quantum state. However, the justification of the mutual information as a measure of total correlation is sometimes questioned. Also, there were attempts at separating the quantum correlation from the classical correlation in the quantum mutual information via the introduction of the measure of quantum correlation called quantum discord. Another approach was proposed to quantify the total correlation in a quantum state via the transformation of the entanglement of Bell pairs via local operation and classical communication. This measure is called the entanglement of purification. Its many properties were studied in [24]. Later, a holographic quantity was proposed as a counterpart of this entanglement of purification using the concept of entanglement wedge cross-section as a conjecture. This quantity is called the holographic entanglement of purification. The definition of entanglement of purification in nonrelativistic quantum mechanics is as follows. If we have a given mixed quantum state ρ_{AB}, then at first, we purify the mixed quantum state ρ_{AB} to $|\Psi\rangle_{AA'BB'}$, and then, the entanglement of purification of the mixed quantum state ρ_{AB} is given as the follows

$$E_P(\rho_{AB}) = \min_{A'B'} E_f(AA' : BB') \tag{9}$$

In the holographic entanglement of purification, the authors have considered the quantity E_W defined as the minimal cross-section of the entanglement wedge in AdS/CFT. The authors have shown there that they have observed that their properties coincide with those of the quantity called entanglement of purification, which measures the total correlation between two subsystems for a mixed quantum state comprising both classical and quantum correlations. Based on their observations and calculations on the entanglement wedge cross-section, the authors have conjectured that the entanglement wedge coincides with the entanglement of purification in holographic conformal field theories. The authors also have given a heuristic argument for this identification. The open question remains whether this conjecture is true or not. Several works are ongoing to check this conjecture in mathematical terms.

Another very important and promising open direction along this line of research is finding an operational interpretation of the definition of entanglement wedge cross-section; similarly, the entanglement of purification was motivated and operationally found in nonrelativistic quantum mechanics. The entanglement of purification was motivated operationally in terms of the conversion of the purely quantum correlation called entanglement in the maximally entangled state called the Bell pairs into the total correlation measure in the arbitrary quantum state using local operation and classical communication. Therefore, to settle the conjecture, another direction could be to try to find an operational footing of the entanglement wedge cross-section as well.

5. Conclusions

In this section, we summarize what we have discussed in the previous sections and open questions. Developing the fundamentals of understanding nature and natural phenomena with a robust mathematical construct and subsequent reproducible experimental verifications has been one of the strongest pillars of physics as we know it today. Many technological applications have stemmed from the robust structure of physical phenomena developed in theories in physics. Thus, we can imagine that such further developments will open up immense possibilities in future technologies that have a high potential for finding solutions for persistent problems in lives of people and such. Thus, it is clear that from the point of view of understanding nature, technological developments, and even resource allocation, the development of fundamental theories of nature is an area of research with immense potential.

Given the above motivation, we have covered a few aspects of physics that are important for developing quantum information theory in relativistic regimes; that is, in high-energy physics, it is paramount to use the relativistic effects via relativistic quantum mechanics and quantum field theory. It is well known that spin is an ill-understood concept in relativistic quantum physics. Prior approaches to spin in quantum physics have been discussed, and some recent promising approaches have been presented. It has also been discussed how a robust formulation of the concept of spin is crucial for developing quantum information in high-energy physics. We also reviewed the problem of spin quantum correlations and have presented the resources in the scientific literature that are trying to verify this experimentally in recent years. This issue is still unresolved, and a consistent resolution can open doors for the applications of relativistic quantum information and its extension of in space-based quantum technologies.

With the above open problem in mind, we also note that quantum field theory tools can be leveraged to develop quantum information protocols in relativistic quantum information in a high-energy arena, especially in the regime of noninertial motion. Such a protocol of relativistic quantum secret sharing has been discussed in detail. Similar techniques can be leveraged to develop such protocols further in high-energy physics.

In the last section of this chapter, we have covered a section on developing the definition of a total correlation measure in the language of quantum field theory and conformal field theory. The entanglement of purification, an active and open area of research, has been discussed here. Another open area of research has been pointed out in this arena, which has a high potential for development in the future. This area is the operational characterization of the quantum correlation measures defined in terms of tools as in quantum field theory and conformal theory. This area of enquiry is based on experimental verification in tabletop setups. This can be an active area of research in the future that might be challenging yet highly promising and with potential for rich dividends.

Acknowledgements

SB acknowledges funding from the Korea Institute of Science and Technology. S. B. acknowledges support from the National Research Foundation of Korea (2020M3E4A1079939 and 2022M3K4A1094774) and the KIST institutional program (2E31531).

Author details

Shrobona Bagchi
Korea Institute of Science and Technology, Seoul, South Korea

*Address all correspondence to: mailshrobona@gmail.com

References

[1] Adesso G, Bromley TR, Cianciaruso M. Measures and applications of quantum correlations. Journal of Physics A: Mathematical and Theoretical. 2016;**49**:473001

[2] Horodecki R, Horodecki P, Horodecki M, Horodecki K. Quantum entanglement. Reviews of Modern Physics. 2009;**81**:865

[3] Modi K, Brodutch A, Cable H, Paterek T, Vedral V. The classical-quantum boundary for correlations: Discord and related measures. Reviews of Modern Physics. 2012;**84**:1655

[4] Horodecki R. Quantum information. Acta Physica Polonica A. 2021;**139**:197

[5] Mann RB, Ralph TC. Relativistic quantum information. Classical and Quantum Gravity. 2012;**29**:22

[6] Fuentes I. Diversities in quantum computation and quantum information. In: Lecture Series in Quantum Information. World Scientific. 2012. pp. 107-147

[7] Tamburini F, Licata I, editors. Relativistic Quantum Information. Entropy, MDPI books; 2020

[8] Tjoa E, Gallock-Yoshimura K. Channel capacity of relativistic quantum communication with rapid interaction. Physical Review D. 2022;**105**:085011

[9] Tjoa E. Quantum teleportation with relativistic communication from first principles. Physical Review A. 2022;**106**: 032432

[10] Saldanha PL, Vedral V. Spin quantum correlations of relativistic particles. Physical Review A. 2012;**85**: 062101

[11] Streiter LF, Giacomini F, Brukner Č. Relativistic bell test within quantum reference frames. Physical Review Letters. 2021;**126**:230403

[12] Mikusch M, Barbado LC, Brukner Č. Transformation of spin in quantum reference frames. Physical Review Research. 2021;**3**:043138

[13] Giacomini F, Castro-Ruiz E, Brukner Č. Relativistic quantum reference frames: The operational meaning of spin. Physical Review Letters. 2019;**123**:090404

[14] Terno DR. Two roles of relativistic spin operators. Physical Review A. 2003; **67**:014102

[15] Zou L, Zhang P, Silenko AJ. Position and spin in relativistic quantum mechanics. Physical Review A. 2020;**101**:032117

[16] Wlodarczyk M, Caban P, Ciborowski J, Dragowski M, Rembielinski J. Quantum spin correlations in Moller scattering of relativistic electron beams. Physical Review A. 2017;**95**:022103

[17] Ciborowski J, Caban P, Dragowski M, Enders J, Fritzsche Y, Poliszczuk A, et al. A project to measure quantum spin correlations of relativistic electron pairs in Moller scattering. In: EPJ Web of Conferences. EDP Sciences. Vol. 164. 2017. p. 01004

[18] Rick Perche T, Martín-Martínez E. Role of quantum degrees of freedom of relativistic fields in quantum information protocols. Physical Review A. 2023;**107**:042612

[19] Benincasa DMT, Borsten L, Buck M, Dowker F. Quantum information processing and relativistic quantum

fields. Classical and Quantum Gravity. 2014;**31**:075007

[20] Bruschi DE, Lee AR, Fuentes I. Time evolution techniques for detectors in relativistic quantum information. Journal of Physics A: Mathematical and Theoretical. 2013;**46**:165303

[21] Calabrese P, Cardy J. Entanglement entropy and quantum field theory. Journal of Statistical Mechanics: Theory and Experiment. 2004;**2004**:P06002

[22] Takayanagi T. Entanglement entropy from a holographic viewpoint. Classical and Quantum Gravity. 2012;**29**: 153001

[23] Ahmadi M, Ya-Dong W, Sanders BC. Relativistic (2,3)-threshold quantum secret sharing. Physical Review D. 2017; **96**:065018

[24] Bagchi S, Pati AK. Monogamy, polygamy, and other properties of entanglement of purification. Physical Review A. 2015;**91**:042323

[25] Umemoto K, Takayanagi T. Entanglement of purification through holographic duality. Nature Physics. 2018;**14**:573-577

Chapter 3

Perspective Chapter: On Entanglement Measures – Discrete Phase Space and Inverter-Chain Link Viewpoint

Felix A. Buot

Abstract

In contrast to abstract statistical analyses in the literature, we present a concrete physical diagrammatic model of entanglement characterization and measure with its underlying discrete phase-space physics. This paper serves as a pedagogical treatment of this complex subject of entanglement measures. We review the important inherent concurrence property of entangled qubits, as well as underscore its emergent qubit behavior. From the discrete phase space point of view, concurrence translates to translation symmetry of entangled binary systems in some quantitative measure of entanglement. Although the focus is on bipartite system, the technique is readily extendable to multi-partite system of qubits, as can easily be deduced from the physical inverter-chain link model. A diagrammatic analysis of the entanglement of formation for any multi-partite qubit system is given. We show that quantum mechanical joint distribution are entanglements.

Keywords: discrete phase space, inverter-chain link model, Hadamard transform, entanglement of formation, concurrence, ergodic distance

1. Introduction

Quantum entanglement has developed from a mere intellectual curiosity [1] of the fundamental structure of quantum mechanics[1] to become an important and practical resource for quantum information processing in the evolving theory of quantum information and ultra-fast computing. Thus, the quantitative measure of entanglement has developed into one of the most active fields of theoretical and experimental research. Here we will try to shed more light on some of the important concepts in the quantification of quantum entanglement by using a concrete simple mechanical model

[1] Note that although Bell's theorem asserts the nonlocality of quantum mechanics, the EPR inquiry is still not resolved, i.e., what is still left unanswered is the mysterious 'link' between qubits corresponding to our "see-saw" or mechanical inverter-chain link. (Note: Bell's inequality theorem is widely discussed in the literature and websites, we prefer not to cite specific reference).

of a bipartite system of qubits or chain of qubits. This treatment will be in contrast with mostly abstract and statistical treatment of entanglement measure in the literature.

We will focus on the so-called entanglement of formation and concurrence, two of the most important concepts to characterize entanglement resource. Here we consider a qubit as a two-state system. Moreover, we also consider an entangled qubit as effectively a two-state system, an *emergent qubit* as depicted in our physical inverter-chain link model. A two-state system has a unity entropy. Thus, a maximally entangled state has entropy equal to 1. From the discrete phase space point of view, any two-state system can be considered to possess two lattice-site states. Discrete Fourier transform or Hadamard transform implements unitary superposition of the two *lattice-site states* (also referred to as '*Wannier functions*') to yield a sort of *crystal-momentum states* (also referred to as '*Bloch functions*'). For example, take the $|00\rangle$ and $|11\rangle$ '*lattice-site states*', then the Hadamard bijective discrete transformation gives the'*crystal-momentum states*', Φ^+ and Φ^-, which are two of the Bell basis states (or'*Bloch function*'states).

2. Bell basis deduced from inverter-chain link model

We sketch here the derivation of the Bell basis states from our physical model, as depicted in **Figure 1**. Clearly, the entanglement of two *bare* qubits is divided into two orthogonal spaces of triplet[2] and singlet entanglement states.

Since the Hadamard transform is unitary, the inverse transform is well-defined, i.e., the transformation is bijective. We refer to $\{|00\rangle, |11\rangle\}$ as the "*Wannier functions*"

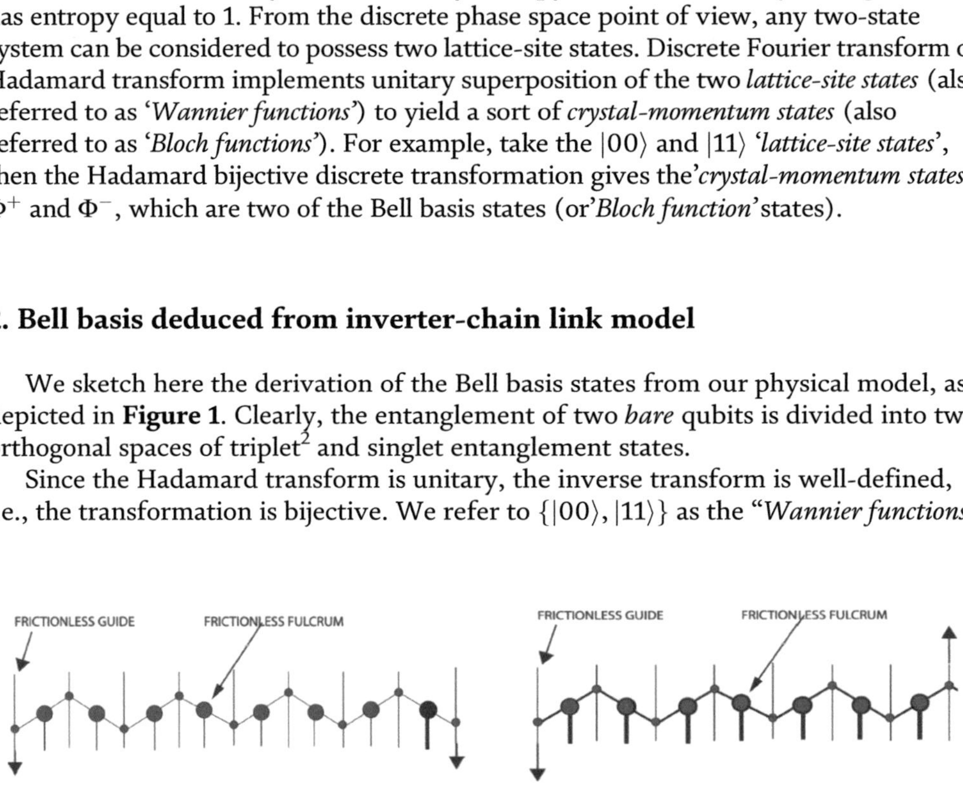

$$\begin{pmatrix} |\Phi^+\rangle \\ |\Phi^-\rangle \end{pmatrix} = \frac{1}{\sqrt{2}} \begin{pmatrix} 1 & 1 \\ 1 & -1 \end{pmatrix} \begin{pmatrix} |0\rangle|0\rangle \\ |1\rangle|1\rangle \end{pmatrix}$$

$$|\Phi^+\rangle = \frac{1}{\sqrt{2}}|00\rangle + \frac{1}{\sqrt{2}}|11\rangle$$

$$|\Phi^-\rangle = \frac{1}{\sqrt{2}}|00\rangle - \frac{1}{\sqrt{2}}|11\rangle$$

$$\begin{pmatrix} |\Psi^+\rangle \\ |\Psi^-\rangle \end{pmatrix} = \frac{1}{\sqrt{2}} \begin{pmatrix} 1 & 1 \\ 1 & -1 \end{pmatrix} \begin{pmatrix} |0\rangle|1\rangle \\ |1\rangle|0\rangle \end{pmatrix}$$

$$|\Psi^+\rangle = \frac{1}{\sqrt{2}}|01\rangle + \frac{1}{\sqrt{2}}|10\rangle$$

$$|\Psi^-\rangle = \frac{1}{\sqrt{2}}|01\rangle - \frac{1}{\sqrt{2}}|10\rangle$$

Figure 1.
Physical diagrammatic model of "triplet" (left) and singlet (right) entanglement. By construction, each diagram is viewed as a two-state system, respectively. The Bell basis are readily derived below each diagram, using the Hadamard transformation. The actual physical implementation of the chain of inverters may need frictionless male/female sliding tube coupling for large-angle swing, but this is beside the point. We assume a rigid coupling model for simultaneity of events at both ends.

---

[2] The use of the term "triplet" is actually a misnomer here since the entangled system is not free to assume a singlet or zero spin state. Thus, this term is used here only as a label.

space and the $\{|\Phi^+\rangle, |\Phi^-\rangle\}$ as the corresponding "*Bloch functions*" space of a two-state *triplet* entanglement system. Likewise, $\{|01\rangle, |10\rangle\}$ as the "*Wannier functions*" space and the $\{|\Psi^+\rangle, |\Psi^-\rangle\}$ as the corresponding "*Bloch functions*" space of a two-state singlet entanglement system.

By virtue of this bijective relationship, any function of Wannier functions will have a corresponding function of Bloch functions. For example, a maximally *mixed* Wannier function states will generate a maximally mixed Bloch function states or the so-called maximally *mixed* entanglement states. Mixed and pure states will be further discussed below.

3. Entangled qubits as an emergent qubit

The virtue of our inverter-chain link model is that the emergent two-state property of entangled qubits is very transparent, since changing the state of one of the qubits also *immediately* change the states of the rest of the entangled partner(s)[3]. Here, we will rigorously justify our claim that entangled qubits behave, as a whole, as an *emergent* qubit.

Here, we employ the matrix representation of states and operators. We now represent $|\Phi^+\rangle$ and$|\Phi^-\rangle$ as,

$$|\Phi^+\rangle = \frac{1}{\sqrt{2}}\begin{pmatrix}1\\1\end{pmatrix}, \tag{1}$$

$$|\Phi^-\rangle = \frac{1}{\sqrt{2}}\begin{pmatrix}1\\-1\end{pmatrix}. \tag{2}$$

Then we have

$$\langle\Phi^+|\sigma_x|\Phi^+\rangle = \frac{1}{2}(1\ \ 1)\begin{pmatrix}0 & 1\\1 & 0\end{pmatrix}\begin{pmatrix}1\\1\end{pmatrix} = 1, \tag{3}$$

$$\langle\Phi^-|\sigma_x|\Phi^-\rangle = \frac{1}{2}(1\ \ -1)\begin{pmatrix}0 & 1\\1 & 0\end{pmatrix}\begin{pmatrix}1\\-1\end{pmatrix} = -1, \tag{4}$$

proving the two eigenvalues, like a qubit property of the triplet system.

For the singlet the most straightforward manipulation is to recognize that the singlet system is an independent system with its own two states which is orthogonal to the triplet states. In this sense we can also represent the singlet states just like that of the triplet states, namely,

---

[3] The idea that entanglement is due to conservation of momentum does not hold for triplet entanglement since its two states give opposing spin-angular momentum. However, one may interpret that quantum superposition of the two-opposing angular momentum states conserves the overall zero net spin-angular momentum, as supported by their eigenvalues similar to a single qubit. On the other hand, for singlet entanglement, the zero angular momentum is conserve in both two states. This seemingly apparent physical difference of the triplet and singlet entanglements underscores the importance of resolving the "mysterious link" in the EPR inquiry, [1] in order to further advance theoretical physics.

$$|\Psi^+\rangle = \frac{1}{\sqrt{2}}\begin{pmatrix}1\\1\end{pmatrix}, \tag{5}$$

$$|\Psi^-\rangle = \frac{1}{\sqrt{2}}\begin{pmatrix}1\\-1\end{pmatrix}. \tag{6}$$

Note that in the 'Bloch-state" space, σ_z in the "Wannier-state" space is transform to σ_x in entangled states,

$$\sigma_x = H\sigma_z H = \begin{pmatrix}0 & 1\\1 & 0\end{pmatrix}. \tag{7}$$

Thus, we also have for the singlet states

$$\langle\Psi^+|\sigma_x|\Psi^+\rangle = \frac{1}{2}(1 \quad 1)\begin{pmatrix}0 & 1\\1 & 0\end{pmatrix}\begin{pmatrix}1\\1\end{pmatrix} = 1, \tag{8}$$

$$\langle\Psi^-|\sigma_x|\Psi^-\rangle = \frac{1}{2}(1 \quad -1)\begin{pmatrix}0 & 1\\1 & 0\end{pmatrix}\begin{pmatrix}1\\-1\end{pmatrix} = -1. \tag{9}$$

Following this notion, Eqs. (8) and (9) can easily be extended to all entangled qubits, bipartite or multi-partite qubit systems to yield an emergent qubit. It is far more simpler to analyze the inverter-chain link diagrams, i.e., employ diagrammatic analyses.

4. Translational/shift invariance of Bell basis

The translational invariance of maximally entangled Bell basis states demonstrates a unique characteristic of entangled qubits. This unique property has been used to detect or measure how much entanglement is present in arbitrary pure and mixed states. This is implemented in terms of the notion of concurrence, to be elaborated below.

The translational or shift operation (also referred to as the *flip* operation in the literature) is demonstrated here for the Bell Basis states and their superpositions. First, let us consider all the Bell basis states. We have,

$$\Phi^+ = \frac{1}{\sqrt{2}}(|0\rangle|0\rangle + |1\rangle|1\rangle). \tag{10}$$

Upon applying the translation (shift by $+1(mod 2)$) operator, $T(+1)$, we obtain

$$\begin{aligned} T(+1)\Phi^+ &= \frac{1}{\sqrt{2}}(|0+1\rangle|0+1\rangle + |1+1\rangle|1+1\rangle),\\ &= \frac{1}{\sqrt{2}}(|1\rangle|1\rangle + |0\rangle|0\rangle) = \Phi^+, \end{aligned} \tag{11}$$

where addition obeys modular arithmetic (*mod* 2). The 'plus' entangled basis states are even upon the operation of $T(+1)$, whereas the 'minus' entangled basis states are odd when operated by $T(+1)$. We still consider the shited, $-\Phi^-$ and $-\Psi^-$, as invariant since it differs from the unshifted Bell states by a global phase factor. In the literature this translation operation is the so-called flipping operation first used by Wooters, et al., [2, 3].

4.1 Translational property of maximal superposition of Bell basis

Now let us consider the superposition of maximally entangled Bell basis states. We can easily see that a superposition of 'plus' entangled basis states with 'minus' entangled basis states yield *unentangled* basis states. For example, we have,

$$\frac{1}{\sqrt{2}}(\Phi^{+} + \Phi^{-}) = |0\rangle|0\rangle, \tag{12}$$

$$\frac{1}{\sqrt{2}}(\Phi^{+} - \Phi^{-}) = |1\rangle|1\rangle. \tag{13}$$

The thing to notice is that although the superposition is made up of two maximally entangled Bell basis, the results are not shift invariant (i.e., generation of another *Wannier* state yields *Wannier* state located at another *site, which is* orthogonal to the unshifted one). This means that the superposition given above yields unentangled states (product states) in Eq. (13). In what follows, we will see that only the following combination of entangled basis states yields another entangled basis states as long as they both belong to either 'even' or 'odd' spaces. We have, by diagrammatic construction,

$$triplet^{\pm} \otimes triplet^{\pm} = triplet^{\pm} \;\text{ or }\; singlet^{\pm}, \tag{14}$$

$$triplet^{\pm} \otimes singlet^{\pm} = singlet^{\pm} \;\text{ or }\; triplet^{\pm}, \tag{15}$$

$$singlet^{\pm} \otimes singlet^{\pm} = triplet^{\pm} \;\text{ or }\; singlet^{\pm}. \tag{16}$$

Equation (15) is quite interesting because it holds on a complete expansion of a direct product of two qubit states. These relations can easily be deduced from the physical diagrammatic model, see **Figure 1**. For example, if we define the operation as a superposition such as,

$$\frac{1}{\sqrt{2}}(\Phi^{+} + \Psi^{+}) = \left[\frac{1}{\sqrt{2}}(|1\rangle + |0\rangle) \otimes \frac{1}{\sqrt{2}}(|1\rangle + |0\rangle)\right]. \tag{17}$$

We see that Eq. (17) is a direct product state of two qubits. We will see in what follows that this is an entangled state and corresponds to Eq. (15) of the $singlet^{+}$ or $triplet^{+}$ of the model diagrams depending on the actual linkage. Similarly, we have,

$$\frac{1}{\sqrt{2}}(\Phi^{-} + \Psi^{-}) = \left[\frac{1}{\sqrt{2}}(|1\rangle + |0\rangle) \otimes \frac{1}{\sqrt{2}}(|1\rangle - |0\rangle)\right], \tag{18}$$

corresponds to Eq. (15) of the *singlet* or *triplet* of the model diagrams, depending on the actual linking.

However, the following combinations of 'even' and 'odd' entangled states result in *unentangled* states, namely,

$$\frac{1}{\sqrt{2}}(\Phi^{+}+\Psi^{-})=\frac{1}{2}[(|1\rangle|1\rangle+|0\rangle|0\rangle)+(|0\rangle|1\rangle-|1\rangle|0\rangle)], \tag{19}$$

and

$$\frac{1}{\sqrt{2}}(\Phi^{-}+\Psi^{+})=\frac{1}{2}[(|1\rangle|1\rangle-|0\rangle|0\rangle)+(|0\rangle|1\rangle+|1\rangle|0\rangle)], \tag{20}$$

by virtue of the failure to have global sign factors. All these claims are justified through the concept of concurrence, an inherent property of entangled qubits, to be discussed in what follows. Moreover, this feature of failing to have global sign factor is also reflected in the failure to represent by our inverter-chain link diagrams.

5. Bipartite system

Let the Hilbert spaces of a bipartite system consisting of A and B be denoted by $\mathscr{H}_A$ and $\mathscr{H}_B$, respectively. A bipartite system is a system with Hilbert space equal to the direct product of $\mathscr{H}_A$ and $\mathscr{H}_B$, i.e.,

$$\mathscr{H}_{AB}=\mathscr{H}_A\otimes\mathscr{H}_B \tag{21}$$

Let the density matrix for the whole system be denoted by ρ. Then the reduced density matrix of a subsystem A is given by the partial trace,

$$\rho_A=Tr_B\ \rho. \tag{22}$$

The entanglement entropy, S_A, is the *von Newmann entropy* of the reduced density matrix, ρ_A,

$$S_A=-Tr\ \rho_A\log\rho_A. \tag{23}$$

Example 1 Let Ω be the number of distinct states in subsystem A. assume a uniform distribution among states, hence ρ_A has eigenvalues $\frac{1}{\Omega}$, i.e., ρ_A can be represented by a diagonal $\Omega\times\Omega$ matrix with identical matrix elements given by $\frac{1}{\Omega}$. Thus, in taking the trace we can use the eigenvalues of the reduced density matrix operator, ρ_A. Therefore, we have

$$\begin{aligned}S_A&=-Tr\frac{1}{\Omega}\log\frac{1}{\Omega},\\&=\log\ \Omega.\end{aligned} \tag{24}$$

Upon multiplying by the Boltzmann constant, k_B, we obtain

$$k_BS_A=k_B\log\ \Omega, \tag{25}$$

which is the Boltzmann thermodynamic entropy, based on ergodic theorem.
Example 2. Two qubit system.

A qubit is simply a quantum bit whose number of distinct eigenstates is 2. We denote these eigenstates as $|0\rangle$ and $|1\rangle$, i.e., a two-state system. If each subsystem A or

B is a single qubit, then the Hilbert space of the whole system is span by the following 4 direct product states, namely,

$$|00\rangle, |01\rangle, |10\rangle, |11\rangle. \tag{26}$$

Here, the first bit refers to subsystem A and the second bit refers to subsystem B. Now, the density matrix of a pure state is,

$$\rho = |\psi\rangle\langle\psi|, \tag{27}$$

then

$$\begin{aligned} \rho^2 &= |\psi\rangle|\psi\rangle\langle\psi||\psi\rangle\langle\psi|, \\ &= \rho. \end{aligned} \tag{28}$$

An operator whose square is equal to itself must have an eigenvalue equal to unity. Let us write for pure state of the two qubits as,

$$|\psi\rangle = \frac{1}{\sqrt{2}}(|00\rangle + |11\rangle), \tag{29}$$

$$\langle\psi| = \frac{1}{\sqrt{2}}(\langle 00| + \langle 11|). \tag{30}$$

We have,

$$\langle\psi||\psi\rangle = 1. \tag{31}$$

Thus, indeed,

$$\begin{aligned} \rho^2 &= |\psi\rangle\langle\psi||\psi\rangle\langle\psi|, \\ &= |\psi\rangle\langle\psi| = \rho. \end{aligned} \tag{32}$$

We refer to $|\psi\rangle$ as a maximally entangled state in the sense that the first qubit is exactly "link" to the second qubit. The reduced density matrix for subsystem A is

$$\rho_A = Tr_B\ \rho = \frac{1}{2}(|0_A\rangle\langle 0_A| + |1_A\rangle\langle 1_A|). \tag{33}$$

Now clearly

$$\begin{aligned} \rho_A^2 &= \frac{1}{4}(|0_A\rangle\langle 0_A| + |1_A\rangle\langle 1_A|)(|0_A\rangle\langle 0_A| + |1_A\rangle\langle 1_A|), \\ &= \frac{1}{4}(|0_A\rangle\langle 0_A| + |1_A\rangle\langle 1_A|) = \frac{1}{2}\rho_A, \end{aligned} \tag{34}$$

so that ρ_A is not a pure state but mixed, i.e., a mixture of two pure states, $|0_A\rangle$ and $|1_A\rangle$. Note that the mixed states density matrix do not possess *off-diagonal elements*. From the last identity, the eigenvalues of ρ_A is $\frac{1}{2}$ and since the subsystem A is a single qubit with two distinct states, then eigenvalues of ρ_A corresponds to $\frac{1}{\Omega}$ of our first example above. Thus we refer to ρ_A as "uniformly'" mixed often referred to as "*maximally mixed*" with the initial state $|\psi\rangle$ as "*maximally entangled*".

6. Entanglement of formation of multi-partite qubit systems

The entanglement entropy of subsystem A can be calculated using the eigenvalues of the 2×2 matrix of ρ_A, which is $\lambda_A = \frac{1}{2}$. Therefore, we have for the entanglement entropy or entanglement of formation given by,

$$\begin{aligned} S_A &= -Tr\rho_A \log \rho_A, \\ &= -Tr\frac{1}{2}\log\frac{1}{2} = \log 2^1, \end{aligned} \tag{35}$$

where exponent base 2, here 1, is the number of qubits that is entangled with system B. This will be made clear in the next example.

Example 3 A four qubit system:

If each subsystem A or B has two qubits, then the Hilbert space of the whole system is span by the following 16 direct product states, i.e., $2^4 = 16$ states. The maximally entangled pure state is determined by the following eight diagrams [4], see **Figure 2**, and their flipped or translated states,

Any combination or superposition of both 'even' triplet and singlet states comprised a maximally entangled state, i.e.,

$$\begin{pmatrix} \Phi_1^+ \\ \Phi_1^- \end{pmatrix} = \frac{1}{\sqrt{2}}\begin{pmatrix} 1 & 1 \\ 1 & -1 \end{pmatrix}\begin{pmatrix} |0\rangle|0\rangle|0\rangle|0\rangle \\ |1\rangle|1\rangle|1\rangle|1\rangle \end{pmatrix}, \tag{36}$$

$$\begin{pmatrix} \Phi_8^+ \\ \Phi_8^- \end{pmatrix} = \frac{1}{\sqrt{2}}\begin{pmatrix} 1 & 1 \\ 1 & -1 \end{pmatrix}\begin{pmatrix} |0\rangle|1\rangle|0\rangle|1\rangle \\ |1\rangle|0\rangle|1\rangle|0\rangle \end{pmatrix}. \tag{37}$$

Thus, from the states given above, we have for example the maximally entangled state,

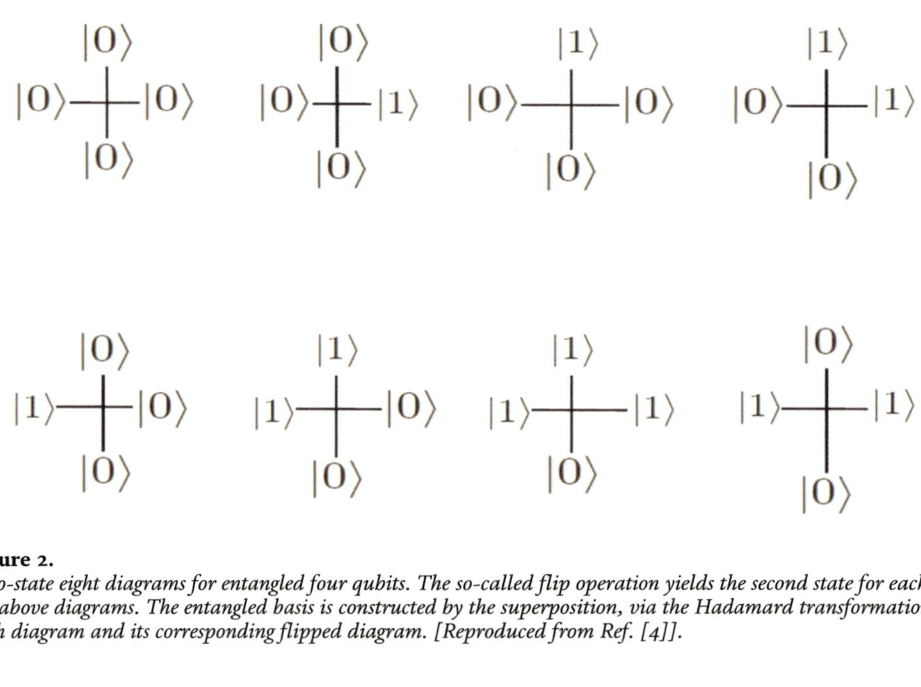

Figure 2.
Two-state eight diagrams for entangled four qubits. The so-called flip operation yields the second state for each of the above diagrams. The entangled basis is constructed by the superposition, via the Hadamard transformation, of each diagram and its corresponding flipped diagram. [Reproduced from Ref. [4]].

$$|\psi\rangle = \frac{1}{\sqrt{2}}\left(\Phi_1^+ + \Phi_8^+\right), \tag{38}$$

and similarly, we can also form another entangled state,

$$|\psi\rangle = \frac{1}{\sqrt{2}}\left(\Phi_1^- + \Phi_8^-\right), \tag{39}$$

where the first two bits belongs to subsystem A and the second two bits belong to subsystem B. We note that

$$\langle\psi||\psi\rangle = 1. \tag{40}$$

Therefore

$$\begin{aligned} \rho^2 &= (|\psi\rangle\langle\psi|)^2, \\ &= |\psi\rangle\langle\psi| = \rho. \end{aligned} \tag{41}$$

So $|\psi\rangle$ is a pure state.

In general, the following maximally entangled state corresponds to a chain of entangled basis states, namely,

$$|\psi^+\rangle = \frac{1}{\sqrt{8}}\left(\Phi_1^+ + \Phi_2^+ + \Phi_3^+ + \Phi_4^+ + \Phi_5^+ + \Phi_6^+ + \Phi_7^+ + \Phi_8^+\right), \tag{42}$$

$$|\psi^-\rangle = \frac{1}{\sqrt{8}}\left(\Phi_1^- + \Phi_2^- + \Phi_3^- + \Phi_4^- + \Phi_5^- + \Phi_6^- + \Phi_7^- + \Phi_8^-\right), \tag{43}$$

of a four qubit system.

The density matrix operator for the whole 4-qubit system can be written as,

$$\rho = \frac{1}{\sqrt{2}}\left(|\Phi_1^- + \Phi_8^-\rangle\right)\frac{1}{\sqrt{2}}\left(\langle\Phi_1^- + \Phi_8^-|\right), \tag{44}$$

$$\rho = \frac{1}{4}[|00,00\rangle + |01,01\rangle + |10,10\rangle + |11,11\rangle][\langle 00,00| + \langle 01,01| + \langle 10,10| + \langle 11,11|]. \tag{45}$$

The reduced density matrix operator for subsystem A is again obtained by taking partial trace with respect to subsystem B.

$$\begin{aligned} \rho_A &= Tr_B\ \rho, \\ &= \langle 00_B|\rho|00_B\rangle + \langle 01_B|\rho|01_B\rangle + \langle 10_B|\rho|10_B\rangle + \langle 11_B|\rho|11_B\rangle, \\ &= \frac{1}{4}|00_A\rangle\langle 00_A| + |01_A\rangle\langle 01_A| + |10_A\rangle\langle 10_A| + |11_A\rangle\langle 11_A|. \end{aligned} \tag{46}$$

To determine the eigenvalues for ρ_A, we take its square,

$$\begin{aligned} \rho_A^2 &= \frac{1}{16}(|00_A\rangle\langle 00_A| + |01_A\rangle\langle 01_A| + |10_A\rangle\langle 10_A| + |11_A\rangle\langle 11_A|), \\ &\times(|00_A\rangle\langle 00_A| + |01_A\rangle\langle 01_A| + |10_A\rangle\langle 10_A| + |11_A\rangle\langle 11_A|), \\ &= \frac{1}{16}(|00_A\rangle\langle 00_A| + |01_A\rangle\langle 01_A| + |10_A\rangle\langle 10_A| + |11_A\rangle\langle 11_A|), \\ &= \frac{1}{4}\rho_A. \end{aligned} \tag{47}$$

So we have

$$\rho_A^2 = \frac{1}{4}\rho_A \tag{48}$$

and the eigenvalues of ρ_A is $\frac{1}{4}$. We can now calculate the entanglement entropy of subsystem A. We have

$$\begin{aligned} S_A &= -Tr\rho_A \log \rho_A, \\ &= \log 2^2. \end{aligned} \tag{49}$$

The exponent 2 correspond to the number of qubits that is entangled with subsystem B. In general, for maximally entangled bipartite system A and B, each having k number of qubits S_A is given by,

$$S_A = \log 2^k. \tag{50}$$

Now of course, for this bipartite system,

$$S_A = S_B, \tag{51}$$

which simply means a complete matching of configurations of each system A and B, respectively, otherwise some degrees of freedom will be hanging, not matched or cannot be entangled.

Example 4. Tripartite system of three qubits:

The following diagrams represent the entangled tripartite system of qubits (**Figure 3**). The entangled basis states are as follows,

$$\begin{pmatrix} \Xi^+ \\ \Xi^- \end{pmatrix} = \frac{1}{\sqrt{2}}\begin{pmatrix} 1 & 1 \\ 1 & -1 \end{pmatrix}\begin{pmatrix} |0\rangle|1\rangle|1\rangle \\ |1\rangle|0\rangle|0\rangle \end{pmatrix}, \tag{52}$$

$$\begin{pmatrix} \Theta_3^+ \\ \Theta_3^{--} \end{pmatrix} = \frac{1}{\sqrt{2}}\begin{pmatrix} 1 & 1 \\ 1 & -1 \end{pmatrix}\begin{pmatrix} |0\rangle|0\rangle|0\rangle \\ |1\rangle|1\rangle|1\rangle \end{pmatrix}, \tag{53}$$

$$\begin{pmatrix} \Omega^+ \\ \Omega^- \end{pmatrix} = \frac{1}{\sqrt{2}}\begin{pmatrix} 1 & 1 \\ 1 & -1 \end{pmatrix}\begin{pmatrix} |0\rangle|1\rangle|0\rangle \\ |1\rangle|0\rangle|1\rangle \end{pmatrix}, \tag{54}$$

$$\begin{pmatrix} \Gamma^+ \\ \Gamma^- \end{pmatrix} = \frac{1}{\sqrt{2}}\begin{pmatrix} 1 & 1 \\ 1 & -1 \end{pmatrix}\begin{pmatrix} |0\rangle|0\rangle|1\rangle \\ |1\rangle|1\rangle|0\rangle \end{pmatrix}. \tag{55}$$

A chain or superposition of any choice of entangled *even* basis states also form a maximally entangled state, e.g.,

$$|\Psi^4\rangle = \frac{1}{\sqrt{4}}\left[\Xi^+ + \Theta_3^+ + \Omega^+ + \Gamma^+\right], \tag{56}$$

$$|\Psi^3\rangle = \frac{1}{\sqrt{3}}\left[\Xi^+ + \Theta_3^+ + \Omega^+\right], \tag{57}$$

$$|\Psi^2\rangle = \frac{1}{\sqrt{2}}\left[\Xi^+ + \Theta_3^+\right], \tag{58}$$

$$|\Psi^1\rangle = \Xi^+, \tag{59}$$

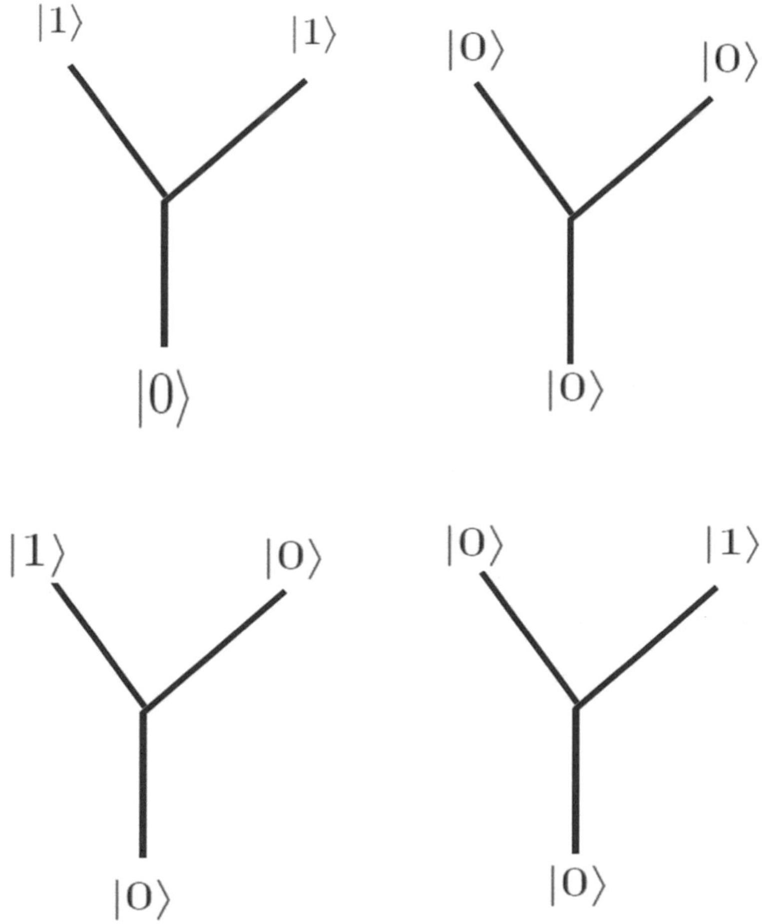

Figure 3.
Two-state four diagrams for entangled three qubits of a tripartite system. Flip operations yield the respective second states. [Reproduced from Ref. [4]].

are all, respectively, maximally entangled states with emergent two-state or qubit properties.

Example 5 Multi-partite systems:

It is important to point out that any entangled qubits, irrespective of their number, identically behave like a single qubit, i.e., behaving exactly like a two-state system. Thus, the natural order of disentangling is as follows: First, one entangles one qubit from the remaining two entangled qubits. The entanglement of formation is equivalent to one qubit. Next, one disentangled the remaining entangled two qubits. This further give an entanglement of formation of one qubit. Thus, the total entanglement of formation is $1 + 1 = 2$ qubits

6.1 Monogamy inequality of entanglement of formation

The reasoning we have given above yields exact equality of the monogamy, usually given as *inequality* in the literature as deduced from statistical analysis, e.g., one given

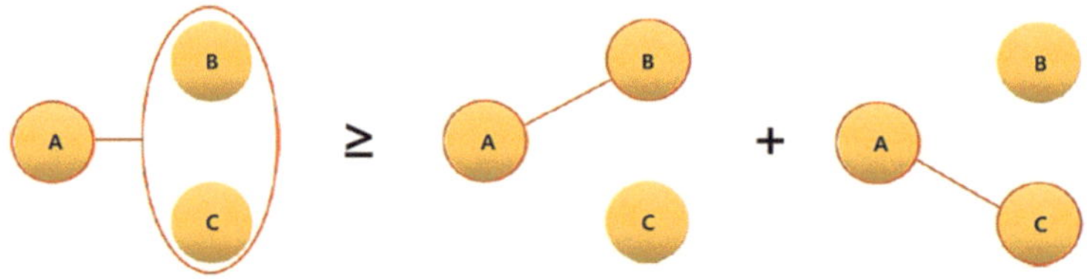

Figure 4.
Figure reproduced from Ref. [5]. In our diagrammatic analysis the right-hand side of the figure yields $1 + 1 = 2$ *as the entanglement of formation, being the sum of two 2-qubit entanglements. Thus, in our diagrammatic analysis, we obtained exact monogamy equality not inequality.*

by Kim [5], The exact equality relation for the entanglement of formation is diagrammatically shown in **Figure 4**.

The exact equality comes about by construction, since on both sides of **Figure 4**, two 2-qubit entangled states are being untangled to obtain the entanglement of formation of this multi-partite system. This comes about through the knowledge that any multi-partite entangled qubits behave as an emergent qubit.

Similarly, for an entangled 4-partite system, the entanglement of formation is determined schematically by the diagrams of **Figure 5**, where the unentangling operations to be done on the left side are itemized on the unentangling operations of *three* entangled 2-qubits on the right side of the equality sign,

This sort of diagrammatic analysis of the entropy of entanglement of formation can straightforwardly be employed to all multi-partite qubit systems by a simple counting argument as is done here. For example, for 18 multi-partite entangled qubits in **Figure 6**, we have the entanglement of formation schematically depicted by the equality where there is 17 number of monogamy in the right-hand side yielding 17 qubits of entropy of entanglement of formation.

7. Joint distributions as entanglements

In this section, we will explicitly show the discrete phase-space physics of qubit entanglements. We will demonstrate that a *joint "Bloch-function"* distributions in *"momentum space"* transforms into entangled qubits (*"Wannier states"*) in *"lattice-site space"*[4]. To the author's knowledge, this demonstrates for the first time the entanglement properties of joint distributions in quantum mechanics, which is ubiquitous in quantum transport theory of topological insulators discussed in Ref. [6].

Consider the generalized Fourier transformation between two Hilbert spaces,

$$|p\rangle = \sum_q \langle q|p\rangle \; |q\rangle, \tag{60}$$

where the $\langle q|p\rangle$ is the transformation function. This is particularly given by the discrete Fourier transform function,

$$\langle q|p\rangle = \frac{1}{\sqrt{N}} \exp\left(-\frac{i}{\hbar} p \cdot q\right). \tag{61}$$

[4] We maintained the *Bloch function/Wannier function* analogy for convenience, and to stress the wide-ranging impact of the discrete phase-space physics.

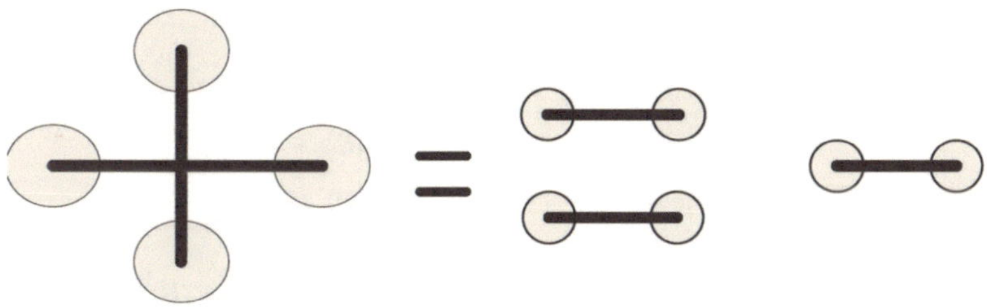

Figure 5.
In our diagrammatic analysis the right-hand side of the figure yields $1 + 1 + 1 = 3$ *as the entanglement of formation, being the sum of three 2-qubit entanglements. Thus, in our diagrammatic analysis, we obtained exact equality not inequality.*

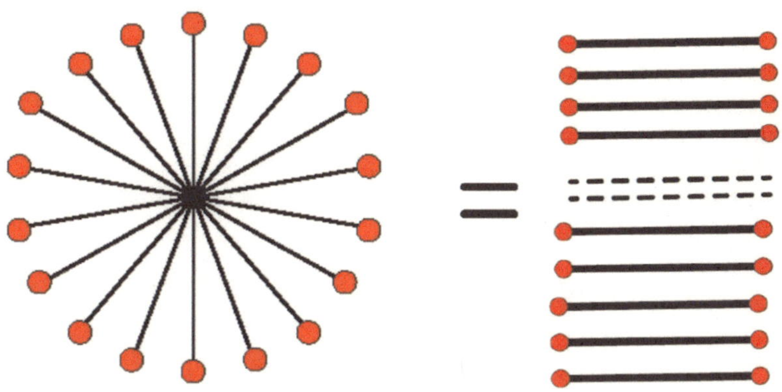

Figure 6.
In our diagrammatic analysis the right hand side of the figure yields $1 + 1 + 1, \ldots\ldots 1 = 17$ *qubits as the entanglement of formation, being the sum of 17 two-qubit entanglements of formation. There are 18 multi-party entangled qubits in the left-hand side.*

where N is the number of discrete points in bijective transformations (this number N is conveniently taken as prime integers).

7.1 Two-point joint distributions

We have a joint distribution in p-space as,

$$|p'\rangle|p''\rangle = \frac{1}{\sqrt{N}}\sum_{q'} \exp\left(-\frac{i}{\hbar}p' \cdot q'\right) \; |q'\rangle \frac{1}{\sqrt{N}}\sum_{q''} \exp\left(-\frac{i}{\hbar}p'' \cdot q''\right)|q''\rangle. \tag{62}$$

Therefore, for a qubit or a two-state system, we write,

$$|p\rangle = \frac{1}{\sqrt{2}}\sum_{q=0}^{1} \exp\left(-\frac{i}{\hbar}p \cdot q\right) \; |q\rangle. \tag{63}$$

Consider the *joint distribution* of two qubits in "*Bloch-function*" space. We can form 4 joint distributions of *Bloch-function* in what follows. To derive these, first, we write the following matrix-representation and Hadamard-transformation identity,

$$\pi\hbar\begin{pmatrix}|0\rangle\\|1\rangle\end{pmatrix}\otimes\pi\hbar\begin{pmatrix}|0\rangle\\|1\rangle\end{pmatrix}$$
$$=\frac{1}{\sqrt{2}}\frac{1}{\sqrt{2}}\begin{pmatrix}(|0\rangle+|1\rangle)\\(|0\rangle-|1\rangle)\end{pmatrix}\otimes\begin{pmatrix}(|0\rangle+|1\rangle)\\(|0\rangle-|1\rangle)\end{pmatrix}. \tag{64}$$

We will denote the joint distribution in p-space as $\mathcal{D}(p_1, p_2, \ldots, p_n)$. We will denote these 4 "*Bloch-function*" distributions by the standard notations for the Bell basis in the literature, namely, Φ^+, Φ^-, Ψ^+ and Ψ^-. Then, we have

$$\mathcal{D}(0,0)=\frac{1}{\sqrt{2}}\{\Phi^+ + \Psi^+\},\quad \mathcal{D}(1,1)=\frac{1}{\sqrt{2}}(\Phi^+ - \Psi^+), \tag{65}$$

$$\mathcal{D}(0,1)=\frac{1}{\sqrt{2}}(\Phi^- - \Psi^-),\quad \mathcal{D}(1,0)=\frac{1}{\sqrt{2}}(\Phi^- + \Psi^-). \tag{66}$$

7.2 Three-point joint distributions

Similarly, for three-fold joint distribution in p-space, we have,

$$|p_1\rangle|p_2\rangle|p_3\rangle=\frac{1}{\sqrt{N}}\sum_{q_1}\exp\left(-\frac{i}{\hbar}p_1\cdot q_1\right)\;|q_1\rangle$$
$$\times\frac{1}{\sqrt{N}}\sum_{q_2}\exp\left(-\frac{i}{\hbar}p_2\cdot q_2\right)\;|q_2\rangle \tag{67}$$
$$\times\frac{1}{\sqrt{N}}\sum_{q_3}\exp\left(-\frac{i}{\hbar}p_3\cdot q_3\right)\;|q_3\rangle.$$

Again, in matrix form, we express the above equation as

$$\pi\hbar\begin{pmatrix}|0\rangle\\|1\rangle\end{pmatrix}\otimes\pi\hbar\begin{pmatrix}|0\rangle\\|1\rangle\end{pmatrix}\otimes\pi\hbar\begin{pmatrix}|0\rangle\\|1\rangle\end{pmatrix}$$
$$=\frac{1}{\sqrt{2}}\begin{pmatrix}1&1\\1&-1\end{pmatrix}\begin{pmatrix}|0\rangle\\|1\rangle\end{pmatrix}\otimes\frac{1}{\sqrt{2}}\begin{pmatrix}1&1\\1&-1\end{pmatrix}\begin{pmatrix}|0\rangle\\|1\rangle\end{pmatrix}\otimes\frac{1}{\sqrt{2}}\begin{pmatrix}1&1\\1&-1\end{pmatrix}\begin{pmatrix}|0\rangle\\|1\rangle\end{pmatrix}, \tag{68}$$
$$=\frac{1}{\sqrt{2}}\frac{1}{\sqrt{2}}\frac{1}{\sqrt{2}}\begin{pmatrix}(|0\rangle+|1\rangle)\\(|0\rangle-|1\rangle)\end{pmatrix}\otimes\begin{pmatrix}(|0\rangle+|1\rangle)\\(|0\rangle-|1\rangle)\end{pmatrix}\otimes\begin{pmatrix}(|0\rangle+|1\rangle)\\(|0\rangle-|1\rangle)\end{pmatrix}.$$

From Eq. (68), we can define 8 joint distributions in p-space, namely, $\mathcal{D}(0,0,0)$, $\mathcal{D}(0,0,1)$, $\mathcal{D}(0,1,0)$, $\mathcal{D}(0,1,1)$, $\mathcal{D}(1,0,0)$, $\mathcal{D}(1,0,1)$, $\mathcal{D}(1,1,0)$, and $\mathcal{D}(1,1,1)$. We obtained,

$$\mathcal{D}(0,0,0)=\frac{1}{\sqrt{4}}\{(\Theta_3^+ + \Gamma^+ + \Xi^+ + \Omega^+)\}, \tag{69}$$

$$\mathcal{D}(0,0,1)=\frac{1}{\sqrt{4}}\{(\Theta_3^- + \Gamma^- + \Xi^- + \Omega^-)\}, \tag{70}$$

$$\mathcal{D}(0,1,0)=\frac{1}{\sqrt{4}}\{\Theta_3^- - \Xi^- + \Gamma^- - \Omega^-\}, \tag{71}$$

$$\mathcal{D}(0,1,1) = \frac{1}{\sqrt{4}}\{\Theta_3^+ - \Gamma^+ - \Omega^+ + \Xi^+\}, \tag{72}$$

$$\mathcal{D}(1,0,0) = \frac{1}{\sqrt{4}}\{\Theta_3^- + \Omega^- + \Gamma^- + \Xi^-\}, \tag{73}$$

$$\mathcal{D}(1,0,1) = \frac{1}{\sqrt{4}}\{\Theta_3^+ - \Xi^+ + \Omega^+ - \Gamma^+\}, \tag{74}$$

$$\mathcal{D}(1,1,0) = \frac{1}{\sqrt{4}}\{\Theta_3^+ - \Xi^+ - \Omega^+ + \Gamma^+\}, \tag{75}$$

$$\mathcal{D}(1,1,1) = \frac{1}{\sqrt{4}}\{\Theta_3^- + \Xi^- - \Omega^- - \Gamma^-\}. \tag{76}$$

In Eqs. (69)–(76), we made use of the following relations:

$$\frac{1}{\sqrt{2}}(|0\rangle|0\rangle|0\rangle + |1\rangle|1\rangle|1\rangle) = \Theta_3^+, \quad \frac{1}{\sqrt{2}}(|0\rangle|0\rangle|1\rangle + |1\rangle|1\rangle|0\rangle) = \Gamma^+, \tag{77}$$

$$\frac{1}{\sqrt{2}}(|0\rangle|1\rangle|0\rangle + |1\rangle|0\rangle|1\rangle) = \Omega^+, \frac{1}{\sqrt{2}}(|0\rangle|1\rangle|1\rangle + |1\rangle|0\rangle|0\rangle) = \Xi^+, \tag{78}$$

$$\frac{1}{\sqrt{2}}(|0\rangle|0\rangle|0\rangle - |1\rangle|1\rangle|1\rangle) = \Theta_3^-, \quad \frac{1}{\sqrt{2}}(|0\rangle|0\rangle|1\rangle - |1\rangle|1\rangle|0\rangle) = \Gamma^-, \tag{79}$$

$$\frac{1}{\sqrt{2}}(|0\rangle|1\rangle|0\rangle - |1\rangle|0\rangle|1\rangle) = \Omega^-, \quad \frac{1}{\sqrt{2}}(|0\rangle|1\rangle|1\rangle - |1\rangle|0\rangle|0\rangle) = \Xi^-. \tag{80}$$

A crucial observation is that only even $(+)$ combinations of entangled basis states correspond to joint distributions and entanglements, and similarly only odd $(-)$ combinations of entangled basis states correspond to joint distributions and entanglements as well. This claim will become clear when we discuss the basic characterization of entanglements and of entanglement measure, which have concurrency[5] equals 1.

8. Chiral degrees of freedom and entangled qubits

Observe that the pseudo-spin variables in semiconductor Bloch equation is defined by the following expressions,

$$S_0 = \rho_{cc} + \rho_{vv}, \tag{81}$$

$$S_x = \rho_{vc} + \rho_{cv}, \tag{82}$$

$$S_y = -i(\rho_{vc} - \rho_{cv}), \tag{83}$$

$$S_z = \rho_{cc} - \rho_{vv}. \tag{84}$$

This maps to

[5] Concurrency is one of the most important characterization of entangled qubits, to be discussed in more details below.

$$e_0 = |\Phi^+\rangle = \frac{1}{\sqrt{2}}(|1\rangle|1\rangle + |0\rangle|0\rangle), \tag{85}$$

$$e_x = |\Psi^+\rangle = \frac{1}{\sqrt{2}}(|0\rangle|1\rangle + |1\rangle|0\rangle), \tag{86}$$

$$ie_y = i|\Psi^-\rangle = \frac{1}{\sqrt{2}}(|0\rangle|1\rangle - |1\rangle|0\rangle), \tag{87}$$

$$e_z = |\Phi^-\rangle = \frac{1}{\sqrt{2}}(|1\rangle|1\rangle - |0\rangle|0\rangle). \tag{88}$$

We refer to the basis (e_0, e_x, e_y, e_z) as the pseudo-spin component basis, which differ from the Bell basis in $|\Psi^-\rangle$ by the factor i in $|\Psi^-\rangle$. An important observation that follows from this is that the entangling of chiral degrees of freedom creates another chiral degrees of freedom, e.g., Eqs. (69)–(76).

This is a very important observation which will aid in understanding the *entanglement-induced delocalization,* in topological insulators treated by the quantum transport approach in Ref. [6].

9. The concurrence concept and emergent qubit

By virtue of our diagrammatic construction, coupled with discrete phase space Hadamard transformation, in deriving the entangled basis states, the concurrence concept defined by Wooters [2, 3], as well as the two-state (qubit) properties of entangled multi-partite qubits, naturally coincides with the mathematical description of our physical model of qubit entanglement. In other words, the essence of the quantum description of an entangled qubits in **Figure 1** consists of the superposition of product 'site states' and their corresponding translated 'site states' of all qubits. This means a superposition of the two states of the entangled two qubits of **Figure 1**. One observe that irrespective of how many qubits are entangled, the resulting entangled state is an *emergent* two-state system and therefore behave just like one qubit with two-states, namely, the first state and its *translated* state [4]. Therefore its associated entropy is just one[6].

The concept of concurrence is basically contained by construction of our physical model of qubit entanglement. It is defined by

$$\left|\langle\Psi||\tilde{\Psi}\rangle\right| = C, \tag{89}$$

where C is the quantitative value of concurrence. This number lies between 0 and 1, $0 \leq C \leq 1$. Here $\tilde{\Psi}$ is the corresponding translated qubits of Ψ.

10. A natural measure of entanglement

The two properties of any entangled qubits, namely, concurrence and its *emergent* qubit behavior, lead Wooters [2, 3] to introduce a measure of entanglement, as incorporated in the two formulas,

---

[6] This is also deduced when concurrence $C = 1$ [3].

$$E(C) = H\left(\frac{1}{2} + \frac{1}{2}\sqrt{1 - C^2}\right), \tag{90}$$

$$H(x) = -x\ln x - (1-x)\ln(1-x). \tag{91}$$

Equation (90) basically say that if there is complete concurrence, i.e., $C = 1$, then Eq. (91) says that the system behave as an emergent qubit or two-state system, as depicted clearly in our diagrams. Thus, for maximally entangled multi-partite qubits, we have $C = 1$,

$$E(C) = H\left(\frac{1}{2}\right), \tag{92}$$

$$H\left(\frac{1}{2}\right) = \frac{1}{2}\ln 2 + \frac{1}{2}\ln 2, \ = 1. \tag{93}$$

affirming that entangled qubits behave as a two-state system or as an *emergent qubit*, yielding entropy equals one.

10.1 Entropic distance in entanglement measure

From the above developments, one can introduce an entanglement entropic distance by the formula

$$\mathscr{E} = |Tr(\rho\ln\rho) - 1|, \tag{94}$$

where $Tr(\rho\ln\rho)$ is evaluated using Eqs. (90) and (91). This distance has a range: $0 \le \mathscr{E} \le 1$. When the concurrence, $C = 1$, $Tr(\rho\ln\rho) = 1 = H(\frac{1}{2})$, then we have the entropy-distance from maximally entangled state, $\mathscr{E} = 0$. When $C = 0$, $Tr(\rho\ln\rho) = 0 = H(1)$, then we have the entropy-distance from maximally entangled state, $\mathscr{E} = 1$.

An example where the $C < 1$ may occur in the following singlet state, with less entanglement,

$$|\psi\rangle = \alpha|01\rangle + \beta|10\rangle, \tag{95}$$

where

$$\alpha^2 + \beta^2 = 1. \tag{96}$$

Then

$$\begin{aligned} C &= |\langle\psi||\tilde{\psi}\rangle| = (\alpha\langle 01| + \beta\langle 10|)(\alpha|10\rangle + \beta|01\rangle), \\ &= \alpha\beta + \beta\alpha \le 1. \end{aligned} \tag{97}$$

11. Mixed states and pure states

In discussing mixed and pure states, one makes use of density-matrix operators. This is the domain of abstract statistical treatment usually found in the literature, perhaps following the statistical tradition of Bell's theorem. To elucidate the basic physics, we will here avoid abstract statistical treatment and only discuss specific situations and examples.

11.1 Mixed states and mixed entanglements

Consider the maximally mixed state of a triplet system,

$$\hat{\rho}_W = \frac{1}{2}(|00\rangle\langle 00| + |11\rangle\langle 11|). \tag{98}$$

We obtain the mixed entanglements given by,

$$\hat{\rho}_B = \frac{1}{2}(|\Phi^+\Phi^+\rangle\langle\Phi^+\Phi^+| + |\Phi^-\Phi^-\rangle\langle\Phi^-\Phi^-|). \tag{99}$$

Similarly, consider the maximally mixed state of a singlet system,

$$\hat{\rho}_W = \frac{1}{2}(|01\rangle\langle 01| + |10\rangle\langle 10|). \tag{100}$$

We obtain the mixed entanglements given by

$$\hat{\rho}_B = \frac{1}{2}(|\Psi^+\Psi^+\rangle\langle\Psi^+\Psi^+| + |\Psi^-\Psi^-\rangle\langle\Psi^-\Psi^-|). \tag{101}$$

11.2 Mixed state from pure state (entangled state)

Consider the example of a bipartite of two qubit system. Consider a pure entangled state,

$$|\psi\rangle = \frac{1}{\sqrt{2}}(|00\rangle + |11\rangle). \tag{102}$$

Then the density matrix is,

$$\rho = \frac{1}{2}(|00\rangle + |11\rangle)(\langle 00| + \langle 11|), \tag{103}$$

where the first qubit belongs to party A and the second qubit belongs to party B.[7] We have

$$\rho^2 = \rho, \tag{104}$$

so that Eq. (103) is a pure state. Again, we have, by tracing the party B we obtain,

$$\begin{aligned}\rho_A &= Tr_B\ \rho, \\ &= \frac{1}{2}(|0_A\rangle\langle 0_A| + |1_A\rangle\langle 1_A|).\end{aligned} \tag{105}$$

[7] What we mean by party A and B is in the general sense since any entangled number of qubits behave as an emergent qubit.

Now clearly

$$\rho_A^2 = \frac{1}{2}\rho_A, \tag{106}$$

so that ρ_A is not a pure state but mixed, i.e., a mixture of two pure states, $|0_A\rangle$ and $|1_A\rangle$.[8] Note that the mixed states density matrix do not possess *off-diagonal elements*. The tracing operation basically eliminates the contribution of the "*off-diagonal*" terms. From the last identity, the eigenvalues of ρ_A is $\frac{1}{2}$ and since the subsystem A is an emergent qubit, in general with two distinct states, then eigenvalues of ρ_A corresponds to $\frac{1}{\Omega}$ of our first example above. Thus we refer to ρ_A as "uniformly'" mixed often referred to as "*maximally mixed*" with the initial state $|\psi\rangle$ as "*maximally entangled*".

12. Concluding remarks

The inverter-chain rigid-coupling mechanical model of entanglement link has been demonstrated to faithfully implement the discrete phase space viewpoint [4]. The crucial observation that arise from this inverter-chain link model is that any multi-partite qubit entangled system has the property of an *emergent* qubit, which has been rigorously justified. This readily lead us to the equality relations of the entropy of entanglement formation in the so-called monogamy inequality of entanglement formation, using statistical arguments [5] discussed in the literature. In this paper, mixed states and mixed entanglements are related by the Hadamard transformations [7]. In addition, we show that mixed state can be extracted from pure entangled state, where the party A or B need not be a single qubit themselves but can each be an entangled multi-partite qubit system, respectively, by virtue of *emergent* qubit property of entangled qubit systems. We find that joint distributions in quantum mechanics have inherent entanglement properties.

The natural measure of entanglement is based on entropy of entanglement formation, concurrence, and entropic distance from maximally entangled reference.

---

8 If the parties A and B are entangled states, then what we have obtain are also mixed entanglement through the inverse Hadamard transformation.

Author details

Felix A. Buot[1,2]

1 C&LB Research Institute, Carmen, Cebu, Philippines

2 Department of Physics, LCFMNN, TCSE Group, University of San Carlos, Cebu City, Philippines

*Address all correspondence to: felixa.buot@gmail.com

References

[1] Einstein A, Podolsky B, Rosen N. Can quantum-mechanical description of physical reality be considered complete? Physics Review. 1935;**47**:777

[2] Hill S, Wooters WK. Entanglement of a pair of quantum bits. Physical Review Letters. 1977;**78**:5022

[3] Wooters WK. Entanglement of a pair of quantum bits. Physical Review Letters. 1998;**80**:2245

[4] Buot FA, Elnar AR, Maglasang G, Galon CM. A mechanical implementation and diagrammatic calculation of entangled basis states. arXiv:2112.10291 [cond-mat.mes-hall]. 2021

[5] Kim JS. Entanglement of formation and monogamy of multi-party quantum entanglement. Scientific Reports. 2021; **11**:2364. DOI: 10.1038/s41598-021-82052-3

[6] Buot FA, Rivero KB, Otadoy RES. Generalized nonequilibrium quantum transport of spin and pseudospins: Entanglements and topological phases. Physica B. 2019;**559**:42-61

[7] Buot FA. Nonequilibrium Quantum Transport Physics in Nanosystem. Hackensack, NJ, USA: World Scientific; 2009, and references therein

Chapter 4

Perspective Chapter: On the Contradiction between Special Relativity and Quantum Entanglement

Yoram Kirsh

Abstract

Demonstrations of quantum entanglement (QE), which confirm the violation of Bell's inequality, indicate that under certain conditions action at a distance is possible. This consequence seems to contradict the relativistic principle of causality, which asserts that an effect never precedes its cause, in any reference frame. By analyzing a numerical example of Bell's experiment with entangled pairs of photons, we show how observers in two inertial reference frames can disagree about the causality relation between two events. One observer claims that event 1 is the cause of event 2, while the other claims that event 1 is the result of event 2. The solution we suggest to the paradox is that in entangled systems, one can find pairs of "entangled events" which have symmetrical causality relations. Each of the events can serve as a cause or as an effect, depending on the frame of reference in which they are observed.

Keywords: quantum entanglement, EPR paradox, special relativity, Bell's inequality, entangled events

1. Introduction

Recently, several independent "loophole-free Bell violation" experiments were reported (*e.g.* [1–3]). These works, as well as previous demonstrations of quantum entanglement (QE) since the pioneering work of Freedman and Clauser [4], have shown that under certain conditions, action at a distance between entangled systems is possible. A straightforward interpretation of the results is that faster-than-light (FTL) or *superluminal* communication is feasible. This consequence may seem to contradict an established tenet of Special Relativity (SR) that claims that FTL transfer of information violates causality and is therefore impossible. The purpose of this chapter was to examine in detail where the conflict arises and to propose a possible solution to the paradox.

In fact, the paradox was already hidden in the Einstein-Podosky-Rosen (EPR) thought experiment [5]. If two identical particles, A and B, move in opposite directions after a brief interaction which ensured that $\mathbf{v}_B = -\mathbf{v}_A$, and if at a specific moment, t_0, when A and B are far apart, the position of A and the momentum of B are measured, one can know both the position and the momentum of A at t_0, since $\mathbf{p}_A = -\mathbf{p}_B$. The authors claimed that this thought experiment contradicted Heisenberg's uncertainty principle, and consequently that quantum mechanics (QM) is incomplete.

In addition to this explicit criticism against QM, another claim can be based on the EPR thought experiment, concerning an inconsistency between QM and SR. Suppose we perform a very accurate measurement of the magnitude of the momentum of A and find that it is $\mathbf{p}_A$. According to the uncertainty principle, the measurement changes the state of A so that the uncertainty in its location becomes infinite, or very large (since $\Delta p \times \Delta x \approx \hbar$). But since A and B are entangled, the measurement should change the state of B as well. If we now measure $\mathbf{p}_B$ we should certainly get $\mathbf{p}_B = -\mathbf{p}_A$ while the uncertainty in the location of B becomes infinite, or very large. This is a substantial change in the wave function Ψ_B, which prior to the measurement on A, could be presented by a wave packet in which both Δx_B and Δp_B were finite.

The question is, how is it possible that an operation made on A instantly influences the state of B which, in principle, could be thousands of kilometers away from A. The effect of the measurement on A on the state of B is immediate, since the time interval Δt between the measurements on A and on B can be arbitrarily short. It seems to contradict SR which claims that no interaction can travel faster than the speed of light.

2. The EPR paradox in the spin version

An alternative version of the EPR paradox is based on measuring the spin direction, instead of momentum and location [6]. In this version, A and B are particles with spin 1/2, which were created with opposite spins (e.g., by the decay of a particle with spin 0). They arrive at two detectors D_A and D_B where their spin directions are measured with respect to an arbitrary z-axis. Due to the conservation of angular momentum, if the spin of A is found to be positive ($|+\frac{1}{2}\rangle$), then the spin of B must be negative ($|-\frac{1}{2}\rangle$), and vice versa. This is truly unrelated to the distance that separates the particles at the moment of measurement.

If we repeat the measurement but measure both spins along the x-axis (perpendicularly to the z-axis), the result will be the same. If the spin of A is found to be $+\frac{1}{2}$, the measured spin of B will be $-\frac{1}{2}$, and vice versa. The results can be explained in two different ways.

1. The particles are entangled in such a way that when one spin is measured, the other spin becomes its opposite. The quantum state of the system, which could be a combination of $|+\frac{1}{2}\rangle$ and $|-\frac{1}{2}\rangle$ for both particles prior to the measurement, collapses into $|+\frac{1}{2}\rangle$ for particle A and $|-\frac{1}{2}\rangle$ for particle B (or vice versa) because of the measurement. This is the QM or the Copenhagen explanation.

2. The two particles were created with definite (opposite) spins around any axis in space we may choose. This is "the hidden variable" explanation.

Both explanations seem to be problematic. In order to illustrate the problem with the first explanation, let us assume that particle A arrives at the detector D_A and after a short time interval, Δt, particle B arrives at the detector D_B. The information about the measured spin of A should travel from D_A to D_B faster than the speed of light, since the distance between them can be many kilometers, while Δt can be arbitrarily small. This seems to violate SR.

According to the second explanation, no information is transferred from one detector to the other. However, we have to assume that each of the two particles is created with definite eigenvalues of both S_z and S_x. But, according to QM, two perpendicular components of the angular momentum cannot be simultaneously in defined states. If a particle has a definite spin direction relative to the z-axis, its spin direction on the x-axis should be a superposition of $|+½\rangle$ and $|-½\rangle$ so that a measurement can give each state with equal probabilities. If the information about the results of potential measurements of the spin along any arbitrary axis exists prior to the measurement, then that information must somehow be concealed. Hence, this model was dubbed "hidden variables."

3. Bell's experimental setup for photons

Until the article of John Stewart Bell [7], no experiment was made in order to find out which of the two explanations is true, since they were thought to be indistinguishable in terms of experimental results. Bell pointed out that the two explanations predict the same results if D_A and D_B are oriented in the same direction in space. However, the results predicted by the two explanations may differ, if the detectors measure the spins in different directions. Bell adopted the spin example advocated by Bohm and Aharonov [6], but since most experiments were made with photons (where the polarization of the photon substitutes the spin component), it will be more convenient in the ensuing discussion to present Bell's thought experiment with photons. The following scheme, depicted in **Figure 1**, is a simplified version of the two-channel experimental setup first employed by Aspect et al. [8].

A source (S) emits two photons, A and B, which travel in opposite directions and have the same polarization state. The photons reach two detectors (e.g., sensitive photomultipliers), D_A and D_B, which are able to detect single photons. In front of each detector, there is a polarizer (P_A and P_B). If the photon passes the polarizer and arrives at the detector, the event is registered as 1. If the photon is stopped by the polarizer and does not reach the detector, the event is recorded as 0.

Let us assume first that the axes of both polarizers are oriented in the z direction, as shown in **Figure 1**. Since the two photons were emitted with the same polarization, we can assume that there will be perfect correspondence between the results recorded by

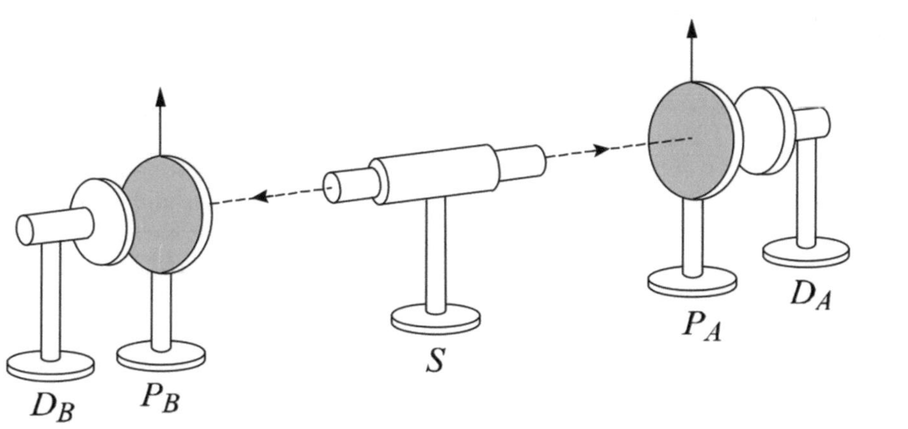

Figure 1.
The setup for Bell's experiment with photons.

the two detectors: if one detector records the series: 1, 0, 0, 1, 0, 1..., the second detector will record exactly the same series. On the other hand, if polarizer P_A is oriented in the z direction while polarizer P_B is oriented in the x direction, there will be a perfect mismatch between the results: if D_A records 1 then D_B will record 0, and vice versa.

Suppose that there is an angle θ between the two polarizers. We define a "matching function" $F(\theta)$, which is the ratio of the number of matches between the two detectors to the total number of readings, in a long series of measurements. We also define a "mismatch function" $E(\theta)$ as the percentage of mismatches between the two detectors. It's easy to see that:

$$E(\theta) = 1\text{–}F(\theta)\ (\text{for any } \theta) \tag{1}$$

$$F(\theta = 0^0) = 1; \quad E(\theta = 0^0) = 0 \tag{2}$$

$$F(\theta = 90^0) = 0; \quad E(\theta = 90^0) = 1 \tag{3}$$

Let us consider the case of $\theta = 0^0$. The correlation between D_A and D_B (Eq. (2)) comes as no surprise if the initial polarization of the pair of photons is parallel to the z-axis ($|z\rangle$) or to the x-axis ($|x\rangle$). However, we expect to get the same correlation, even when the initial polarization of the photons forms an arbitrary angle ϕ ($0^0 < \phi < 90^0$) with the z-axis. In this case, the initial polarization prior to the measurement can be considered as a superposition of $|z\rangle$ and $|x\rangle$, and the measurement actually causes the collapse of the wave function of each photon to one of the eigenstates, $|z\rangle$ or $|x\rangle$. As mentioned above, the fact that in two remote locations, the collapse is to the same eigenstate can be explained in two alternative ways: the QM interpretation (which involves an immediate action at a distance) or the hidden-variable model.

4. Bell's inequality and Bell's theorem

Let us explore the following four-stage thought experiment, which will lead us to Bell's Inequality.

Stage 1: P_A and P_B both point in the z direction ($\theta = 0$), as shown in **Figure 2**-1. Two entangled photons are emitted from the source. Photon A arrives at P_A while photon B arrives at P_B. The readings of the two detectors will be the same all the time, therefore $E(\theta) = 0$.

Stage 2: P_B is rotated counterclockwise at an angle θ ($0^0 < \theta < 45^0$, e.g., $\theta = 15^0$), as shown in **Figure 2**-2. Once more, two photons are emitted from the source. Let us assume that photon A passes through P_A and reaches D_A. The probability that photon B will reach D_B is no longer 100%; sometimes it will arrive at D_B and sometimes it will not. In some cases, photon A will be blocked while photon B arrives at D_B. Therefore, $E(\theta)$, which corresponds to the average mismatch between D_A and D_B, is no longer 0.

Stage 3: P_B is restored to its previous position (parallel to the z-axis) while P_A is rotated clockwise at the same angle θ, as shown in **Figure 2**-3. The situation is symmetrical to **Figure 2**-2, and we expect the average mismatch to be $E(\theta)$ as in Stage 2.

Stage 4: We leave P_A as in Stage 3, i.e., skewed at an angle θ clockwise. We rotate P_B anticlockwise at the same angle θ. The angle between the two polarizers is now 2θ, as shown in **Figure 2**-4. We repeat the series of measurements that were performed in the previous stages. The mismatch between D_A and D_B is now $E(2\theta)$.

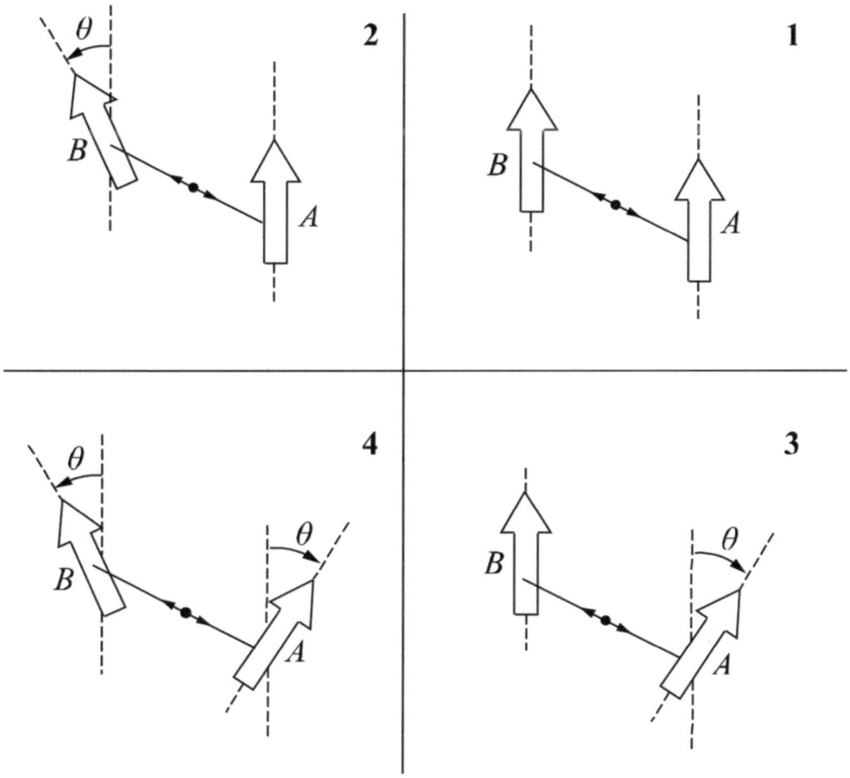

Figure 2.
A four-stage thought experiment that leads to Bell's inequality.

We shall now see that the two models, which explain the correlation between the readings of D_A and the readings of D_B, provide different predictions about the relationship between $E(2\theta)$ and $E(\theta)$. Therefore, by measuring $E(\theta)$ and $E(2\theta)$ for various values of θ, it can be determined as to which of the two models is correct.

According to the hidden-variable model—that rejects action at a distance—there is no connection between the two polarizers. There is no way that polarizer P_A could sense the state of polarizer P_B. Therefore, the rotation of polarizer P_B between Stage 3 and Stage 4 cannot affect polarizer P_A. Consequently, at Stage 4, polarizer P_A will continue to pass or block photons with the same rate of mismatch which we would get if P_B stayed upright. In other words, at Stage 4, the mismatch between the series a (which shows the readings of D_A) and a hypothetical series c (which represents the results we would get from D_B at Stage 4, if P_B remained upright) is still $E(\theta)$.

By the same token, the mismatch between c and b (the actual readings of D_B at Stage 4) will also be $E(\theta)$. One might conclude that the mismatch between a and b will be the sum of the mismatch between a and c and the mismatch between b and c, namely, that $E(2\theta) = 2E(\theta)$. However, this conclusion is too hasty. In some cases, an element in a will be the opposite of the corresponding element in c, and the same element in b will also be the opposite of the element in c. In this case, the elements in a and in b will be identical. Therefore, the average mismatch between a and b at Stage 4, namely $E(2\theta)$, can be $2E(\theta)$ but can also be smaller than $2E(\theta)$. We can write:

$$E(2\theta) \leq 2E(\theta) \; \left[\text{Bell's inequality}\right] \tag{4}$$

We can demonstrate the relation between $E(2\theta)$ and $2E(\theta)$ by the following example. Let us assume that.

$$c = 0, 0, 1, 0, 1, 0, 1, 1, 0, 1, 1, 0 \tag{5}$$

In order to create series a, we duplicate c but randomly change 4 of the 12 digits. To create series b, we once more duplicate c and randomly change 4 of the 12 digits. The mismatch between c and a as well as between c and b will be 1/3.

$$E(c,a) = E(c,b) = 1/3 \tag{6}$$

If the elements of c, which were changed to create b, were different from the elements of c which were changed to create a, the mismatch between a and b will be 2/3. However, if the same elements were changed in both cases, the mismatch between a and b will be 0. In the general case, we can write:

$$E(a,b) \leq 2/3 \tag{7}$$

In correspondence with Eq. (4). Eq. (4) is Bell's inequality for the specific experimental procedure described above. According to Bell's theorem, Eq. (4) would be verified experimentally if the hidden-variable model is true. On the other hand, if Bell's inequality is violated, then the QM model is true.

5. *E(θ)* according to quantum mechanics

We can go on and evaluate $E(\theta)$ according to the traditional interpretation of QM. Let us assume that in Stage 2 of the thought experiment described in Section 4, photon A passes through P_A and reaches D_A. This means that the polarization of photon A is parallel to the z-axis. Since the two photons are entangled, this is also the polarization of photon B. The angle between the axes of P_A and P_B is θ. Therefore, the polarization of photon B is skewed relatively to P_B at an angle θ.

According to Malus's law, when a polarized beam of light hits a perfect polarizer, the intensity of the light that passes through the polarizer is given by:

$$I = I_0 \cos^2\theta \tag{8}$$

where I_0 is the initial intensity and θ is the angle between the light's direction of polarization and the axis of the polarizer.

When we regard the beam as a stream of photons, we can ascribe to each photon a defined direction of polarization which creates an angle θ with P_B. According to Eq. (8), out of N photons, approximately $N\cos^2\theta$ will pass P_B and about $N\sin^2\theta$ will be blocked. The probability that a single photon will reach D_A or D_B, while its companion will be blocked at the other detector is therefore $\sin^2\theta$, and this will be the average mismatch rate in a long series of measurements[1]. Thus, according to QM, at Stages 2 and 3:

[1] If photon A is blocked, the probability that photon B will pass is $\cos^2(90^0-\theta) = \sin^2\theta$

$$E(\theta) = \sin^2\theta \text{ (Stages 2 and 3)} \tag{9}$$

In Stage 4, the angle between the polarizers is 2θ and, according to QM, the mismatch rate will be:

$$E(2\theta) = \sin^2 2\theta \tag{10}$$

If, for example, $\theta = 30^0$, we shall get in Stages 2 and 3 a mismatch rate of:

$$E(30^0) = \sin^2(30^0) = 0.5^2 = 0.25 \tag{11}$$

while in Stage 4 the mismatch rate will be:

$$E(60^0) = \sin^2(60^0) = 0.75 \tag{12}$$

In this case, $E(2\theta) > 2E(\theta)$ in contrast to Bell's inequality (Eq. (4)). Indeed, it's easy to prove that $E(2\theta) > 2E(\theta)$ for any θ in the range: $0^0 < \theta < 45^0$.

To prove this, we notice that:

$$\sin^2(2\theta) = 4 \sin^2(\theta) \times \cos^2(\theta) \tag{13}$$

In the range $0^0 \leq \theta \leq 45^0$, the function $\cos^2(\theta)$ is a monotonically descending function, which has a minimum at $\theta = 45^0$, where $\cos^2(\theta) = 0.5$. At that point:

$$\sin^2(2\theta) = 4\sin^2(\theta) \times 0.5 = 2 \sin^2(\theta) \ (\theta = 45°) \tag{14}$$

namely, for $\theta = 45°$, $E(2\theta) = 2E(\theta)$. If $0^0 < \theta < 45°$ then $\cos(2\theta) > 0.5$ and $\sin^2(2\theta) > 2\sin^2(\theta)$, which means that $E(2\theta) > 2E(\theta)$, in contrast with Bell's inequality.

Figure 3 depicts $E(\theta)$ according to the QM description ($E(\theta) = \sin^2\theta$) in the range $0^0 \leq \theta \leq 90^0$. On the straight line (described by $E(\theta) = \theta/90^0$), the relation between $E(\theta)$ and $E(2\theta)$ is: $E(2\theta) = 2E(\theta)$. One can see that in the range $0^0 < \theta < 45°$, the function $E(\theta) = \sin^2\theta$ (which represents QM's results) lies below the straight line, while Bell's inequality Eq. (4) is valid only for points that are above the straight line. Thus, by performing the experiment described in Section 4, one can find out whether the QM model or the hidden-variable model is correct.

In a pioneering work of Freedman and Clauser [4], an experiment similar to the thought experiment described above was performed. They measured the polarization correlation of two entangled photons emitted in an atomic cascade of calcium. The wavelengths of the photons were 551.3 nm and 422.7 nm. The measurements were made at nine different angles in the range $0^0 \leq \theta \leq 90°$. The results were in agreement with quantum mechanics and violated Bell's inequality to a high statistical accuracy. Actually, the curve describing the results (**Figure 3** in [4]) is similar to $E(\theta) = \sin^2\theta$ in **Figure 3** above, except for the values on the vertical axis which reflect the fact that the efficiency of the detectors was less than 100%. The authors considered the results as strong evidence against local hidden-variable theories.

Over the years, additional experiments demonstrated clearly the violation of Bell's inequality. It was established that in experiments such as those described in the EPR and Bell's papers, a measurement performed on one particle does affect the other particle which can be far away. The effect is immediate, indicating that a single

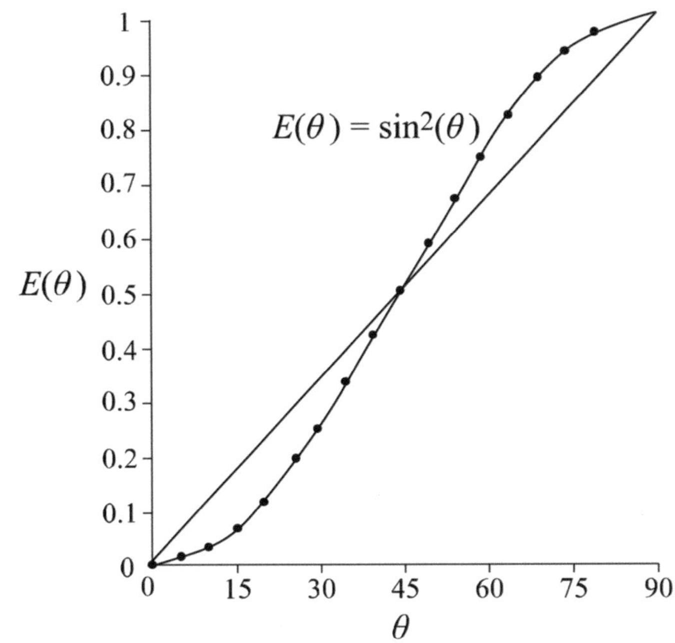

Figure 3.
The points depict $E(\theta)$ *according to quantum mechanics (QM)* ($E(\theta) = \sin^2\theta$). *On the straight line (described by* $E(\theta) = \theta/90°$), $E(2\theta) = 2E(\theta)$.

unified wave function continues to describe the two particles even when they are far apart. Thus, a measurement performed on one particle causes the collapse of the wave function in the other one as well. In order to resolve the contradiction between this action at a distance and SR, we have to see why the contradiction arises in the first place.

6. Special relativity and action at a distance

Let S and S' denote two inertial reference frames. S' is moving with respect to S at a constant velocity v along the x-axis. At time $t = 0$, the spatial axes and clocks of S and S' coincide. Suppose that an event that occurred in S, at point x_1 in time t_1 creates a signal that travels at a velocity u. The signal arrives at point x_2 in time t_2 and creates a second event there (e.g., turning on a lamp). The coordinates and times of the two events in S' are (x'_1, t'_1) and (x'_2, t'_2). According to the Lorentz transformation:

$$t'_1 = \gamma\left(t_1 - \frac{vx_1}{c^2}\right); \qquad t'_2 = \gamma\left(t_2 - \frac{vx_2}{c^2}\right) \tag{15}$$

Where $\gamma = 1/(1-v^2/c^2)^{1/2}$. From (15), we get:

$$\Delta t' = \gamma\left(\Delta t - v\Delta x/c^2\right) \tag{16}$$

Where $\Delta t' = t'_2 - t'_1$; $\Delta t = t_2 - t_1$; $\Delta x = x_2 - x_1$.
Since $\Delta x = u\Delta t$, we can write:

$$\Delta t' = \gamma\ \Delta t\ \left(1\text{–}uv/c^2\right) \tag{17}$$

If the signal velocity u is greater than the speed of light ($u > c$), we can define a reference frame S' that moves at a velocity v which is smaller than c but close enough to c so that $uv\ /c^2 > 1$ (in order for that to happen, v should meet the condition $c > v > c^2/u$). The expression $(1 - uv/c^2)$ in Eq. (17) will be negative, and therefore if Δt is positive, $\Delta t'$ should be negative and vice versa. Thus, in S', event 1 will occur after event 2 and an observer in S' would see the effect precede its cause. For example, let $u = 1.1c$ and $\Delta x = c \times 1$ s while $v = 0.98c$ ($\gamma = 5.025$). With these values:

$\Delta t = \Delta x/u = 0.9091$ s.

$\Delta t' = 5.025 \times 0.9091(1\text{–}0.98 \times 1.1) = -\ 0.356$ s.

The minus sign indicates that the order of events in S and S' is reversed. According to an observer in S', event 1 (which is the cause of event 2) occurred 0.356 s after event 2. The assumption that a signal can travel faster than the speed of light leads to a violation of the relativistic principle of causality, which asserts that an effect never precedes its cause, in any reference frame. This is why SR forbids instantaneous action at a distance as well as traveling of matter, energy, or information at speeds greater than the speed of light.

In order to demonstrate the inconsistency of Bell-like experiments with SR, let us return to Stage 1 in the experiment described in Section 4. Both polarizers P_A and P_B are oriented in the z direction ($\theta = 0$), as shown in **Figure 2**-1. Thus, the readings of the two detectors are the same all the time and $E(\theta) = 0$. Let us assume that D_A, D_B, and the source are on the x-axis of a rest frame S. The source is at $x_0 = 0$ and the coordinates of D_A and D_B are:

$$x_1 = x(D_A) = 15\ \text{m}; x_2 = x(D_B) = -15.3\ \text{m} \tag{18}$$

In order to facilitate the calculations, we redefine our unit of length so that $c = 3 \times 10^8$ m/s exactly. We denote by t_1 and t_2 the times of two events. **Event 1**: "Photon A reached P_A and then D_A registered 0 or 1," **Event 2**: "Photon B reached P_B and then D_B registered 0 or 1." It's easy to see that:

$$t_1 = 50\ \text{ns}; t_2 = 51\ \text{ns} \tag{19}$$

Before checking the times in another frame, S', let us discuss the following question: Can an observer in S consider event 2 as the result of event 1? I claim that she can. The two events can be considered a combination of cause and effect for the following reasons.

1. In frame S, event 1 precedes event 2 by 1 ns.

2. Prior to the occurrence of event 1, the reading of D_B could be either 0 or 1 in equal probabilities. After event 1 takes place, the reading of D_B is definitely determined.

3. If D_A was removed after the photons left the source, and before they reached the polarizers, the reading of D_B could be either 0 or 1 in equal probabilities. The fact that D_A operated and registered the arriving photon influenced the reading of D_B.

Therefore, event 2 can be considered the result of event 1. An argument against this claim is that the two events are separated by a space-like interval. According to SR, only if two events are separated by a time-like or light-like interval can one event influence the other. However, it can be claimed that the concept of causality is metaphysically prior to the relativistic restrictions. Actually, it was argued that many standard philosophical theories would treat the relationship between such two events as causal, despite the contradiction with SR [9].

Let us assume a second frame S' which is moving with respect to S in the negative direction of the x-axis at a velocity $v = -0.6c$. Let us assume that when the two photons are ejected from the source, the clocks at S and S' show $t = t' = 0$, and the origins coincide. By using the Lorentz transformation (Eq. 15), we find:

$$t_1' = 100 \text{ ns}; \quad t_2' = 25.5 \text{ ns} \tag{20}$$

Thus, in S' event 2 occurs before event 1, although we defined event 1 as the cause of event 2. Actually, since the velocity of the signal which carries the information between D_A and D_B is infinite, the paradox appears for v as small as $|v| \approx 0.01c$ for the numerical values of the example above. In general, it occurs whenever:

$$-v \geq \frac{t_2 - t_1}{x_1 - x_2} c^2 \tag{21}$$

7. Solutions to the paradox

Several solutions can be offered for the contradiction, which was demonstrated in the previous section between SR and Bell-like experiments. Ballentine and Jarrett [10] suggested a distinction between a "strong" locality principle and a "weak" one that is needed to satisfy the demands of relativity. They claimed that QM satisfies the latter and therefore there is no contradiction between QM and SR. Instead, one can argue that we do not have two separate events here but only one spatially extended but indivisible event which is "the collapse of the wave function which represents the polarization of the two photons" ([9], p. 41). Another alternative is to formulate a theory of causation which requires some conditions which two events need to fulfill in order to represent a cause-and-effect relationship, and then show that these conditions are not realized here ([9], p. 42).

The solution, which I suggest to the paradox, is based on the following principles:

1. Event 1 and event 2, in the example discussed in Section 6, are two distinct and separate events that occur at different points in space–time.

2. There is a cause-and-effect connection between the two events.

3. According to an observer in S, event 2 is a result of event 1.

4. According to an observer in S', on the other hand, event 1 is a result of event 2.

5. The disagreement between the two observers does not violate the causality principle of SR, since in this particular case the cause-and-effect relationship

between the two events is symmetrical: each of them can be regarded as a result of the other, depending on the frame of reference in which they are observed.

Usually, when there is a causal relationship between two remote events, they are physically different. That's why the effect cannot precede its cause in any reference frame. For example, if event 1 is the ejection of a signal from point (x_1, y_1, z_1, t_1) and event 2 is the arrival of the signal to point (x_2, y_2, z_2, t_2) where it turns on a lamp, we demand that, in any reference frame, $t_2 > t_1$. This demand is fulfilled only if the velocity of the signal does not exceed the speed of light, as shown in Section 6.

However, in Bell-like experiments, like the one described in Section 4, there is no physical difference between the two events: they are totally symmetrical. Each of them can serve as a cause or as an effect, depending on the frame of reference in which they are observed. If event 1 is observed before event 2, event 1 is the cause and event 2 is the result. If the order of times is reversed, then event 2 is the cause and event 1 is the result.

It is customary to think that the causal relation between two events can be one of the four types:

1. Event 2 is a result of event 1.

2. Event 1 is a result of event 2.

3. Both events have a common cause.

4. There is no causal relation between the two events.

The analysis of the ostensible contradiction between Bell's theorem and SR indicates that there is an additional possibility. In entangled systems, one can find pairs of "entangled events" which have symmetrical cause-and-effect relations. Each of them can appear to be the cause of the other, depending on the frame of reference in which they are observed. This fifth possibility solves the paradox which the action at a distance creates. Experimental results (e.g., [11]) can be interpreted as supporting this suggestion.

Author details

Yoram Kirsh
The Open University of Israel, Raanana, Israel

*Address all correspondence to: yoramk@openu.ac.il

References

[1] Hensen B et al. Loophole-free Bell inequality violation using electron spins separated by 1.3 kilometres. Nature. 2015;**526**:682-686. Available from: https://www.nature.com/articles/nature 15759

[2] Giustina M et al. A significant-loophole-free test of Bell's theorem with entangled photons. Physical Review Letters. 2015;**115**:250401

[3] Shalm LK et al. A strong loophole-free test of local realism. Physical Review Letters. 2015;**115**:250402

[4] Freedman SJ, Clauser JF. Experimental test of local hidden-variable theories. Physical Review Letters. 1972;**28**:938-941. DOI: 10.1103/ PhysRevLett.28.938

[5] Einstein A, Podolsky B, Rosen N. Can quantum-mechanical description of physical reality be considered complete? Physical Review. 1935;**47**:777-780. DOI: 10.1103/PhysRev.47.777

[6] Bohm D, Aharonov Y. Discussion of experimental proof for the paradox of Einstein, Rosen, and Podolsky. Physical Review. 1957;**108**:1070-1076

[7] Bell JS. On the Einstein-Podolsky-Rosen paradox. Physics. 1964;**1**:195-200. Available from: https://cds.cern.ch/rec ord/111654/files/vol1p195-200_001.pdf

[8] Aspect A, Grangier P, Roger G. Experimental realization of Einstein-Podolsky-Rosen-Bohm gedankenexperiment: A new violation of Bell's inequalities. Physical Review Letters. 1982;**49**:91-94

[9] Butterfield J. David Lewis meets John Bell. Philosophy of Science. 1992;**59**: 26-43

[10] Ballentine LE, Jarrett JP. Bell's theorem: Does quantum mechanics contradict relativity? American Journal of Physics. 1987;**55**:696-701. DOI: 10.1119/1.15059

[11] Goswami K et al. Indefinite causal order in a quantum switch. Physical Review Letters. 2018;**121**:090503

Chapter 5

Phenomenology of Heavy Quark at the LHC

Rachid Benbrik and Mohammed Boukidi

Abstract

In the Standard Model, the Higgs boson is responsible for giving particles mass through a process called the Higgs mechanism. However, the interactions between the Higgs boson and usual quarks are determined by Yukawa couplings, which means that the strength of their interactions depends on their mass. In contrast, vector-like quarks are expected to have a more "vector" coupling with the Higgs boson, which would make their interactions independent of their mass. This property could potentially make them easier to detect experimentally. In this chapter, we develop the theoretical framework of such exotic heavy quarks both in the standard model and beyond it such as the two Higgs doublets model (2HDM). This chapter introduces a Lagrangian which follows the most generic concept of the Standard Model which is consistent with the basics, and then we explore the consequences of identifying the eigenstates and interactions with SM-VLQ. We also provide a detailed description of the full interaction in the 2HDM mass basis. We shortly comment on the production of heavy quark pairs at the LHC.

Keywords: exotic particles, vector-like quark, LHC, Higgs, SM

1. Introduction

The introduction of vector quarks into the realm of particle physics represents a departure from the conventional characteristics of Standard Model (SM) quarks on multiple intriguing fronts. In stark contrast to their SM counterparts, vector quarks typically do not acquire masses through Yukawa couplings with a Higgs doublet, a feature that resonates with the experimental observations of the Higgs boson [1, 2]. This distinct trait opens up new avenues for exploration. What further sets vector quarks apart is their remarkable capacity for mixing with SM quarks. This mixing dynamic has profound implications, altering their couplings with essential bosons like the Z, W, and Higgs bosons [3]. Such mixing represents a straightforward mechanism for breaking the Glashow-Iliopoulos-Maiani mechanism, a tantalizing prospect that holds the potential for ushering in novel phenomena in low-energy physics [4]. However, the allure of vector quarks goes beyond these unique traits. Two primary theoretical motivations underpin their extensive study. Firstly, they play a pivotal role in the intricate dance of electroweak symmetry breaking, offering a promising avenue for explaining the observed light mass of the Higgs boson [5]. This becomes particularly compelling within theoretical frameworks that posit the Higgs as a pseudo Goldstone boson [6].

Secondly, vector-like quarks emerge organically as fermion resonances within theories that delve into the nuanced realm of partial flavor composition within the Standard Model. In both of these theoretical scenarios, substantial mixing between the novel vector quarks and the third family of SM quarks, often dubbed "top partners," takes center stage [7–10]. Notably, the presence of vector quarks is not confined to a single theoretical framework but extends its reach into diverse models. These encompass small Higgs and composite Higgs models, as well as their holographic counterparts [11]. Consequently, the study of vector quarks not only enriches our understanding of particle physics but also offers a tantalizing glimpse into the potential extensions and augmentations of the Standard Model, unearthing the profound intricacies that underlie the fundamental forces shaping the universe's particle interactions.

Simultaneously, it is worth noting that testing quantum entanglement directly in proton-proton (pp) collisions at the Large Hadron Collider (LHC) poses a formidable challenge due to the macroscopic and complex nature of these interactions [12]. Quantum entanglement typically manifests at the microscopic scale, involving the correlation of properties between individual or pairs of particles. Nevertheless, researchers at the LHC can indirectly explore aspects related to quantum entanglement [13–15]. They may investigate the correlations and angular momentum conservation among particles produced in these high-energy collisions [16]. While not direct evidence of entanglement, these studies shed light on the quantum dynamics governing the particle interactions. Additionally, the search for new, exotic particles beyond the Standard Model is a primary mission of the LHC [17]. These hypothetical particles might exhibit quantum properties, including entanglement, which could be uncovered through their distinctive decay patterns and behaviors. Although the LHC primarily serves as a platform for fundamental particle physics research, it contributes to our broader comprehension of quantum mechanics and its role in the universe's fundamental fabric [18, 19].

1.1 From weak eigenstates to mass eigenstates and mixing angles

Let us consider a general 2×2 Hermitian matrix of the form:

$$\mathcal{M} = \begin{pmatrix} a & b \\ \overline{b} & M \end{pmatrix}, \tag{1}$$

Where a, b, c, and d are complex numbers. In order to diagonalize the above matrix with the aid of the unitary matrix $\mathcal{U}$, we seek a diagonal matrix $\mathcal{M}_{\text{diag}}$ such that:

$$\mathcal{U}\mathcal{M}\mathcal{U}^{\dagger} = \mathcal{M}_{diag} \tag{2}$$

Here $\mathcal{U}^{\dagger}$ denotes the conjugate transpose of $\mathcal{U}$. We adopt the following form for the unitary matrix $\mathcal{U}$:

$$\mathcal{U} = \begin{pmatrix} \cos\ \theta & -\sin\ \theta e^{i\varphi} \\ \sin\ \theta e^{-i\varphi} & \cos\ \theta \end{pmatrix}, \tag{3}$$

where θ and φ are real parameters. We determine these parameters by requiring that $\mathcal{U}$ diagonalizes $\mathcal{M}$. Specifically, we choose θ and φ such that:

$$\tan(2\theta) = \frac{2|b|}{M-a} \quad \text{and} \quad \varphi = \arg(b), \tag{4}$$

We proceed to substitute the matrices $\mathcal{U}$, $\mathcal{U}^\dagger$, and $\mathcal{M}$ into the equation for diagonalization, which we then simplify as follows:

$$\begin{aligned}
\mathcal{U}\mathcal{M}\mathcal{U}^\dagger &= \begin{pmatrix} c & -se^{i\varphi} \\ se^{-i\varphi} & c \end{pmatrix} \begin{pmatrix} a & b \\ \overline{b} & M \end{pmatrix} \mathcal{U}^\dagger \\
&= \begin{pmatrix} ca - s|b| & cb - sMe^{i\varphi} \\ sae^{-i\varphi} + c\overline{b} & s|b| + cM \end{pmatrix} \begin{pmatrix} c & se^{i\varphi} \\ -se^{-i\varphi} & c \end{pmatrix} \\
&= \begin{pmatrix} c^2a - sc|b| - sc|b| + s^2M & scae^{i\varphi} - s^2b + c^2b - scMe^{i\varphi} \\ scae^{-i\varphi} + c^2\overline{b} - s^2\overline{b} - scMe^{-i\varphi} & s^2a + sc|b| + sc|b| + c^2M \end{pmatrix}
\end{aligned} \tag{5}$$

Here, $c = \cos\theta$, $s = \sin\theta$, and $\overline{b}$ denotes the complex conjugate of b.

To simplify the (1,2) entry, we utilize the fact that $\mathcal{M}$ is Hermitian along with the following relationship derived from Eq. (1):

$$\frac{2sc}{c^2 - s^2} = \frac{2|b|}{M-a} \tag{6}$$

This yields the following equations:

$$\begin{aligned}
(1,\ 2) &= sc(a-M)e^{i\varphi} + (c^2 - s^2)b \\
&= \frac{b}{M-a}(a-M)\,(c^2 - s^2)e^{i\varphi} + (c^2 - s^2)b \\
&= 0
\end{aligned} \tag{7}$$

$$(2,\ 1) = (1,\ 2)^* = 0 \tag{8}$$

$$(1,\ 1) = c^2a + s^2M - 2sc|b| \tag{9}$$

$$(2,\ 2) = s^2a + c^2M + 2sc|b| \tag{10}$$

We shall invoke the definition of the matrix $\mathcal{M}$, previously established as follows:

$$\mathcal{M} = \begin{pmatrix} y_{33} & y_{34} \\ 0 & M \end{pmatrix} \tag{11}$$

Utilizing the aforementioned definition of $\mathcal{M}$, the products $\mathcal{M}\mathcal{M}^\dagger$ and $\mathcal{M}^\dagger\mathcal{M}$ can be computed as:

$$\mathcal{M}\mathcal{M}^\dagger = \begin{pmatrix} |y_{33}|^2 + |y_{34}|^2 & y_{34}M \\ y_{34}M & M^2 \end{pmatrix} \quad \text{where} \quad \mathcal{U}_L = \begin{pmatrix} \cos\theta_L & -\sin\theta_L e^{i\varphi_L} \\ \sin\theta_L e^{-i\varphi_L} & \cos\theta_L \end{pmatrix} \tag{12}$$

$$\mathcal{M}^\dagger\mathcal{M} = \begin{pmatrix} |y_{33}|^2 & y_{33}^* y_{34} \\ y_{33} y_{34}^* & M^2 + |y_{34}|^2 \end{pmatrix} \quad \text{where} \quad \mathcal{U}_R = \begin{pmatrix} \cos\theta_R & -\sin\theta_R e^{i\varphi_R} \\ \sin\theta_R e^{-i\varphi_R} & \cos\theta_R \end{pmatrix} \tag{13}$$

The mixing angles θ_L and θ_R can be expressed as follows:

$$\begin{cases} \tan(2\theta_L) = \dfrac{2|y_{34}|M}{M^2 - |y_{33}|^2 - |y_{34}|^2} & \text{where} \quad \varphi_L = arg(y_{34}) \\ \tan(2\theta_R) = \dfrac{2|y_{33}y_{34}|}{M^2 + |y_{34}|^2 - |y_{33}|^2} & \text{where} \quad \varphi_R = arg(y_{33}^* y_{34}) \end{cases} \tag{14}$$

Let us define a new matrix $\mathcal{M}$ with a distinct element configuration from the previous one:

$$\mathcal{M} = \begin{pmatrix} y_{33} & 0 \\ y_{43} & M \end{pmatrix} \tag{15}$$

Upon calculating the products $\mathcal{M}^{\mathcal{M}}$ and $\mathcal{M}^\dagger \mathcal{M}$, we obtain two matrices as a result:

$$\mathcal{M}\mathcal{M}^\dagger = \begin{pmatrix} |y_{33}|^2 & y_{33}y_{43}^* \\ y_{33}^* y_{43} & M^2 + |y_{43}|^2 \end{pmatrix} \qquad \mathcal{M}^\dagger \mathcal{M} = \begin{pmatrix} |y_{33}|^2 + |y_{43}|^2 & y_{43}^* M \\ y_{43}M & M^2 \end{pmatrix} \tag{16}$$

From the above expressions, we can extract the mixing angles θ_L and θ_R, as well as the phases φ_L and φ_R.

$$\begin{cases} \tan(2\theta_L) = \dfrac{2|y_{33}y_{43}|}{M^2 + |y_{43}|^2 - |y_{33}|^2} & \text{where} \quad \varphi_L = arg(y_{33}y_{43}^*) \\ \tan(2\theta_R) = \dfrac{2|y_{43}|M}{M^2 - |y_{33}|^2 - |y_{43}|^2} & \text{where} \quad \varphi_R = arg(y_{33}^*) \end{cases} \tag{17}$$

2. Framework of SM + VLQs singlet

In the basis of weak eigenstates (t, T), where t represents the standard model top quark, the matrix describing the masses of the top quark and vector-like quarks (VLQ) [20] can be expressed as follows

$$\mathcal{M} = \begin{pmatrix} y_{33}\dfrac{v}{\sqrt{2}} & y_{34}\dfrac{v}{\sqrt{2}} \\ 0 & M \end{pmatrix} \tag{18}$$

The aforementioned mass matrix clearly indicates that the physical mass of the heavy top quark, m_T, differs from the mass parameter M due to the mixing of the weak eigenstates t and T. To obtain the physical states $(t_{L,R}, T_{L,R})$ in terms of the gauge eigenstates $(\tilde{t}_{L,R},\ \tilde{T}_{L,R})$, the mass matrix $\mathcal{M}$ can be diagonalized through a bi-unitary transformation, as described in Section 1.

$$\begin{pmatrix} t_{L,R} \\ T_{L,R} \end{pmatrix} = \mathcal{U}_{L,R} \begin{pmatrix} \tilde{t}_{L,R} \\ \tilde{T}_{L,R} \end{pmatrix} \tag{19}$$

Where the rotation matrices $\mathcal{U}_L$ and $\mathcal{U}_R$ corresponding to left-handed and right-handed rotations, respectively, can be defined as follows:

$$\mathcal{U}_{L,R} = \begin{pmatrix} \cos(\theta_{L,R}) & -\sin(\theta_{L,R})e^{i\varphi_{L,R}} \\ \sin(\theta_{L,R})e^{-i\varphi_{L,R}} & \cos(\theta_{L,R}) \end{pmatrix} \tag{20}$$

In regards to the mixing angles, it is important to note that θ_L and θ_R are not considered independent parameters. This can be deduced from the bi-unitary transformations, which yield the following relationships:

$$\tan\ 2\theta_L = \left(\frac{2|y_{34}|\frac{v}{\sqrt{2}}M}{M^2 - |y_{33}|^2\frac{v^2}{2} - |y_{34}|\frac{v^2}{2}}\right), \qquad \tan\ 2\theta_R = \left(\frac{|y_{33}y_{34}|v^2}{M^2 + |y_{34}|^2\frac{v^2}{2} - |y_{33}|\frac{v^2}{2}}\right) \tag{21}$$

Here y_{33} is assumed to be real, $\varphi_L = \arg(y_{34})$ and $\varphi_R = arg\left(y_{33}^* y_{34}\right)$..

Based on the aforementioned relation in Section 1, namely $\mathcal{U}_L \mathcal{M}\mathcal{U}_R^\dagger = \mathcal{M}_{diag}$, we can introduce the following notations:

$$\overline{u}_L^0 \mathcal{M} u_R^0 = \overline{u}_L^0 \mathcal{U}_L^\dagger \mathcal{M}_{diag} \mathcal{U}_R u_R^0 \Rightarrow \quad \begin{cases} u_{L,R} = \mathcal{U}_{L,R} u_{L,R}^0 \\ u_{L,R}^0 = \mathcal{U}_{L,R}^\dagger u_{L,R} \end{cases} \tag{22}$$

2.1 Modified gauge boson couplings

The couplings of the W boson to the top quark (t) and its heavy partner (T) with the bottom quark (b) can be expressed in terms of the left-handed up-type quark mixing matrix $\mathcal{U}_L$, as follows (**Figure 1**).

$$\left(\overline{u}_{L3}^0\ \overline{u}_{L4}^0\right)\ (1\,0)\ b_L = \left(\bar{t}_L \quad \overline{T}_L\right) \mathcal{U}_L \begin{pmatrix} 1 \\ 0 \end{pmatrix} b_L \tag{23}$$

where

$$\mathcal{U}_L \begin{pmatrix} 1 \\ 0 \end{pmatrix} = \begin{pmatrix} c_L \\ s_L e^{-i\varphi_L} \end{pmatrix} \tag{24}$$

Using these expressions, we obtain the following couplings:

$$V_{tb} = c_L, \quad \text{and} \quad V_{Tb} = s_L e^{-i\varphi_L} \tag{25}$$

Figure 1.
Typical Feynman diagram of the W boson couplings to the top quark (t) and its heavy partner (T) with the bottom quark (b).

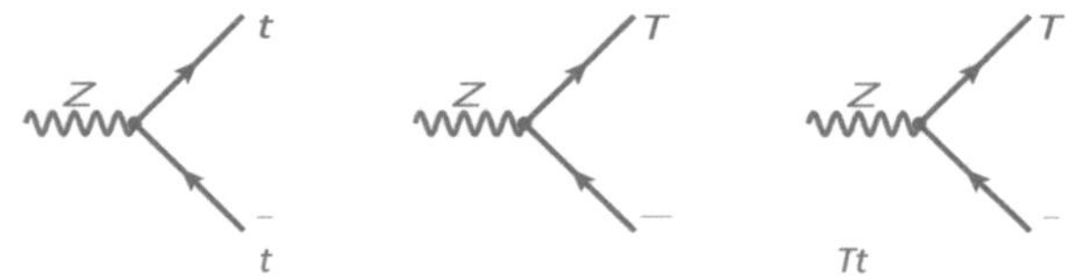

Figure 2.
Typical Feynman diagram of the Z boson couplings to pairs of top (t) *and top partner* (T)*.*

The neutral current expression can be employed to describe the neutral couplings of the Z boson to pairs of top (t) and top partner (T) (**Figure 2**).

$$\left(\bar{u}{L_3}^0 \quad \bar{u}{L_4}^0\right) \begin{pmatrix} 1 & 0 \\ 0 & 0 \end{pmatrix} \begin{pmatrix} u_{L_3}^0 \\ u_{L_4}^0 \end{pmatrix} = \left(\bar{t}_L \quad \overline{T}_L\right) \mathcal{U}_L \begin{pmatrix} 1 & 0 \\ 0 & 1 \end{pmatrix} \mathcal{U}_L^{\dagger} \begin{pmatrix} t_L \\ T_L \end{pmatrix} \tag{26}$$

The computation of the product yields the following relationships:

$$\left(\bar{u}{L_3}^0 \quad \bar{u}{L_4}^0\right) \begin{pmatrix} 1 & 0 \\ 0 & 0 \end{pmatrix} \begin{pmatrix} u_{L_3}^0 \\ u_{L_4}^0 \end{pmatrix} = \left(\bar{t}_L \quad \overline{T}_L\right) \mathcal{U}_L \begin{pmatrix} 1 & 0 \\ 0 & 1 \end{pmatrix} \mathcal{U}_L^{\dagger} \begin{pmatrix} t_L \\ T_L \end{pmatrix} \tag{27}$$

The aforementioned equations can be employed to define the following couplings:

$$X_{tt} = c_L^2, \quad X_{TT} = s_L^2, \quad X_{Tt} = c_L s_L e^{-i\varphi_L} \tag{28}$$

2.2 Modified Higgs couplings

The neutral Higgs couplings to top (t) and top partner (T) pairs can be derived from the following equations (**Figure 3**):

$$\mathcal{M} = \begin{pmatrix} \frac{y_{33}}{\sqrt{2}} & \frac{y_{34}}{\sqrt{2}} \\ 0 & 0 \end{pmatrix} = \begin{pmatrix} \frac{y_{33}}{\sqrt{2}} & \frac{y_{34}}{\sqrt{2}} \\ 0 & \frac{M^0}{v} \end{pmatrix} - \begin{pmatrix} 0 & 0 \\ 0 & \frac{M^0}{v} \end{pmatrix} = \frac{M}{v} - \frac{M^0}{v} \begin{pmatrix} 0 & 0 \\ 0 & 1 \end{pmatrix} \tag{29}$$

Then we compute:

$$\left(\bar{u}{L_3}^0 \quad \bar{u}{L_4}^0\right)\mathcal{M}\begin{pmatrix} u_{R_3}^0 \\ u_{R_4}^0 \end{pmatrix} = \left(\bar{u}{L_3}^0 \quad \bar{u}{L_4}^0\right)\frac{M}{v}\begin{pmatrix} u_{R_3}^0 \\ u_{R_4}^0 \end{pmatrix} - \frac{M^0}{v}\left(\bar{u}{L_3}^0 \quad \bar{u}{L_4}^0\right)\begin{pmatrix} 0 & 0 \\ 0 & 1 \end{pmatrix}\begin{pmatrix} u_{R_3}^0 \\ u_{R_4}^0 \end{pmatrix} \tag{30}$$

Further simplifying the above equation, we get:

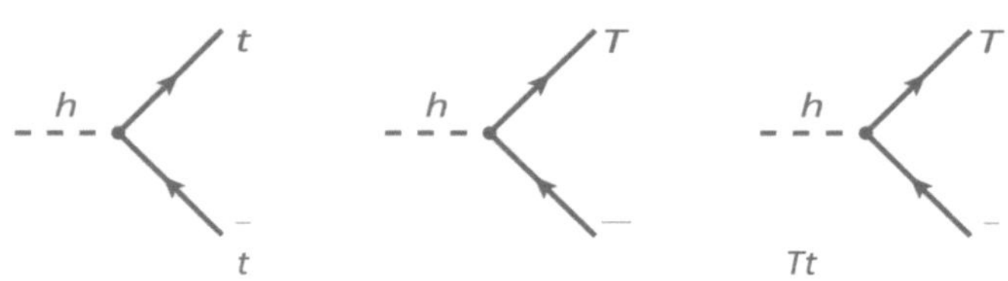

Figure 3.
Typical Feynman diagram of the Higgs boson h couplings to pairs of top (t) *and top partner* (T)*.*

$$\begin{pmatrix}\bar{t}_L & \overline{T}_L\end{pmatrix}\frac{1}{v}\begin{pmatrix}m_t & 0\\ 0 & M\end{pmatrix}\begin{pmatrix}t_R\\ T_R\end{pmatrix}-\frac{M^0}{v}\begin{pmatrix}\bar{t}_L & \overline{T}_L\end{pmatrix}\mathcal{U}_L\begin{pmatrix}0 & 0\\ 0 & 1\end{pmatrix}\mathcal{U}_R^{+}\begin{pmatrix}t_R\\ T_R\end{pmatrix} \tag{31}$$

$$\begin{pmatrix}c_L & -s_Le^{i\varphi_L}\\ s_Le^{-i\varphi_L} & c_L\end{pmatrix}\begin{pmatrix}0 & 0\\ 0 & 1\end{pmatrix}\mathcal{U}_R^{+}=\begin{pmatrix}0 & -s_Le^{i\varphi_L}\\ 0 & c_L\end{pmatrix}\begin{pmatrix}c_R & s_Re^{i\varphi_R}\\ -s_Re^{-i\varphi_R} & c_R\end{pmatrix} \tag{32}$$

$$=\begin{pmatrix}s_Ls_Re^{i(\varphi_L-\varphi_R)} & -s_Lc_Re^{i\varphi_L}\\ -c_Ls_Re^{-i\varphi_R} & c_Lc_R\end{pmatrix} \tag{33}$$

Applying the aforementioned transformation yields the following couplings:

$$y_{ttH}=c_L^2, \tag{34}$$

$$y_{TTH}=s_L^2, \tag{35}$$

$$y_{TtH}^L=\frac{m_t}{m_T}s_Lc_Le^{i\phi},\qquad y_{TtH}^R=s_Lc_Le^{i\phi}. \tag{36}$$

2.3 Framework of SM + VLQs doublet

Let us start with the Yukawa Lagrangian:

$$\mathcal{L}=-y_{4j}^u\begin{pmatrix}\overline{u}_{L4}^0 & \overline{d}_{L4}^0\end{pmatrix}u_{Rj}^0\tilde{\Phi}-y_{4j}^d\begin{pmatrix}\overline{u}_{L4}^0 & \overline{d}_{L4}^0\end{pmatrix}d_{Rj}^0\Phi+hc. \tag{37}$$

The matrix describing the masses the vector-like quarks (VLQ) can be expressed as follows:

$$\mathcal{M}^u=\begin{pmatrix}y_{33}^u\frac{v}{\sqrt{2}} & 0\\ y_{43}^u\frac{v}{\sqrt{2}} & \mathcal{M}^0\end{pmatrix}\equiv\begin{pmatrix}W_{33}^u & 0\\ W_{43}^u & \mathcal{M}^0\end{pmatrix}\quad;\quad\mathcal{U}_L^u\mathcal{M}^u\mathcal{U}_R^{u+}=\mathcal{M}_{diag}^u \tag{38}$$

$$\mathcal{M}^d=\begin{pmatrix}y_{33}^d\frac{v}{\sqrt{2}} & 0\\ y_{43}^d\frac{v}{\sqrt{2}} & \mathcal{M}^0\end{pmatrix}\equiv\begin{pmatrix}W_{33}^d & 0\\ W_{43}^d & \mathcal{M}^0\end{pmatrix}\quad;\quad\mathcal{U}_L^d\mathcal{M}^d\mathcal{U}_R^{d+}=\mathcal{M}_{diag}^d \tag{39}$$

The rotation matrices $\mathcal{U}_L^q$ and $\mathcal{U}_R^q$ corresponding to left-handed and right-handed rotations, respectively, can be defined as follows:

$$\mathcal{U}_{L,R}^q=\begin{pmatrix}c_{L,R}^q & -s_{L,R}^qe^{i\varphi_{L,R}^q}\\ s_{L,R}^qe^{-i\varphi_{L,R}^q} & c_{L,R}^q\end{pmatrix} \tag{40}$$

$$\begin{cases}u_{L,R}=\mathcal{U}_{L,R}^u u_{L,R}^0, & u_{L,R}^0=\mathcal{U}_{L,R}^{u\dagger}u_{L,R}\\ d_{L,R}=\mathcal{U}_{L,R}^d d_{L,R}^0, & d_{L,R}^0=\mathcal{U}_{L,R}^{d\dagger}d_{L,R}\end{cases} \tag{41}$$

2.3.1 Charged currents

The Yukawa Lagrangian of charged currents may be expressed as follows:

$$\mathcal{L} = \frac{-g}{\sqrt{2}} \left[\overline{u}^0_{L4} \gamma^\mu d^0_{L4} + \overline{u}^0_{R4} \gamma^\mu d^0_{R4} + \overline{u}^0_{L3} \gamma^\mu d^0_{L3} \right] W^+_\mu + hc. \tag{42}$$

The Left-Handed (LH) terms can be represented as follows:

$$\begin{pmatrix} \overline{u}^0_{L3} & \overline{u}^0_{L4} \end{pmatrix} \begin{pmatrix} 1 & 0 \\ 0 & 1 \end{pmatrix} \gamma^\mu \begin{pmatrix} d^0_{L3} \\ d^0_{L4} \end{pmatrix} = \begin{pmatrix} \bar{t}_L & \overline{T}_L \end{pmatrix} \mathcal{U}^u_L \begin{pmatrix} 1 & 0 \\ 0 & 1 \end{pmatrix} \mathcal{U}^{d+}_L \gamma^\mu \begin{pmatrix} b_L \\ B_L \end{pmatrix} \tag{43}$$

$$\begin{pmatrix} c^u_L & -s^u_L e^{i\varphi^u_L} \\ s^u_L e^{-i\varphi^u_L} & c^u_L \end{pmatrix} \begin{pmatrix} c^d_L & c^d_L e^{i\varphi^d_L} \\ -s^d_L e^{-i\varphi^d_L} & c^d_L \end{pmatrix} = \begin{pmatrix} c^u_L c^d_L + s^u_L s^d_L e^{i(\varphi^u_L - \varphi^d_L)} & c^u_L s^d_L e^{i\varphi^d_L} - s^u_L c^d_L e^{i\varphi^u_L} \\ s^u_L c^d_L e^{-i\varphi^u_L} - c^u_L s^d_L e^{-i\varphi^d_L} & s^u_L s^d_L e^{i(\varphi^d_L - \varphi^u_L)} + c^u_L c^d_L \end{pmatrix} \tag{44}$$

Applying the aforementioned transformation yields the following couplings:

$$y^L_{tbW^+} = c^u_L c^d_L + s^u_L s^d_L e^{i(\varphi^u_L - \varphi^d_L)}, \tag{45}$$

$$y^L_{TbW^+} = s^u_L c^d_L e^{-i\varphi^u_L} - c^u_L s^d_L e^{-i\varphi^d_L}, \tag{46}$$

$$y^L_{tBW^+} = c^u_L s^d_L e^{i\varphi^d_L} - s^u_L c^d_L e^{i\varphi^u_L}, \tag{47}$$

$$y^L_{TBW^+} = c^u_L c^d_L + s^u_L s^d_L e^{-i(\varphi^u_L - \varphi^d_L)}. \tag{48}$$

The Right-Handed (RH) terms can be represented as follows:

$$\begin{pmatrix} \overline{u}^0_{R3} & \overline{u}^0_{R4} \end{pmatrix} \begin{pmatrix} 0 & 0 \\ 0 & 1 \end{pmatrix} \gamma^\mu \begin{pmatrix} d^0_{R3} \\ d^0_{R4} \end{pmatrix} = \begin{pmatrix} \bar{t}_R & \overline{T}_R \end{pmatrix} \mathcal{U}^u_R \begin{pmatrix} 0 & 0 \\ 0 & 1 \end{pmatrix} \mathcal{U}^{d+}_R \gamma^\mu \begin{pmatrix} b_L \\ B_L \end{pmatrix} \tag{49}$$

$$\begin{aligned} \begin{pmatrix} c^u_R & -s^u_R e^{i\varphi^u_R} \\ s^u_R e^{-i\varphi^u_R} & c^u_R \end{pmatrix} \begin{pmatrix} 0 & 0 \\ 0 & 1 \end{pmatrix} \mathcal{U}^{d+}_R &= \begin{pmatrix} 0 & -s^u_R e^{i\varphi^u_R} \\ 0 & c^u_R \end{pmatrix} \begin{pmatrix} c^d_R & s^d_R e^{i\varphi^d_R} \\ -s^d_R e^{-i\varphi^d_R} & c^d_R \end{pmatrix} \\ &= \begin{pmatrix} s^u_R s^d_R e^{i(\varphi^u_R - \varphi^d_R)} & -s^u_R c^d_R e^{i\varphi^u_R} \\ -c^u_R s^d_R e^{-i\varphi^d_R} & c^u_R c^d_R \end{pmatrix} \end{aligned} \tag{50}$$

Upon performing the calculations, the aforementioned transformation yields the expression for the following couplings:

$$y^R_{tbW^+} = s^u_R s^d_R e^i \left(\varphi^u_R - \varphi^d_R\right), \tag{51}$$

$$y^R_{TbW^+} = -c^u_R s^d_R e^{-i\varphi^d_R}, \tag{52}$$

$$y^R_{tBW^+} = -s^u_R c^d_R e^{i\varphi^u_R}, \tag{53}$$

$$y^R_{TBW^+} = c^u_R c^d_R. \tag{54}$$

2.3.2 Neutral currents

The Yukawa Lagrangian of neutral currents may be expressed as follows:

$$\mathcal{L} = \frac{-g}{2c_W} \left[\overline{u}^0_{Li} \gamma^\mu u^0_{Li} - \overline{d}^0_{Li} \gamma^\mu d^0_{Li} + \overline{u}^0_{R4} \gamma^\mu u^0_{R4} - \overline{d}^0_{R4} \gamma^\mu d^0_{R4} \right] z_\mu + J^\mu_{EM}, \quad \text{with} \quad i = 3, 4 \tag{55}$$

The Left-Handed (LH) terms can be represented as follows:

$$\begin{aligned}\left(\overline{u}^0_{L3} \quad \overline{u}^0_{L4}\right)\begin{pmatrix}1 & 0\\0 & 1\end{pmatrix}\gamma^\mu\begin{pmatrix}u^0_{L3}\\u^0_{L4}\end{pmatrix} &= \left(\bar{t}_L \quad \overline{T}_L\right)\mathcal{U}^u_L\begin{pmatrix}1 & 0\\0 & 1\end{pmatrix}\mathcal{U}^u_L\gamma^\mu\begin{pmatrix}t_L\\T_L\end{pmatrix}\\ &= \left(\bar{t}_L \quad \overline{T}_L\right)\gamma^\mu\begin{pmatrix}t_L\\T_L\end{pmatrix} etc\end{aligned} \tag{56}$$

Applying the aforementioned transformation yields the following couplings:

$$y^L_{ttZ} = y^R_{bbZ} = 1, \tag{57}$$

$$y^L_{TTZ} = y^R_{TTZ} = 1 \tag{58}$$

$$y^L_{tTZ} = y^L_{bBZ} = 0. \tag{59}$$

The Right-Handed (RH) terms can be represented as follows:

$$\begin{aligned}\mathcal{U}^u_R\begin{pmatrix}0 & 0\\0 & 1\end{pmatrix}\mathcal{U}^{u\dagger}_R &= \begin{pmatrix}c^u_R & -s_R e^{i\varphi^u}\\ s^u_R e^{-i\varphi^u} & c^u_R\end{pmatrix}\begin{pmatrix}0 & 0\\0 & 1\end{pmatrix}u^{u\dagger}_R\\ &= \begin{pmatrix}0 & -s^u_R e^{i\varphi^u}\\0 & c^u_R\end{pmatrix}\begin{pmatrix}c^u_R & s^u_R e^{i\varphi^u}\\ -s^u_R e^{-i\varphi^u} & c^u_R\end{pmatrix}\\ &= \begin{pmatrix}s^{u^2}_R & -s^u_R c^u_R e^{i\varphi^u}\\ -s^u_R c^u_R e^{-i\varphi^u} & c^{u^2}_R\end{pmatrix} \equiv X^u\end{aligned} \tag{60}$$

After performing the requisite calculations, the aforementioned transformation leads to the expression of the following couplings:

$$y^R_{ttZ} = s^{u^2}_R \tag{61}$$

$$y^R_{TTZ} = c^{u^2}_R \tag{62}$$

$$y^R_{tTZ} = -s^u_R c^u_R e^{i\varphi^u_R} \tag{63}$$

$$y^R_{bbZ} = s^{d^2}_R, \tag{64}$$

$$y^R_{BBZ} = c^{d^2}_R, \tag{65}$$

$$y^R_{bBZ} = -s^d_R c^d_R e^{i\varphi^d_R}. \tag{66}$$

2.3.3 Modified Higgs couplings

$$\mathcal{L} = -\left(\overline{u}^0_{L3} \quad \overline{u}^0_{L4}\right)\begin{pmatrix}y^u_{33} & 0\\ y^u_{34} & 0\end{pmatrix}\frac{H}{\sqrt{2}} + (u \to d) + hc \tag{67}$$

$$\begin{pmatrix}\frac{y^u_{33}}{\sqrt{2}} & 0\\ \frac{y^u_{43}}{\sqrt{2}} & 0\end{pmatrix} = \frac{1}{v}\begin{pmatrix}\frac{y^u_{33}}{\sqrt{2}} & 0\\ \frac{y^u_{43}}{\sqrt{2}} & M^0\end{pmatrix}\begin{pmatrix}1 & 0\\0 & 0\end{pmatrix} = \frac{1}{v}\mathcal{M}^u\begin{pmatrix}1 & 0\\0 & 0\end{pmatrix} \tag{68}$$

$$
\begin{aligned}
\mathcal{U}_L^u \frac{1}{v}\mathcal{M}^u \begin{pmatrix}1 & 0\\ 0 & 0\end{pmatrix} \mathcal{U}_R^{u+} &= \frac{1}{v}\mathcal{U}_L^u\mathcal{M}^u\mathcal{U}_R^{u+}\mathcal{U}_R^u \begin{pmatrix}1 & 0\\ 0 & 0\end{pmatrix} \mathcal{U}_R^{u+} \\
&= \frac{1}{v}\mathcal{M}_{diag}^u\,\mathcal{U}_R^u \left[1 - \begin{pmatrix}0 & 0\\ 0 & 1\end{pmatrix}\right] \mathcal{U}_R^{u+} \\
&= \frac{1}{v}\mathcal{M}_{diag}^u\left(1 - X^u\right)
\end{aligned} \tag{69}
$$

Upon completing the necessary calculations, the resulting transformation yields the following couplings:

$$y_{ttH}^R = c_R^{u^2} \tag{70}$$

$$y_{TTH}^R = s_R^{u^2} \tag{71}$$

$$y_{tTH}^L = s_R^u c_R^u e^{i\phi_u}, \quad y_{tTH}^R = \frac{m_t}{m_T}\left(s_R^u c_R^u e^{i\phi_u}\right) \tag{72}$$

$$y_{bbH}^R = c_R^{d^2}, \tag{73}$$

$$y_{BBH}^R = s_R^{d^2}, \tag{74}$$

$$y_{bBH}^R = s_R^d c_R^d u e^{i\phi_d}, \quad y_{bBH}^R = \frac{m_t}{m_T}\left(s_R^d c_R^d e^{i\phi_d}\right). \tag{75}$$

In **Figure 4**, we depict the branching ratios ($\mathcal{BR}s$) of three decay channels, namely $T \rightarrow W^{+}b$, tZ, and th, as a function of the top quark partner mass m_T, at a fixed s_L for SM + T representation (left) and at a fixed s_R^u and s_R^d for SM + TB representation (right). It can be observed from both panels that the $T \rightarrow W^{+}b$ decay mode exhibits dominance over other modes throughout the entire mass spectrum of T, ranging from 300 to 3000 GeV.

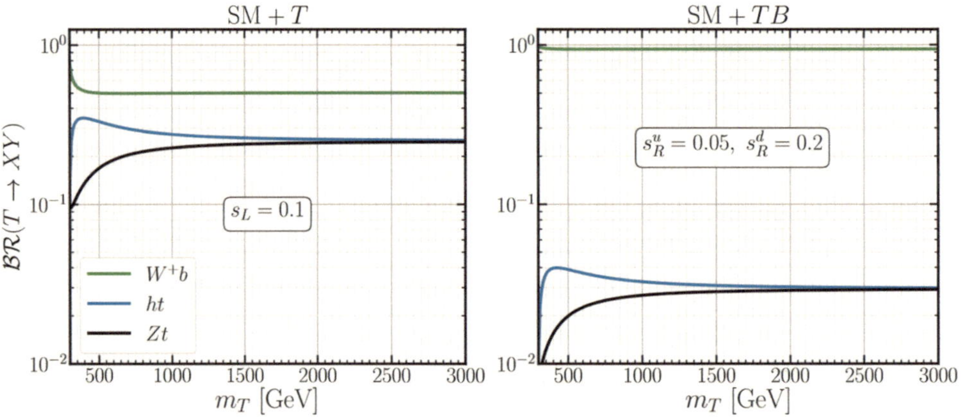

Figure 4.
$BR(T \rightarrow XY)$ as a function of m_T for SM + T (left) and SM + TB (right).

3. 2HDM + VLQs singlet: concepts and formulations

The most general renormalizable potential, of 2HDM + VLQ [21, 22], which is invariant under $\mathcal{SU}(2) \otimes \mathcal{U}(1)$ can be written as

$$\begin{aligned} V(\phi_1, \phi_2) = & m_1^2 \phi_1^\dagger \phi_1 + m_2^2 \phi_2^\dagger \phi_2 + \left(m_{12}^2 \phi_1^\dagger \phi_2 + h.c\right) + \lambda_1 \left(\phi_1^\dagger \phi_1\right)^2 + \lambda_2 \left(\phi_2^\dagger \phi_2\right)^2 \\ & + \lambda_3 \left(\phi_1^\dagger \phi_1\right)\left(\phi_2^\dagger \phi_2\right) + \lambda_4 \left(\phi_1^\dagger \phi_2\right)\left(\phi_2^\dagger \phi_1\right) + \left[\lambda_5 \left(\phi_1^\dagger \phi_2\right)^2 + h.c\right] \\ & + \phi_1^\dagger \phi_1 \left[\lambda_6 \left(\phi_1^\dagger \phi_2\right) + h.c\right] + \phi_2^\dagger \phi_2 \left[\lambda_7 \left(\phi_1^\dagger \phi_2\right) + h.c\right] \end{aligned} \tag{76}$$

where ϕ_i, i = 1, 2 are complex $\mathcal{SU}(2)$ doublets

$$\phi_1 = \begin{pmatrix} G^+ \\ \dfrac{v + \varphi_1^0 + iG^0}{\sqrt{2}} \end{pmatrix}, \quad \phi_2 = \begin{pmatrix} H^+ \\ \dfrac{\varphi_2^0 + iA}{\sqrt{2}} \end{pmatrix} \tag{77}$$

The physical CP-even scalars h and H are mixtures of $\varphi_{1,2}^0$:

$$\begin{pmatrix} h \\ H \end{pmatrix} = \begin{pmatrix} s_{\beta\alpha} & s_{\beta\alpha} \\ c_{\beta\alpha} & -s_{\beta\alpha} \end{pmatrix} \begin{pmatrix} \varphi_1^0 \\ \varphi_2^0 \end{pmatrix} \tag{78}$$

where: $s_{\beta\alpha} = \sin(\beta - \alpha)$, $c_{\beta\alpha} = \cos(\beta - \alpha)$.
we write:

$$\tilde{\phi}_1 = \begin{pmatrix} \dfrac{v + \varphi_1^0 - iG^0}{\sqrt{2}} \\ -G^- \end{pmatrix}, \quad \tilde{\phi}_2 = \begin{pmatrix} \dfrac{\varphi_2^0 - iA}{\sqrt{2}} \\ -H^- \end{pmatrix} \tag{79}$$

where $\tilde{\phi}_i = i\tau_2 \phi_i^* \quad (\mathrm{i} = 1, 2),$.
we add the Lagrangian of Yukawa interaction for the heavy Quark $T_{L,R}^0$ and $B_{L,R}^0$:

$$\begin{aligned} -\mathcal{L}_Y^{II} \supset & y_t \overline{Q_L^0} \tilde{\phi}_2 T_R^0 + \lambda_t \overline{Q_L^0} \tilde{\phi}_1 T_R^0 + M_T \overline{T_L^0} T_R^0 + y_b \overline{Q_L^0} (\sigma_1 \tilde{\phi}_2) B_R^0 \\ & + \lambda_b \overline{Q_L^0} (\sigma_1 \tilde{\phi}_1) B_R^0 + M_B \overline{B_L^0} B_R^0 + Hc \end{aligned} \tag{80}$$

where $\overline{Q_L^0} = \left(\overline{t_L^0}, \overline{b_L^0}\right)$, and $\sigma_1 = \begin{pmatrix} 0 & 1 \\ 1 & 0 \end{pmatrix}$.

3.1 Neutral Higgs couplings

- For the Light-Light couplings of Top quarks to the triplets Higgs h, H, and A (**Figure 5**)

$$y_{htt} = \left(s_{\beta\alpha} + c_{\beta\alpha} \cot\beta\right) c_L^2 \tag{81}$$

$$y_{Htt} = \left(c_{\beta\alpha} - s_{\beta\alpha} \cot\beta\right) c_L^2 \tag{82}$$

$$y_{Att} = -\cot\beta c_L^2 \tag{83}$$

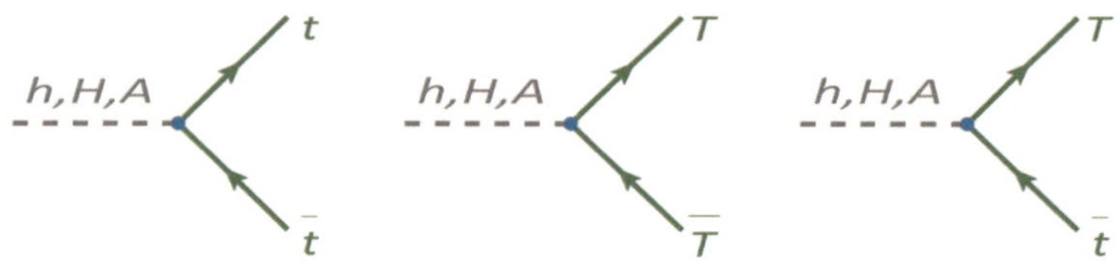

Figure 5.
Typical Feynman diagram of the Higgs bosons h, H and a couplings to pairs of top (t) and top partner (T).

- For the heavy-heavy couplings of Top quarks to the triplets Higgs h, H, and A

$$y_{hTT} = (s_{\beta\alpha} + c_{\beta\alpha}\cot\beta)s_L^2 \tag{84}$$
$$y_{HTT} = (c_{\beta\alpha} - s_{\beta\alpha}\cot\beta)s_L^2 \tag{85}$$
$$y_{ATT} = -\cot\beta s_L^2 \tag{86}$$

- For the Light-heavy left and right couplings of Top quarks to the triplets Higgs h, H, and A

$$\begin{aligned} y^L_{htT} &= (s_{\beta\alpha} + c_{\beta\alpha}\cot\beta)\frac{m_t}{m_T}c_L s_L e^{i\phi} y^R_{htT} = (s_{\beta\alpha} + c_{\beta\alpha}\cot\beta)c_L s_L e^{i\phi} \\ y^L_{HtT} &= (c_{\beta\alpha} - s_{\beta\alpha}\cot\beta)\frac{m_t}{m_T}c_L s_L e^{i\phi} y^R_{HtT} = (c_{\beta\alpha} - s_{\beta\alpha}\cot\beta)c_L s_L e^{i\phi} \\ y^L_{AtT} &= -\cot\beta\frac{m_t}{m_T}c_L s_L e^{i\phi} y^R_{AtT} = -\cot\beta c_L s_L e^{i\phi} \end{aligned} \tag{87}$$

3.2 Charged Higgs couplings

The couplings of the charged Higgs boson$H^{\pm}$ to the top quark (t) and its heavy partner (T) with the bottom quark (b) can be expressed as follows (**Figure 6**):

$$y^L_{tbH^+} = c_L, \quad y^R_{tbH^+} = 1 \tag{88}$$
$$y^L_{TbH^+} = s_L, \quad y^R_{tbH^+} = 0 \tag{89}$$

4. 2HDM + VLQs doublet case

4.1 Neutral Higgs couplings

Following the methodology outlined in the preceding section, the Higgs coupling can be determined within the framework of the 2HDM + *TB* [23, 24] as follows:

Figure 6.
Typical Feynman diagram of the charged Higgs boson H⁺ couplings to the top quark (t) and its heavy partner (T) with the bottom quark (b).

- For the Light-Light couplings of Top quarks to the triplets Higgs h, H, and A

$$y_{htt} = \left(s_{\beta\alpha} + c_{\beta\alpha}\cot\beta\right)c_R^{u^2} \tag{90}$$

$$y_{Htt} = \left(c_{\beta\alpha} - s_{\beta\alpha}\cot\beta\right)c_R^{u^2} \tag{91}$$

$$y_{Att} = -\cot\beta c_R^{u^2} \tag{92}$$

- For the heavy-heavy couplings of Top quarks to the triplets Higgs h, H, and A

$$y_{hTT} = \left(s_{\beta\alpha} + c_{\beta\alpha}\cot\beta\right)s_R^{u^2} \tag{93}$$

$$y_{HTT} = \left(c_{\beta\alpha} - s_{\beta\alpha}\cot\beta\right)s_R^{u^2} \tag{94}$$

$$y_{ATT} = -\cot\beta s_R^{u^2} \tag{95}$$

- For the Light-heavy left and right couplings of Top quarks to the triplets Higgs h, H, and A

$$\begin{aligned}
y_{htT}^L &= \left(s_{\beta\alpha} + c_{\beta\alpha}\cot\beta\right)s_R^u c_R^u e^{i\phi_u} y_{htT}^R = \left(s_{\beta\alpha} + c_{\beta\alpha}\cot\beta\right)\frac{m_t}{m_T} s_R^u c_R^u e^{i\phi_u} \\
y_{HtT}^L &= \left(c_{\beta\alpha} - s_{\beta\alpha}\cot\beta\right)s_R^u c_R^u e^{i\phi_u} y_{HtT}^R = \left(c_{\beta\alpha} - s_{\beta\alpha}\cot\beta\right)\frac{m_t}{m_T} s_R^u c_R^u e^{i\phi_u} \\
y_{AtT}^L &= -\cot\beta s_R^u c_R^u e^{i\phi_u} y_{AtT}^R = -\cot\beta \frac{m_t}{m_T} s_R^u c_R^u e^{i\phi_u}
\end{aligned} \tag{96}$$

4.2 Charged Higgs couplings

Let us start with the Yukawa Lagrangian (**Figure 7**):

$$\begin{aligned}
\mathcal{L} &= -y_{ij}^{u*}\overline{q}_{Li}u_{Rj}\epsilon H_u^* + y_{ij}^{d*}\overline{q}_{Li}d_{Rj}\epsilon H_d^* - y_{4j}^{u*}\overline{Q}_L u_{Rj}\epsilon H_u^* + y_{4j}^{d*}\overline{Q}_L d_{Rj}\epsilon H_d^* + h.c. \\
&= -y_{ij}^{u*}\begin{pmatrix}\overline{u}_{Li}^0 & \overline{d}_{Li}^0\end{pmatrix}\begin{pmatrix}H_u^{0*} \\ -H_u^+\end{pmatrix}u_{Rj}^0 + y_{ij}^{d*}\begin{pmatrix}\overline{u}_{Li}^0 & \overline{d}_{Li}^0\end{pmatrix}\begin{pmatrix}H_d^- \\ -H_d^{0*}\end{pmatrix}u_{Rj}^0 \\
&\quad - y_{4j}^{u*}\begin{pmatrix}\overline{u}_{L4}^0 & \overline{d}_{L4}^0\end{pmatrix}\begin{pmatrix}H_u^{0*} \\ -H_u^+\end{pmatrix}u_{Rj}^0 + y_{4j}^{d*}\begin{pmatrix}\overline{u}_{L4}^0 & \overline{d}_{L4}^0\end{pmatrix}\begin{pmatrix}H_d^- \\ -H_d^{0*}\end{pmatrix}u_{Rj}^0 \\
&= -y_{ij}^{u*}\begin{pmatrix}\overline{u}_{Li}^0 & \overline{d}_{Li}^0\end{pmatrix}\begin{pmatrix}\frac{1}{\sqrt{2}}\left[\cos\ \alpha h^0 + \sin\alpha H_1^0 - i\ \left(\sin\beta G^0 + \cos\beta P_1^0\right)\right] \\ -\ \left(\sin\beta G^+ + \cos\beta H^+\right)\end{pmatrix}u_{Rj}^0 \\
&\quad + y_{ij}^{d*}\begin{pmatrix}\overline{u}_{Li}^0 & \overline{d}_{Li}^0\end{pmatrix}\begin{pmatrix}-\ \left(\cos\beta G^- + \sin\beta H^-\right) \\ -\frac{1}{\sqrt{2}}\left[-\sin\alpha h^0 + \cos\ \alpha H_1^0 + i\left(\cos\beta G^0 - \sin\beta P_1^0\right)\right]\end{pmatrix}u_{Rj}^0 \\
&\quad - y_{4j}^{u*}\begin{pmatrix}\overline{u}_{L4}^0 & \overline{d}_{L4}^0\end{pmatrix}\begin{pmatrix}\frac{1}{\sqrt{2}}\left[\cos\ \alpha h^0 + \sin\alpha H_1^0 - i\left(\sin\beta G^0 + \cos\beta P_1^0\right)\right] \\ -\ \left(\sin\beta G^+ + \cos\beta H^+\right)\end{pmatrix}u_{Rj}^0 \\
&\quad + y_{4j}^{d*}\begin{pmatrix}\overline{u}_{L4}^0 & \overline{d}_{L4}^0\end{pmatrix}\begin{pmatrix}-\ \left(\cos\beta G^- + \sin\beta H^-\right) \\ -\frac{1}{\sqrt{2}}\left[-\sin\alpha h^0 + \cos\ \alpha H_1^0 + i\ \left(\cos\beta G^0 - \sin\beta P_1^0\right)\right]\end{pmatrix}u_{Rj}^0
\end{aligned} \tag{97}$$

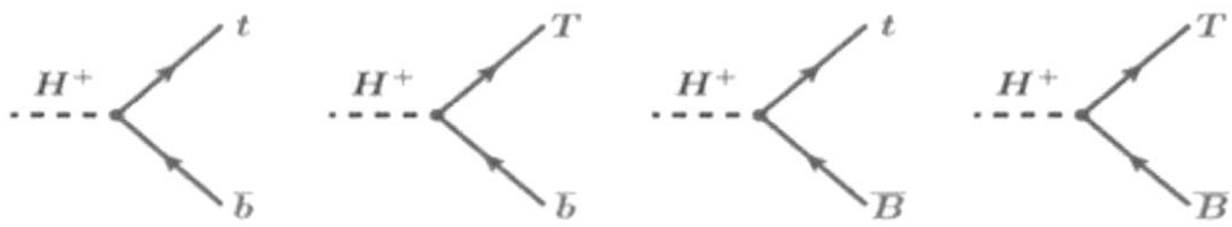

Figure 7.
Typical Feynman diagram of the charged Higgs boson H⁺ couplings to the top quark (t) and its heavy partner (T) with the bottom quark (b) and its heavy partner (B).

If we select only the $H^{\pm}$ particle, the result is:

$$\begin{aligned}
\mathcal{L} \supset\, & y_{ij}^{u*} \begin{pmatrix} \overline{u}_{Li}^{0} & \overline{d}_{Li}^{0} \end{pmatrix} \begin{pmatrix} 0 \\ \cos\beta H^{+} \end{pmatrix} u_{Rj}^{0} \\
& + y_{ij}^{d*} \begin{pmatrix} \overline{u}_{Li}^{0} & \overline{d}_{Li}^{0} \end{pmatrix} \begin{pmatrix} \sin\beta H^{-} \\ 0 \end{pmatrix} u_{Rj}^{0} \\
& + y_{4j}^{u*} \begin{pmatrix} \overline{u}_{L4}^{0} & \overline{d}_{L4}^{0} \end{pmatrix} \begin{pmatrix} 0 \\ \cos\beta H^{+} \end{pmatrix} u_{Rj}^{0} \\
& + y_{4j}^{d*} \begin{pmatrix} \overline{u}_{L4}^{0} & \overline{d}_{L4}^{0} \end{pmatrix} \begin{pmatrix} \sin\beta H^{-} \\ 0 \end{pmatrix} u_{Rj}^{0} \\
\supset\, & y_{ij}^{u*} \cos\beta H^{\pm} \overline{d}_{Li}^{0} u_{Rj}^{0} + y_{ij}^{d*} \sin\beta H^{\pm} \overline{u}_{Li}^{0} d_{Rj}^{0} \\
& + y_{4j}^{u*} \cos\beta H^{\pm} \overline{d}_{L4}^{0} u_{Rj}^{0} + y_{4j}^{d*} \sin\beta H^{\pm} \overline{u}_{L4}^{0} d_{Rj}^{0} \\
\supset\, & y_{ij}^{u*} \cos\beta H^{\pm} \overline{d}_{Li}^{0} u_{Rj}^{0} + y_{4j}^{u*} \cos\beta H^{\pm} \overline{d}_{L4}^{0} u_{Rj}^{0} \\
& + y_{ij}^{d*} \sin\beta H^{\pm} \overline{u}_{Li}^{0} d_{Rj}^{0} + y_{4j}^{d*} \sin\beta H^{\pm} \overline{u}_{L4}^{0} d_{Rj}^{0}
\end{aligned} \tag{98}$$

Subsequently, we can express the Yukawa Lagrangian of the charged Higgs boson as follows:

$$\mathcal{L} = \begin{pmatrix} \overline{u}^{0}{}_{L3} & \overline{u}^{0}{}_{L4} \end{pmatrix} \begin{pmatrix} y_{33}^{d} & 0 \\ y_{43}^{d} & 0 \end{pmatrix} \begin{pmatrix} d_{R3}^{0} \\ d_{R4}^{0} \end{pmatrix} \sin\beta H^{\pm} + \begin{pmatrix} \overline{d}^{0}{}_{L3} & \overline{d}^{0}{}_{L4} \end{pmatrix} \begin{pmatrix} y_{33}^{u} & 0 \\ y_{43}^{u} & 0 \end{pmatrix} \begin{pmatrix} u_{R3}^{0} \\ u_{R4}^{0} \end{pmatrix} \cos\beta H^{\pm} \tag{99}$$

$$\begin{pmatrix} y_{33}^{d} & 0 \\ y_{43}^{d} & 0 \end{pmatrix} = \frac{\sqrt{2}}{v} \left[\begin{pmatrix} \frac{y_{33}^{d} v}{\sqrt{2}} & 0 \\ \frac{y_{43}^{d} v}{\sqrt{2}} & \frac{y_{44}^{d} v}{\sqrt{2}} \end{pmatrix} - \begin{pmatrix} 0 & 0 \\ 0 & M^{0} \end{pmatrix} \right] = \frac{\sqrt{2}}{v} \mathcal{M}^{d} - \frac{M^{0}\sqrt{2}}{v} \begin{pmatrix} 0 & 0 \\ 0 & 1 \end{pmatrix} \tag{100}$$

$$\begin{aligned}
& \mathcal{U}_{L}^{u} \left[\frac{\sqrt{2}}{v} \mathcal{M}^{d} - \frac{M^{0}\sqrt{2}}{v} \begin{pmatrix} 0 & 0 \\ 0 & 1 \end{pmatrix} \right] \mathcal{U}_{R}^{d\dagger} = \frac{\sqrt{2}}{v} \mathcal{U}_{L}^{u} \left[\mathcal{U}_{L}^{d\dagger} \mathcal{U}_{L}^{d} \mathcal{M}^{d} \mathcal{U}_{R}^{d\dagger} \mathcal{U}_{R}^{d} - M^{0} \begin{pmatrix} 0 & 0 \\ 0 & 1 \end{pmatrix} \right] \mathcal{U}_{R}^{d\dagger} \\
& = \frac{\sqrt{2}}{v} \mathcal{U}_{L}^{u} \mathcal{M}_{diag}^{d} \underbrace{\mathcal{U}_{R}^{d} \mathcal{U}_{R}^{d\dagger}}_{=1} - M^{0} \frac{\sqrt{2}}{v} \mathcal{U}_{L}^{u} \begin{pmatrix} 0 & 0 \\ 0 & 1 \end{pmatrix} \mathcal{U}_{R}^{d\dagger} \\
& = \frac{\sqrt{2}}{v} \left(A \times \mathcal{M}_{diag}^{d} - m_{B} B \right)
\end{aligned} \tag{101}$$

with

$$\begin{aligned} A &= \mathcal{U}_L^u \mathcal{U}_L^{d\dagger} \\ &= \begin{pmatrix} c_L^u & -s_L^u e^{i\phi_u} \\ s_L^u e^{-i\phi_u} & c_L^u \end{pmatrix} \begin{pmatrix} c_L^d & s_L^d e^{i\phi_d} \\ -s_L^d e^{-i\phi_d} & c_L^d \end{pmatrix} \\ &= \begin{pmatrix} c_L^u c_L^d + s_L^u s_L^d e^{i(\phi_u-\phi_d)} & c_L^u s_L^d e^{i\phi_d} - c_L^d s_L^u e^{i\phi_u} \\ c^d s_L^u e^{-i\phi_u} - c_L^u s_L^d e^{-i\phi_d} & s_L^u s_L^d e^{i(\phi_d-\phi_u)} + c_L^u c_L^d \end{pmatrix} \end{aligned} \tag{102}$$

$$\begin{aligned} B &= \mathcal{U}_L^u \begin{pmatrix} 0 & 0 \\ 0 & 1 \end{pmatrix} \mathcal{U}_R^{d\dagger} \\ &= \begin{pmatrix} c_L^u & -s_L^u e^{i\phi_u} \\ -s_R^d e^{-i\phi_u} & c_L^u \end{pmatrix} \begin{pmatrix} 0 & 0 \\ 0 & 1 \end{pmatrix} \mathcal{U}_R^{d\dagger} \\ &= \begin{pmatrix} 0 & -s_L^u e^{i\phi_u} \\ 0 & c_L^u \end{pmatrix} \begin{pmatrix} c_R^d & s_R^d e^{i\phi_d} \\ -s_R^d e^{i\phi_d} & c_R^d \end{pmatrix} \\ &= \begin{pmatrix} s_L^u s_R^d e^{i(\phi_u-\phi_d)} & -c_R^d s_L^u e^{i\phi_u} \\ -c_L^u s_R^d e^{-i\phi_d} & c_L^u c_R^d \end{pmatrix} \end{aligned} \tag{103}$$

Now we can define the Matrix M^u as follows:

$$M^u = A \times \mathcal{M}_{diag}^d - m_B B$$

$$= \begin{pmatrix} \left[c_L^u c_L^d + s_L^u s_L^d e^{i(\phi_u-\phi_d)}\right] m_b - mBs_L^u s_R^d e^{i(\phi_u-\phi_d)} & \left[c_L^u s_L^d e^{i\phi_d} - c_L^d s_L^u e^{i\phi_u}\right] m_B + m_B c_R^d s_L^u e^{i\phi_u} \\ \left[c_L^d s_L^u e^- i\phi_u - c_L^u s_L^d e^{-i\phi_d}\right] m_b + m_B c_L^u s_R^d e^{-i\phi_d} & \left[s_L^u s_L^d e^{i(\phi_d-\phi_u)} + c_L^u c_L^d\right] m_B - m_B c_L^u c_R^d \end{pmatrix} \tag{104}$$

Similarly, the matrix M^d can be defined in the same manner as M^u.

$$M^d = \begin{pmatrix} \left[c_L^d c_L^u + s_L^d s_L^u e^{i(\phi_d-\phi_u)}\right] m_t - mTs_L^d s_R^u e^{i(\phi_d-\phi_u)} & \left[c_L^d s_L^u e^{i\phi_u} - c_L^u s_L^d e^{i\phi_d}\right] m_T + m_T c_R^u s_L^d e^{i\phi_d} \\ \left[c_L^u s_L^d e^{-i\phi_d} - c_L^d s_L^u e^{-i\phi_u}\right] m_t + m_T c_L^d s_R^u e^{-i\phi_u} & \left[s_L^d s^u d_L e^{i(\phi_u-\phi_d)} + c_L^d c^u\right] m_T - m_T c_L^d c_R^u \end{pmatrix} \tag{105}$$

Finally, the charged Higgs couplings can be defined as follows:

- For the coupling $H^{\pm}tb$:

$$y_{H^+tb}^L = c_L^d c_L^u + \frac{s_L^d}{s_L^u} \left(s_L^{u^2} - s_R^{u^2}\right) e^{i(\phi_u-\phi_d)} \tag{106}$$

$$y_{H^+tb}^R = \frac{m_b}{m_t} \left[c_L^u c_L^d + \frac{s_L^u}{s_L^d} \left(s_L^{d^2} - s_R^{d^2}\right) e^{i(\phi_u-\phi_d)}\right] \tag{107}$$

- For the coupling H^+Tb:

$$y^L_{H^+Tb} = c^d_L c^u_L + \frac{s^d_L}{s^u_L}\left(s^{u^2}_L - s^{u^2}_R\right) e^{-i\phi_d *} \tag{108}$$

$$y^R_{H^+Tb} = \frac{m_b}{m_T}\left[c^u_L c^d_L + \frac{s^u_L}{s^d_L}\left(s^{d^2}_L - s^{d^2}_R\right) e^{-i\phi_d}\right] \tag{109}$$

- For the coupling H^+tB

$$y^L_{H^+tB} = \frac{m_t}{m_B}\left[c^u_L c^d_L e^{i\phi_d} + \frac{c^d_L}{s^u_L}\left(s^{d^2}_L - s^{d^2}_R\right) e^{i\phi_u}\right] \tag{110}$$

$$y^R_{H^+tB} = c^d_L c^u_L + \frac{s^d_L}{s^u_L}\left(s^{u^2}_L - s^{u^2}_R\right) e^{\phi_u} \tag{111}$$

- For the coupling $H^{\pm}TB$:

$$y^L_{H^+tb} = s^d_L s^u_L e^{i(\phi_d - \phi_u)} + \frac{c^d_L}{c^u_L}\left(s^{u^2}_R - s^{u^2}_L\right) \tag{112}$$

$$y^R_{H^+tb} = \frac{m_B}{m_T}\left[s^u_L s^d_L e^{i(\phi_d - \phi_u)} + \frac{c^u_L}{c^d_L}\left(s^{d^2}_R - s^{d^2}_L\right) e^{i(\phi_u - \phi_d)}\right] \tag{113}$$

Figure 8 presents the branching ratios ($\mathcal{BR}s$) for all open decay channels, including $T \to W^+b$, tZ, th, H^+b, Ht, and At, as a function of the mass of the top quark partner m_T. The left panel illustrates the singlet representation 2HDM + T, with fixed values for m_h, m_H, m_A, $\tan\beta$, and s_L, while the right panel depicts the 2HDM + TB representation with the same fixed 2HDM parameters in addition to s^u_R and s^d_R.

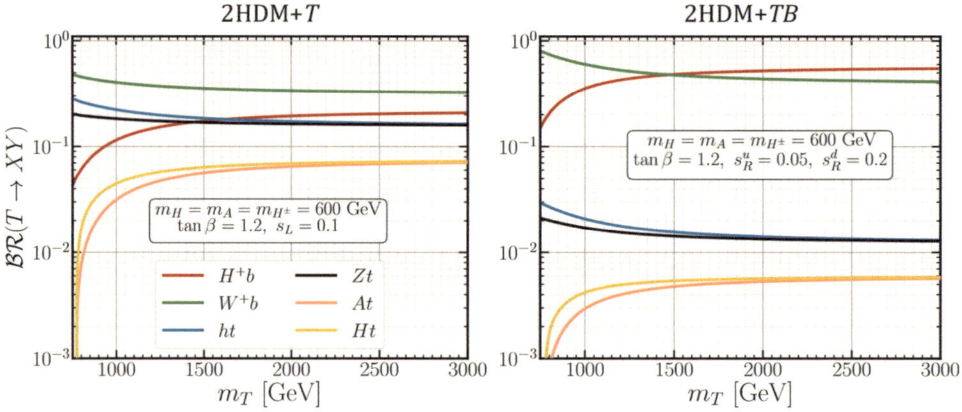

Figure 8.
$BR(T \to XY)$ as a function of m_T for 2HDM + T (left) and 2HDM + TB (right).

5. Summary of gauge interactions

5.1 Light–light interactions

$$
\begin{aligned}
\mathcal{L}_W &= -\frac{g}{\sqrt{2}}\bar{t}\gamma^\mu\left(y^L_{tb}P_L + y^R_{tb}P_R\right)bW^+_\mu + H.c.,\\
\mathcal{L}_Z &= -\frac{g}{2c_W}\bar{t}\gamma^\mu\left(y^L_{tt}P_L + y^R_{tt}P_R - 2Q_t s^2_W\right)tZ_\mu\\
&\quad -\frac{g}{2c_W}\bar{b}\gamma^\mu\left(-y^L_{Zbb}P_L - y^R_{bb}P_R - 2Q_b s^2_W\right)bZ_\mu + H.c.,\\
\mathcal{L}_h &= -\frac{gm_t}{2M_W}y^h_{tt}\bar{t}th^0 - \frac{gm_b}{2M_W}y^h_{bb}\bar{b}bh^0 + H.c.,\\
\mathcal{L}_H &= -\frac{gm_t}{2M_W}y^H_{tt}\bar{t}tH^0 - \frac{gm_b}{2M_W}y^H_{bb}\bar{b}bH^0 + H.c.,\\
\mathcal{L}_A &= -\, i\frac{gm_t}{2M_W}y^A_{tt}\bar{t}\gamma_5 tA + i\frac{gm_b}{2M_W}y^A_{bb}\bar{b}\gamma_5 bA + H.c.,\\
\mathcal{L}_{H^+} &= -\, \frac{gm_t}{\sqrt{2}M_W}\bar{t}\left(\cot\beta y^L_{tb}P_L + \tan\beta y^R_{tb}P_R\right)bH^+ + H.c.
\end{aligned}
\tag{114}
$$

5.2 Heavy-heavy interactions

$$
\begin{aligned}
\mathcal{L}_W &= -\, \frac{g}{\sqrt{2}}\overline{Q}\gamma^\mu\left(y^L_{QQ'}P_L + y^R_{QQ'}P_R\right)Q'W^+_\mu + H.c.,\\
\mathcal{L}_Z &= -\, \frac{g}{2c_W}\overline{Q}\gamma^\mu\left(\pm y^L_{QQ}P_L \pm y^R_{QQ}P_R - 2Q_Q s^2_W\right)QZ_\mu\\
\mathcal{L}_h &= -\, \frac{gm_Q}{2M_W}y^h_{QQ}\overline{Q}Qh^0 + H.c.,\\
\mathcal{L}_H &= -\, \frac{gm_Q}{2M_W}y^H_{QQ}\overline{Q}QH^0 + H.c.,\\
\mathcal{L}_A &= \pm\, i\frac{gm_Q}{2M_W}y^A_{QQ}\overline{Q}\gamma_5 QA + H.c.,\\
\mathcal{L}_{H^+} &= -\, \frac{gm_Q}{\sqrt{2}M_W}\overline{Q}\left(\cot\beta y^L_{QQ}P_L + \tan\beta y^R_{QQ}P_R\right)QH^+ + H.c.
\end{aligned}
\tag{115}
$$

5.3 Light-heavy interactions

$$
\begin{aligned}
\mathcal{L}_W &= -\, \frac{g}{\sqrt{2}}\overline{Q}\gamma^\mu\left(y^L_{Qq}P_L + y^R_{Qq}P_R\right)qW^+_\mu\\
&\quad -\, \frac{g}{\sqrt{2}}\bar{q}\gamma^\mu\left(y^L_{qQ}P_L + y^R_{qQ}P_R\right)QW^+_\mu + H.c.\\
\mathcal{L}_Z &= -\, \frac{g}{2c_W}\bar{q}\gamma^\mu\left(\pm y^L_{qQ}P_L \pm y^R_{qQ}P_R\right)QZ_\mu + H.c\\
\mathcal{L}_h &= -\, \frac{gm_T}{2M_W}\bar{t}\left(Y^L_{htT}P_L + y^R_{htT}P_R\right)Th^0\\
&\quad -\frac{gm_B}{2M_W}\bar{b}\left(y^L_{hbB}P_L + y^R_{hbB}P_R\right)Bh^0 + H.C.\\
\mathcal{L}_H &= -\, \frac{gm_T}{2M_W}\bar{t}\left(y^L_{HtT}P_L + y^R_{HtT}P_R\right)TH^0\\
&\quad -\frac{gm_B}{2M_W}\bar{b}\left(y^L_{HbB}P_L + y^R_{HbB}P_R\right)Bh^0 + H.C.\\
\mathcal{L}_A &= i\frac{gm_T}{2M_W}\bar{t}\left(y^L_{AtT}P_L - Y^R_{AtT}P_R\right)TA\\
&\quad -\, i\frac{gm_B}{2M_W}\bar{b}\left(y^L_{AbB}P_L - y^R_{AbB}P_R\right)BA + H.C.\\
\mathcal{L}_{H^+} &= -\, \frac{gm_T}{\sqrt{2}M_W}\overline{T}\left(\cot\beta y^L_{Tb}P_L + \tan\beta y^R_{Tb}P_R\right)bH^+\\
&\quad -\frac{gm_B}{\sqrt{2}M_W}\bar{t}\left(\cot\beta y^L_{tB}P_L + \tan\beta y^R_{tB}P_R\right)BH^+ + H.c,
\end{aligned}
\tag{116}
$$

6. Vector-like quarks pair production at the LHC

The production process of vector-like quark pairs (VLQs) is an essential mechanism for investigating the properties of these hypothetical particles and for understanding the behavior of high energy interactions. A key benefit of VLQ pair production is its model independence, since it depends only on the masses of the VLQs and the energy of the colliding particles. At hadron colliders [25, 26], the extra-heavy quarks are able to get paired production because of their unique gauge couplings with the gluons.

In **Figure 9**, a typical Feynman diagram of pair-produced vector-like quarks $Q(\equiv T,\ B,\ X,\ Y)$ is shown.

The expression for the total inclusive $t\bar{t}$ production cross-section can be written as follows, as described in [27]:

$$\sigma_{\text{tot}}\ \left(pp \to T\overline{T}\right) = \sum_{ij} \int_0^{\beta_{\max}} d\beta \Phi_{ij}\ \left(\beta,\ \mu_F^2\right)\ \hat{\sigma}_{ij}\ \left(\beta,\ m^2,\ \mu_F^2,\ \mu_R^2\right) \tag{117}$$

Here, the indices i and j enumerate all possible initial state particles; $\beta_{\max}$ is defined as $\sqrt{1-4m^2/s}$, where $\sqrt{s}$ is the centre-of-mass energy of the hadron collider; and $\beta = \sqrt{1-\rho}$, where $\rho \equiv 4m^2/s$, is the relative velocity of the final state top quarks with pole mass m and partonic centre-of-mass energy $\sqrt{s}$.

The function Φ is the partonic flux and is given by:

$$\Phi_{ij}\ \left(\beta,\ \mu_F^2\right) = \frac{2\beta}{1-\beta^2} \mathcal{L}ij\ \left(\frac{1-\beta \max}{1-\beta}, \mu_F^2\right) \tag{118}$$

Here, $\mathcal{L}$ is the usual partonic luminosity, defined as:

$$\mathcal{L}\ \left(x,\ \mu_F^2\right) = x\ \left(f_i \otimes f_j\right)\ \left(x,\ \mu_F^2\right) \tag{119}$$

The renormalization and factorization scales are denoted by μ_R and μ_F, respectively. Setting $\mu_F = \mu_R = m$, the NNLO (next-to-next-to-leading order) partonic cross-section can be expanded as follows:

$$\hat{\sigma}_{ij}\ (\beta) = \frac{\alpha_s^2}{m^2}\ \left(\sigma_{ij}^{(0)} + \alpha_s \sigma_{ij}^{(1)} + \alpha_s^2 \sigma_{ij}^{(2)} + \mathcal{O}\ \left(\alpha_s^3\right)\right) \tag{120}$$

In the above equation, α_s represents the renormalized $\overline{\text{MS}}$ coupling, considering the presence of $N_L = 5$ active flavors at scale $\mu_R^2 = m^2$. Moreover, the functions $\sigma_{ij}^{(n)}$ solely depend on the value of β.

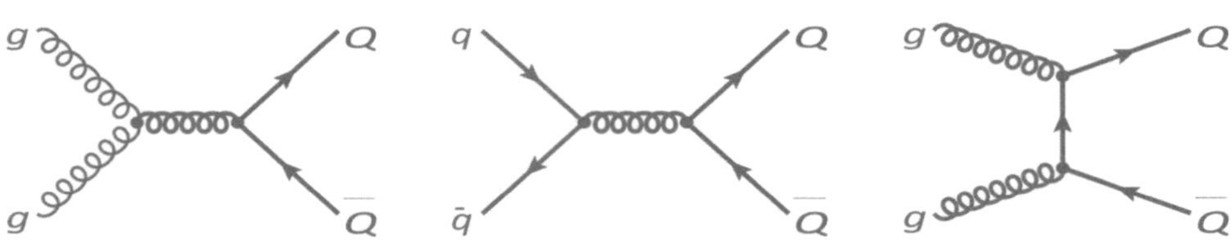

Figure 9.
Typical Feynman diagram of pair-produced vector-like quarks $Q(\equiv T,\ B,\ X,\ Y)$.

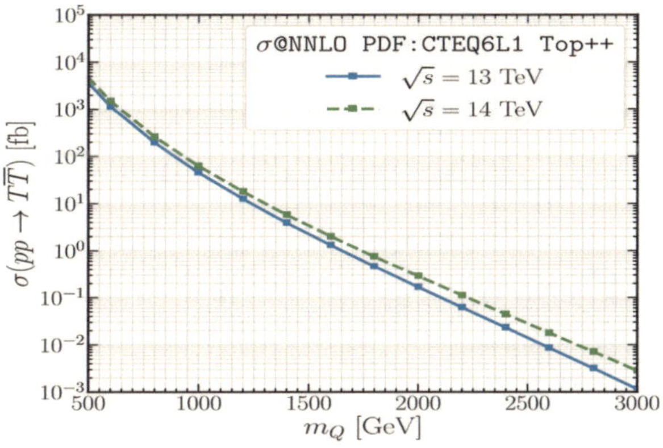

Figure 10.
Production cross section for pair-produced vector-like quarks T at the LHC. In green (blue) we show the NNLO cross section for 14 TeV (13 TeV) Centre of mass energy computed by top++ [28].

In **Figure 10**, we present the production cross-section at the next-to-next-to-leading order (NNLO) of vector-like quark T pairs produced at the Large Hadron Collider (LHC), as a function of their masses m_Q. The curves are depicted in green and blue, corresponding to centre of mass energies of 14 and 13 TeV, respectively.

7. Conclusions

In this chapter, we focus on the Lagrangian framework that adheres to the fundamental principles of the standard model, exploring its consequences by determining eigenstates and interactions with SM-VLQ. We provide a comprehensive examination of the full interaction within the mass basis of 2HDM. Additionally, we touch upon the pair production of heavy quarks at the LHC, operating at energies of 13 and 14 TeV, underlining the importance of these endeavors in advancing our understanding of the fundamental forces and particles shaping the universe.

Simultaneously, the pursuit of heavy quarks, which share quantum characteristics like spin and electric charge with their standard model counterparts, is highly motivated. These quarks are abundantly produced within the LHC, and their discovery would mark a monumental advancement in particle physics. Although current experimental searches for vector-like quarks at CERN's LHC have not yielded conclusive evidence, the quest continues, holding the potential to usher in a new era of particle physics research. In this context, researchers at the LHC grapple with the challenge of directly testing quantum entanglement in proton-proton collisions, which is complicated by the macroscopic nature of these interactions. Quantum entanglement, typically a phenomenon observed at the microscopic scale, involves correlations between particles or pairs of particles. While the LHC's experiments may not provide direct proof of entanglement, they indirectly explore related aspects by studying correlations and angular momentum conservation among particles generated during collisions.

Appendix: heavy quark decay widths

The partial widths for T decays, including all possible mixing terms, are

$$\begin{aligned}
\Gamma(T \to W^+ b) &= \frac{g^2}{64\pi} \frac{m_T}{M_W^2} \lambda(m_T,\ m_b,\ M_W)^{1/2} \left\{ \left(|y_{Tb}^L|^2 + \ |y_{Tb}^R|^2 \right) \right. \\
&\quad \times \left. \left[1 + r_W^2 - 2r_b^2 - 2r_W^4 + r_b^4 + r_W^2 r_b^2 \right] - 12 r_W^2 r_b\, |y_{Tb}^L\, |y_{Tb}^{R*} \right\}, \\
\Gamma(T \to Zt) &= \frac{g}{128\pi c_W^2} \frac{m_T}{M_Z^2} \lambda(m_T,\ m_t,\ M_Z)^{1/2} \left\{ \left(|y_{tT}^L|^2 + |y_{tT}^R|^2 \right) \right. \\
&\quad \times \left. \left[1 + r_Z^2 - 2r_t^2 - 2r_Z^4 + r_t^4 + r_Z^2 r_t^2 \right] - 12 r_Z^2 r_t\, |y_{tT}^L\, |y_{tT}^{R*} \right\}, \\
\Gamma(T \to Ht) &= \frac{g^2}{128\pi} \frac{m_T}{M_W^2} \lambda(m_T,\ m_t,\ M_H)^{1/2} |y_{tT}^H|^2 \ \left(1 + 6r_t^2 - r_H^2 + r_t^4 - r_t^2 r_H^2 \right),
\end{aligned} \tag{121}$$

and for the 2HDM-T quark they are completely analogous, with $r_x \equiv m_x/m_Q$, where $x = t, b, W, Z, H$ and Q is the heavy quark, and

$$\lambda(x, y, z) \equiv \left(x^4 + y^4 + z^4 - 2x^2y^2 - 2x^2z^2 - 2y^2z^2 \right). \tag{122}$$

Author details

Rachid Benbrik*† and Mohammed Boukidi†
Polydisciplinary Faculty, Laboratory of Fundamental and Applied Physics, Cadi Ayyad University, Safi, Morocco

*Address all correspondence to: r.benbrik@uca.ac.ma

† These authors contributed equally.

References

[1] Aad G et al. Observation of a new particle in the search for the standard model Higgs boson at the large hadron collider. Physics Letters B. 2012;**716**(1): 1-29

[2] Chatrchyan S et al. Observation of a new boson at a mass of 125 GeV with the CMS experiment at the LHC. Physics Letters B. 2012;**716**(1):30-61

[3] Buchkremer M et al. Vector-like quarks and Higgs effective couplings. Journal of High Energy Physics. 2012; **2012**(9):1-34

[4] Glashow SL, Iliopoulos J, Maiani L. Weak interactions with lepton-hadron symmetry. Physical Review D. 1970;**2** (7):1285-1292

[5] Chatrchyan S et al. Study of the mass and spin-parity of the Higgs boson candidate via its decays to Z boson pairs. Physical Review Letters. 2012;**110**(8): 081803

[6] Contino R, Servant G. Discovering the top partners at the LHC using same-sign dilepton final states. Journal of High Energy Physics. 2010;**2010**(6): 1-30

[7] ATLAS collaboration, Aad G, et al. Search for pair production of heavy top-like quarks decaying to a high-pT W boson and a b quark in the lepton plus jets final state at $\sqrt{s} = 7$ TeV with the ATLAS detector. Physics Letters. 2013; **B718**:1284-1302, 1210.5468, arXiv: 1210.5468 [hep-ex]

[8] CMS collaboration, Chatrchyan S, et al. Search for pair produced fourth-generation up-type quarks in pp collisions at $\sqrt{s} = 7$ TeV with a lepton in the final state. Physics Letters. 2012; **B718**:307-328, arXiv:1209.0471 [hep-ex]

[9] Matsedonskyi O, Panico G, Wulzer A. On the interpretation of top partners searches. JHEP. 2014;**12**:97

[10] Aguilar-Saavedra JA. Identifying top partners at LHC. JHEP. 2009;**11**:030, arXiv:0907.3155 [hep-ex]

[11] Contino R et al. Holography for fermions. Journal of High Energy Physics. 2006;**2006**(5):1-27

[12] Aaboud M et al. Measurement of the correlation between the polar angles of leptons from top quark decays in the helicity basis at 7 TeV using the ATLAS detector. Physics Letters B. 2016;**757**: 236-256

[13] Varma M, Baker OK. Quantum Entanglement in Top Quark Pair Production. [arXiv:2306.07788 [hep-ph]]

[14] Aguilar-Saavedra JA. Post-decay quantum entanglement in top pair production. [arXiv:2307.06991 [hep-ph]]

[15] Dong Z, Gonçalves D, Kong K, Navarro A. When the Machine Chimes the Bell: Entanglement and Bell Inequalities with Boosted $t^{-}t$,. [arXiv: 2305.07075 [hep-ph]]

[16] CMS Collaboration. Measurement of long-range near-side two-particle angular correlations in pp collisions at 7 TeV. Physical Review Letters. 2013; **111**(12):122301

[17] CERN. Large Hadron Collider (LHC). n.d. Available from: https:// home.cern/science/accelerators/large-hadron-collider] (https://home.cern/science/accelerators/large-hadron-collider)

[18] Aaboud M et al. Measurement of the W boson polarisation in tt^{-} events from

pp collisions at 8 TeV in the lepton + jets channel with ATLAS. The European Physical Journal C. 2017;**77**(4):264

[19] Aad G et al. (ATLAS Collaboration). Measurements of the Higgs boson production and decay rates and constraints on its couplings from a combined ATLAS and CMS analysis of the LHC pp collision data at 7 and 8 TeV. Journal of High Energy Physics. 2015; **2015**(8):1

[20] Aguilar-Saavedra JA, Benbrik R, Heinemeyer S, Pérez-Victoria M. Handbook of vectorlike quarks: Mixing and single production. Physics Review. 2013;**D88**(9):094010, arXiv:1306.0572 [hep-ex]

[21] Arhrib A, Benbrik R, King SJD, Manaut B, Moretti S, Un CS. Phenomenology of 2HDM with vectorlike quarks. Physics Review. 2018; **D97**:095015, arXiv:1607.08517 [hep-ex]

[22] Benbrik R, Kuutmann EB, Buarque Franzosi D, Ellajosyula V, Enberg R, Ferretti G, et al. Signatures of vector-like top partners decaying into new neutral scalar or pseudoscalar bosons. JHEP. 2020;**5**:28, arXiv:1907.05929 [hep-ex]

[23] Benbrik R, Boukidi M, Moretti S. Probing light charged Higgs bosons in the 2-Higgs doublet model type-II with vector-like quarks. arXiv:2211.07259 [hep-ex].

[24] Arhrib A, Benbrik R, Boukidi M, Enberg R, Manaut B, Moretti S, et al. Anatomy of vector-like top-quark models in the alignment limit of the 2-Higgs doublet model type-II

[25] ATLAS collaboration, Aaboud M, et al. Combination of the searches for pair-produced vector-like partners of the third-generation quarks at $\sqrt{s} = 13$ TeV with the ATLAS detector. Physical Review Letters. 2018;**121**(21):211801, arXiv:1808.02343 [hep-ex]

[26] CMS Collaboration. Search for pair production of vector-like quarks in leptonic final states in proton-proton collisions at $\sqrt{s} = 13$ TeV. arXiv: 2209.07327 [hep-ex]

[27] Czakon M, Fiedler P, Mitov A. Total top-quark pair-production cross section at hadron colliders through $O(\alpha_S^4)$. Physical Review Letters. 2013;**110**: 252004, arXiv:1303.6254 [hep-ex]

[28] Czakon M, Mitov A. Top++: A program for the calculation of the top-pair cross-section at hadron colliders. Computer Physics Communications. 2014;**185**:2930, arXiv:1112.5675 [hep-ex]

Chapter 6

The Electromagnetic Inter-Nucleon Quark-to-Quark Bond and Its Effect on the Nuclear Force

Nancy Lynn Bowen

Abstract

This chapter discusses the electromagnetic forces inside a nucleus. Previous nuclear theories have ignored the electromagnetic force or treated it as a minor component, considering only Coulomb forces between protons. Since quarks are the centers for both the electric charge and the magnetic dipole moments within a nucleon, such assumptions are not valid. Since the electromagnetic interaction between inter-nucleon quarks may be formidable, electromagnetism can, indeed, be the force holding the nucleons together in a nucleus. Thus, the electromagnetic forces within a nucleus should not be ignored, but rather given the foremost of consideration—specifically with regard to nuclear behavior. New understandings are gained by applying the laws of electromagnetism to the nuclear structure inside an atomic nucleus. In this chapter, historic misunderstandings are debunked and clarified—including the supposed limitations of the electromagnetic force, the miscalculated violations of uncertainty principles, and the misconceived lowest-energy shape of the nuclides.

Keywords: quarks, electromagnetics, nuclear structure, nuclear binding energy, nuclear behavior

1. Introduction

The development of a proper theory of the Nuclear Force has occupied nuclear physicists for over eight decades as one of the main topics of physics research in the twentieth and twenty-first centuries. Currently, there is no singular model of the Nuclear Force that can explain all nuclear behaviors [1, 2]. A successful model of the Nuclear Force should explain the more salient nuclear behaviors: the shape of binding energy curve, particle decay, large quadrupole moments, and excited energy states.

The emphasis of this chapter is electromagnetics; however, the concepts of quantum physics are not ignored or contradicted in any way, and they are properly included in this examination of nuclear behavior.

1.1 Definition and clarification of terms

The Nuclear Force binds the nucleons inside the nucleus. Historically, the Nuclear Force was called the "Strong Nuclear Force," one of four forces of nature, along with the Gravitational Force, the Electromagnetic Force, and the Weak Nuclear Force. After the discovery of quarks, the "Chromodynamic Force" was identified as the force holding the quarks together inside a nucleon. Later, the Chromodynamic Force was renamed the "Strong Nuclear Force," considered to be the force responsible for binding the sub-nucleon particles together inside a nucleon. The force holding the nucleons together in a nucleus was renamed the "Nuclear Force."

To add to this confusion, there is a model of the Nuclear Force called the "Residual Chromodynamic Force," also known as the "Residual Strong Force." Because of this model, and its association with chromodynamics, the Nuclear Force was thought to be a subset of the Chromodynamic Force. The Strong Nuclear Force is now considered to have two parts: the Chromodynamic Force, holding the quarks inside a nucleon, and the Nuclear Force, holding the nucleons inside a nucleus.

To avoid confusion regarding these forces, the term "Strong Nuclear Force" is not used in this chapter. Rather the term "Chromodynamic Force" is used to describe the force holding the quarks together in a nucleon, and the term "Nuclear Force" is used to describe the force holding the nucleons together in a nucleus.

2. Brief review of other models of the nuclear force

There are a number of models of the Nuclear Force, each of which can explain a limited range of nuclear behavior. The current state of the theory for the Nuclear Force is simply a study of the various models, with no one single model being able to explain the majority of the observed behaviors in an atomic nucleus.

The Liquid Drop Model, developed in 1929, claims that the nucleons bind to their closest neighbors, similar to a drop of liquid. This concept is based on empirical data for the binding energy curve, wherein the binding energy per nucleon is relatively constant. Used in conjunction with the Liquid Drop Model is the Weizsäcker formula [3]. This formula is a mathematical equation that is simply a curve-fitting estimate for the experimental binding energy. It uses five parameters and conditional logic to obtain the best fit to the experimental binding curve, duplicating the experimental binding energies to within a few percent. The Liquid Drop Model assumes that the nucleus is spherical in shape, without any consideration for other shapes. It does not explain excited states, large quadrupole moments, or nuclear particle decay.

The Nuclear Shell Model was developed in 1949 [4], and it mimics the Electronic Shell Model for electrons around an atom. The Nuclear Shell Model attempts to explain the minor deviations that were observed, in the 1940s, between the experimental data and the Weizsäcker formula. At that time, with only a limited amount of experimental data, there were notable deviations that were observed at "magic" numbers. However, when utilizing the current data, these minor deviations are revealed to be either non-existent, relatively minor, or related to obscure behavior [5]. The Shell Model uses the Pauli Exclusion Principle to predict the spins of the various atomic nuclides; however, these predictions are accurate only when the Nilsson terms are included [6]. The Nilsson terms, often referred to as "spaghetti plots" due to their complicated natures, provide two or three different empirical terms per nuclide in order to match the experimental spin data to the theory [7]. The Shell Model does not explain particle decay or large quadrupole moments.

There are several Independent Particle Models [8] of the Nuclear Force. These models hypothesize that nucleons are confined inside a 3-D potential energy well, with no interaction among the nucleons. Although such a concept is unrealistic, it is employed to make the mathematics of the Schrödinger equation easier to solve. Depending on the atomic nuclide being studied, the Independent Particle Models alter the shape of the 3-D potential energy well within the Schrödinger equation, using numerous empirically selected variables to allow the models to duplicate the excited states of the nuclide. These models replicate large quadrupole moments only if the potential energy well is modeled as a 3-D ellipsoid, rather than a sphere. These models do not attempt to explain particle decay or the binding energy curve.

The Alpha-Cluster Model of 1938 [9] explained the stronger binding energy, observed for the nuclei with an integer number of alpha segments, claiming that alpha particles were formed as clusters inside a nucleus. This model was recently extended beyond the alpha nuclei and renamed the Clustering Model [10], with much recent research in the field of a nuclear clustering. According to this model, clusters of nucleons link together to form a chain-like structure. There is much experimental and theoretical evidence that atomic nuclides are made of clusters of alpha particles, as well as other types of segments. Recent research regarding the Clustering Model has indicated that clustering structures do indeed exist within nuclei [11, 12].

The Residual Color Model (also known as the Residual Chromodynamic Model or Residual Strong Force Model) is a theoretical model of the Nuclear Force, postulating the exchange of virtual pi-mesons as the binding force between nucleons [13]. In this model, the Nuclear Force is related to the Chromodynamic Force, but with four notable exceptions: it acts outside rather than inside the nucleons, it is a short-range force, it has different mediating particles, and it has a significantly lower strength. The Residual Color Force bonds a quark in one nucleon to a quark in another nucleon, thereby binding the two nucleons together. The Residual Color Model asserts that each nucleon is bonded to its nearest neighbors, similar to the Liquid Drop Model. The derivation of this residual force is mathematically difficult, even for only two nucleons. Only the smallest atomic nuclides have been simulated; however, these simulations resulted in significantly large errors. Thus, this model is unable to predict nuclear behavior. It should be noted that the computer simulations for the Residual Color Model used inflated quark masses, much larger than the theoretical mass of a "naked" quark. This mass inflation of the quarks was done to aid in the convergence of the simulations, since the simulations with small quark masses would not converge. Also, the use of inflated quark masses avoids a violation of Copenhagen interpretation of the Heisenberg uncertainty principle. Fortunately, there is some validity to this mass inflation of the quarks, related with the Constituent Quark Model.

In the Constituent Quark Model [14], the nucleons consist of three constituent quarks that are "dressed" with gluons and quark-antiquark pairs, thereby significantly increasing the effective mass of the quarks. If the quarks could be isolated from each other, then their chromodynamic binding energy could be experimentally measured. The quark rest masses would be the sum of the chromodynamic binding energy plus the rest mass of the nucleon. Unfortunately, the chromodynamic binding energy of the quarks is unknown. However, it is known to be at least as large as the amount of energy required to make a nucleon spontaneously emit a meson. (The actual binding energy could be much larger than this amount.) From this minimum estimate of the chromodynamic binding energy, the minimum value for the quark mass can be determined: the minimum mass of each dressed quark is between 300 and 400 MeV/ c^2. The inflated quark masses used in the Residual Color Model are consistent with the minimum values for a dressed quark.

3. A brief review of quarks

In 1964, the concept of quarks was introduced by Gell-Mann and Zweig [15, 16], stating that the proton and neutron are non-elementary particles. A proton is made of two up quarks and one down quark. A neutron is made of two down quarks and one up quark. The terms "up" and "down" are simply quantum attributes, unrelated to vertical orientation. Up-quarks have an electrical charge that is +2/3 of an elementary charge, and down-quarks have an electric charge that is −1/3 of an elementary charge. The electrical charge and magnetic moments of a nucleon are confined to the quarks, rather than being homogeneously distributed throughout the nucleon. Quarks have the attributes of mass, spin, electric charge, magnetic moments, and a quantum attribute called "color," which is unrelated to visual color.

Illustrated in **Figure 1** are shorthand symbolic representations of the proton and neutron, showing the up- and down-quarks. In this representation, the up-quarks have a "++" charge, and the down-quarks have a "−" charge (2/3 and −1/3 of an elementary charge, respectively). The colors in these illustrations do not relate to the Chromodynamic Force, rather the colors are used to distinguish between the up- and down-quarks. The dots for the quarks do not represent the relative quark size, but rather are sized for easy viewing. The magnetic dipole moments of the quarks, not shown, are perpendicular to the page.

Gell-Mann stated that quark charges are localized, and Richard Feynman asserted that high-energy experiments proved the reality of quarks. Feynman surmised that the quarks have a distribution of position and momentum, like any other particle, and he correctly believed that the diffusion of quark momentum explained diffractive scattering. James Bjorken proposed that point-like partons would imply certain relations in deep inelastic scattering of electrons and protons, relations that were verified in experiments at Stanford Linear Accelerator in 1969 [17].

The deep inelastic scattering experiments indicate that the quarks have a confined three-dimensional spatial distribution for their position. Specifically, the quarks are not moving with a simple random motion throughout the entire volume of the nucleon, rather there is a physical separation of the quarks, as illustrated in **Figure 2a**.

The quarks exist as individual and definitive particles, as in **Figure 2a**, with each quark having a relative charge distribution in three-dimensional space and each quark

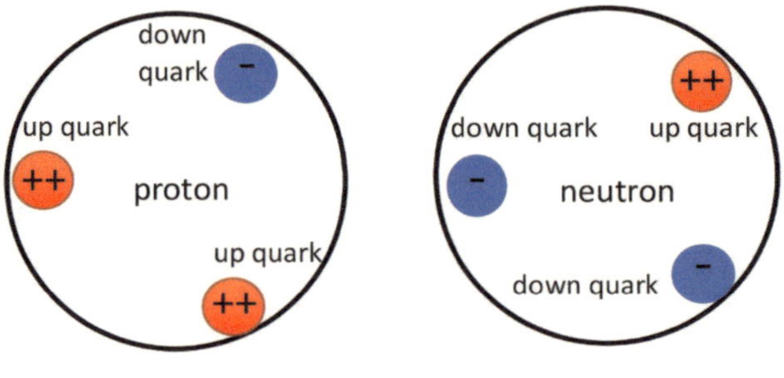

Figure 1.
Shorthand simplified representations of a proton and neutron.

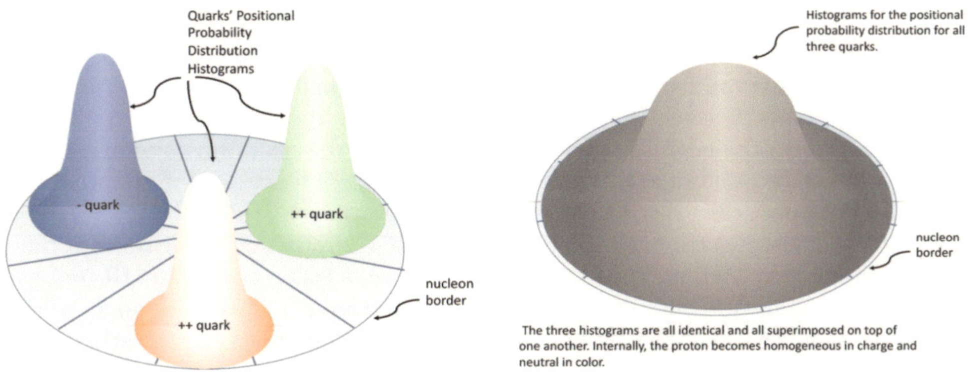

Figure 2.
In 2a, (left) the quark probability distributions are distinct and separate. In 2b, (right) is an illustration of an incorrect probability distribution of the three quarks, superimposed on one another and coincident with the nucleon boundary, causing a homogeneous charge and a neutral color.

being separate and distinct from the other quarks. Nucleons do not have a homogeneous charge or neutral gray color throughout their volumes. This incorrect hypothetical situation is illustrated in **Figure 2b**, shown as a comparison with **Figure 2a**.

A quote about the Laureates, who won the 1990 Nobel Prize for their deep elastic experiments at the Stanford Linear Accelerator, describes this concept.

> *"Earlier investigations of the proton at low energies had shown that this ought to be 'soft' with a relatively even internal distribution of its electrical charge. This year's Laureates therefore had reason to believe there would be a decline in the probability of photon absorption (low number of events). But they found instead a high probability level (many events), i.e. there seemed to be something small and "hard" inside the proton."*
> *"Thus the new investigations gave the surprising result that the electrical charge within the proton is concentrated to smaller components of negligible size."*
> *"This unexpected discovery by the 1990 Laureates was noted immediately by certain skilled theoreticians, chiefly R.P. Feynman and J.D. Bjorken. The result was first interpreted within the framework of what is termed the parton model, which, however, soon came to be identified with the quark model."*

Thus, the quark probability distributions are not merged together, but rather they are individual and distinct distributions, as shown in **Figure 2a**.

3.1 The chromodynamic, nuclear, and electromagnetic forces on quarks

The Chromodynamic Force is much stronger than the Nuclear Force, by several orders of magnitude. However, an actual definitive measurement of its strength has not been made, because particle physicists are unable to separate the quarks from a nucleon. The Chromodynamic Force is thought to be similar to the force of a spring—the further it is stretched from its neutral point, the stronger it pulls back. Presently, there are numerous models of what comprises the mass of a nucleon. These models of the nucleon describe: the number of quarks inside a nucleon, what constitutes the mass of a nucleon, and what gives the nucleon its spin. Some models claim there are several hundred non-valence quarks and gluons inside a nucleon. For the Electromagnetic Model of the Nuclear Force, only the three valence quarks are considered. The Chromodynamic Force is the force inside of the protons and neutrons, and it confines and contains the quarks therein. The Chromodynamic Force abruptly neutralizes itself, falling to zero near the edge of the nucleon. Therefore, the quantum color of a quark strongly affects the other quarks inside that same nucleon; however, the quantum color of a quark does not affect quarks within other nucleons.

Conversely, the Electromagnetic Force does not abruptly neutralize itself at the edge of the nucleon. Rather, the Electromagnetic Force is completely unaffected by the nucleon boundary. These Electromagnetic Forces have an influence on all of the other quarks within the entire atomic nucleus, attracting or repulsing the other quarks within an atomic nucleus, depending on the polarity of the associated electric charges and magnetic dipoles. As a result, a quark inside of one nucleon feels the electromagnetic influences of a quark from another nucleon. Those same two quarks are contained and confined within their respective nucleons by the Chromodynamic Force; however, outside of their respective nucleons, the quarks are able to influence one another due to their electromagnetic forces.

3.2 Quarks and the electromagnetic bond

For the Electromagnetic Model, the nuclear bond is between two quarks forming the bond—a positively charged up quark from one nucleon and a negatively charged down quark from another nucleon. The concept of a nuclear bond is similar to the electronic bonds among the atoms of a molecule. In electronic molecular bonding of atoms, one atom cannot bond to an indefinite number of other atoms. Similarly, one nucleon cannot bond to an indefinite number of other nucleons. The number of bonds is limited by the number of quarks available for bonding. Thus with only three valence quarks, each nucleon can bond only to three other nucleons.

Prior to the 1960s, the proton was incorrectly thought to be homogeneously charged, and to be an elementary particle, with no sub-particles inside of it. As a result, the strongest electrical energy between two such protons was thought to be about 10^{-13} joules. The experimental energy required to free a single nucleon from an atomic nucleus is much larger than this. For this reason, the Nuclear Force was believed to be much stronger than the Electromagnetic Force. As a result, this incorrect concept of a homogenously charged elementary proton created the erroneous limitation for the maximum strength of the Electromagnetic Force. Unfortunately, this incorrect concept is still often perpetuated.

Mathematically, in the limit as the distance goes to zero, the electromagnetic energy goes to infinity. Hence, the electromagnetic energy can be extremely large if the quarks are close enough to each other. Since the electric charge of the nucleons resides only within the quarks, a quark from one nucleon can bond electromagnetically with a quark in another nucleon, and the resulting force between two such quarks can be large enough to be the Nuclear Force.

3.3 Slight conceptual change to the inter-nucleon quark-to-quark bonding

If the Nuclear Force is dependent on the up and down flavor of quarks, rather than on the color of the quarks, then many problems are resolved, and many questions can be answered [18]. This one simple conceptual change to the Residual Color Model makes an extremely important difference in the understanding of the Nuclear Force.

Specifically, the concept is that a down-quark is attracted to an up-quark, but not to another down-quark. Similarly, an up-quark is attracted to down-quark, but not to another up-quark. The Nuclear Force is simply the attraction between an up-quark in one nucleon and a down-quark in another nucleon. When a bond is made between these two quarks in two different nucleons, the nucleons themselves become bonded. This bond lowers the overall energy of the nuclide as compared with its constituent parts, thereby giving the nuclide a higher binding energy.

This one simple conceptual change, that the Nuclear Force is dependent upon the up/down polarity rather than the quantum color of the quark, explains why a system of six protons and six neutrons is at a lower energy (and at a higher binding energy) than five protons and seven neutrons. It is because six protons and six neutrons form one bond for every pair of up-down quarks. There are 18 up-quarks and 18 down-quarks in the system of six protons and six neutrons; thus, there are 18 pairs of up-down quarks and 18 bonds. However, for five protons and seven neutrons, there are 19 down-quarks and 17 up-quarks. Thus, only 17 bonds are formed, and two quarks remain unbonded. The nucleus with five protons and seven neutrons is at a higher overall energy (and at a lower binding energy) than the system of six protons and six neutrons. With this simple change, it is now understood why a nucleon only bonds to its nearest neighbors—a nucleon can only bond three times because it only has three quarks available for bonding.

This concept also explains the first term of the Weizsäcker formula; it is because of the limited number of possible bonds for each nucleon. Also, this simple change of concept explains the asymmetry force of the Weizsäcker formula—the greatest number of bonds occurs when there are equal numbers of up-quarks and down-quarks, which means an equal number of protons and neutrons. The Coulomb energy term of the Weizsäcker formula is also easily explained as being related to the electrical energy of the net positive charges within a nucleus.

Furthermore, due to the inter-nucleon up-to-down quark bonding with three possible bonds per nucleon, there is a predilection for the formation of an alpha segment, made of two neutrons and two protons. This predilection explains the pairing term in the Weizsäcker formula, giving a higher binding energy (and a lower overall energy) when N and Z are even. The highest binding energy occurs when there are an even numbers of protons and an even number of neutrons, allowing for the formation of alpha segments. The "surface" term of the Weizsäcker formula is related to the two unbonded up quarks on the ends of the chain-like configuration and the two unbonded down quarks in the middle open-alpha segment. (The reason for these unbonded quarks will be explained later.)

Thus, for this one simple change—that the inter-nucleon quark-to-quark bond is related to up/down flavor rather than to color—the five terms of the Weizsäcker formula can be understood as a direct consequence of the inter-nucleon up-to-down quark bonding. Simple calculations show that the concept of inter-nucleon up-to-down quark bonding reproduces the binding energy curve surprisingly well [18]. If the force between the inter-nucleon quarks is assumed to be the Electromagnetic Force, then more rigorous calculations can be made, as done in Section 5.

4. Quarks and the Copenhagen interpretation of uncertainty principles

It is shown here that there is no violation of the Copenhagen interpretation of the Heisenberg uncertainty principle. An example calculation is done to illustrate this. However, some definitions, explanations, and equations are discussed first.

The original derivation of the Heisenberg uncertainty principle states that if the two variables, position and momentum, are measured simultaneously, then the uncertainty of the measurement must be greater than $\hbar/2$, shown in Eq. (1).

$$\sigma_x \eta_p \geq \hbar/2 \tag{1}$$

Here, σ_x is the noise in the error of the measurement of position, and η_p is the resultant disturbance in the momentum, as a result of the measurement [19]. This equation and these definitions of the variables represent the original form for the Heisenberg uncertainty principle. The Copenhagen interpretation, however, does not involve laboratory measurements. The Copenhagen interpretation claims that the existence of a quantum system—whether observed or not—cannot exist in a state that is more precise than the Heisenberg limit. This changes the original Heisenberg uncertainty principle from that of Eq. (1) to that of Eq. (2),

$$\sigma_x \sigma_p \geq \hbar/2 \tag{2}$$

where σ_x is the standard deviation of the position probability distribution of the particle, and σ_p is the standard deviation of the momentum probability distribution of the particle.

There are probability distributions for position and momentum for each quark. Associated with these probability distributions are the quantum Expectation Values for position and momentum, which are essentially the mean position and momentum of a quark. The mean value for position is a summation of all the possible values that the position can have, with each value being weighted according to the probability $P(x)$ of that value occurring. To normalize these probability distributions, this is divided by the total number of events, which is the summation of $P(x)$. For a smooth distribution, the summation is replaced by an integral, as shown in Eq. (3):

$$Mean\ of\ x = \frac{\int_{-\infty}^{\infty} xP(x)dx}{\int_{-\infty}^{\infty} P(x)dx} \tag{3}$$

In a quantum mechanical calculation, the probability function $P(x)$ of the particle is the complex conjugate of the wave function times the original wave function. The wave function $\Psi(x,t)$ of the particle is found by solving the Schrödinger equation with the appropriate boundary conditions. For the Expectation Value of position, the probability distribution is multiplied by x, then integrated. Since the overall probability of the particle's position within infinity is unity, the integral in the denominator of Eq. (3) is defined to be one. Thus, for quantum mechanics, the Expectation Value of the position x is as shown in Eq. (4):

$$Expectation\ Value\ of\ x = \int_{-\infty}^{\infty} \Psi^*(x,t)\, x\, \Psi(x,t)dx \tag{4}$$

Since the denominator is defined as one, it is no longer explicitly necessary for the equation. A similar operation is done to find for the Expectation Value of momentum.

As mentioned previously, in the computer simulations of the Residual Color Force Model, the mass of the bare "naked" quarks was inflated to avoid non-convergence. This inflated mass also avoided a violation of the Copenhagen interpretation of the Heisenberg uncertainty principle—since for "dressed" quarks, the quantum uncertainty principles are not inherently violated. This larger mass is thought to be due to either relativistic speeds or to a quantum energy associated with the gluons. If relativistic speeds are involved, then the dressed quarks would also have a significant uncertainty in their momentum probability distribution. Regardless, the Chromodynamic Force is considered to be strong enough to confine the quarks inside the nucleus, despite relativistic speeds.

Due to their vibrations and quantum fluctuations, the quarks have a quantum probability distribution in three dimensions for both space and momentum. Thus, the quarks are not in a fixed or static position relative to each other. Rather there are probability distributions for their relative positions and momentums. When the positions are averaged over spacetime, the Expectation Values of the three quarks appear to be in point-like spatial positions, relative to each other.

This is an important concept: the Expectation Value of the three-dimensional spatial positions of the three quarks inside a nucleon can be modeled as point-like positions with respect to each other. At any given instant in time, there is an uncertainty in the position and momentum of the quarks. Although the spatial and momentum probability distributions are not point-like, the positional Expectation Values do appear as point-like positions.

Several experimental studies have been conducted to measure the radial charge distribution inside a proton and neutron [20–22]. The resulting radial charge densities are shown in **Figure 3**.

As seen in the **Figure 3**, these experiments indicate that there is zero charge density at the center of both the proton and neutron—consistent with the quarks having a probability distribution situated more toward the periphery. Using these experimentally determined charge distributions, the quark probability distributions can be found by solving for the up and down probability distributions necessary to achieve these same charge distributions [23]. These quark probability distributions are shown in **Figure 4**.

As can be seen, the quarks are not homogeneously distributed throughout the entire volume of the nucleon. Their mean position is approximately 0.8 femtometers

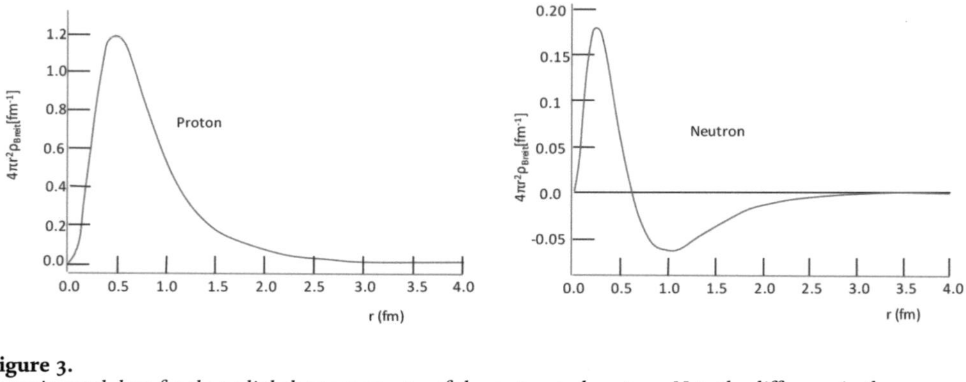

Figure 3.
Experimental data for the radial charge symmetry of the proton and neutron. Note the difference in the vertical axis.

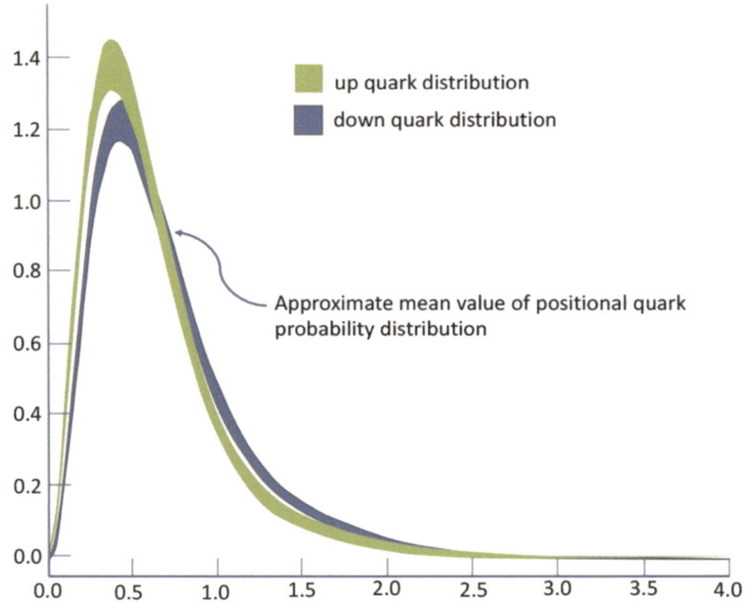

Figure 4.
Up-quark and down-quark orbital radial distributions. The mean value of the quarks position is roughly at 0.8 femtometers from the center.

from the center, in agreement with the calculations of this paper. (Note, these distributions are radial densities.)

While it is not possible to definitively calculate the combined uncertainty in space and momentum, by using reasonable values for these standard deviations, some estimates can be made to determine if the quarks fit within the confines of quantum uncertainty principles.

If the quarks inside a proton cannot be unbound by the addition of 139.6 MeV of energy, then the binding mass is at least this 139.6 MeV/c^2. The chromodynamic binding mass is subtracted from the masses of the three isolated quarks. In other words, the total mass of the three quarks is equal to the rest mass of the nucleon plus the chromodynamic binding mass. This information gives us a lower limit to the masses of the three constituent quarks. This calculation is shown in Eq. (5), where the mass of a proton is 938.3 MeV/c^2.

$$\sum Mass_{consituent\ parts} \geq 938.3\frac{MeV}{c^2} + 139.6\frac{MeV}{c^2} \geq 1077.9\ \frac{MeV}{c^2} \qquad (5)$$

Assuming there are no other massive particles, as yet undiscovered, inside the proton, this gives the sum of the masses of the three quarks to be at least 1077.9 MeV/c^2. This indicates that each of the effective masses of the three constituent quarks is at least 350 MeV/c^2 = 6.4×10^{-28} kg in mass. (Most likely, they are more massive than this.) Regardless of whether this constituent quark mass is the rest mass of the three "dressed" quarks or a relativistically inflated mass of the "naked" quarks, this value of 6.4×10^{-28} kg is a reasonable estimate for the lower limit of the quark mass. Also, this value is in good agreement with the Constituent Quark Model and the computer simulations used in the Residual Quark Model.

A reasonable estimate for the standard deviation of the quark vibrational velocity distribution is about 0.8 times the speed of light. Thus, $\sigma_v = 0.8c = 2.4 \times 10^8$ m/sec. This gives, σ_p, the standard deviation for the quark momentum as: $\sigma_p = (2.4 \times 10^8$ m/sec)$(6.4 \times 10^{-28}$ kg) $= 1.54 \times 10^{-19}$ kg-m/sec. A reasonable estimate for the standard deviation of the spatial distribution, σ_x, is about ¼ the diameter of the nucleon. Thus, σ_x = (¼) (1.68 fm) = 0.42×10^{-15} m.

Using these reasonable estimates and the lower limit of the quark mass, the product of the two standard deviations falls within the limits of the Heisenberg uncertainty principle, as shown in Eq. (6).

$$\sigma_x \sigma_p = \left(0.42 \times 10^{-15}\, m\right)\left(1.54 \times 10^{-19}\, kg\, m/sec\right) = 6.45 \times 10^{-35} \geq \hbar/2 \qquad (6)$$

Thus, using these reasonable estimates for the uncertainties, the product is above the Heisenberg limit; there is no violation of the Copenhagen interpretation. Furthermore, the binding mass is very likely to be more than 139.6 MeV/c^2, which means that the product of the uncertainties is even larger. Also, the vibrational speed of the quarks could easily exist within the realm of relativistic speeds, again increasing the uncertainty product. Both considerations make a violation of the Copenhagen interpretation less likely. Hence, the concept of a quark that "appears" to be a point-like electric charge and magnetic moment is fundamentally allowed.

5. Calculations of the energies within the nucleus

For this calculation, the energies that are taken into consideration are the centrifugal energy, electric energy, and magnetic energy. The electric charges and magnetic dipole moments of the nucleons are contained in the quarks. Since the electric charges and magnetic dipole moments reside in the quarks, it can easily be realized that the electromagnetic forces between an up-quark and a down-quark are strong enough to bind the two nucleons together. No other force, different from the electromagnetic force, is needed to account for the strong bond between nucleons. **Figure 5** shows this electromagnetic bond. Also present is the magnetic force, perpendicular to the page, but not shown.

The force binding these two nucleons need not be anything other than the electromagnetic forces between the up- and down-quarks.

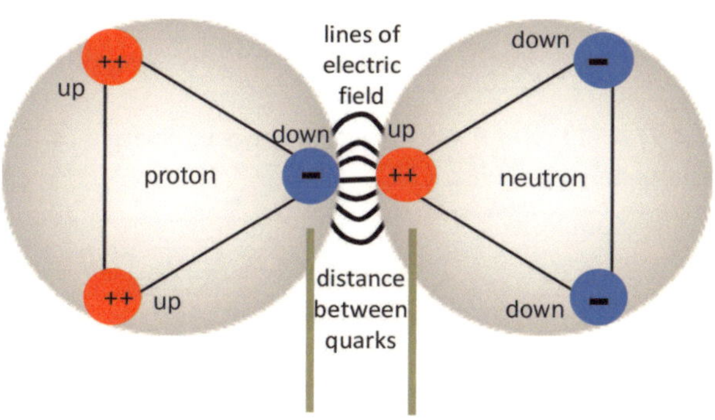

Figure 5.
The electromagnetic bond between quarks.

5.1 The electric energy

The electric energy [24] between two charged particles is shown in Eq. (7):

$$Energy_{12} = \frac{q_1 q_2}{(4\pi\varepsilon_0)(distance_{12})} \tag{7}$$

where $distance_{12}$ is the distance between particles 1 and 2, and q_1 and q_2 are the electric charges on particles 1 and 2. For additional charges, the total electric energy is the double summation over all pairs of charges, as shown in Eq. (8).

$$E_{electric\_total} = \sum_{i=1}^{n-1} \sum_{j=i+1}^{n} \frac{q_i q_j}{(4\pi\varepsilon_0)(distance_{ij})} \tag{8}$$

5.2 The magnetic energy

The magnetic energy [25, 26] between two magnets has vector and positional dependence. Given two magnets, with magnetic moments μ_1 and μ_2, the magnetic field of $magnet_1$ is determined at the location of the $magnet_2$. This vector field is symbolized as B_{12}, as shown in Eq. (9).

$$\overrightarrow{B_{12}} = \frac{\mu_0}{4\pi} \left\{ \frac{3(\overrightarrow{\mu_2} \cdot \overrightarrow{r_{21}})\overrightarrow{r_{21}} - r_{21}^2(\overrightarrow{\mu_2})}{r_{21}^5} \right\} \tag{9}$$

The resultant energy, $U_{magnetic12}$, is the negative dot product of the vector of the magnetic moment μ_2 with the vector of B_{12}, as shown in Eq. (10).

$$U_{magnetic12} = -\overrightarrow{\mu_2} \cdot \overrightarrow{B_{12}} \tag{10}$$

For a collection of magnets, the total magnetic energy is the double summation over all pairs of magnets, as shown in Eq. (11).

$$U_{magnetic\ total} = \sum_{i-1}^{n-1} \sum_{j=i+1}^{n} -\overrightarrow{\mu_i} \cdot \overrightarrow{B_{ij}} \tag{11}$$

where $\overrightarrow{B_{ij}}$ is the vector magnetic field established by the ith magnet at the location of the jth magnet.

Combining Eqs. (9)–(11), the total electromagnetic energy of a distribution of charges and magnets is shown in Eq. (12):

$$U_{EM\ total} = \sum_{i=1}^{n-1} \sum_{j=i+1}^{n} \frac{q_i q_j}{(4\pi\varepsilon_0)(r_{ij})} + \sum_{i=1}^{n-1} \sum_{j=i+1}^{n} \frac{\mu_0}{4\pi} \left\{ \frac{3\left(\overrightarrow{\mu_j} \cdot \overrightarrow{r_{ji}}\right)\left(\overrightarrow{\mu_i} \cdot \overrightarrow{r_{ji}}\right) - r_{ji}^2\left(\mu_i \cdot \mu_j\right)}{r_{ji}^5} \right\} \tag{12}$$

Due to the vector properties of this energy, the lowest energy configuration for two magnets is a stacked bond, in which the magnetic moments of the magnets are oriented in the same direction, stacked one atop the other, and as close as physically possible. A side-by-side bond is with anti-parallel magnetic moments, oriented side by side, and as a close as physically possible. An angled bond gives intermediate results. From quantum field theory, it is known that quarks behave as point-like Dirac particles, each having their own inherent magnetic moment [27].

5.3 The centrifugal energy

The kinetic energy of spin must be properly included in the total energy of a nuclide. A rotating rigid object comprises numerous individual particles that orbit about a central axis of rotation. The kinetic energy of this orbital movement, E_{orbital}, is related to the angular velocity, ω, and to the moment of inertia, I_{orbital}, shown in Eq. (13):

$$E_{orbital} = ½I\omega^2 = ½L^2/I_{orbital} \tag{13}$$

where L is the orbital angular momentum of the object. Note that $L = \omega I_{\text{orbital}}$. Putting these concepts into the quantum realm of nuclear physics [28], the equation for the quantum orbital angular momentum, L, is shown in Eq. (14):

$$L = \hbar\sqrt{l(l+1)} \tag{14}$$

where l is the value of the quantum number for the orbital angular momentum. Combining Eqs. (13) and (14) yields Eq. (15).

$$E_{orbital} = \frac{½L^2}{I_{orbital}} = \frac{½\left(\hbar^2[l(l+1)]\right)}{I_{orbital}} \tag{15}$$

5.4 The total binding energy

The total binding energy can be calculated using these equations. There is only one parameter that must be selected, and that is the minimum distance between the inter-nucleon quarks, as shown previously in **Figure 5**, between two bonded quarks of two different nucleons. The value of 2.11082×10^{-16} meters is used to match the experimental data. This value gives excellent matching to the experimental binding energy, as seen in Section 7. The calculated binding energy is the difference of the total energies of the configuration and the total energies of the isolated constituent parts, shown in Eq. (16).

$$U_{binding\ energy} = U_{total\ of\ configuration} - U_{total\ of\ constituent\ parts} \tag{16}$$

Note that a positive binding energy lowers the total energy and mass of the configuration, as compared with the individual constituent parts.

5.5 Nuclear bonds and the quantum hard-core repulsion

The Pauli Exclusion principle states that nucleons cannot overlap in their physical dimensions; this is also known as the hard-core repulsion. For this reason, a pair of double-bonded nucleons, defined as two nucleons bonded twice to each other, is not

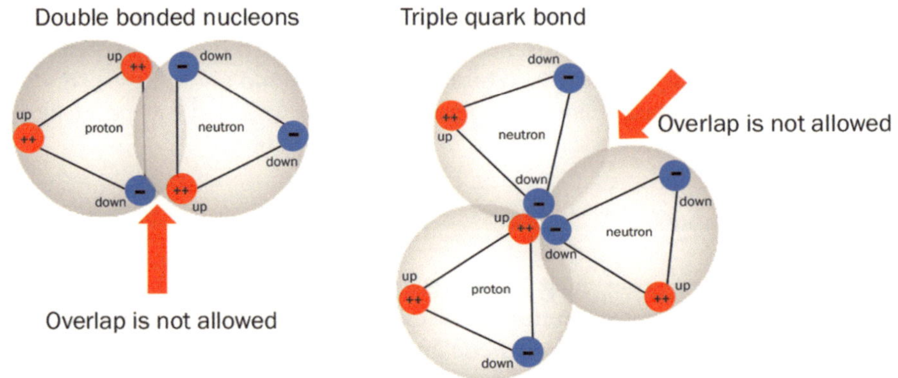

Figure 6.
Left is an illustration of a double bond between a proton and neutron. Right is a triple-quark bond.

allowed. Similarly, a triple-quark bond, defined as three quarks attempting to bond together, is not allowed. These conditions are illustrated in **Figure 6**.

6. The determination of the lowest energy configurations

The lowest energy configurations of the atomic nuclei are determined by using the laws and equations for electromagnetics and angular momentum, as applied to the quarks within a structured nucleus. The position of each quark is defined in a matrix with xyz spatial coordinates and an electric charge value of either −1/3 or +2/3 of an elementary charge. The value of the magnetic moment and the three-dimensional vector direction of each magnetic moment are also included in the matrix. Based on the physical constraints of the configuration, a determination is made as to whether the magnetic bond is stacked, angled, or side by side.

The various possible configurations are tested to determine which configuration is the lowest energy state. As a result, every nucleus in this paper is in its lowest energy state in accordance with the rules for electromagnetic and spin energies. Also, double-bonded nucleons or triple-quark bonds are not allowed.

The atomic nuclides from hydrogen ^2H up to Copernicium ^{283}Cn have been simulated and placed in their lowest energy configuration. For this lowest energy configuration, a pattern emerges from ^{12}C upward, as discussed below.

The nucleons bond together to form the various basic clusters or segments; these segments subsequently bond together to form the larger atomic nuclides. For the lowest energy configuration, these basic segments bond together in a chain-like configuration, in strong agreement with the Cluster Model.

For stable atomic nuclides, the alpha segment predominates. When there are only three possible bonds per nucleon, the alpha segment is the lowest energy configuration for two protons and two neutrons.

6.1 Alpha segments predominate

The nuclear bond is an attractive electromagnetic bond between the up-quark in one nucleon and the down-quark in another nucleon. This bond lowers the overall energy of the atomic nucleus, similar to the way that chemical bonds between the atoms lower the overall energy of a molecule. The number of times that an atom can bond to another atom is limited by the number of valence electrons available for bonding. Similarly, the number of bonds for each nucleon is limited by the number of valence quarks available for bonding. Thus, with only three valance quarks, each nucleon can only bond three times. Two quarks, one from each of the two nucleons, are needed to form one bond. If every quark were fully bonded in an atomic nucleus, there would be three bonds for every two nucleons.

The alpha segment is predominant because there are only three bonds for each nucleon. Each nucleon can bond a maximum of three times, causing the nucleons to cluster into alpha segments. If there were only two bonds within each nucleon, they would simply string together as shown in **Figure 7a**. As can be seen in that illustration, each hypothetical proton and neutron can bond only twice. A structure similar to a string of beads is formed.

With three bonds allowed for each nucleon, the protons and neutrons form alphas segments, as shown in **Figure 7b**. This results in clusters of alpha segments, linked together in a chain-like structure.

If there were more bonds within each proton or neutron, the predominant cluster would be something larger than the alpha segment. For example, with five possible bonds per nucleon, the predominant structure inside such a hypothetical atomic nucleus would be ^{6}Li, as seen in **Figure 7c**.

Experimentally, it is known that segments, such as a ^{6}Li segment, are not the predominant structure of atomic nuclei, but rather the alpha particle predominates. This can be seen in the scallop-like pattern of the curve of binding energy and in the prevalence of alpha decay for the larger radioactive nuclides. The predominance of the alpha segment strongly implies that there are only three bonds per nucleon, as is expected for a nucleon having three valance quarks.

An important observation to make can be seen in the representation of the alpha segments, shown in **Figure 7b**. This nuclide represents oxygen ^{16}O. The left side of the oxygen ^{16}O configuration would have an unbonded positively charged up-quark on

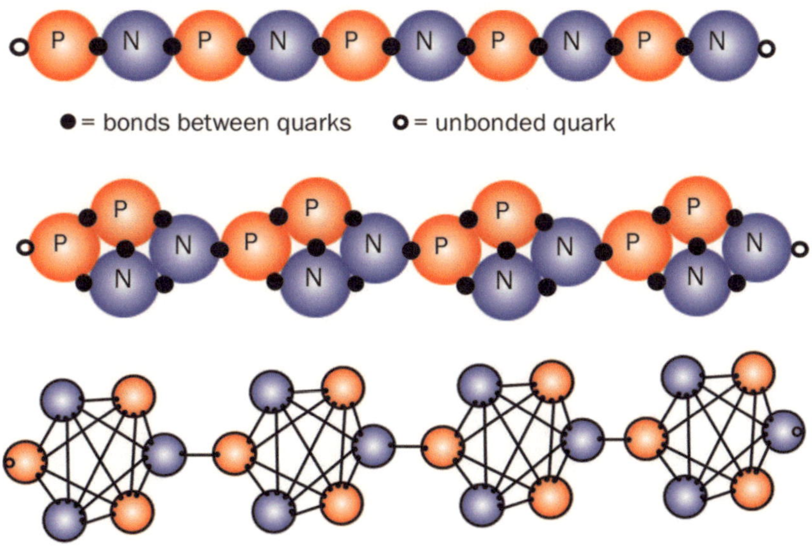

Figure 7.
(a) Hypothetical nuclear bonding if there were only two bonds for each proton and neutron. (b) Nuclear bonding with three bonds for each proton and neutron. (c) Hypothetical nuclear bonding if there were five bonds for each proton and neutron.

the end, and the right side would have an unbonded negatively charged down-quark on the end. This charge polarity causes an electric dipole moment in the configuration. However, quantum mechanics states there can be no net electric dipole moment within a nuclide [29].

Thus, to prevent this dipole, a bond is broken in one of the alpha segments. This broken bond creates an alpha segment, called an open-alpha segment, which bonds to the rest of the chain via its positive quarks, causing there to be two unbonded down-quarks in its center. This allows a negative charge to be situated near the center of the configuration and also spreads out the net positive charge of the nuclide. Both ends of the chain terminate with a positive charge, and there is no dipole moment.

The lowest energy configuration is one that maximizes the number of up-to-down quark bonds. To reduce the Coulomb energy, the lowest energy configuration spreads out the net positive charge as much as possible, while still maintaining the bonds. Any negative charge, such as an unbonded down-quark of an extra neutron, is situated in what would otherwise be the highest concentration of positive charge.

6.2 Spherical nuclides and the deformation parameter

Thus, the lowest energy configuration of any atomic nuclide is not a sphere. Rather, the electric force of the net positive charge within a nucleus attempts to spread apart that net positive charge as much as possible, without breaking the bonds. This forms a clustered, chain-like structure. When considering the electric force, a spherical shape for a net positively charged nuclide is exactly the opposite of its lowest energy shape.

There is much experimental evidence to substantiate this non-spherical shape. For example, the deformation parameter is a measure of an atomic nuclide's non-spherical shape. The experimentally known deformation parameters are shown in **Figure 8**. If the nuclides were spherical, then all of the deformation parameters seen in

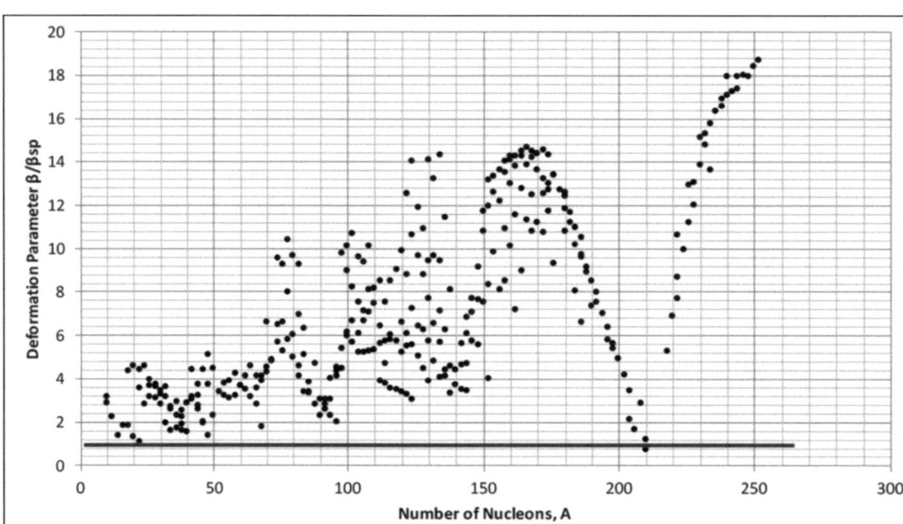

Figure 8.
The ratio of the experimental nuclear deformation parameter divided by the predicted deformation parameter of the Shell Model, indicated by the blue line at the value 1.

Figure 8 should be less than 1, indicated by the blue line. However, they are much higher. (All data for **Figure 8** were extracted from reference [30].)

The deformation parameters are much larger than the Shell Model can explain. This theoretical discrepancy with the empirical data is also true of electric quadrupole moments. Note that the distortion from a spherical shape does not appear in just a few small isolated regions of the nuclear table; rather it covers the entire range of the table. These large experimentally measured deformation parameters for the vast majority of atomic nuclides invalidate the concept of spherical atomic nuclides, in conflict with any model that pre-assumes a spherical shape, as is commonly done in many previous models [31].

For example, when researchers interpret the scattering data from an experimentally probed atomic nucleus, the shape of the atomic nucleus is pre-assumed to be a spherical shape, prior to interpreting the data. Furthermore, only the electric monopole moment is considered. For this pre-assumed spherical shape, the best value of its radius is forced-fit to the data, without any consideration that a different shape may be possible [32].

6.3 The patterns of the lowest energy configurations

For stable atomic nuclides with $A \geq 12$, the lowest-energy configurations follow a pattern, as might be expected. Nuclides below $A = 12$ cannot follow the pattern, simply because there are not enough nucleons to do so. For stable nuclides, with $A \geq 12$, the pattern is as follows.

- Alpha segments predominate within the configuration. For every two protons and two neutrons, an alpha segment is formed.

- There is an open-alpha segment near the center of the configuration.

- There is a triton segment, made of two neutrons and one proton, with an unbonded down quark, included in all stable nuclides with odd Z (with the exception of ^{14}N).

- For stable atomic nuclides with $N > Z$, the extra neutrons are positioned in the location that would otherwise have the highest positive electric charge density.

- For stable atomic nuclides smaller than bismuth ^{209}Bi, the excess neutrons are positioned between the alpha segments.

For unstable nuclides, the pattern is as follows:

- For unstable nuclides larger than ^{209}Bi, the extra neutrons are forced to double up between the alpha segments. There are no stable nuclides with doubled-up neutrons in their configuration.

- In the lowest energy state for radioactive nuclides above the nuclear drip line (either having excess protons or not having enough neutrons), the net positive charge of the nuclide is spread out as much as possible.

- For radioactive nuclides below the nuclear drip line, the excess neutrons, with their negatively charged unbonded down quarks, are spread out as much as possible.

- These chain-like configurations may also curve or curl depending on the orientation of the constraints on the magnetic bonds between the segments. For example, a helix shape may form.

7. Summary of the electromagnetic force within the nucleon

- The electromagnetic force is valid inside the atomic nucleus.

- The electric charge and the magnetic dipole moment of the nucleons are contained within the quarks.

- Within a single nucleon, the three quarks have a spatial quantum probability distribution relative to each other. (An equilateral triangle is assumed for simplicity.)

- There are three possible bonds per nucleon, one for each quark.

- An up-quark and a down-quark, one from two different nucleons, are needed for one bond.

- The configuration, with the lowest electromagnetic energy plus spin kinetic energy, is assumed for the ground state.

- The kinetic energy of the quantum angular momentum of the nucleus [33, 34] is properly taken into account.

- There is a minimum distance between two quarks of two different nucleons, consistent with the hard core repulsion.

- The hard core repulsion of the nucleons prevents a nucleon from bonding more than once to another nucleon. Double-bonded nucleons and triple-quark bonds are not allowed.

Once the lowest energy configurations are determined, there remains only one parameter to be selected for the best fit to the binding energy data—namely, the "minimum quark-to-quark distance." This model has only one parameter to determine, whereas the Weizsäcker formula uses five empirically fit parameters and a conditional logic statement to achieve its mathematical curve fitting. The electromagnetic model of the Nuclear Force is able to get comparable results with only one parameter.

Over a thousand different configurations for different atomic nuclides have been computer modeled using this method to calculate the binding energy based on electromagnetics. These include stable and unstable atomic nuclides; large, medium, and small atomic nuclides; and ground and excited states. These detailed calculations have been done for every atomic nuclide shown in **Figure 9**. Each atomic nuclide is placed in the lowest-energy configuration; then the electromagnetic energy of the configuration is calculated, and the binding energy is determined.

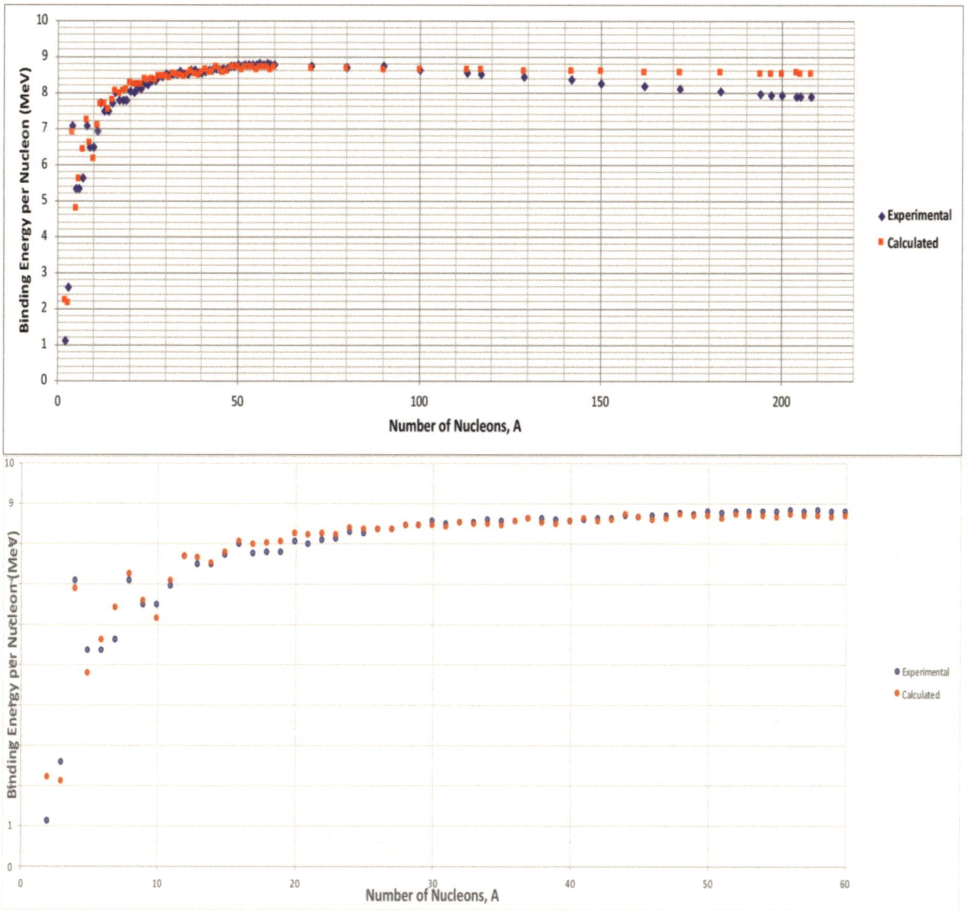

Figure 9.
(a) Binding energy per nucleon versus A, for both calculated and experimental data, from A = 2 to A = 208.
(b) Binding energy per nucleon versus A, for both calculated and experimental data, from A = 2 to A = 60.

The value for the minimum distance between quarks is 2.11082×10^{-16} meters. The resulting binding energy curve is shown in **Figure 9**. The top pane shows all the points from A = 2 to A = 208. The lower pane shows only the points from A = 2 to A = 60, for ease of viewing the details. As can be seen, there is excellent agreement in the reproduction of the experimental data.

Most of the atomic nuclides fall within a 1 or 2% error. For the atomic nuclides with large A, the predicted downward slope is not as severe as is experimentally observed. For these larger atomic nuclides, the worst error is 8.32% for ^{204}Pb. (Experimental data are extracted from reference [35].) Using electromagnetic equations, atomic nuclides as large as copernicium ^{283}Cn have been easily modeled.

By applying the electromagnetic forces to the quarks and nucleons within a nucleus, new insights about many other behaviors of the Nuclear Force can be achieved. For example, by examining the electromagnetic forces that could possibly cause violations of the hard-core repulsion—such as a nuclide attempting to form either a double-bonded nucleons or triple-quark bonds—this model can potentially explain the nuclear behaviors of neutron ejection, proton ejection, and alpha decay.

Also, by considering configurations that are not the lowest energy state, new insights can be provided about the nuclear behaviors of these excited states and their isomeric transitions. Understanding how the electromagnetic forces within a nuclide might change after the occurrence of beta decay also allows for many interesting and discerning insights.

8. Conclusions

This chapter acts as an introduction to the electromagnetic behaviors inside the atomic nucleus. It also serves to clarify the important role of electromagnetics and to debunk the historical misconceptions regarding the electromagnetic forces inside an atomic nucleus. Rather than disregarding the electromagnetic forces and energies of the quarks, when taken into full account and understanding, the electromagnetic forces inside a nucleus can explain much about nuclear behavior. It is the intent of this chapter to highlight the role of electromagnetics, as a topic of nuclear physics that deserves further analysis and serious consideration. By recognizing the Electromagnetic Forces within the nuclear structure, a better understanding of nuclear behavior can be obtained.

The Electromagnetic Model of the Nuclear Force asserts that the electromagnetic properties of the quarks are the force binding the nucleons together, via an inter-nucleon quark-to-quark bond. The ground state configurations of hundreds of atomic nuclides, from ^{2}H to ^{283}Cn, have been determined and computer-simulated. The calculated binding energies agree with the experimental binding energies to within a few percent. These computations are done by using only one selected parameter. No previous theoretical model of the Nuclear Force is able to demonstrate such an accurate prediction of binding energy with only one parameter.

With this understanding that the electromagnetic properties of the quarks are what that bind the nucleons together in a nucleus, the Nuclear Force is directly unified to the Electromagnetic Force.

Author details

Nancy Lynn Bowen
Colorado Mountain College, Glenwood Springs, CO, USA

*Address all correspondence to: nbowen@coloradomtn.edu

References

[1] Eisberg R, Resnick R. Quantum Physics of Atoms, Molecules, Solids, Nuclei, and Particles. New York: Wiley; 1985. p. 509

[2] Lilley J. Nuclear Physics Principles and Applications. Chichester: Wiley; 2001. p. 35

[3] Von Weizsäcker CF. Zur theorie der kernmassen. Zeitschrift für Physik. 1935; **96**(7–8):431-458. (in German)

[4] Mayer MG, Jensen JHD. Elementary Theory of Nuclear Shell Structure. New York: Wiley; 1955

[5] Bowen NL. An examination of the updated empirical data in support of the shell model. Journal of Condensed Matter Nuclear Science. 2020;**33**:224

[6] Nilsson SG. Binding states of individual nucleons in strongly deformed nuclei. Det Kongelige Danske Videnskabernes Selskab. Matematisk-fysiske Meddelelser. 1955; **29**:16

[7] Brown BA, Wildenthal BH. Status of the nuclear shell model. Annual Review of Nuclear and Particle Science. 1988;**38**: 29-66

[8] Mukhin K. Experimental Nuclear Physics, Volume 1, Physics of Atomic Nucleus. Moscow: Mir Publishers; 1987. p. 133

[9] Hafstad LR, Teller E. The alpha-particle model of the nucleus. Physics Review. 1938;**54**:681

[10] Brink DM. History of cluster structure in nuclei. Journal of Physics: Conference Series. 2008;**111**:012001. DOI: 10.1088/1742-6596/111/1/012001

[11] Ichikawa T, Maruhn JA, Itagaki N, Ohkubo S. Linear chain structure of four-α clusters in 16O. Physics Review Letters. 2011;**107**:112501

[12] Beck C. Present status of nuclear cluster physics and experimental perspectives. Journal of Physics: Conference Series. 2014;**569**:012002. DOI: 10.1088/1742-6596/569/1/012002

[13] Vayenas G, Souentie S. Gravity, Special Relativity and the Strong Force. New York: Springer; 2012. pp. 25-27

[14] Griffiths D. Introduction to Elementary Particles. New Jersey: Wiley-VCH; 2008. p. 135

[15] Gell-Mann M. A schematic model of baryons and mesons. Physics Letters. 1964;**8**(3):214-215. DOI: 10.1016/S0031-9163(64)92001-3

[16] Zweig G. An SU(3) Model for Strong Interaction Symmetry and Its Breaking. CERN-TH-401. 1964. Available from: https://cds.cern.ch/record/570209

[17] Friedman JI, Kendall HW. Deep inelastic electron scattering. Annual Review of Nuclear Science. 1972;**22**:203

[18] Bowen NL. The inter-nucleon up-to-down quark bond and its implications for nuclear binding. In: Szadkowski ZP, editor. Quantum Chromodynamic. London: IntechOpen; 2020. DOI: 10.5772/intechopen.94377

[19] Heisenberg W. Quantum theory and measurement. In: Wheeler JA, Zurek WH, editors. Princeton, NJ: Princeton University Press. pp. 62-84 [Translation into English by J. A. Wheeler and W. H. Zurek, 1981. Originally published under the title,

"Uber den anschaulichen Inhalt der quantemtentheoretischen Kinematik und Mechanik," Zeitschrift für Physik, 43, 172-98 (1927)]

[20] Hasell DK et al. The BLAST experiment. Nuclear Instrument Mathematics. 2009;**A603**:247. DOI: 10.1016/j.nima.2009.01.131

[21] Hasell DK et al. Spin-dependent electron scattering from polarized protons and deuterons with the BLAST experiment at MIT-Bates. Annual Review of Nuclear Particle Science. 2011; **61**:409-433

[22] Hata H, Murata M, Yamato S. Chiral currents and static properties of nucleons in holographic QCD. Physical Review D. 2008;**78**:086006. DOI: 10.1103/PhysRevD.78.086006

[23] Smith TP. The anatomy of a neutron. American Scientist. 2010;**98**:478-485. Available from: http://www.jstor.org/stable/25766725. [Accessed: 21 June 2022]

[24] Lorrain P, Corson D. Electromagnetic Fields and Waves. San Francisco: Freeman and Company; 1970. pp. 72-76

[25] Owen GE. Electromagnetic Theory. Boston: Allen and Bacon, Inc.; 1963. p. 205

[26] Yosida K. Theory of Magnetism. New York: Springer; 1996. p. 13

[27] Peskin ME, Schroeder DV. An Introduction to Quantum Field Theory. Reading, MA: Perseus Books; 1995. pp. 175-198

[28] Krane KS. Introductory Nuclear Physics. New York: Wiley; 1988. pp. 34-37

[29] Bertulani CA. Nuclear Physics in a Nutshell. Princeton, New Jersey: Princeton University Press; 2007. p. 105

[30] Raman S, Nestor CW, Tikkanen P. Transition probability from the ground to the first 2+ state of even-even nuclides. In: Pritychenko B, editor. Atomic Data and Nuclear Data Tables. Vol. 78. Cambridge, MA: Elsevier Inc.; 2001. pp. 1-128

[31] Krane KS. Introductory Nuclear Physics. New York: Wiley; 1988. pp. 49-57

[32] De Vries H, De Jager CW, De Vries C. Nuclear charge-density-distribution parameters from elastic electron scattering. Atomic Data and Nuclear Data Tables. 1987;**36**:495-536

[33] Krane KS. Introductory Nuclear Physics. New York: Wiley; 1988. p. 144

[34] Bertulani CA. Nuclear Physics in a Nutshell. Princeton, NJ: Princeton University Press; 2007. p. 155

[35] National Nuclear Data Center. Available from: The NuDat 2 database, http://www.nndc.bnl.gov/nudat2/ [Accessed: 04 October 2021]

Chapter 7

Exploring Strange Entanglement: Experimental and Theoretical Perspectives on Neutral Kaon Systems

Nahid Binandeh Dehaghani, A. Pedro Aguiar and Rafal Wisniewski

Abstract

This chapter provides an in-depth analysis of the properties and phenomena associated with neutral K-mesons. Kaons are quantum systems illustrating strange behaviors. We begin by examining the significance of strangeness and charge parity violation in understanding these particles. The concept of strangeness oscillations is then introduced, explaining oscillations between K^0 and $\overline{K}^0$ states. The regeneration of K_S is investigated, uncovering the underlying mechanisms involved. The discussion moves on to quasi-spin space, exploring its bases and their implications. The entangled states of kaon pairs $\left(K^0, \overline{K}^0\right)$ are considered, with a focus on maximally entangled neutral kaons and nonmaximally entangled states. Decoherence effects on entangled kaons are examined, utilizing the density matrix description to capture the dynamics. A dedicated decoherence parameter is introduced to quantify the impact of decoherence. Furthermore, the chapter investigates the loss of entanglement through measures such as von Neumann entanglement entropy, entanglement of formation, and concurrence. These measures provide insights into quantifying and characterizing entanglement in the context of neutral kaons. Through this comprehensive exploration of properties, phenomena, and entanglement dynamics, this chapter aims to pointing out recent works on neutral kaons, contributing to advancements in particle physics.

Keywords: decoherence, entanglement, K-meson, open quantum system, particle physics, strange particles

1. Introduction

Quantum entanglement, being among the most counterintuitive and subtle foundational elements of quantum mechanics, pertains to the correlations observed between distant components of certain composite systems. This intriguing phenomenon was brought to light by the pioneering work of Einstein, Podolsky and Rosen (EPR) in [1] and Schrödinger in [2], who uncovered a "spooky" characteristic of quantum machinery, better featured as nonlocality in the correlations of an EPR pair. Well-known and valuable tools for exploring this nonlocality are obtained by means of the subsequent development of initial Bell inequalities [3], and their subsequently reformulated variations [4, 5]. Experimental tests have consistently demonstrated the violation of Bell inequalities [6, 7], indicating the failure of local realistic theories and affirming the nonlocal nature of the universe.

Consequently, there is considerable interest in exploring the EPR-Bell correlations of measurements in various branches of physics, including particle physics. As a result, several pioneering researchers in particle physics have proposed investigating EPR-entangled massive particles, such as neutral kaons [8–10]. They referred to the unique characteristics of individual neutral kaon states, which exhibit various rare phenomena such as strangeness oscillation, small mass splitting, different lifetimes between the physical states, violations of two primary symmetries: charge parity (CP) and time reversal (T), regeneration when traversing a slab of material, and most notably, strange entanglement. Neutral kaons exhibit a unique form of entanglement known as strange entanglement, referring to the specific entanglement between two neutral kaons [11].

Numerous studies have been conducted to test quantum mechanics in the neutral kaon systems and search for CPT violation through neutral-meson oscillations. Notably, a significant focus on CP, T, and CPT violation in the neutral kaon system was first conducted at CERN in the CPLEAR experiment [12]. Additional contributions came from experiments such as NA48 and NA62 [13], which played key roles in discovering direct CP violation, yielding crucial experimental results. Moreover, experiments like KLOE [14], conducted at the DAΦNE collider, and its successor KLOE-2 [15], at the Frascati National Laboratory, achieved enhanced precision in investigating CPT violations and conducting quantum decoherence tests. The LHCb experiment, located at CERN's Large Hadron Collider (LHC) [16], and KOTO, performed at the Japan Proton Accelerator Research Complex (J-PARC) [17], were among the other experiments which significantly contributed to our understanding of strange entanglement in neutral kaon system, providing valuable insights into the properties and behavior of entangled kaon states.

The investigation of the evolution of an entangled kaon system subjected to decoherence is a crucial aspect. This analysis is carried out by studying the system's behavior over time using the so-called master equation as time progresses, the level of decoherence in the initially entangled kaon system increases, leading to a loss in the system's entanglement. This reduction in entanglement can be accurately assessed in the field of quantum information, where the degree of entanglement in a state is quantified using specific measures, such as entropy of entanglement, concurrence, and entanglement of formation. These measures are widely employed for quantifying quantum entanglement. The unique characteristics of strange entanglement exhibited by neutral kaons, distinct from any other system, enable the exploration of novel phenomena. Considering the high importance of strange entanglement, a significant portion of the chapter is dedicated to exploring the stability of the entangled quantum system and examining the potential occurrence of decoherence due to interactions with its surrounding environment. We aim to understand the extent of these effects and their impact on entanglement by focusing on the correlation between decoherence and the loss of entanglement.

The structure of the chapter is as follows: In Section II, we review the strange behavior of neutral kaons through several phenomena, that is, strangeness and CP violation, strangeness oscillation, and regeneration. Section III introduces the bases in quasi-spin formalism, including the strangeness basis, free-space basis, and inside-matter basis. In Section IV, entangled states of kaon pairs are studied in two main subsections, that is, maximally and nonmaximally entangled states. In the next section, the effects of decoherence on entangled kaons are studied. The density matrix description of entangled kaon system is introduced, and its evolution is studied through the Gorini-Kossakowski-Sudarshan-Lindblad equation. In section VI, the decline in entanglement is quantitatively measured, and explicit evidence of the loss of entanglement is provided. The chapter ends with a conclusion and an overview of prospective research challenges.

1.1 Notation

The ket symbol is used to denote a column vector in a Hilbert space, indicating the quantum state of the particle, that is, $|K^0\rangle$ represents the state vector of a neutral kaon particle. The bra vector is used to describe the dual space or the bra state in quantum mechanics. In the case of a neutral kaon, $\langle K^0|$ represents the bra vector corresponding to the quantum state of the neutral kaon particle. We use the superscript † to show the conjugate transpose of a matrix (or vector). The symbol $\otimes$ indicates the tensor product, and $\oplus$ the direct sum. For A and B being operators, we use the notation $[A, B] = AB - BA$, and $\{A, B\} = AB + BA$. The expression $A(t = t_i; t_r)$ represents the state (or probability) A at a specific time, where the time t is equal to t_i, and t_r represents a specific reference time. For a generic function $f(t)$, its derivative with respect to time is denoted by $\dot{f}(t)$. For a composite system consisting of subsystem A and subsystem B, the partial trace over subsystem B is denoted as tr_B. The imaginary unit is shown by $i = \sqrt{-1}$.

2. Properties of the neutral kaons

Kaons are the lightest strange mesons whose quark content is understood as $u\bar{s}$, $s\bar{u}$, $d\bar{s}$ for the charged kaons $K^+(494MeV/c^2)$ and $K^-(494MeV/c^2)$, and the neutral kaon $K^0(498MeV/c^2)$, respectively [18]. Neutral K-mesons exhibit fascinating quantum phenomena that showcase their peculiar and intriguing behavior. In the following, we study notable instances that exemplify this strangeness.

2.1 Strangeness and charge parity violation

The neutral kaon and its antiparticle $\overline{K}^0(498MeV/c^2) = s\bar{d}$ are distinguished by a quantum number $\mathcal{S}$, known as strangeness, such that $\mathcal{S}|K^0\rangle = +|K^0\rangle$, and $\mathcal{S}|\overline{K}^0\rangle = -|\overline{K}^0\rangle$. Kaons are pseudoscalar particles with a total spin of 0 and parity $P = -1$ ($J^P = 0^-$). They exhibit charge conjugation symmetry (C), which corresponds to the transformation $K^0 \rightleftarrows \overline{K}^0$. Hence, for the joint transformation, one can write

$$CP|K^0\rangle = -\left|\overline{K}^0\right\rangle, \qquad CP\left|\overline{K}^0\right\rangle = -|K^0\rangle \tag{1}$$

From a theoretical point of view, in order to obtain the CP eigenstates, one can implement the superposition of $\overline{K}^0$ and K^0 as

$$|K_1^0\rangle = \frac{1}{\sqrt{2}}\left(|K^0\rangle - \left|\overline{K}^0\right\rangle\right), \qquad |K_2^0\rangle = \frac{1}{\sqrt{2}}\left(|K^0\rangle + \left|\overline{K}^0\right\rangle\right). \tag{2}$$

where $CP|K_1^0\rangle = +|K_1^0\rangle, CP|K_2^0\rangle = -|K_2^0\rangle$. The state K_1 principally decays to two pions while K_2 primarily decays to three pions, which occurs around 600 times slower in comparison to the decay of K_1 into two pions. The reason why this decay occurs so slowly is due to the fact that the mass of K_2 is a bit greater than the total masses of the three pions. Strangeness is not conserved in weak interactions; moreover, such interactions are CP-violating. The observation of these two modes of decay led to the establishment of the existence of two weak eigenstates of the neutral kaons, called K_L (K-long, T) and K_S (K-short, θ). The weak eigenstates are slightly different in mass, $\Delta m = m(K_L) - m(K_S) = 3.49 \times 10^{-12} MeV$, however, they differ considerably in their lifetimes and decay modes. The state (K_S) is a combination of K_2 (K_1) with a small portion of K_1 (K_2), expressed as

$$|K_S\rangle = \frac{1}{\sqrt{|p|^2 + |q|^2}}\left(p|K^0\rangle - q\left|\overline{K}^0\right\rangle\right) \qquad |K_L\rangle = \frac{1}{\sqrt{|p|^2 + |q|^2}}\left(p|K^0\rangle + q\left|\overline{K}^0\right\rangle\right) \tag{3}$$

where $p = 1 + \varepsilon$ and $q = 1 - \varepsilon$, with ε being the complex CP-violating parameter that is the same for K_L and K_S, that is, $\varepsilon_L = \varepsilon_S = \varepsilon$. According to the CPT Theorem [19], since CP symmetry is violated, time reversal (T) symmetry must also be violated to maintain the overall CPT symmetry. The decay of K_L (long-lived neutral kaon) is dominantly governed by CP violation, similar to the decay of K_2. The primary decay mode of K_L is the three-pion decay, represented as $K_L \to 3\pi$, with a lifetime of approximately $\tau_L = 5.17 \times 10^{-8}$ seconds. Similarly, K_S (short-lived neutral kaon) predominantly decays via the strong interaction, similar to the decay of K_1. The main decay mode of K_S is the two-pion decay, denoted as $K_S \to 2\pi$, with a lifetime of approximately $\tau_S = 8.954 \times 10^{-11}$ seconds. It is important to note that while the dominant decay mode of K_L is $K_L \to 3\pi$, there is also a small amount of CP-violating decay observed, specifically $K_L \to 2\pi$ [20].

2.2 Strangeness oscillation

The two kaons K^0 and $\overline{K}^0$ transfer to common states, and subsequently, they mix, meaning that they oscillate between K^0 and $\overline{K}^0$ before they decay. Consider the time-dependent Schrödinger equation for the state vector $|\psi(t)\rangle$ as

$$i\hbar|\dot{\psi}(t)\rangle = H|\psi(t)\rangle \tag{4}$$

where $\hbar$ is the plank constant, and H is the non-Hermitian effective mass Hamiltonian describing the decay characteristics and strangeness oscillations of kaons, defined as

$$H = M - \frac{i}{2}\Gamma \tag{5}$$

whose eigenstates are K_S and K_L. The matrices M, related to mass, and Γ, a decay-matrix, are 2×2 Hermitian expressed as

$$M = \begin{pmatrix} M_{11} & M_{12} \\ M_{12}^* & M_{11} \end{pmatrix}, \qquad \Gamma = \begin{pmatrix} \Gamma_{11} & \Gamma_{12} \\ \Gamma_{12}^* & \Gamma_{11} \end{pmatrix}, \tag{6}$$

in which $\begin{cases} M_{11} = \frac{1}{2}(m_L + m_S) \\ M_{12} = \frac{1}{2}(m_L - m_S) \end{cases}$ and $\begin{cases} \Gamma_{11} = \frac{\hbar}{2}(\Gamma_L + \Gamma_S) \\ \Gamma_{12} = \frac{\hbar}{2}(\Gamma_L - \Gamma_S) \end{cases}$, with $\Gamma_j = \tau_j^{-1}, j = S, L$.

The eigenvalues of H satisfy $\begin{cases} H|K_S(t)\rangle = \left(m_S - \frac{i\hbar}{2}\Gamma_S\right)|K_S(t)\rangle \\ H|K_L(t)\rangle = \left(m_L - \frac{i\hbar}{2}\Gamma_L\right)|K_L(t)\rangle \end{cases}$. The system evolves exponentially; that is, the solutions to the Hamiltonian are obtained as

$$|K_S(t)\rangle = e^{-\left(\frac{i}{\hbar}m_S + \frac{\Gamma_S}{2}\right)t}|K_S(t=0)\rangle, \qquad |K_L(t)\rangle = e^{-\left(\frac{i}{\hbar}m_L + \frac{\Gamma_L}{2}\right)t}|K_L(t=0)\rangle \tag{7}$$

Subsequently, one can find the solution for K^0 and $\overline{K}^0$ as

$$\begin{aligned} \left|K^0(t)\right\rangle &= \frac{1}{2}\left(e^{-\left(\frac{i}{\hbar}m_S + \frac{\Gamma_S}{2}\right)t} + e^{-\left(\frac{i}{\hbar}m_L + \frac{\Gamma_L}{2}\right)t}\right)\left|K^0\right\rangle + \frac{q}{2p}\left(-e^{-\left(\frac{i}{\hbar}m_S + \frac{\Gamma_S}{2}\right)t} + e^{-\left(\frac{i}{\hbar}m_L + \frac{\Gamma_L}{2}\right)t}\right)\left|\overline{K}^0\right\rangle \\ \left|\overline{K}^0(t)\right\rangle &= \frac{p}{2q}\left(-e^{-\left(\frac{i}{\hbar}m_S + \frac{\Gamma_S}{2}\right)t} + e^{-\left(\frac{i}{\hbar}m_L + \frac{\Gamma_L}{2}\right)t}\right)\left|K^0\right\rangle + \frac{1}{2}\left(e^{-\left(\frac{i}{\hbar}m_S + \frac{\Gamma_S}{2}\right)t} + e^{-\left(\frac{i}{\hbar}m_L + \frac{\Gamma_L}{2}\right)t}\right)\left|\overline{K}^0\right\rangle \end{aligned} \tag{8}$$

Suppose an experiment in which a beam of pure K^0 is produced at $t = t_i$, where t_i is the initial time, via the strong interaction. The probability of observing a K^0 in the beam at a subsequent time t is determined by

$$\begin{aligned} \left|\langle K^0|K^0(t)\rangle\right|^2 &= \frac{1}{4}\left(e^{\left(\frac{i}{\hbar}m_S - \frac{\Gamma_S}{2}\right)t} + e^{\left(\frac{i}{\hbar}m_L - \frac{\Gamma_L}{2}\right)t}\right)\left(e^{-\left(\frac{i}{\hbar}m_S + \frac{\Gamma_S}{2}\right)t} + e^{-\left(\frac{i}{\hbar}m_L + \frac{\Gamma_L}{2}\right)t}\right) \\ &= \frac{1}{4}\left(e^{-\Gamma_S t} + e^{-\Gamma_L t} + 2e^{-\frac{t}{2}(\Gamma_S + \Gamma_L)}\cos\left(\frac{t(m_L - m_S)}{\hbar}\right)\right) \end{aligned} \tag{9}$$

where the third term shows interference, which is the reason for an oscillation in the K^0 beam [18]. By following the same procedure, one can compute the probability of observing $\overline{K}^0$ particles in a beam at a later time, given that the beam initially consists of K^0 particles. Therefore,

$$\left|\left\langle \overline{K}^0|K^0(t)\right\rangle\right|^2 = \frac{1}{4}\frac{|q|^2}{|p|^2}\left(e^{-\Gamma_S t} + e^{-\Gamma_L t} - 2e^{-\frac{t}{2}(\Gamma_S + \Gamma_L)}\cos\left(\frac{t(m_L - m_S)}{\hbar}\right)\right) \tag{10}$$

The K^0 beam oscillates with frequency $f = \frac{(m_L - m_S)}{2\pi}$, with $(m_L - m_S)\tau_S = 0.47$. The probability of finding a K^0 or $\overline{K}^0$ from an initially pure K^0 beam is shown in **Figure 1**.

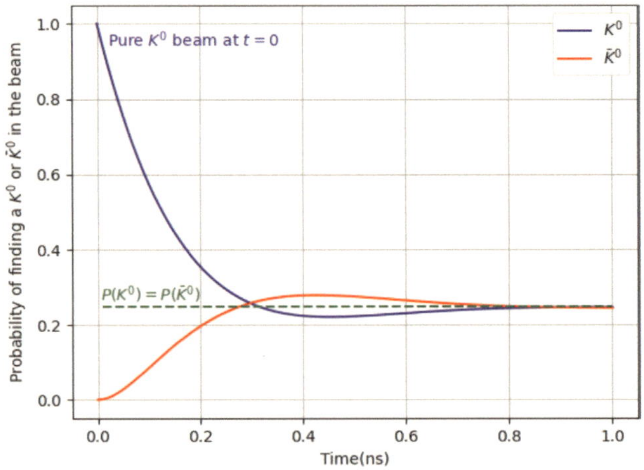

Figure 1.
Probability of finding a K° or $\overline{K}^{\circ}$ state in an initially produced K° beam over time.

The plank constant $\hbar$ is considered as a unit for convenience. The oscillation becomes apparent when considering times on the order of a few τ_S, before all K_S mesons have decayed and only K_L mesons remain in the beam. Therefore, in a beam initially consisting of only K^0 mesons at $t = 0$, the presence of the $\overline{K}^0$ is observed at a distance from the production source through its equal probability of being found in the K_L meson. A similar phenomenon occurs when starting with a $\overline{K}^0$ beam [21].

Over time, the composition of the beam undergoes variations in strangeness due to the different nature of K^0 and $\overline{K}^0$ particles. This intriguing phenomenon is commonly referred to as strangeness oscillations, reflecting the oscillating strangeness content within the beam. In a broader context, this fascinating occurrence is known as flavor oscillations.

2.3 Regeneration of K_S

A beam of K-meson decays in flight after a few centimeters, so the short-lived kaon state K_S disappears, and only a pure beam of long-lived K_L is left. By shooting the K_L beam into a block of matter, which is usually regarded as a composition of protons and neutrons for all practical purposes, then the K^0 and $\overline{K}^0$ components of the beam interact dissimilarly with matter, which also causes the loss of quantum coherence between them. The K^0 particle engages in quasi-elastic scattering interactions with nucleons, while $\overline{K}^0$ has the ability to produce hyperons. Since the emerging beam contains various different linear combinations of K^0 and $\overline{K}^0$, that is, a mixture of K_L and K_S, the K_S would eventually be regenerated in the beam [22].

3. Bases in quasi-spin space

The "quasi-spin" picture for kaons, initially proposed by Lee and Wu [23] and later developed by Lipkin [24], offers notable advantages when compared to spin-$\frac{1}{2}$ particles or photons with vertical (V)/horizontal (H) polarization. The strangeness eigenstates K^0 and $\overline{K}^0$ are regarded as members of a quasi-spin doublet, where $K^0 = \begin{pmatrix} 1 \\ 0 \end{pmatrix}$ (or V polarized photon) and $\overline{K}^0 = \begin{pmatrix} 0 \\ 1 \end{pmatrix}$ (or H polarized photon) are considered as the quasi-spin states up $|\Uparrow\rangle_z$ and down $|\Downarrow\rangle_z$, respectively. All operators acting in the quasi-spin space can be expressed by Pauli matrices, that is, σ_x, σ_y, σ_z. The strangeness operator $\mathcal{S}$ is identified by σ_z, that is,

$$\sigma_z |K^0\rangle = +|K^0\rangle, \qquad \sigma_z \left|\overline{K}^0\right\rangle = -\left|\overline{K}^0\right\rangle, \tag{11}$$

the CP operator with $-\sigma_x$, and the CP violation is relative to σ_y. This formalism is suitable for all two-level quantum systems. In this regard, the Hamiltonian in Eq. (5) can be implemented as

$$H = \alpha I + \beta(\sin\theta\sigma_x + \cos\theta\sigma_y) \tag{12}$$

where $\alpha = \frac{1}{2}\left(m_L + m_S - \frac{i}{2}(\Gamma_L + \Gamma_S)\right)$, $\beta = \frac{1}{2}\left(m_L - m_S - \frac{i}{2}(\Gamma_L - \Gamma_S)\right)$, and the phase θ corresponds to the CP parameter ε such that $e^{i\theta} = \frac{1-\varepsilon}{1+\varepsilon}$.

Overall, in the quasi-spin formalism, we may work with one of the following bases [25]:

- Strangeness basis $\left\{K^0, \overline{K}^0\right\}$: This basis is well-suited for examining electromagnetic and strong interaction processes that conserve strangeness, including the formation of $K^0\overline{K}^0$ systems from nonstrange initial states, for instance $e^+e^- \to \phi(1020) \to K^0\overline{K}^0$ or $p\overline{p} \to K^0\overline{K}^0$, and the detection of neutral kaons through strong kaon-nucleon interactions. This is an orthonormal basis, that is, $\left\langle K^0 | \overline{K}^0 \right\rangle = 0$.

- Free-space basis: $\{K_S, K_L\}$: In the quasi-spin space, the weak interaction eigenstates are similar to the CP eigenstates $|K_1\rangle$ and $|K_2\rangle$. However, the K_S, K_L basis provides a useful framework for analyzing the propagation of particles in free space while the CP basis is particularly suited for studying weak kaon decays. This basis is quasi–orthonormal with $\langle K_S|K_S\rangle = \langle K_L|K_L\rangle = 1$, and $\langle K_S|K_L\rangle = \langle K_L|K_S\rangle = \frac{\varepsilon+\varepsilon^*}{1+|\varepsilon|^2} \simeq 0$.

- Inside-matter basis: $\{K'_S, K'_L\}$: The behavior of neutral kaons as they travel through a homogeneous medium of nucleonic matter, serving as both a regenerator and an absorber is determined by the medium Hamiltonian, which includes an extra strong interaction term as

$$H_{medium} = H - \frac{2\pi\nu}{m_K}\begin{pmatrix} f_0 & 0 \\ 0 & \overline{f}_0 \end{pmatrix} \tag{13}$$

where ν indicates the nucleonic density of the homogeneous medium, m_K is the mean value of $K_{S,L}$ mass, f_0 and $\overline{f}_0$ show the forward scattering amplitudes for K^0 and $\overline{K}^0$, respectively. The $|K'_L\rangle$ and $|K'_S\rangle$ are the eigenstates of H_{medium} expressed as

$$|K'_L\rangle = \frac{1}{\sqrt{1+|r\overline{\rho}|^2}}\left(|K^0\rangle + r\overline{\rho}\left|\overline{K}^0\right\rangle\right), \quad |K'_S\rangle = \frac{1}{\sqrt{1+\left|r(\overline{\rho})^{-1}\right|^2}}\left(|K^0\rangle - r(\overline{\rho})^{-1}\left|\overline{K}^0\right\rangle\right) \tag{14}$$

where the dimensionless regenerator parameter ρ, the auxiliary parameter $\overline{\rho}$ and its inverse $(\overline{\rho})^{-1}$ are introduced as

$$\rho \equiv \frac{\pi\nu}{m_K}\frac{f_0 - \overline{f}_0}{m_L - m_S - \frac{i}{2}(\Gamma_L - \Gamma_S)} \tag{15}$$
$$\overline{\rho} \equiv \sqrt{1+4\rho^2} + 2\rho, \quad (\overline{\rho})^{-1} = \sqrt{1+4\rho^2} - 2\rho$$

and $r = \frac{1-\varepsilon}{1+\varepsilon}$. This basis is also quasi–orthonormal.

$$\langle K'_S|K'_L\rangle = \langle K'_L|K'_S\rangle^* = \frac{1 - |r|^2(\overline{\rho}^*/\overline{\rho})}{\sqrt{1+|r\overline{\rho}|^2}\sqrt{1+|r/\overline{\rho}|^2}} \tag{16}$$

Two limiting cases exist:

1. For a very low density medium: $|K'_S\rangle \to |K_S\rangle$ and $|K'_L\rangle \to |K_L\rangle$

2. For extremely high density media: $|K'_L\rangle \to \left|\overline{K}^0\right\rangle$ and $|K'_S\rangle \to |K^0\rangle$

4. Entangled states of kaon pairs

In general terms, we classify a state as entangled when it cannot be expressed as a convex combination of product states, otherwise it is separable [26]. Quantum entanglement, as a central feature of quantum mechanic, is a phenomenon where two or more particles can become correlated in such a way that the properties of one particle are immediately affected by the properties of the other particle, regardless of the distance between them. In quantum information, entanglement is regarded as a resource. Hence, one is interested in maximally entangled quantum states. To this end, we investigate the entangled states of kaon pairs in two main class, that is, maximally and nonmaximally entangled states.

4.1 Maximally entangled neutral kaons

The spin-singlet states, initially proposed by Bohm, are the most commonly studied and simplest form of bipartite states. These states involve a pair of spin-1/2 particles. In analogy to the standard Bohm state, we consider entangled states of $K^0\overline{K}^0$ [21, 27]. In both cases of Φ−resonance decays and s-wave proton–antiproton annihilation, the process begins at time, $t = 0$ with an initial state denoted as $|\phi(0)\rangle$ with global spin, charge conjugation, and parity $J^{PC} = 1^{--}$ expressed as

$|\phi(t=0)\rangle = \frac{1}{\sqrt{2}}\left(|K^0\rangle_l \otimes \left|\overline{K}^0\right\rangle_r - \left|\overline{K}^0\right\rangle_l \otimes |K^0\rangle_r\right)$, which can further be written in free-space basis as

$$|\phi(t=0)\rangle = \frac{1+|\varepsilon|^2}{\sqrt{2}|1-\varepsilon^2|}\left[|K_S\rangle_l \otimes |K_L\rangle_r - |K_L\rangle_l \otimes |K_S\rangle_r\right] \tag{17}$$

The neutral kaons separate and can be observed both to the left (l) and right (r) of the source. The weak interactions, which violate CP symmetry, come into play only in Eq. (17). It is worth noting that this state is both antisymmetric and maximally entangled in the two observable bases. Consequently, any measurements performed will consistently yield left–right anticorrelated outcomes. After production, the left-moving and right-moving kaons undergo evolution as described by Eq. (7) for respective proper times t_l and t_r. This formal evolution results in the formation of the "two-times" state. Therefore,

$$|\phi(t_l,t_r)\rangle = \frac{1}{\sqrt{2}}e^{-(\Gamma_S t_l+\Gamma_L t_r)/2}\left(|K_S\rangle_l \otimes |K_L\rangle_r - e^{(i\Delta m+\Delta\Gamma/2)\Delta t}|K_L\rangle_l \otimes |K_S\rangle_r\right) \tag{18}$$

where $\Delta t = t_l - t_r$, $\Delta m = m_L - m_S$, $\Delta\Gamma = \Gamma_L - \Gamma_S$, and $\varepsilon \to 0$. Equivalently, Eq. (18) can be written in strangeness basis

$$\begin{aligned}|\phi(t_l,t_r)\rangle = &\frac{1}{2\sqrt{2}}e^{-(\Gamma_S t_l+\Gamma_L t_r)/2}\Big(\left(1-e^{(i\Delta m+\Delta\Gamma/2)\Delta t}\right)\left(|K^0\rangle_l \otimes |K^0\rangle_r - \left|\overline{K}^0\right\rangle_l \otimes \left|\overline{K}^0\right\rangle_r\right)\\ &+\left(1-e^{(i\Delta m+\Delta\Gamma/2)\Delta t}\right)\left(|K^0\rangle_l \otimes \left|\overline{K}^0\right\rangle_r - \left|\overline{K}^0\right\rangle_l \otimes |K^0\rangle_r\right)\Big)\end{aligned} \tag{19}$$

Typically, it is common to examine two-kaon states at a unique time, that is, $t \equiv t_r = t_l$. In this scenario, we have the following equation

$$\begin{aligned}|\phi(t,t)\rangle &= \frac{1}{\sqrt{2}}e^{-(\Gamma_S+\Gamma_L)t/2}\left(|K^0\rangle_l \otimes \left|\overline{K}^0\right\rangle_r - \left|\overline{K}^0\right\rangle_l \otimes |K^0\rangle_r\right)\\ &= \frac{1}{\sqrt{2}}e^{-(\Gamma_S+\Gamma_L)t/2}\left(|K_S\rangle_l \otimes |K_L\rangle_r - |K_L\rangle_l \otimes |K_S\rangle_r\right)\end{aligned} \tag{20}$$

exhibiting similar maximal entanglement and anticorrelations over time.

4.2 Nonmaximally entangled states

In addition to the previously discussed maximally entangled state of kaons, there is interest in exploring other nonmaximally entangled states for testing the local realism versus Quantum Mechanics theories. To prepare these states, we begin with the initial state described in Eq. (17). A thin and homogeneous regenerator is positioned along the right beam, as close as possible to the source of the two-kaon state. If the regenerator is placed in close proximity to this origin and the proper time (Δt) required for the right-moving neutral kaon to pass through the regenerator is sufficiently short, that is, much smaller than τ_S, weak decays can be neglected, and the resulting state after traversing the thin regenerator is obtained as

$$|\phi(\Delta t)\rangle = \frac{1}{\sqrt{2}}(|K_S\rangle \otimes |K_L\rangle - |K_L\rangle \otimes |K_S\rangle + \eta(|K_S\rangle \otimes |K_S\rangle - |K_L\rangle \otimes |K_L\rangle)) \tag{21}$$

The regeneration effects are designated by $\eta = i\rho\left(\Delta m - \frac{i}{2}\Delta\Gamma\right)\Delta t$. One may note the difference between Eqs. (20) and (21) at made by the terms linear in η. To intensify that difference, let the state Eq. (21) propagate in free space up to a proper time $\tau_S \leq T \leq \tau_L$, so

$$\begin{aligned} |\phi(T)\rangle = & \frac{e^{-(\Gamma_L\tau_l + \Gamma_S\tau_r)/2}}{\sqrt{2}}(|K_S\rangle \otimes |K_L\rangle - |K_L\rangle \otimes |K_S\rangle \\ & -\eta\left(e^{-(i\Delta m+\Delta\Gamma/2)T}|K_L\rangle \otimes |K_L\rangle - e^{(i\Delta m+\Delta\Gamma/2)T}|K_S\rangle \otimes |K_S\rangle\right)\Big) \end{aligned} \tag{22}$$

Eq. (22) shows that the $|K_L\rangle \otimes |K_L\rangle$ component has exhibited remarkable resilience against weak decays compared to the accompanying terms $|K_S\rangle \otimes |K_L\rangle$ and $|K_L\rangle \otimes |K_S\rangle$, resulting in its significant enhancement. Conversely, the $|K_S\rangle \otimes |K_S\rangle$ component has experienced substantial suppression and can therefore be disregarded provided that $T > > \tau_S$. By normalizing Eq. (22) to the surviving pairs, one obtains

$$|\Phi\rangle = \frac{1}{\sqrt{2 + |R_L|^2 + |R_S|^2}}(|K_S\rangle \otimes |K_L\rangle - |K_L\rangle \otimes |K_S\rangle + R_L|K_L\rangle \otimes |K_L\rangle + R_S|K_S\rangle \otimes |K_S\rangle) \tag{23}$$

in which $R_L = -re^{-(i\Delta m+\Delta\Gamma/2)T}$, $R_S = re^{(i\Delta m+\Delta\Gamma/2)T}$. The state Φ, which is nonmaximally entangled, encompasses all pairs of kaons wherein both the left and right partners persist until the common proper time T. Due to the specific normalization of Φ, kaon pairs exhibiting decay of one or both members prior to time T need to be identified and excluded. This exclusion occurs before any measurement utilized in a Bell-type test, rendering this approach a "preselection" procedure rather than a "postselection" one, thereby avoiding any conflicts between local realism and quantum mechanics. Upon the establishment of the state $|\Phi\rangle$, it becomes essential to examine alternative joint measurements on each corresponding pair of kaons when conducting a Bell-type test.

5. Decoherence effects on entangled kaons

Exploring the factors that could potentially lead to decoherence of entangled kaons is of high importance [21, 28–31]. Besides, the decoherence allows us to gather insights into the quality of the entangled state. In the subsequent analysis, we explore potential decoherence effects that may arise from interactions between the quantum "system" and its surrounding "environment," as shown in **Figure 2**. Decoherence effects are mainly divided into two groups: standard and nonstandard [21]. Examples of "standard" decoherence effects can be found in various sources, including:

- Strong interaction scatterings of kaons with nucleons
- Weak interaction decays
- Noise from the experimental setup

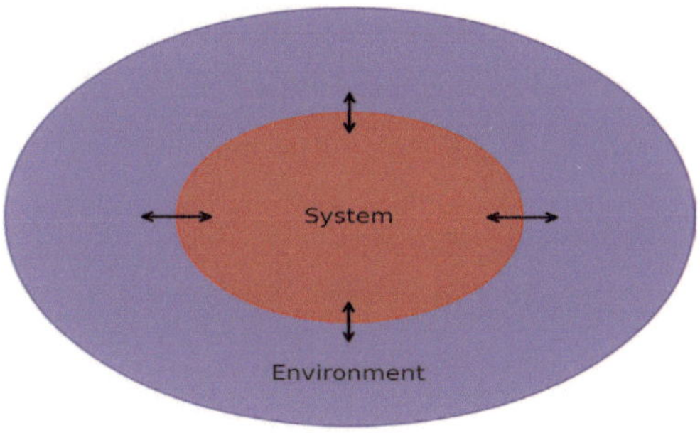

Figure 2.
The overall system can be divided into two components: the system of interest, referred to as the "system", and the surrounding "environment".

Nonstandard decoherence effects are due to the fundamental modifications of quantum mechanics, such as:

- Influence of quantum gravity [32–34]
- Quantum fluctuations in the space-time structure at Planck mass scale [35]
- Dynamical state-reduction theories [36]

5.1 Density matrix description of entangled kaon system

We will now delve into the decoherence model within the Hilbert space $\mathcal{H} = \mathbb{C}^2$, which represents a two-dimensional complex vector space. Our analysis will specifically focus on the usual effective mass Hamiltonian, as denoted by Eq. (5). Here, it is supposed that CP invariance is not violated as in the case of CPLEAR experiment [12], whose data are not sensitive to the impacts of CP violation. Therefore, $p = q = 1$ meaning that

$$|K_1^0\rangle \equiv |K_S\rangle, \quad |K_2^0\rangle \equiv |K_L\rangle, \quad \langle K_S|K_L\rangle = 0 \tag{24}$$

We can effectively track the evolution of the density operator ρ by employing the so-called Gorini-Kossakowski-Sudarshan-Lindblad equation representing the dynamics of a subsystem within a Markovian system as the entire system expressed as [37],

$$\dot{\rho}(t) = \mathcal{L}\rho(t) = -i\left(H\rho(t) - \rho(t)H^\dagger\right) + \underbrace{\sum_j \left(L_j\rho(t)L_j^\dagger - \frac{1}{2}\left\{L_j^\dagger L_j, \rho(t)\right\}\right)}_{\mathcal{D}(\rho)} \tag{25}$$

in which $\mathcal{L}$ is the Liouville superoperator, and L_j is the Lindblad (or jump) operator. The effect of decoherence is added by the dissipation term $\mathcal{D}(\rho)$, for which we consider the following ansatz [30],

$$\mathcal{D}(\rho) = \frac{\lambda}{2}\sum_j \left[P_j, \left[P_j, \rho\right]\right] \quad \text{with} \quad P_j = |K_j\rangle\langle K_j|, \quad j = S, L \tag{26}$$

where $\lambda \geq 0$ is the decoherence parameter. Eq. (26) shows a special case of dissipation where $L_j = \sqrt{\lambda} P_j$. Therefore, for the elements of the density operator $\rho(t) = \sum_{i,j=S,L} \rho_{ij}(t)|K_i\rangle\langle K_j|$, we attain

$$\begin{aligned} \rho_{SS}(t) &= \rho_{SS}(0)e^{-\Gamma_S t} \\ \rho_{LL}(t) &= \rho_{LL}(0)e^{-\Gamma_L t} \\ \rho_{LS}(t) &= \rho_{LS}(0)e^{-(i(m_L - m_S) - \Gamma - \lambda)t} \end{aligned} \tag{27}$$

where $\Gamma = \frac{1}{2}(\Gamma_S + \Gamma_L)$.

Consider the maximally entangled state Eq. (20) at initial time $t = 0$ as

$$|\psi^-\rangle = \frac{1}{\sqrt{2}}(|e_1\rangle - |e_2\rangle) \tag{28}$$

where $|e_1\rangle = |K_S\rangle_l \otimes |K_L\rangle_r$ and $|e_2\rangle = |K_L\rangle_l \otimes |K_S\rangle_r$. The total system Hamiltonian is then described by a tensor product of the one-particle Hilbert spaces as $H = H_l \otimes I_r + I_l \otimes H_r$ with l and r denoting the direction of the moving particles. The state in Eq. (28) is a Bell state [38–40] and is equivalently expressed by the density operator

$$\rho(0) = \frac{1}{2}(|e_1\rangle\langle e_1| + |e_2\rangle\langle e_2| - |e_1\rangle\langle e_2| - |e_2\rangle\langle e_1|) \tag{29}$$

In this case, the projectors are $P_1 = |e_1\rangle\langle e_1|$ and $P_2 = |e_2\rangle\langle e_2|$, which project to the eigenstates of two-particle Hamiltonian. Therefore, the element-wise time evolution obtained from Eq. (25) with the ansatz Eq. (26) is expressed as

$$\begin{aligned} \dot{\rho}_{ij}(t) &= -2\Gamma\rho_{ij}(t) & \text{for} \, i = j : \rho_{ij}(t) = \rho_{ij}(0)e^{-2\Gamma t} \\ \dot{\rho}_{ij}(t) &= -(2\Gamma + \lambda)\rho_{ij}(t) & \text{for} \, i \neq j : \rho_{ij}(t) = \rho_{ij}(0)e^{-(2\Gamma + \lambda)t} \end{aligned} \tag{30}$$

From Eq. (29), we already know $\rho_{11}(0) = \rho_{22}(0) = \frac{1}{2}$ and $\rho_{12}(0) = \rho_{21}(0) = -\frac{1}{2}$. As a result, we acquire the time-varying density operator as the following:

$$\rho(t) = \frac{1}{2}e^{-2\Gamma t}(|e_1\rangle\langle e_1| + |e_2\rangle\langle e_2| - e^{-\lambda t}(|e_1\rangle\langle e_2| + |e_2\rangle\langle e_1|)) \tag{31}$$

From Eq. (25), it follows that $\rho(t) = e^{\mathcal{L}t}\rho(0)$. This means that the initial state $\rho(0)$ is transformed to $\rho(t)$ by the completely positive and trace-preserving (CPTP) map $\mathcal{V}(t) = e^{\mathcal{L}t}$ generated by the superoperator $\mathcal{L}$ [41, 42]. Note that while $\mathcal{V}(t)$ should satisfy trace-preserving characteristics, the non-Hermitian nature of the system Hamiltonian results in a deviation from the property of trace preservation. Specifically, we observe that $tr(\rho(0)) = 1$ but $tr(\rho(t)) = e^{-2\Gamma t}$, in other words, $tr(\dot{\rho}(t)) \neq 0$. The issue arising from the non-Hermitian Hamiltonian in this particular system has been addressed in previous studies [43, 44]. By introducing certain modifications to the Hilbert space and the dynamical equation, one can effectively work with this Hamiltonian. Therefore, the Hilbert space $\mathcal{H}$ of the system is extended by adding the Hilbert space $\mathcal{H}_0$ which corresponds to the decay states resulting from the dissipation, so $\mathcal{H}_{tot} = \mathcal{H} \oplus \mathcal{H}_0$. By using the effective mass Hamiltonian defined in Eq. (5), the dynamical equation is then expressed as

$$\dot{\rho}(t) = -i[M, \rho(t)] - \left\{\frac{1}{2}\Gamma, \rho(t)\right\} \tag{32}$$

Let define $B : \mathcal{H} \to \mathcal{H}_0$, and $\Gamma = B^\dagger B$, then we obtain $tr(\dot{\rho}) = -tr\left(B^\dagger B\rho(t)\right) \neq 0$. By adding $B\rho(t)B$ to Eq. (32), such that $\dot{\rho}(t) = -i[M, \rho(t)] - \left\{\frac{1}{2}\Gamma, \rho(t)\right\} + B\rho(t)B$, then $tr(\dot{\rho}(t)) = 0$, meaning that the trace of $\rho(t)$ is preserved.

5.1.1 Purity

Decoherence arises exclusively from the influence of the factor $e^{-\lambda t}$ on the off-diagonal elements. Hence, for $t = 0$, the density operator corresponds to a pure state; however, for $t > 0$ and $\lambda \neq 0$, the state $\rho(t)$ does not exhibit a pure state anymore. As time elapses and environmental factors come into play, the quantum effects, characterized by coherences, gradually fade away, giving rise to the phenomenon of decoherence.

In quantum mechanics, and particularly in the field of quantum information theory, the purity of a quantum state described by the density operator is defined as $\mathcal{P}(t) = tr(\rho^2(t))$. The purity expresses a measure of quantum states and provides information regarding the degree of mixture in a given state. Here, the purity of $\rho(t)$ is [31],

$$\mathcal{P}(t) = \frac{1}{2}e^{-4\Gamma t}\left(1 + e^{-2\lambda t}\right) \tag{33}$$

that is, for $t = 0$, $\mathcal{P}(t) = 1$, and for $t > 0$, $\mathcal{P}(t) < 1$. We can conclude that the evolution of $\rho(t)$ transfers from a pure state to a mixed state is due to the occurrence of decoherence phenomenon. In addition, the purity of a normalized quantum state satisfies $\frac{1}{2} \leq \mathcal{P}(t) \leq 1$ for a state defined upon a two-dimensional Hilbert space. However, in this case, the value of $\mathcal{P}(t)$ can be less than $\frac{1}{2}$. It happens when one of the conditions of the density operator is not satisfied for $t > 0$, that is, $tr(\rho(t)) \neq 1$.

5.2 Decoherence parameter associated with entangled kaon system

In the CPLEAR experiment, as described in [12], entangled kaons are generated. Subsequently, the strangeness content (S) of the right-moving and left-moving particles is measured at time $t = t_r$ and $t = t_l$, respectively. Consider a specific scenario where the detection reveals that a $\overline{K}^0$ is observed at the right side at time $t = t_r$, while a K^0 is detected at the left side at time $t = t_l$, where $t_r \leq t_l$. We indicate two operators S_r^+ and S_l^- to represent the measurement of strangeness at the right and left sides, respectively (see **Figure 3**). After the measurement occurred at the right side, the density operator of the left-moving particle turned out to be.

$$\rho_l(t = t_r; t_r) = tr_r\left(S_r^+\rho(t_r)\right) \tag{34}$$

From here, the probability of this case becomes [31],

$$P\left(\overline{K}^0, t_l; K^0, t_r\right) = tr(S_l^-\rho_l(t_l; t_r)) = tr\left(tr_r\left(S_r^+\rho_l(t_r)\right)\right) \tag{35}$$

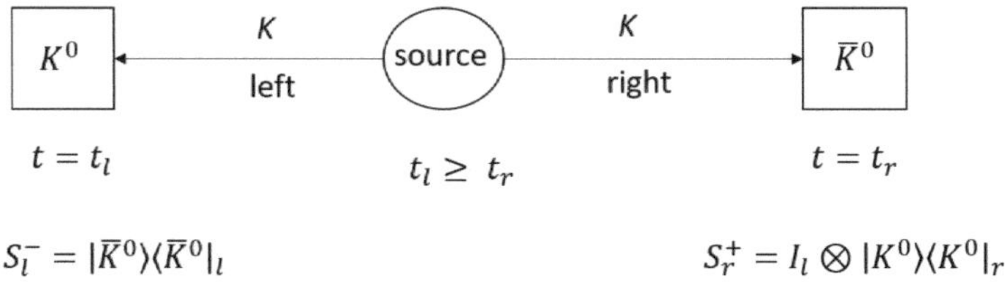

Figure 3.
We consider a case that $\overline{K}^0$ and K^0 are detected at the right side at $t = t_r$ and left side at $t = t_l$, respectively, where $t_l \geq t_r$. The operators S_r^+ and S_l^- correspond to strangeness measurement at the right and left sides.

In the following, we obtain $P\left(\overline{K}^0, t_l; K^0, t_r\right)$. For the other cases, the same procedure can be employed. In order to determine Eq. (35), one can use $|K_S\rangle$ and $|K_L\rangle$ as the basis and write Eq. (34) as

$$\rho_l(t = t_r; t_r) = \frac{1}{4}e^{-2\Gamma t_r}\left(|K_S\rangle\langle K_S|_l + |K_L\rangle\langle K_L|_l - e^{-\lambda t_r}\left(|K_S\rangle\langle K_L|_l + |K_L\rangle\langle K_S|_l\right)\right) \quad (36)$$

Suppose the density operator related to a left-moving particle is expressed as

$$\begin{aligned}\rho_l(t;t_r) = \rho_{SS}(t;t_r)|K_S\rangle\langle K_S|_l + \rho_{SL}(t;t_r)|K_S\rangle\langle K_L|_l \\ +\rho_{LS}(t;t_r)|K_L\rangle\langle K_S|_l + \rho_{LL}(t;t_r)|K_L\rangle\langle K_S|_l\end{aligned} \quad (37)$$

which is supposed to evade decoherence for the time interval $t > t_r$. Therefore, it evolves according to $\dot{\rho}_l(t;t_r) = -i\left(H\rho_l(t;t_r) - \rho_l(t;t_r)H^\dagger\right)$. Hence, the following equation is obtained

$$\begin{aligned}\dot{\rho}_l(t;t_r) = -\Gamma_S\rho_{SS}(t;t_r)|K_S\rangle\langle K_S|_l + (i\Delta m - \Gamma)\rho_{SL}(t;t_r)|K_S\rangle\langle K_L|_l \\ -(i\Delta m + \Gamma)\rho_{LS}(t;t_r)|K_L\rangle\langle K_S|_l - \Gamma_L\rho_{LL}(t;t_r)|K_L\rangle\langle K_L|_l\end{aligned} \quad (38)$$

Let us assume that C_{ij} with $i,j = S, L$ is constant, so

$$\begin{aligned}\rho_{SS}(t;t_r) = C_{SS}e^{-\Gamma_S t}, \qquad \rho_{LL}(t;t_r) = C_{LL}e^{-\Gamma_L t}, \\ \rho_{SL}(t;t_r) = C_{SL}e^{(i\Delta m - \Gamma)t}, \qquad \rho_{LS}(t;t_r) = C_{LS}e^{-(i\Delta m + \Gamma)t}\end{aligned} \quad (39)$$

From our knowledge of Eqs. (36) and (37), the values of C_{ij} and subsequently $\rho_{ij}(t;t_r)$ can be obtained. Finally, by replacing $t = t_l$, we attain $P\left(\overline{K}^0, t_l; K^0, t_r\right)$. Explicitly, by assuming that $\Delta t = t_l - t_r$, we have the following results

$$\begin{aligned}P\left(K^0, t_l; \overline{K}^0, t_r\right) &= P\left(\overline{K}^0, t_l; K^0, t_r\right) \\ &= \frac{1}{8}e^{-2\Gamma t_r}\left(e^{-\Gamma_S\Delta t} + e^{-\Gamma_L\Delta t} + 2e^{-\lambda t_r}\cos(\Delta m\Delta t)e^{-\Gamma\Delta t}\right)\end{aligned} \quad (40)$$

$$\begin{aligned}P\left(K^0, t_l; K^0, t_r\right) &= P\left(\overline{K}^0, t_l; \overline{K}^0, t_r\right) \\ &= \frac{1}{8}e^{-2\Gamma t_r}\left(e^{-\Gamma_S\Delta t} + e^{-\Gamma_L\Delta t} - 2e^{-\lambda t_r}\cos(\Delta m\Delta t)e^{-\Gamma\Delta t}\right)\end{aligned} \quad (41)$$

Let consider $\Delta t = 0$, that is, $t_l = t_r = t$, then from Eq. (40), one obtains

$$P\left(K^0, t_l; \overline{K}^0, t_r\right) = P\left(\overline{K}^0, t_l; K^0, t_r\right) = \frac{1}{4} e^{-2\Gamma t}\left(1 - e^{-\lambda t}\right) \tag{42}$$

which is in contradiction to the pure quantum mechanical EPR-correlations. The asymmetry of probabilities is the captivating factor of interest, as it directly responds to the interference term and can be quantified by means of experimental measurements. In the realm of pure quantum mechanics, where a system does not experience decoherence, we encounter this phenomenon by A^{QM} expressed as

$$\begin{aligned} A^{QM}(t_l, t_r) &= \frac{P\left(K^0, t_l; \overline{K}^0, t_r\right) + P\left(\overline{K}^0, t_l; K^0, t_r\right) - P\left(K^0, t_l; K^0, t_r\right) - P\left(\overline{K}^0, t_l; \overline{K}^0, t_r\right)}{P\left(K^0, t_l; \overline{K}^0, t_r\right) + P\left(\overline{K}^0, t_l; K^0, t_r\right) + P\left(K^0, t_l; K^0, t_r\right) + P\left(\overline{K}^0, t_l; \overline{K}^0, t_r\right)} \\ &= \frac{\cos(\Delta m \Delta t)}{\cosh\left(\frac{1}{2}\Delta\Gamma\Delta t\right)} \end{aligned} \tag{43}$$

in which $\Delta\Gamma = \Gamma_L - \Gamma_S$. In the decoherence model of entangled kaon system, since it is not known which particle will first be detected, t_r in Eqs. (40) and (41) need to be replaced by $\tau = min\,(t_r, t_l)$. By inserting Eqs. (40) and (41), we obtain

$$A^{\lambda}(t_l, t_r) = A^{QM}(t_l, t_r) e^{-\lambda\tau} \tag{44}$$

which indicates that the decoherence effect represented by $e^{-\lambda\tau}$ is dependent on the time of the first detected kaon.

The decoherence model in Eq. (44) can be introduced in a phenomenological way, where the decoherence parameter λ corresponds to an effective decoherence parameter ζ as $\zeta(t_l, t_r) = 1 - e^{-\lambda\tau}$. Apparently, the value of $\zeta = 0$ represents pure quantum mechanics and $\zeta = 1$ corresponds to complete decoherence or spontaneous factorization of the wave function (Schrödinger-Furry hypothesis). By means of standard least squares method [45, 46], $\zeta = 0.13 \pm 0.865$ is obtained in [31], which is in agreement with the results obtained from the effective variance method, where $\zeta = 0.13^{+0.16}_{-0.15}$ [47, 48] correspondent to $\lambda = \left(1.84^{-2.17}_{+2.50}\right) \times 10^{-12} MeV$. The value of both parameters is compatible with quantum mechanics, that is, $\zeta = 0$ and $\lambda = 0$ far away from the total decoherence, that is, $\zeta = 1$ or $\lambda = \infty$. It indicates that the interaction between the system and its environment has negligible influence on the system. As a result, the quantum properties related to the entanglement of the strangeness are preserved without significant alteration.

6. Loss of entanglement

As time progresses, the degree of decoherence in the initially fully entangled $K^0\overline{K}^0$ system increases, leading to a decrease in the system's entanglement. This loss of entanglement, defined as the disparity between an entanglement value and its maximum unity, can be precisely measured [21, 49]. In the realm of quantum information, the quantification of entanglement in a state is assessed by employing specific measures designed for quantifying entanglement. In this context, entropy plays a pivotal role. The entropy serves as a measure of the level of uncertainty or lack of knowledge associated with a quantum state. If the quantum state is pure, maximum information about the system is provided; however, mixed states only offer partial information. The entropy quantifies the extent to which maximal information is absent. In the following, we focus on the three most important and wildly used entanglement measures: Von Neumann entanglement entropy, entanglement of formation, and concurrence. When the interest is focused on the effect of decoherence, one needs to compensate for the decay up to time t to attain a proper density operator for the kaon system. Therefore, we divide the state Eq. (31) with its trace, that is, $\rho_N(t) = \frac{\rho(t)}{tr(\rho(t))}$.

6.1 Von Neumann entanglement entropy

Von Neumann's entropy of the quantum state is expressed as

$$S(\rho_N(t)) = -tr\big(\rho_N(t)\, log\,_d \rho_N(t)\big) \tag{45}$$

where d is the dimension of the Hilbert space, that is, for the Hilbert space of a qubit $d = 2$, hence $0 < S(\rho_N(t)) < 1$. For $t = 0$, the entropy is zero, meaning that the state is pure and also maximally entangled. However, as $t \to \infty$, the entropy approaches to one, that is, the system becomes mixed. The von Neumann entropy is a good criterion of entanglement, specifically for pure quantum states [21, 30, 50].

The bipartite von Neumann entanglement of its reduced states [51]. This entropy can be defined as the von Neumann entropy of any of its reduced states. This definition holds true because both reduced states have the same value, as can be demonstrated through the Schmidt decomposition of the state with respect to the bipartition. Consequently, the outcome remains unchanged regardless of the specific reduced state chosen. Generally, two subsystems are maximally entangled when their reduced density operators are maximally mixed. In our case, the left- (subsystem l) and right- (subsystem r) propagating kaons are our subsystems. Therefore, the reduced density operators are defined as

$$\rho_N^l(t) = tr_r(\rho_N(t)), \qquad \rho_N^r(t) = tr_l(\rho_N(t)) \tag{46}$$

The von Neumann entropy of $\rho_N^l(t)$ (or $\rho_N^r(t)$) provides the uncertainty in the subsystem l (or r) before measuring the subsystem r (or l). In the case of the kaon system, we have

$$S\big(\rho_N^l(t)\big) = S\big(\rho_N^r(t)\big) = 1 \quad \forall t \geq 0 \tag{47}$$

which are independent of the decoherence parameter λ, meaning that the correlation stored in the entire system is lost to the environment and not to the subsystems.

6.2 Entanglement of formation

Entanglement entropy is also known as the entanglement of formation for pure states. It is possible to express any density matrix as a collection of pure states forming an ensemble $\rho_i = |\psi_i\rangle\langle\psi_i|$, with each pure state having a corresponding probability p_i, that is $\rho = \sum_i p_i \rho_i$. For mixed states, entanglement of formation can be generalized by defining a quantity minimized over all the ensemble realizations of the mixed state. For the kaon system, we have

$$E_f(\rho) = \min \sum_i p_i S(\rho_i^l) \tag{48}$$

The entanglement of formation can be simplified to $E_f(\rho) \geq \mathcal{E}(f(\rho))$ by introducing the lower bound of $E_f((\rho))$, given by

$$\begin{aligned} \mathcal{E}(f(\rho)) &= H\left(\frac{1}{2} + \sqrt{f(\rho)(1-f(\rho))}\right) \quad \text{for} \quad f(\rho) \geq \frac{1}{2} \\ \mathcal{E}(f(\rho)) &= 0 \qquad\qquad\qquad\qquad\qquad\qquad \text{for} \quad f(\rho) < \frac{1}{2} \end{aligned} \tag{49}$$

where $f(\rho) = \max\langle e|\rho|e\rangle$, known as the fully entangled fraction of ρ, is the maximum overall completely entangled states $|e\rangle$, and

$$H(x) = -x log_2 x - (1-x)\, log\,_2(1-x). \tag{50}$$

For our model, the lower bound expressed for $E_f(\rho)$ is saturated, that is, $E_f(\rho) = \mathcal{E}(f(\rho))$. Therefore, one can calculate the entanglement of formation simply by computing $\mathcal{E}(f(\rho))$ [21].

The fully entangled fraction of $\rho_N(t)$ is $f(\rho_N(t)) = \frac{1}{2}\left(1 + e^{-\lambda t}\right) \geq 0$. Therefore, the entanglement of formation for the $K^0\overline{K}^0$ is assessed in terms of $E_f(\lambda)$ as

$$E_f(\lambda) = -\frac{1+\sqrt{1-e^{-2\lambda t}}}{2}\log_2\frac{1+\sqrt{1-e^{-2\lambda t}}}{2} - \frac{1-\sqrt{1-e^{-2\lambda t}}}{2}\log_2\frac{1-\sqrt{1-e^{-2\lambda t}}}{2} \tag{51}$$

From which one can obtain the entanglement loss $L_E(t)$ as

$$\begin{aligned} L_E(t) &= 1 - E(\rho_N(t)) \\ &\simeq \frac{\lambda}{\ln 2} t = \frac{1}{\ln 2}\xi(t) = 1.44\xi(t) \end{aligned} \tag{52}$$

approximated for small values of λ. Eq. (52) shows that the entanglement loss equals the weighted amount of decoherence.

6.3 Concurrence

Wootters and Hill, in their research publications [52–54], discovered a relation between entanglement of formation and a measure known as concurrence. This connection allows the expression of entanglement of formation for a general mixed state ρ of two qubits in terms of the concurrence as

$$E_f(\rho) = \mathcal{E}(C(\rho)) = H\left(\frac{1}{2} + \frac{1}{2}\sqrt{1 - C^2(\rho)}\right) \tag{53}$$

with $0 \leq C(\rho) \leq 1$. As The function $\mathcal{E}(C(\rho))$ is monotonically increasing from 0 to 1 as $C(\rho)$ goes from 0 to 1.

The concurrence $C(\rho)$ in defined as $C(\rho) \equiv max\{0, \lambda_1 - \lambda_2 - \lambda_3 - \lambda_4\}$ with λ_i's representing the square roots of the eigenvalues, in decreasing order, of $R = \rho\tilde{\rho}$ matrix where $\tilde{\rho}$ is the spin-flipped state of ρ defined as

$$\tilde{\rho} = (\sigma_y \otimes \sigma_y)\rho^* (\sigma_y \otimes \sigma_y) \tag{54}$$

The complex conjugate of ρ, that is, ρ^* is taken in the basis $\{|\Uparrow\Uparrow\rangle, |\Downarrow\Downarrow\rangle, |\Uparrow\Downarrow\rangle, |\Downarrow\Uparrow\rangle\}$. In our model, since $\rho_N(t)$ is not variant under spin flip, hence $R = \rho_N^2$, and the concurrence is

$$C(\rho_N(t)) \equiv max\{0, e^{-\lambda t}\} = e^{-\lambda t} \tag{55}$$

Hence, $L_C(t)$ is computed as

$$L_C(t) = 1 - C(\rho_N(t)) = 1 - e^{-\lambda t} = \xi(t) \tag{56}$$

The value of $L_C(t)$ is precisely equivalent to decoherence $\xi(t)$, describing the factorization of the initial spin-singlet state to the state $|K_S\rangle_l \otimes |K_L\rangle_r$ or $|K_S\rangle_l \otimes |K_L\rangle_r$. In both cases, Eqs. (52) and (56) show that the loss of entanglement is equivalent to decoherence and increases linearly with time [30]. In **Figure 4**, we show the loss of information by the von Neumann entropy $S(\lambda)$ in comparison with the loss of entanglement of formation $L_E(t)$ depending on the normalized time τ of the propagating $K^0\overline{K}^0$ system for the experimental mean value $\lambda = 1.84 \times 10^{-12} MeV$, and the upper bound $\lambda = 4.34 \times 10^{-12} MeV$ of the decoherence parameter. The rate of increase in the loss of entanglement of formation is slower as time progresses, indicating the amount of resources required to create a specific entangled state. In the overall state, the level of entanglement decreases until separability is achieved, which occurs exponentially fast as time approaches infinity. In the CPLEAR experiment, where the propagation of one kaon for 2 cm corresponds

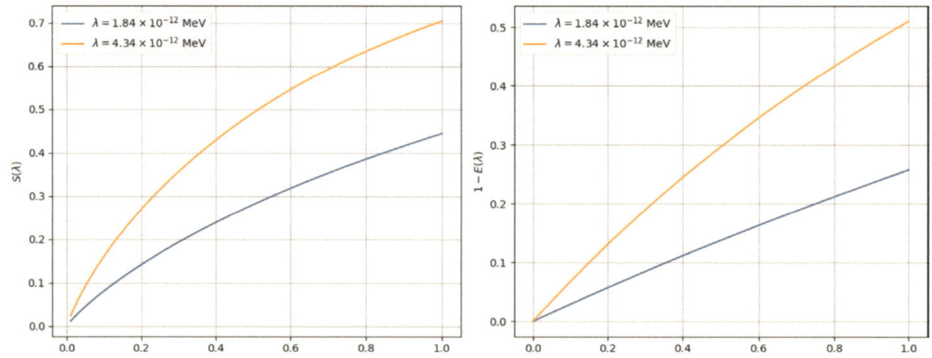

Figure 4.
The time dependence of von Neumann entropy and the loss of entanglement of formation for two values of λ. The time t is normalized versus τ_S, that is, $\tau = \frac{t}{\tau_S}$.

to the propagation time $\tau = 0.55$, until an absorber measures it, the loss of entanglement is approximately 0.18 and 0.38 for the mean value and upper bound of λ, respectively.

7. Conclusions

This chapter has provided a comprehensive analysis of the properties and phenomena associated with neutral K-mesons, highlighting their intriguing and often puzzling behaviors. We began by emphasizing the significance of strangeness and charge parity violation in the understanding of these particles. Next, the concept of strangeness oscillations, exemplified by the oscillations between K^0 and $\overline{K}^0$ states, was introduced and thoroughly explored. We delved into the regeneration of K_S and unraveled the underlying mechanisms that govern these oscillations, shedding light on the intricate dynamics involved. Next, we examined the quasi-spin space and its bases, unraveling their implications and providing insights into the entangled states of kaon pairs, particularly focusing on both maximally and nonmaximally entangled neutral kaons. This exploration has broadened our understanding of the entanglement properties exhibited by these particles. Furthermore, we dedicated significant attention to the study of decoherence effects on entangled kaons. Through the use of density matrix formalism, we captured the dynamic nature of decoherence and introduced a dedicated parameter to quantify its impact. Measures such as von Neumann entanglement entropy, entanglement of formation, and concurrence were employed to explore the loss of entanglement, providing valuable tools for characterizing and quantifying entanglement in the context of neutral kaons. However, several intriguing questions for future research remain open.

8. Outlook

The study of kaon entanglement and decoherence raises a host of intriguing open questions that drive ongoing research and exploration. How do environmental interactions, such as scattering or absorption, impact the entanglement and coherence properties of neutral kaons? Can we develop methods to quantify and control these effects? Additionally, investigating entanglement dynamics in multipartite kaon systems, involving more than two neutral kaons, presents an exciting avenue of research. How does multipartite entanglement and its degradation relate to the underlying interactions and dynamics? Developing novel experimental techniques and theoretical models to explore these phenomena could shed light on the intricate nature of entanglement in kaon systems.

Furthermore, the generation and manipulation of entangled kaon states offer intriguing possibilities. Can we create specific entanglement patterns or preserve entanglement over extended timescales? Understanding and enhancing entanglement in kaon systems could have implications for quantum information processing and communication. The role of entanglement in comprehending and quantifying CP violation in kaon systems is another crucial aspect. How does entanglement contribute to our understanding of the underlying mechanisms driving CP violation? Exploring connections between the entanglement properties of neutral kaons and other areas of physics, such as quantum information theory, quantum field theory, or quantum gravity, holds promise for uncovering fundamental principles and phenomena.

Additionally, the presence of decoherence poses significant challenges. How does decoherence affect the entanglement properties of neutral kaons? Can we develop techniques to mitigate or minimize its detrimental effects and preserve entanglement over longer durations? Exploring the entanglement and coherence properties of neutral kaons within nonstandard models beyond the Standard Model of particle physics could provide insights into new physics and phenomena. Moreover, extending the study of entanglement and decoherence to other meson systems or particles with similar characteristics is a compelling direction. How do the entanglement properties differ or align between different types of particles? Finally, experimental methodologies and theoretical frameworks for directly observing or measuring entanglement in kaon systems, as well as exploring the implications of entanglement for quantum computing and communication, open up exciting possibilities for practical applications in quantum technologies.

These open questions highlight the ongoing exploration and research endeavors, presenting intriguing avenues for further investigation and potential breakthroughs.

Acknowledgements

The authors gratefully acknowledge the financial support provided by the Foundation for Science and Technology (FCT/MCTES) within the scope of the PhD grant 2021.07608.BD, the Associated Laboratory ARISE (LA/P/0112/2020), the R&D Unit SYSTEC through Base (UIDB/00147/2020) and Programmatic (UIDP/00147/2020) funds and project RELIABLE - Advances in control design methodologies for safety-critical systems applied to robotics (PTDC/EEI-AUT/3522/2020), all supported by national funds through FCT/MCTES (PIDDAC). The work has been done in honor and memory of Professor Fernando Lobo Pereira.

Conflict of interest

The authors declare no conflict of interest.

Author details

Nahid Binandeh Dehaghani[1]*, A. Pedro Aguiar[1] and Rafal Wisniewski[2]

1 Research Center for Systems and Technologies (SYSTEC), ARISE and the Electrical and Computer Engineering Department, Faculty of Engineering, University of Porto, Porto, Portugal

2 Department of Electronic Systems, Aalborg University, Aalborg, Denmark

*Address all correspondence to: nahid@fe.up.pt

References

[1] Einstein A, Podolsky B, Rosen N. Can quantum-mechanical description of physical reality be considered complete? Physical Review. 1935; **47**:777-780. DOI: 10.1103/physrev.47.777

[2] Schrödinger E. Die gegenwärtige situation in der Quantenmechanik. Naturwissenschaften. 1935;**23**:844-849. DOI: 10.1007/bf01491987

[3] Bell JS. Physics. Reprinted in JS Bell, Speakable and Unspeakable in Quantum Mechanics. 2nd ed. Vol. 1. Cambridge University Press; 1964, 1987. pp. 195. DOI: 10.1017/CBO9780511815676

[4] Clauser JF, Horne MA. Experimental consequences of objective local theories. Physical Review D. 1974;**10**:526. DOI: 10.1103/physrevd.10.526

[5] Wigner EP. On hidden variables and quantum mechanical probabilities. In: Part I: Particles and Fields. Part II: Foundations of Quantum Mechanics. Berlin, Heidelberg: Springer-Verlag; 1997. pp. 515-523. DOI: 10.1007/978-3-662-09203-3_56

[6] Aspect A, Dalibard J, Roger G. Experimental test of Bell's inequalities using time-varying analyzers. Physical Review Letters. 1982;**49**:1804. DOI: 10.1103/physrevlett.49.1804

[7] Tittel W, Brendel J, Zbinden H, Gisin N. Violation of Bell inequalities by photons more than 10 km apart. Physical Review Letters. 1998;**81**:3563. DOI: 10.1103/physrevlett.81.3563

[8] Feynman RP, Leighton RB, Sands M. The Feynman Lectures on Physics. Reading, MA: Addison-Wesley Pub. Co.; 1963. DOI: 10.1063/1.3051743

[9] Lee TD. Particle Physics and Introduction to Field Theory. Chur, Switzerland: Harwood Academic Publishers; 1981. DOI: 10.1201/b16972-2

[10] Okun LB. Leptons and Quarks. North-Holland: Elsevier; 2013. DOI: 10.1016/b978-0-444-86924-1.50031-3

[11] Bernabeu J, Di Domenico A. Can future observation of the living partner post-tag the past decayed state in entangled neutral K mesons? Physical Review D. 2022;**105**:116004. DOI: 10.1103/physrevd.105.116004

[12] Apostolakis A, Aslanides E, Backenstoss G, Bargassa P, Behnke O, Benelli A, et al. An EPR experiment testing the non-separability of the K0K0 wave function. Physical Review B. 1999; **422**:339-348. DOI: 10.1016/S0370-2693(97)01545-1

[13] Bizzeti A. Recent results from the NA62 and NA48/2 experiments at CERN. Nuclear and Particle Physics Proceedings. 2023;**324**:113-118. DOI: 10.1016/j.nuclphysbps.2023.01.023

[14] Aloisio A, Ambrosino F, Antonelli A, Antonelli M, Bacci C, Bencivenni G, et al. Recent results from KLOE at DAΦNE. Nuclear Physics B – Proceedings Supplements. 2002;**111**: 213-218. DOI: 10.1016/S0920-5632(02)01708-5

[15] Gauzzi P, Perez del Rio E. The KLOE-2 experiment at DAΦNE. EPJ Web Conference. 2019;**212**:01002. DOI: 10.1051/epjconf/201921201002

[16] Aaij R, Beaucourt L, Chefdeville M, Decamp D, Déléage N, et al. LHCb detector performance. International Journal of Modern Physics A. 2015;**30**:

1530022. DOI: 10.1142/S0217751X15300227

[17] Yamanaka T et al. The J-PARC KOTO experiment. Progress of Theoretical and Experimental Physics. 2012;**2012**:02B006. DOI: 10.1093/ptep/pts057

[18] Martin BR, Shaw G. Particle Physics. Chichester, United Kingdom: John Wiley & Sons; 2016. DOI: 10.1063/1.2808907

[19] Greaves H, Thomas T. Studies in History and Philosophy of Science Part B: Studies in History and Philosophy of Modern Physics. Vol. 45. Elsevier; 2014. pp. 46-65. DOI: 10.1016/j.shpsb.2013.10.001

[20] Griffiths D. Introduction to Elementary Particles. Wiley-VCH Verlag GmbH & Co. KGaA; 2004. DOI: 10.1002/9783527618460

[21] Bertlmann RA. Entanglement, Bell inequalities and decoherence in particle physics. Quantum Coherence: From Quarks to Solids. 2006;**2006**:1-45. DOI: 10.1007/11398448_1

[22] Pais A, Piccioni O. Note on the decay and absorption of the θ^0. Physical Review. 1995;**100**:1487-1489. DOI: 10.1103/PhysRev.100.1487

[23] Lee TD, Wu CS. Weak interactions (second section) chapter 9: Decays of neutral K mesons. Annual Review of Nuclear Science. 1966;**16**:511-590. DOI: 10.1007/978-1-4612-5397-6_27

[24] Lipkin HJ. CP violation and coherent decays of kaon pairs. Physical Review. 1968;**176**:1715. DOI: 10.1103/physrev.176.1715

[25] Bramon A, Escribano R, Garbarino G. A review of Bell inequality tests with neutral kaons. Handbook on Neutral Kaon Interferometry at a Φ Factory. 2007;**2007**:217-254

[26] Bruß D. Characterizing entanglement. Journal of Mathematical Physics. 2002;**43**:4237-4251. DOI: 10.1364/icqi.2001.t4

[27] Bramon A, Garbarino G, Hiesmayr BC. Quantum marking and quantum erasure for neutral kaons. Physical Review Letters. 2004; **92**:020405. DOI: 10.1103/physrevlett.92.020405

[28] Di Domenico A. Search for CPT violation and decoherence effects in the neutral kaon system. Journal of Physics: Conference Series. 2009;**171**:012008. DOI: 10.1088/1742-6596/171/1/012008

[29] Di Domenico A. Latest results on kaon physics at KLOE-2. JACoW. 2023; **eeFACT2022**:30-34. DOI: 10.18429/JACoW-eeFACT2022-TUXAS0101

[30] Bertlmann RA, Durstberger K, Hiesmayr BC. Decoherence of entangled kaons and its connection to entanglement measures. Physical Review A. 2003;**68**:012111. DOI: 10.1103/physreva.68.012111

[31] Syahbana A, Zen FP, Dwiputra D. Study of entangled K-meson and its decoherence. Indonesian Journal of Physics. 2022;**33**:39-44. DOI: 10.5614/itb.ijp.2022.33.1.4

[32] Rovelli C. Quantum Gravity. Cambridge, United Kingdom: Cambridge University Press; 2004. DOI: 10.1017/cbo9780511755804

[33] Kiefer C. Why Quantum Gravity? Berlin, Heidelberg: Springer-Verlag; 2007. DOI: 10.1093/acprof:oso/9780199212521.003.01

[34] Gambini R, Porto RA, Pullin J. Fundamental decoherence from quantum gravity: A pedagogical review. General Relativity and Gravitation. 2007;**39**:1143-1156. DOI: 10.1007/s10714-007-0451-1

[35] Sivaram C. What is special about the planck mass? arXiv Preprint. 2007. DOI: 10.48550/arXiv.0707.0058

[36] Pearle P. Combining stochastic dynamical state-vector reduction with spontaneous localization. Physical Review A. 1989;**39**:2277. DOI: 10.1103/physreva.39.2277

[37] Dehaghani NB, Pereira FL, Aguiar AP. Quantum control modelling, methods, and applications. Extensive Reviews. 2022;**2**:75-126. DOI: 10.21467/exr.2.1.5037

[38] Nielsen MA, Chuang IL. Quantum Computation and Quantum Information. Cambridge: Cambridge University Press; 2010. DOI: 10.1017/cbo9780511976667

[39] Kurgalin S, Borzunov S. Concise Guide to Quantum Computing. Switzerland: Springer; 2021. DOI: 10.1007/978-3-030-65052-0

[40] Zaman F, Jeong Y, Shin H. Counterfactual Bell-state analysis. Scientific reports. 2018;**8**:14641. DOI: 10.1038/s41598-018-32928-8

[41] Breuer HP, Petruccione F. The Theory of Open Quantum Systems. USA: Oxford University Press; 2007. DOI: 10.1093/acprof:oso/9780199213900.001.0001

[42] Manzano D. A short introduction to the Lindblad master equation. AIP Advances. 2020;**2020**:10. DOI: 10.1063/1.5115323

[43] Caban P, Rembieliński J, Smoliński KA, Walczak Z. Unstable particles as open quantum systems. Physical Review A. 2005;**72**:032106. DOI: 10.1103/physreva.72.032106

[44] Bertlmann RA, Grimus W, Hiesmayr BC. Open-quantum-system formulation of particle decay. Physical Review A. 2006;**73**:054101. DOI: 10.1103/physreva.73.054101

[45] Barlow RJ. Statistics: A Guide to the Use of Statistical Methods in the Physical Sciences. Chichester: Wiley; 1993. DOI: 10.2307/1269014

[46] Bevan A. Statistical Data Analysis for the Physical Sciences. NewYork: Cambridge University Press; 2013. DOI: 10.1017/CBO9781139342810

[47] Bertlmann RA, Grimus W, Hiesmayr BC. Quantum mechanics, Furry's hypothesis, and a measure of decoherence in the K0 system. Physical Review D. 1999;**60**:114032. DOI: 10.1103/physrevd.60.114032

[48] Bertlmann RA, Zeilinger A. Quantum [Un]Speakables II. The Frontiers Collection. 2017

[49] Hiesmayr BC. The Puzzling Story of the Neutral Kaon System: Or What We Can Learn from Entanglement [Ph.D. thesis]. Austria: University of Vienna; 2002

[50] Varizi AD. Open Quantum System Approach to Neutral Kaon Interferometry [Ph.D. thesis]. Brazil: Universidade Federal de Minas Gerais; 2017

[51] Bennett CH, DiVincenzo DP, Smolin JA, Wootters WK. Mixed-state entanglement and quantum error correction. Physical Review A. 1996;**54**: 3824. DOI: 10.1103/physreva.54.3824

[52] Wootters WK. Entanglement of formation of an arbitrary state of two qubits. Physical Review Letters. 1998;**80**: 2245. DOI: 10.1103/physrevlett.80.2245

[53] Hill SA, Wootters WK. Entanglement of a pair of quantum bits. Physical Review Letters. 1997;**78**:5022. DOI: 10.1103/physrevlett.78.5022

[54] Wootters WK. Entanglement of formation and concurrence. Quantum Information and Computation. 2001;**1**: 27-44. DOI: 10.26421/qic1.1-3

Section 3

Interdisciplinarity

Chapter 8

Perspective Chapter: Why Do We Care about Violating Bell Inequalities?

Christopher G. Timpson

Abstract

High energy experiments present an exciting new regime in which to explore the violation of Bell inequalities by nature. There are two main reasons why one is interested in Bell inequality violation. The first is that—for suitable experimental configurations—Bell inequality violation can indicate the failure of the condition of *Local Causality*, which condition is a natural way of capturing the desideratum of no superluminal action-at-a-distance. The second is that Bell inequality violation is an *Entanglement Witness*. I review both of these reasons for interest, and suggest that high energy experiments plausibly involve the latter rather more than the former, at least as currently configured.

Keywords: Bell inequality, entanglement witness, high energy, local causality, nonlocality

1. Introduction

Recently there have been both a number of proposals for, and actual experimental tests of, Bell inequality violation in high energy experiments (see e.g., [1–3]). This is an important new regime for exploring entanglement and Bell inequality violation, involving very different length-scales from the current state of the art in Bell experiments (compare [4]). It offers the promise of new insights into quantum correlations and entanglement at the femtometre scale, and will help us in our concrete understanding of the behaviour of relativistic quantum fields.

But at the same time, it is noteworthy that despite its now being nearly 60 years since Bell's seminal paper [5], and despite the award of the 2022 Nobel prize to Aspect, Clauser and Zeilinger for their experimental work demonstrating Bell inequality violation, there is still considerable controversy over what Bell inequality violation really means, and why we should care about it [6, 7].

Here I will review the meaning of Bell inequalities and the significance of their violation, distinguishing two distinct broad areas of interest:

1. Tests of whether the world is nonlocal, or more precisely, is not *locally causal* [8], and

2. Tests of whether a multipartite system described by quantum mechanics is in an entangled state or not (that is, the use of Bell inequalities as *entanglement witnesses* [9]).

In the high energy Bell tests, the latter notion is certainly in play and it is important. As we shall see, however, it is an open question the extent to which the criteria for providing a test of the first kind are satisfied in high energy experiments.

2. Bell correlations and entangled states

For Schrödinger, entanglement was *the* characteristic feature of quantum theory, that feature which 'enforces its entire departure from classical lines of thought' [10]. Feynman, by contrast, insisted that it was the concept of superposition, as manifested in the two-slit experiment, that was 'the heart of quantum mechanics. In reality, it contains the only mystery ... ' [11]. Reassuringly, we need not see these two claims as being as much in dispute as one might initially suppose: for entanglement can be understood simply as genuinely multi-particle superposition.

Thus, given some system A with a Hilbert space $\mathcal{H}_A$ and a set of basis states $\{|\phi_i\rangle \in \mathcal{H}_A\}, \left\langle \phi_i|\phi_j \right\rangle = \delta_{ij}$, then A is in a superposition with respect to that basis *iff* its state vector $|\psi\rangle \in \mathcal{H}_A$ is given by:

$$|\psi\rangle = \sum_i c_i|\phi_i\rangle, \tag{1}$$

where more than one of the coefficients c_i is non-zero.

Now consider a second system, B, with $\mathcal{H}_B$; $\left\{|\chi_j\rangle \in \mathcal{H}_B\right\}, \left\langle \chi_j|\chi_k \right\rangle = \delta_{jk}$. The pair of systems AB will be (pure state) entangled *iff* their joint state $|\Psi\rangle \in \mathcal{H}_A \otimes \mathcal{H}_B$ *cannot* be written as a product, i.e.:

$$|\Psi\rangle \neq |\phi\rangle|\chi\rangle, \tag{2}$$

for any $|\phi\rangle, |\chi\rangle$ in $\mathcal{H}_A, \mathcal{H}_B$, respectively. If the joint state *could* be written as a product $|\phi\rangle|\chi\rangle$, then even though $|\phi\rangle$ might be a superposition with respect to some basis of interest for $\mathcal{H}_A$, and $|\chi\rangle$ might be a superposition with respect to some basis of interest for $\mathcal{H}_B$, we would not have *multiparticle* superposition: each particle individually would merely be in its own superposition. By contrast if we have a state of the form:

$$|\Psi\rangle = \sum_i c_i|\phi_i\rangle|\chi_i\rangle, \tag{3}$$

with more than one c_i non-zero, then it cannot be written as a product, and we have genuinely multiparticle superposition: a superposition of the product basis $|\phi_i\rangle|\chi_i\rangle$ for the two systems.

We can immediately see that something interesting is going on with an entangled state like (3).

Take a somewhat simplified case for illustrative purposes: suppose the states $|\phi_i\rangle$ corresponded to states with compact spatial support within a small width Δ around a location which is i units along the x-axis (with a unit length being $> 2\Delta$, say), and similarly, suppose that the states $|\chi_i\rangle$ corresponded to states with the same width but at a location $i + 2$ units along the x-axis. Then if we think that being in an eigenstate of the operator corresponding to some physical quantity is both necessary and sufficient for a system's having a definite value of that property, our pair of systems in the state given by (3) will definitely be two units of length apart from one another, whilst neither system A, nor system B, individually has any definite location of its own. This is a very striking, and non-classical, kind of holism about quantum property possession: the properties possessed by the whole system are not determined by properties possessed by the sub-systems, even for such a simple and important property as position. To take another, very familiar, example, consider the singlet state of a pair of spin-half particles:

$$|\Psi^-\rangle = \frac{1}{\sqrt{2}}(|\uparrow\rangle|\downarrow\rangle - |\downarrow\rangle|\uparrow\rangle). \tag{4}$$

This is of the form (3) and clearly an entangled state. In this case, the two spins are definitely in opposite directions, without there being a fact about which direction, if any, each individual spin is pointing in. Again: we have properties of the whole which are not determined by the properties of the parts, in a way which is not possible classically.

Even if we do not think that a system's being in an eigenstate of the operator corresponding to some quantity implies that the system actually has a definite value of that quantity, but (as in some approaches to understanding quantum theory, e.g. [12]) restrict ourselves to the weaker claim that it merely implies that a suitable measurement interaction will give a particular result with probability one, entangled states remain strikingly non-classical and of great interest. An entangled state such as (3) is a pure state of the theory: it is not composed (in the theory) of a statistical mixture of more fine-grained states. Yet it will both give rise to non-trivial probabilities (i.e. those other than 0 and 1) for all observables for each system taken individually, whilst also providing *correlations* between measurement results on the two sides. This is not possible classically: in a classical probabilistic theory, pure states cannot give rise to correlations. (Consider two classical bits: we can only make a correlated state by producing a mixture of two of the pairs of bit values, e.g. by tossing a fair coin and producing (pure) bit values 01 on heads and 10 on tails, say).

The definition of pure state entanglement above can be generalised in two different directions: we can define entanglement for non-pure, i.e., mixed states, and we can define it for multipartite systems, i.e. those with a number of components $N > 2$, rather than just for bipartite ($N = 2$) systems.

A mixed state of a bipartite system AB, represented by a density operator ρ_{AB}, is an entangled state *iff* it cannot be written in the form:

$$\rho_{AB} = \sum_i \lambda_i \rho_A^i \otimes \rho_B^i, \tag{5}$$

where $\rho^i_{A/B}$ are density operators on $\mathcal{H}_{A/B}$ respectively and the λ_i are convex coefficients: real numbers $0 \leq \lambda_i \leq 1$ which sum to one: $\sum_i \lambda_i = 1$.

A state in the form (5) with only one term in the sum is called a product state; one with two or more terms is called *seperable* (or sometimes *clasically correlated*). If we include pure state density operators ($\rho^2 = \rho$) in our reckoning then this definition of entanglement covers both pure and mixed state cases.

The state $\rho_{1\dots N}$ of an N-party system is N-party entangled *iff* it cannot be expressed as a convex sum of the form:

$$\rho_{1\dots N} = \sum_i \lambda_i \rho^i, \tag{6}$$

where each ρ^i can be factorised into products states of less than N parties. An N-party system may be $M < N$-party entangled, even if it has no M-party subsystem whose reduced state is itself M -party entangled [13].

The GHZ state:

$$|\Psi^{\mathrm{GHZ}}\rangle = \frac{1}{\sqrt{2}}(|\uparrow\rangle|\uparrow\rangle \dots |\uparrow\rangle + |\downarrow\rangle|\downarrow\rangle \dots |\downarrow\rangle) \tag{7}$$

is a familiar example of an N-party state which is fully N-party entangled.

2.1 The Einstein-Podolsky-Rosen argument

Nowadays, entanglement is a workhorse of quantum information science, essential for entanglement-assisted communication (e.g. quantum teleportation), quantum computation, and quantum cryptography [14]. But it was first explicitly put to work by Einstein in his debates with Bohr on the completeness of quantum theory [15].

In these debates, Einstein would formulate thought experiments designed to illustrate that there was some knowable physical fact which a description according to the rules of quantum theory would leave out [16]. Bohr typically responded by arguing that the measurements involved in the putative experiments must be disturbing (due, as he was wont to say, to 'the finiteness of the quantum of action' [15, 17]) and hence they did not after all reveal that there were some knowable physical facts which the quantum description left out.

Einstein brilliantly realised that the correlations encapsulated in entangled states could transform this debate [18]. If we consider an experiment in which the measurement is performed on one half of a spatially separated entangled pair, and we ask about the physical features of the other half, then Bohr would be able to maintain his disturbance-on-measurement doctrine only at the cost of invoking some form of action-at-a-distance. The measurement on one side would have to disturb the physical features of the far side.

Thus the argument of the famous Einstein-Podolsky-Rosen (EPR) paper [19] was to the conclusion of a dilemma: either quantum theory is incomplete—it does not describe all the knowable physical features of systems—or it is nonlocal: it involves instantaneous action-at-a-distance. Einstein, for obvious reasons, preferred the first horn of the dilemma.

As formulated in the EPR paper (Einstein himself would prefer simpler, later, presentations [18, 20]) the argument used an important principle, the *EPR criterion of reality*:

> *'If, without in any way disturbing a system, we can predict with certainty (i.e., with probability equal to unity) the value of a physical quantity, then there exists an element of reality corresponding to that quantity.* '[19]

Consider the singlet state (4) above, and suppose that Alice, in one lab, holds one half of a pair of systems in this state, whilst Bob, in a distant lab, holds the other. If Alice were to measure spin in the z-direction, say, she would get either the outcome spin-up with 0.5 probability, or the outcome spin-down, with 0.5 probability. Suppose she gets spin-up. As soon as she has her measurement result, she knows that a measurement on Bob's system in the z-direction would now with certainty give a spin-down outcome. From the EPR criterion we can infer that there is an element of reality of Bob's system corresponding to this spin value. Alternatively we might simply note that following Alice's measurement, to get the correct probabilities for measurement results on Bob's system, we now need to assign it the state $|\downarrow_z\rangle$. Either way, the supposedly complete quantum state (4) misses something out: it does not include an element of reality corresponding to spin down in the z direction for Bob's system, and it does not assign the state $|\downarrow_z\rangle$. So either the quantum description is missing things out, or the result of Alice's measurement is to change the features of Bob's system, e.g., in the instantaneous change of the state of his system from $\rho_B = 1/2(|\uparrow\rangle\langle\uparrow| + |\downarrow\rangle\langle\downarrow|)$ to $|\downarrow_z\rangle\langle\downarrow_z|$.

Bohr's reply to EPR [21] is generally felt to be somewhat obscure [15, 22, 23] but a summary version would be that—for all its plausibility—he rejected the EPR criterion of reality.

One simple way of taking this is as a rejection of the idea that the changing of the probabilities for the outcomes of measurements on Bob's far system, given Alice's nearby action, need correspond to the changing of any intrinsic physical feature of Bob's system. (Nonlocal action, note, would require changes in the locally defined—the intrinsic—properties of the far system).

In turn, an extreme way of implementing this thought is to step back from the idea that the formalism of quantum theory, and in particular the quantum state, provides much in the way of description of the actual microphysical world in the first place. For example, if one took the (controversial) view that quantum states are no more than compendia of probabilities for the observable macroscopic results of measurement interactions, for ensembles of similarly prepared systems [12], and do not purport to describe microphysical reality, then one would not be troubled by the change in the state of Bob's system from $1/2(|\uparrow\rangle\langle\uparrow| + |\downarrow\rangle\langle\downarrow|)$ to $|\downarrow_z\rangle\langle\downarrow_z|$. Since quantum states, on this view, do not describe the actual microphysical properties of individual quantum systems, such a change in the state would not correspond to a change in the properties of any system, hence not to a nonlocal change. On the other hand, it might be felt that stepping back so far from quantum theory's having descriptive microphysical content would be to throw the baby out with the bathwater.

2.2 Bell's theorem

In any case: the EPR argument was, as we have seen, an argument about the content of quantum theory, over whether the theory was incomplete or was nonlocal. Bell changed things completely. His concern was not whether some particular theory (e.g. standard quantum theory) was nonlocal. His question was rather whether *the world* is nonlocal. That is, does any empirically adequate theory of the world—any

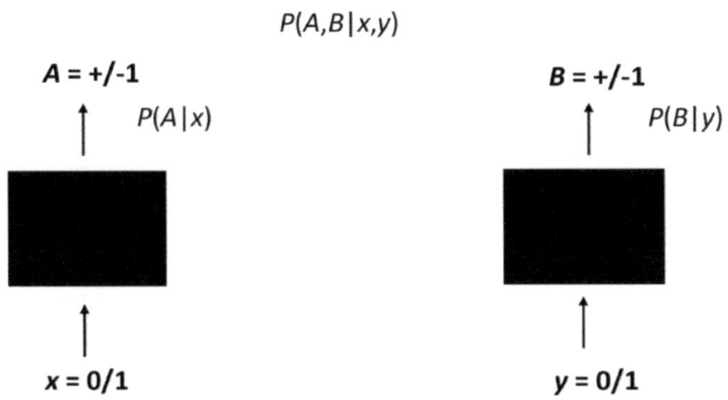

Figure 1.
Correlation boxes. For randomly chosen input values x, y *we gather output values A, B and estimate the individual and joint probabilities* $P(A|x), P(B|y)$ *and* $P(AB|xy)$ *respectively.*

theory which will account for the experimental results—have to include nonlocality within it?[1] He was able to address this question by beginning from a suitably theory-independent starting place.[2]

Thus let us frame the question of the existence of various correlations operationally. Suppose we begin with a pair of black boxes 1 and 2, where box 1 takes an input bit value $x = 0, 1$ and outputs a value $A = \pm 1$ with some probability distribution $P(A|x)$, and box 2 takes an input value $y = 0/1$ and outputs a value $B = \pm 1$ with probability distribution $P(B|y)$ (**Figure 1**).

Suppose that we run a large number of repeated trials using these boxes, choosing input values x and y randomly and independently of each other. Gathering the input and output data, we will be able to estimate the underlying probability distributions. Suppose furthermore that when we gather the statistics we find there to be correlations between the outputs of the boxes, so that:

$$P(AB|xy) \neq P(A|x)P(B|y). \tag{8}$$

We can explore the idea that these correlations are due, and due only, to some further variable λ connecting the boxes. If they are, then when we condition on this further variable, the joint probability for A and B should factorise:

$$P(AB|xy\lambda) = P(A|x\lambda)P(B|y\lambda). \tag{9}$$

Call this the *factorisability condition*. We have been supposing that the input values x and y are chosen independently of each other. If they are, additionally, independent of the new variable λ (too)

$$P(\lambda|xy) = P(\lambda), \tag{10}$$

a condition we may label *λ-independence*, then the joint probability for the outcomes can be expressed as:

[1] Bell's thinking became increasingly general as it developed. By 1975 it had reached its fully mature form. See [24] for an account of the developments.

[2] Of course, there is no *entirely* theory neutral starting place for science; it is a matter of degree. Bell was able to formulate a sufficiently theory-independent framework to show something interesting and important.

$$P(AB|xy) = \int P(A|x\lambda)P(B|y\lambda)P(\lambda)d\lambda. \tag{11}$$

Call a joint probability which can be expressed in the form (11) *Bell correlated*.

If the probabilities for the behaviour of our boxes are Bell correlated, then a number of interesting inequalities—called Bell inequalities [5]—follow. For example, if we define the expectation value (equivalently, the correlation function) for the outputs of our boxes for various inputs as:

$$E(x,y) = \sum_{AB} ABP(AB|xy), \tag{12}$$

then one such is:

$$|E(x=0,y=0) + E(x=0,y=1) + E(x=1,y=0) - E(x=1,y=1)| < 2. \tag{13}$$

This is known as the Clauser-Horne-Shimony-Holt (CHSH) inequality [25]; it is a particular Bell inequality. Nowadays, a menagerie of different Bell inequalities is available, for example allowing different numbers of inputs and outputs, and numbers of boxes greater than two [26]. They all follow from the assumption that the salient joint probabilities can be expressed in the form (11).

But why should we care about this? Why should we care whether our boxes are Bell correlated or not; whether they satisfy a Bell inequality or not?

We care when we embed the boxes in a relativistic spacetime, and the choice of x and the outcome A occur at spacelike separation from the choice of y and the outcome B.

Start with the idea that correlations should be explicable, and start with the idea that:

> *'The direct causes (and effects) [of events] are nearby, and even the indirect causes (and effects) are no further away than permitted by the velocity of light.'*[27]

This is Bell's principle of *Local Causality*. It incorporates both the idea that causal relations are mediated (a causal link between distant events a and b will be made up of a causal chain of events between a and b, so the causal link does not obtain at a distance) and the constraint from relativity theory that the speed of light is the upper bound on the propagation of causal influence.

We need to formulate the principle mathematically to deploy it in our theories. This Bell did as follows [8, 27]:

If we consider an event A in spacetime (an event such as one our boxes producing a particular output on a particular occasion), then a full specification (or a sufficient specification) M of facts on a spacelike hypersurface through the past lightcone of A will render facts N associated with any other region of spacetime probabilistically irrelevant to A (see **Figure 2**). Thus:

$$P(A|MN) = P(A|M). \tag{14}$$

This is the formal statement of Local Causality.

Applying it to the case—call it the standard configuration—in which we have embedded our correlation boxes within spacetime, with the choice of x and the output

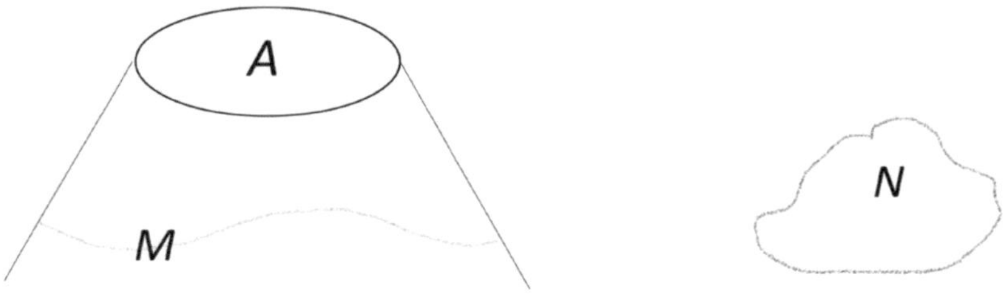

Figure 2.
Local causality. A sufficient specification of facts on the surface M *renders facts within N probabilistically irrelevant to the event A: $P(A|MN) = P(A|M)$.*

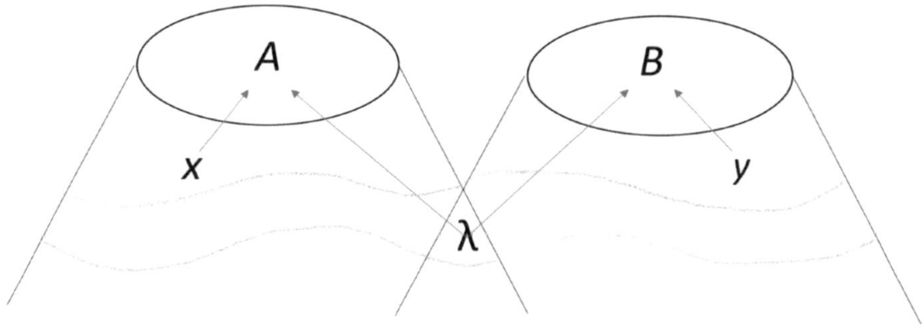

Figure 3.
A spacetime picture of the Bell experiment in standard configuration. λ specifies facts in a spacetime region cutting through the union of the past light cones of A *and* B*, including where they overlap.*

A at spacelike separation from the choice of y and the output B, and with λ a full or sufficient specification of facts in a spacetime region cutting through the past light cones of A, B, including where they overlap (see **Figure 3**), Local Causality will imply the factorisability condition (9).[3]

Given too that x and y can be taken as free variables—or sufficiently free variables (we will return to this)—λ-independence will also hold, and so the CHSH inequality will follow for the outputs of our boxes.

Bell's theorem is the claim that correlations in a standard-configuration experiment satisfying Local Causality and λ-independence will also satisfy a Bell inequality. It follows that any theory which predicts that the results of correlation experiments in standard configuration can violate a Bell inequality must also violate Local Causality, λ-independence, or both.

We know that the CHSH inequality has been shown to be violated for high quality experiments in standard configuration (e.g. [4, 28]), and we very reasonably believe that in these experiments, it was possible to choose the input variables x and y independently. Accordingly it seems we must conclude that *Local Causality fails*.

This is often taken to lead to two further conclusions (very distinct from one another):

---

[3] By coarse-graining λ by integrating over its values in certain regions, we can obtain if we wish variables which correspond just to the part of the hypersurface where the past light cones of A, B, overlap; just to the part within the remainder of the past light cone of A; and just to the part within the remainder of the lightcone of B.

- Quantum mechanics is nonlocal; and
- *The world* is nonlocal.

3. Now for some controversy

These conclusions are remarkable in reach and content, but the path that led to them seems clear enough. How can there be any controversy regarding the meaning of Bell inequality violation?

The simplest place to begin is by noting that certain ways of understanding quantum theory would seem already to provide counterexamples to the last two conclusions of the previous section.

Note to begin that we did not (as Bell was himself very aware) need experimental violation of Bell inequalities to conclude that quantum theory is not locally causal. The quantum probabilities for the correlation experiments are given by:

$$P(AB|xy\lambda) = \left\langle \Psi | P_x^A \otimes P_y^B | \Psi \right\rangle, \tag{15}$$

where P_x^A is the projector (or more generally, positive operator) associated with the value A for measurement of quantity x, and P_y^B similarly, and λ captures that we have prepared state $|\Psi\rangle$ for our bipartite system. It is automatic that this quantity will not factorise whenever $|\Psi\rangle$ is entangled.

Meanwhile, approaches to quantum theory which take its content to be mainly operational—capturing the statistics of macroscopic measurement results, say, rather than describing physical features of quantum sytems—or that otherwise step back from the detailed description of a mind-independent physical world can still count as local. There are even ways—at least one—of taking quantum theory to be descriptive of the detailed microphysical features of the world and of individual quantum systems whilst still remaining local: the Everett (Many Worlds) interpretation of the theory is generally taken to be local [24, 29].

This raises the question of how failure of the specific condition of Local Causality relates to the presence of nonlocality, and especially to the presence of any nonlocality we might find worrying or problematic.

It is helpful to note that we can distinguish between two different notions, each of which can be thought to fall under the more general heading of 'nonlocality' (the distinction goes back to Einstein in fact [20], though my terminology is not quite his):

- *Dynamical nonlocality*—there are changes in *locally defined* (intrinsic) properties in some region due to spacelike happenings elsewhere. (This is a notion of genuine superluminal action).
- *Kinematical nonlocality*—the states assigned to unions of spacetime regions are not determined by the states assigned to the individual regions. (This is often known as the failure of *separability*).

Certainly in quantum mechanics when we have a failure of separability in the sense that we have an entangled state for a pair of systems, and when those two systems are spatially separated, we will also have a failure of separability in the sense of kinematical nonlocality. Perhaps other, post-quantum, theories might similarly be kinematically nonlocal. Kinematical nonlocality is certainly a striking and non-classical feature, but it seems perfectly coherent theoretically (it is an actual feature in quantum theory) and in and of itself it does not give rise to any problems of consistency with relativity (though we should note that a kinematically nonlocal theory will still need to allow a rich-enough range of locally defined properties given by the states of individual regions, as is achieved in quantum theory via use of the reduced density operator [30]). If kinematical nonlocality can be combined with the *absence* of any dynamical nonlocality then we may have a perfectly satisfactory theory, consistent with relativity, and perhaps one which can violate a Bell inequality unproblematically.

On the other hand, Bell's intuitive prose statement of Local Causality did seem to capture our notions of dynamical locality consistent with relativity, and his reasoning from this statement to the factorisability condition seems simple and sound. What, if anything, can have gone wrong with it?

To make this question more pointed, suppose that we did try to appeal to a mere failure of kinematical locality, rather than a failure of dynamical locality, to explain Bell-inequality violation. How would this failure of kinematical locality actually help? We may grant there to be non-separable features in the joint past of Alice and Bob's correlation experiments, but surely we can just condition on these and include them within λ [31]? Then a Bell inequality should still hold, and kinematical nonlocality will not have helped us to violate it.

3.1 Looking at the examples

To unpick these puzzles, let us look a little more closely at how the proposed examples of local understandings of quantum theory work. We can discern three kinds of example.

The first kind of example we are already familiar with from what I called the 'extreme' way of offering a response to EPR in wake of Bohr. It is the kind of view which retreats from the idea that quantum theory, and the quantum state, provide us with means of describing the microphysical world. Instead the formalism of the theory is seen as a device for organising our experimental interactions with the world, and quantum probabilities only pertain to the behaviour of measuring devices which can be characterised non-quantum mechanically and at the macroscopic level. The only role of the quantum state is to furnish these probabilities; it does not describe actual features of quantum systems. Such views [12] can be called *operationalist* (foregrounding operational procedures in the lab) or *anti-realist* in the sense that they depart from a scientific realist conception of our physical theories. In a scientific realist conception, the aim of science is to provide us with a literally true description of the what the mind-independent world is like in both its directly observable and non-directly observable features, and as applied to quantum theory, it would require us to interpret at least a large part of the quantum formalism as directly representing facts about the microphysical systems from which the world is built.

In a sufficiently operationalist or anti-realist interpretation of a theory, the notion of (dynamical) locality will itself be given a suitably operationalist understanding. Since the theory is not in the business of describing microphysical goings-on in individual runs of an experiment, but rather in the business of codifying statistical features, non-microphysically characterised, of repeated runs of experiments, the notion of dynamical locality will be spelt out in statistical, operational, terms: a theory will be (operationally) nonlocal *iff* it is *signalling*, where a theory is signalling *iff* operational procedures confined to one spacetime region produce a change in the statistics for measurement outcomes in a spacelike region.

Quantum theory itself is rather obviously *non-signalling* since the reduced density operator for a far system is unchanged by any operation one can perfom on a nearby system. A sufficiently operationalist interpretation of quantum theory, therefore, will—by the relevant estimation—count as dynamically local despite predicting Bell inequality violation.

Notably, such an approach does not explain how Bell inequality violation comes about in experiments, it just asserts that it does. It gives-up on the idea of explaining physically how correlated measurement results turn out as they do on individual runs of an experiment, as part more broadly of relinquishing descriptive ambitions for quantum theory. Remaining fastidiously aloof from describing microphysical systems, it will remain equally aloof from causal description of events involving those systems, hence Bell's prose statement of the principle of Local Causality above will not apply straightforwardly, if at all. But as remarked before in the context of the EPR argument, taking so stringently thin a notion of the content of quantum theory as this may be depriving ourselves of too much that we need.

The second kind of example may agree with the prose statement of Local Causality, but instead raise trouble for the mathematical expression of the principle in terms of probabilities as in Eq. (14) and Eq. (9). In particular, one might offer a notion of probability according to which changes in the probabilities assigned by Alice to the possible results of measurements on Bob's distant system do not count as changes in any objective, or in any localised, feature of his system. If changes in the probabilities pertaining to a system need not correspond to changes in objective or in localised features of the system, then the inference from changes in the probabilities to causal consequences in the world is rendered shaky.

It is tempting to think that there is a single correct, best, probability distribution for what the results of measurement on a given system should be, a probability distribution fixed by the features of the system and its immediate surroundings (perhaps including the features of any measurement device with which it is about to interact). In the quantum case, it is very tempting to think that this single correct, best, probability distribution is given by what the correct quantum state (pure or mixed) that this system should be assigned is.

But it might be argued, as in [32, 33] that there is no single set of probabilities which should be assigned to a given system; the physical features of a system, or more generally, the physical features of the world in the past light cone of a system, need not determine a single, univocal, probability for the results of measurement on that system. It will follow that there is not a single quantum state that should be assigned to a system either. Thus Alice following her spin measurement on her half of the entangled pair shared with Bob correctly assigns to Bob's system the pure state $| \downarrow_z\rangle$, but this assignment is not made correct by any feature intrinsic to (locally defined for) Bob's system; it is made correct by what Alice has learnt at her location. Bob, correctly, and unimprovably (for now) at his location, not missing any physical feature intrinsic to his system, continues to assign the mixed state to his system.

In this kind of view, the full specification of physical facts on some spacelike hypersurface do not determine *univocal* and ideally best probabilities for each event on future spacelike hypersurfaces, so we cannot straightforwardly draw causal inferences from the probabilities assigned to an occurence changing when conditioned on facts pertaining to a spacelike separated region.

In [32, 33] probabilities (and quantum states) are objective, even if they are not univocal (because relational). A rather different picture is given in QBism [34], where again there are not univocal probabilities, and there are not unique quantum states for systems, but this time because probabilities (and quantum states) are viewed as subjective, in the sense of being expressions of individual agents's degrees of belief. Again, this renders the probabilistic formulation of Local Causality as ineligible to express causal notions.

The third example is the particular concrete example given by the Everettian (Many Worlds) approach to quantum mechanics and it is quite unlike the first two kinds of example. Unlike these, in the Everettian approach, the quantum state (of the universe!) is unique, objective, and intended to be part of the detailed description of individual quantum mechanical systems. It is a thorough-going scientific realist approach to understanding quantum theory. But, it is generally held to be a dynamically local theory, indeed it will satisfy the prose statement of Local Causality, since in Everett, the state of any spacetime region will be fixed by the state on a spacelike surface cutting the past light cone of that region.

A number of tightly-interconnected factors seem to be involved in the Everett approach being able to violate Bell inequalities whilst remaining dynamically local:

1. Though the quantum state is real in the sense of playing a central role in describing real features of the world, it does not collapse on measurement, so there are no dynamically nonlocal changes in the state. Instead the state evolves by unitary dynamics given by a local Hamiltonian.

2. Apart from some background spacetime structure, all the remaining physical ontology (physical features of the world) supervenes on the quantum state and its evolution. (Nothing extra is added, such as hidden variables which might need to be pushed-about dynamically nonlocally.)

3. The theory is emphatically kinematically non-local, since entanglement between different spatially separated regions is generic.

4. There is non-uniqueness of measurement outcomes (different outcomes of a given measurement, in different worlds which have branched from the measurement). We do not need to guarantee single correlated outcomes on a given run of the experiment, at the expense of other possible outcomes: all the outcomes are realised.

It is an interesting open question whether any theory which is suitably scientific realist, which is dynamically local, which is kinematically nonlocal, and yet which can violate a Bell inequality (without λ-dependence), is also one in which uniqueness of measurement outcomes will fail. That is, do the features which co-occur in the Everett interpretation *have* to co-occur?

In any case: the clearest lesson from reflecting on the Everett example seems to be that in a theory which is kinematically non-local, the probabilistic formulation of Local Causality may again not be an appropriate way of expressing the prose formulation of Local Causality [24].

3.2 Summing-up this controversy

Though they incorporate, or express, a range of very different philosophical approaches, perhaps the simplest, quick, way to understand these controversies is to discern an ambiguity (deliberately planted there) in my earlier statement of Bell's theorem. I said that the theorem was the claim that correlations in a standard-configuration experiment satisfying Local Causality and λ-independence will also satisfy a Bell inequality. But which version of Local Causality? The initial, intuitive, prose version? Or the mathematical formulations, Eqs. (9) and (14)? No one could deny the content of the theorem if we meant the latter. But as we have seen, there are a number of routes by which one might endorse the prose version of the principle, but not agree with its mathematical formulation, thus not agree with the claim, if Local Causality in the prose formulation is intended.

Still: the class of theories for which the mathematical formulation is a good way of expressing Local Causality is very broad and important. So the claim that Bell inequality violation in nature shows that Local Causality, as mathematically expressed, is not true, is still very deep, remarkable, and powerful.

3.3 'Local causality' vs. 'local realism'

It is rather common to find Bell's results discussed under the heading of 'Local Realism' instead of 'Local Causality'. This alternative terminology may perhaps be traced back to [35, 36], but it can be highly, and distractingly, controversial. For some, asserting that Bell inequality violation (for standard configuration experiments, etc.) rules out Local Realism invites the prospect that locality might be salvaged by jettisoning a(n allegedly atavistic) realism assumption. For others, such a contention must be quite mistaken, for Bell's reasoning *did not* and *need not* make any appeal to any assumptions beyond locality (Local Causality) itself [37–39].

There is some truth in both contentions, but there is no doubt that 'Local Causality' is the better terminology, for it does not invite the confusions that 'Local Realism' does.

We have already seen that some dialling-down of realist, descriptive, ambitions for physics does allow scope for evading inferences to the presence of dynamical nonlocality (action at a distance) both in the context of the original EPR argument (recalling the tenor of Bohr's response), and in one of the routes we explored for resisting the idea that the mathematical and the prose formulations of Local Causality aptly express conditions for dynamical locality in all contexts. So some realist commitment is involved as part of the background of Local Causality's expressing an interesting physical feature. But what is the realism in Local Realism supposed to be?

For some it is just a generic commitment to the notion of scientific realism along the lines introduced above. Thus Clauser and Shimony: 'Realism is a philosophical view, according to which external reality is assumed to exist and have definite properties, whether or not they are observed by someone.' [35]. On other occasions, realism is taken to be the logically much stronger—or more specific—view that all physical quantities (e.g. all quantities represented by self-adjoint operators in quantum mechanics) have a definite value for what the result of measuring them on some occasion would be, even if they were not in fact measured on that occasion [36]. The latter is a stronger view since it commits to there being hidden variables which are deterministic (fixing definite measurement outcomes), whereas in the former view it could be that there are independently existing properties of the world but that they only (in general) furnish non-deterministic probabilities for measurement outcomes. Further, the latter reading suggests definite values for all physical quantities simultaneously, where one might take the external world to have real mind-independent features at all times, but perhaps only for a subset of the physical quantities.

The strong view (realism as deterministic hidden variables) is too narrow. Local Causality covers a broader range of theories than this (and it needs to, to be applicable to realistic experimental scenarios). But if we relax merely to the broader view (scientific realism as applied in the quantum context) then we find examples of theories which are (dynamically) local and realist in this sense, but which can violate Bell inequalities: the Many Worlds approach. So locality and realism (in this broader sense) do not entail Bell inequalities; while locality and realism (taken in the narrower sense) does not cover all cases. We should simply drop 'Local Realism' and stick with 'Local Causality'.

4. Back to λ-independence

If our correlation experiments were not in standard configuration, then it would be easy for them to produce Bell inequality violating results without there needing to be any failure of local causality, or any other non-classical funny-business. For example, if the setting—x, say—on one side was not at spacelike separation from the setting or outcome on the other, then there would be plenty of time for a classical sub-luminal, or a lightlike, channel to connect the two sides. Bob's box would then be able to take both x and y as inputs, and locally mimic failure of local causality. (The response function which captures the behaviour of Bob's box can now be of the form $P(B|x, y, \lambda) \neq P(B|x, \lambda)$ consistently with local causality.) This additional channel connecting the two sides of the experiment might not be obvious to us, but it could certainly be there. Requiring that the experiment be in the standard configuration rules out there being any such channel, on pain of inconsistency with relativity.

It is also extremely important that there actually be a choice of x and y values when the experiment is run. If x and y took the same fixed values across all runs of the experiment then we would not have any interesting correlations. Any correlations in the outputs ($P(AB) \neq P(A)P(B)$) could readily be accounted for by a purely locally causal mechanism. More subtly, if x and y choices were made on each run, but we happened to know in advance what they would be, it would again be possible to mimic failure of local causality locally: we could adjust the other input to the boxes, via λ, to make it look as if local causality were failing when it was not. This brings us back to the question of λ-independence.

It will be recalled that λ-independence ((10) above) was an essential ingredient in arriving at Bell-correlations ((11) above) and thus in deriving Bell inequalities. We are used to the idea that we can often have a good deal of control over at least some of the parameters which are involved in our experiments. We are also used to the idea that, in general, separate systems will not be correlated with one another unless they have had some explicit interaction in the past, often, in the recent past. Accordingly, if when constructing our correlation experiment we have taken pains that the process by which the x input value is selected is independent of other things, and the process by which the y input value is selected is independent of other things, then it will seem reasonable to assume a) that x and y are not correlated with each other (in fact we can explicitly check whether $P(xy) = P(x)P(y)$) and b) that, crucially, they are not correlated with anything else important for the results of the experiment. With these precautions, then, λ-independence seems a safe assumption.

On the other hand, if this condition fails, then locally causal theories could produce arbitrary correlations—whether the quantum correlations, or even stronger correlations, which maximally violate the CHSH inequality (the Popescu-Rohrlich correlations [40]). To see this, let us consider a simple example.

Suppose we have a theory which is deterministic in the sense that specification of all the facts on a spacelike hypersurface which cuts through the past light cone of some event fixes all the facts to the future of that surface within the lightcone in question. By construction such a theory will be locally causal. It will also automatically violate λ-independence. To see this, consider that in such a theory both x and y will be functions of λ—call them $f(\lambda)$ and $g(\lambda)$ respectively. Then we can consider the set Λ_{xy} of all those values of λ which would give rise to some specific x, y pair:

$$\Lambda_{xy} = \{\lambda | f(\lambda) = x, g(\lambda) = y\}. \tag{16}$$

Given some specific x, y in some run of the experiment, then, λ in that run will be constrained to lie within Λ_{xy}, which is a strict subset of the set Λ of all values of λ, which means that λ-independence fails. Knowing the values of x, y we know that λ must be within a specific subset of values. Letting χ_{xy} be the characteristic function of the set Λ_{xy} (giving value 1 if $\lambda \in \Lambda_{xy}$ and 0 otherwise), we have that:

$$P(\lambda | xy) = \frac{\chi_{xy}(\lambda) P(\lambda)}{P(xy)} \neq P(\lambda). \tag{17}$$

With the λ-dependence of this deterministic theory in hand, we can now see how to produce arbitrary correlations (this kind of construction seems first explicitly to have been noted, for quantum correlations, in [41]). We just need to put an appropriate probability distribution over the initial set of λs.

Note that in this theory, not only x and y but also the outcomes A and B will be functions of λ. We can thus further refine the Λ_{xy} with a partition:

$$\Lambda_{xy} = \bigcup_{AB} \Lambda_{xy}^{AB}, \tag{18}$$

where the Λ_{xy}^{AB} are the sets of all λ values which produce the specific values x, y, A, B. (Note, $\Lambda_{xy}^{AB} \cap \Lambda_{x'y'}^{A'B'} = \emptyset$).

Taking χ_{xy}^{AB} to be the characteristic function of the set Λ_{xy}^{AB}, the joint probability $P(ABxy\lambda)$ will be given by:

$$P(ABxy\lambda) = \chi_{xy}^{AB}(\lambda) P(\lambda). \tag{19}$$

Suppose that it is the quantum correlations which we want to recover. Then we need merely postulate a $P(\lambda)$ which is such that the correct weight gets given to the sets Λ_{xy}^{AB}. Thus we choose $P(\lambda)$ such that:

$$P\left(\lambda \in \Lambda_{xy}^{AB}\right) = \int \chi_{xy}^{AB}(\lambda) P(\lambda) d\lambda = \left\langle \Psi | P_x^A \otimes P_y^B | \Psi \right\rangle P(xy). \tag{20}$$

The correlation probabilities in this theory will be given by:

$$P(AB | xy) = \int \frac{\chi_{xy}^{AB}(\lambda) P(\lambda)}{P(xy)} d\lambda. \tag{21}$$

For the weights on Λ_{xy}^{AB} given by $P(\lambda)$ satisfying (20) we will immediately obtain $\left\langle \Psi | P_x^A \otimes P_y^B | \Psi \right\rangle$ as desired. By choosing different weights on the Λ_{xy}^{AB} we can obtain any desired correlations.

The λ-dependence of this deterministic theory was of a rather strong kind, with λ fixing x and y. It is also possible to pursue the more subtle question of how little λ-dependence is required in order to reproduce certain correlations locally causally. For example [42] shows that at most one bit of correlation (as measured by the mutual information) between λ and one of the measurement settings is sufficient in order to be able to reproduce any correlations which could be generated by quantum measurements on a singlet state (see also [43]). In the other direction, interestingly, [44] shows that if one's theory retains a little λ independence, in the sense that given λs do not absolutely rule out particular x, ys, then there is a Bell inequality which would be violated for a less than maximally entangled quantum state, but which this λ-dependent theory would not be able to violate.

4.1 Effective λ-independence

The discussion so far of the λ-dependence of the deterministic theory leaves us with something of a puzzle. Earlier we noted that, having taken care to apply certain precautions, it would be perfectly reasonable to believe that λ-independence was satisfied for one's experiment in standard configuration. But as we have just seen, if there is an underlying deterministic theory, then λ-dependence is obligatory. And it seems a perfectly reasonable thought that there might be some (perhaps as yet to be discovered) deterministic theory underlying quantum theory.

The key to resolving this conundrum is to recognise that what is important is *effective* λ-independence, as opposed to what might be called *absolute* or *fine-grained* λ-independence. (I take this to be one of the key lessons from—and an area of agreement within—the exchange between Bell and Clauser, Horne and Shimony [45].)

When considering the behaviour of some physical system—for example one of our black boxes—it may well be (at least for the sake of argument) that there is some fully deterministic theory in the background, which gives us a detailed λ which is sufficient entirely to fix the system's behaviour. Yet that physical system may in a certain sense be insensitive to λ. That is, although λ fully fixes the behaviour, the system might behave in much the same way, or exactly the same way, for a range of values of λ. In other words, the behaviour of the system may best be characterised as responsive to some coarse-grained variables, corresponding to some coarse-graining of Λ.

Let us introduce the notation $\bar{\lambda}$ for new variables given by some minimal coarse-graining of Λ (e.g., averaging values of λ within some small ϵ-volume of Λ). Considering now the boxes in our correlation experiment, suppose them to be governed by a locally causal theory; their behaviour will be captured by the response functions:

$$P(A|Bxy\lambda) = P(A|x\lambda), \text{and} \tag{22}$$

$$P(B|Axy\lambda) = P(B|y\lambda). \tag{23}$$

Suppose now that the boxes are insensitive to fine-grained λ. There will be some coarse-grained $\bar{\lambda}$ which screens-off λ in the response functions:

$$P(A|x\bar{\lambda}) = P(A|x\lambda); \tag{24}$$

$$P(B|y\bar{\lambda}) = P(B|y\lambda). \tag{25}$$

Now it may be that at this coarse-grained level, $\bar{\lambda}$-independence will hold —$P(\bar{\lambda}|xy) = P(\bar{\lambda})$—even if λ independence does not. Then we will again have Bell correlated joint probabilities, and Bell inequalities will hold for our locally causal theory:

$$P(AB|xy) = \int P(A|x\bar{\lambda})P(B|y\bar{\lambda})P(\bar{\lambda})d\bar{\lambda}. \tag{26}$$

Say that *effective* λ-independence holds, iff there is some minimally coarse-grained $\bar{\lambda}$ which screens-off λ in the measurement response functions and for which $\bar{\lambda}$-independence holds.

Effective λ-independence can certainly hold in a deterministic theory, and in the circumstances relevant to our experiment in standard configuration we would expect it to do so. Reflect that many and various different processes could be selected which produce the x and y input values for the experiment (and we can check to see whether it makes any difference to the presence of Bell-inequality violation which process is used). These processes could be of the kind where we have very good reason to believe that the dynamics (whatever it is) leading-up to the production of the value is extremely complicated (as for example in a pseudo-random-number-generator), so that an exact specification of the final state of the choosing-device would be necessary to determine its initial state, whilst the value produced itself is a very coarse-grained function of the final state of the choosing-device. In this circumstance, we can envisage that if there is an underlying deterministic theory, its sets Λ_{xy} are highly fibrillated and interwoven with one another. Then knowing the exact λ would still fix which Λ_{xy} we were in. But even a slightly coarse-grained $\bar{\lambda}$ would not pin down any Λ_{xy}: within the small ϵ-ball some (approximately) equal-measured subset of each Λ_{xy} might be present. Thus $\bar{\lambda}$-independence would hold.

Important as it is, this is not enough for effective λ-independence to hold. For that we also need that $\bar{\lambda}$ screens-off λ in the response functions. However, given the enormous variety of ways (ways which we can to some extent choose between), and the enormous complexity of ways (which we can to some extent ensure), in which any fine-grained λs can be connected to x and y values, it would be extraordinary if the response functions were always sensitive to the fineness of grain of λ required to fix x and y. In other words, it would be extraordinary if $\bar{\lambda}$ did not screen-off λ in the response functions, and thus effective λ-indepedence hold. (Plausibly, it would be far more extraordinary than Local Causality's failing.)

Bell in [46] talks about input x and y variables, when they are produced by plausible randomisation processes, being 'sufficiently free variable[s] for the purposes at hand'. I take the notion of effective λ-independence just outlined to be a way of capturinging this notion of 'sufficiently free'. There may be some description (e.g. the envisaged underlying deterministic one) in which x and y are not free variables, since they are correlated with, or caused by, something else. But that need not stop them being effectively, or sufficiently, free; and the degree of coarseness of grain of λ which both screens-off λ in the measurement response functions and delivers $\bar{\lambda}$-independence captures the relevant metric for sufficient freedom.

4.2 The question of λ-independence in the case of high energy experiments

So far, then, we have reassured ourselves that in the standard configuration for the correlation experiment, and if one has taken trouble that the procedures for selection of x and y values are one from some number of suitably randomising procedures—or otherwise give us a high degree of confidence that the selection processes are independent of other aspects of the experiment—we may be confident that λ-independence, or effective λ-independence, holds.

It is sometimes suggested, however, that there may be special reasons even so why (effective) λ-independence might fail, thus allowing us to avoid the conclusion that the world is not locally causal, despite Bell inequality violation. Such special reasons might include the thought that backwards causation might be possible, so that we should not rule out the possibility that Alice and Bob selecting their x and y values could affect the value of λ in the past (e.g. [47, 48]) or that a particular form of deterministic hidden variable theory might naturally give rise to λ-dependence [49]. In each of these cases (as also in [50]) we face the question of whether there is more reason to believe that these special reasons obtain than to believe that effective λ-independence does. Given the strength of the general physical grounds for believing effective λ-indpendence obtains, the general view is against these special reasons.

Throughout our discussion of (effective) λ-independence, however, it was extremely important that the processes which produced the selection of x and y values were separately identifiable and understandable to the extent that we could have confidence that the values produced were (at least effectively) independent of each other and of other features which might be important in the experiment. By controlling and probing the behaviour of these processes, we could come to a very high degree of confidence that such independence obtains. In terms of a causal diagram (rather than just a spacetime diagram) where arrows indicate the presence of causal influence and the absence of arrows its absence, we need a causal structure as given in **Figure 4a**.

Let us now turn (finally) to some more specific aspects of Bell tests in high energy experiments. Here (as in e.g. [1–3]) we envisage a high-energy collision product which itself decays into two parts, where conservation principles will ensure that the two parts are in a suitable entangled state. For example for a Higgs particle (spinless) decaying into a $W^{\pm}$ pair, the spin state of the $W^{\pm}$ pair should be close to a singlet state [1]. (The spin state in top/anti-top quark pair production is also a very

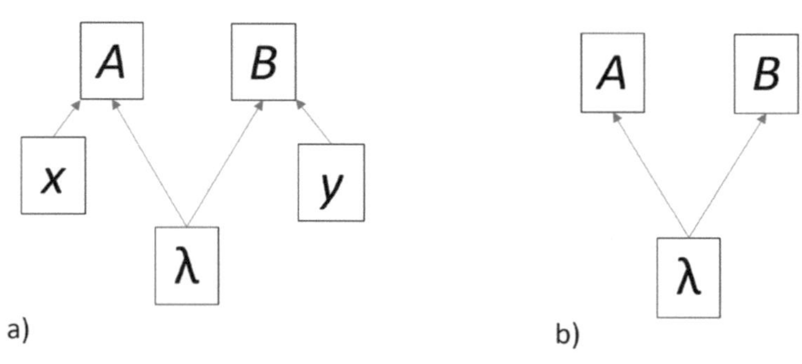

Figure 4.
Causal diagrams for correlation experiments. In a) we have standard configuration with x *and* y *values independently selected, independent of* λ *(and outside the forward light cone of the preparation event for the systems measured). In b) there is no independent selection of* x, y *values.*

interesting case [51, 52]). These entangled decay products separate with opposite momenta, and then themselves decay further, where these final decay products are then detected, and from their angular spread, the spin of the parent (entangled) particle inferred.

It will be noticed immediately that we do not have a part of this process which provides an external choice of measurement on the particles in the entangled state (corresponding to a choice of x and y). The decay of each particle into the finally detected products is spontaneous (or is a matter of the interaction of the particles with the quantum fields immediately surrounding them) rather than being selected or being a matter of choice. We do not have an independently identifiable and controllable process in the experiment which is determining what measurement is made and whose features we can separately study and ascertain. The causal structure of the experiment seems to be of the form of **Figure 4b** rather than **Figure 4a**.

How much of a problem is that? From the point of view of performing an ideal test of whether the world is not locally causal, it is a significant problem. We no longer have a sufficient guarantee that (effective) λ-independence will hold since we no longer have independent access to whatever it might be that is determining x and y values. If we are attempting experimentally to constrain the features of a general class of theories, we need to consider the possibility that there might be some hidden variables underlying our entangled state for the $W^{\pm}$ particles, which were carried along by the two particles as they separated, and which in due course told the particles when and how to decay. In that case we would have strong λ-dependence, no reason to believe that effective λ-independence obtained, and the theory giving us these hidden variables could perfectly well be locally causal while still giving us Bell inequality violation.

It is a different case if we are assuming quantum theory to be the true theory of the world. Then we might see the quantum description of the experiment giving us an analogue of, or perhaps an instance of, a choice of x and y values independent of the features of the particles being measured: we might look to the particles' decay being caused by their interaction with the quantum fields surrounding them, and we might see this as a process which we have good reason to believe is intrinsically random and not such as to give rise to λ-dependence. Indeed, we might see such interactions with local quantum fields as being much like the use of pseudo-random-number-generators in a traditional Bell experiment.[4] However, if, as in an ideal Bell test, we are interested in establishing general truths about the world (about how things have to be in broad classes of theories in order for them to be empirically adequate) then we cannot assume quantum theory as a whole, or even too-large chunks of it. Thus we cannot assume that the process leading to the entangled particles' final decays are driven by suitably random and independent processes just because they would be if quantum theory were true. The main point at issue is: how much can we infer about the world irrespective of whether quantum theory is true? (Recall: quantum theory itself is automatically not locally causal. We want to know how much we can pin down experimentally the different question of whether *the world* is not locally causal.)

Of course: we do use theory in our assessment of the functioning of pseudo-random-number-generators in standard Bell tests too. It is not that no aspect of our ideal Bell experiment can use theory. But so far as possible, the theory used should be independent of the theory being tested, and so far as possible, we should have detailed

[4] I owe this observation to Alan Barr.

experimental access to the functioning of the various devices we use in our experiment, including whatever it is that is selecting x and y values.

To sum-up then. High energy experimental tests of Bell inequalities, at least as currently configured and conceived, do not seem to give us sufficient guarantee that (effective) λ-independence should hold since we do not have independent access to, or independent understanding of, the processes giving rise to the selection of a given measurement. Thus as things stand the experiments may not be in a good position to provide us with a crucial experiment for whether the world is locally causal or not. But this is far from saying that they are not interesting, however! Seeing (if that is what we will see) more instances of strikingly quantum behaviour in new regimes we have not closely studied before experimentally is an important endeavour. But more than this, we can also seek to develop our understanding of various aspects of these high energy experiments, and in particular our understanding of the process of measurement in the sense of the quantum-to-classical transition in the experiments. The quantum-to-classical transition is an important topic in its own right and has been very little studied in the context of realistic high energy experiments. Careful analysis of how classical measurement results come about in this setting may well move us towards the possibility of developing stronger reasons for belief that λ-independence should be thought to hold in these experiments. We may be able to move towards non-question-begging grounds for conviction that λ-independence holds.

5. Bell inequality violation as an entanglement witness

There is a different kind of reason altogether why one might be interested in Bell inequality violation. We saw the definitions of pure and mixed, bipartite and N-party entanglement earlier. These definitions are clear enough. What may be less clear, and what in some cases is a very difficult problem, is whether some given quantum state is entangled or not. Recall—the definition is couched negatively: one's state is entangled if it is not possible to write it as a separable (or a product) state. If we are not already given a state in product or separable form in general it is a difficult (in some cases, unsolved) problem whether it can be put in that form. The device of *entanglement witnesses* was developed to help address this question [9, 53, 54].

An entanglement witness is a self-adjoint operator $\mathcal{W}$ which has a positive expectation value for all separable states, and a negative expectation value for at least one entangled state. Thinking geometrically in the vector space of linear operators, $Tr(\rho\mathcal{W}) \geq 0$ defines a plane, normal to $\mathcal{W}$, on one side of which (the positive side) sit all the separable states, and on the other side of which is at least one entangled state. The set of separable states is a convex set (i.e. convex combinations of separable states gives you back a separable state—the same is not necessarily true for entangled states, we may note: entanglement is not preserved under convex combination). It was first noted in [53] that it is therefore a consequence of the Hahn-Banach theorem that for any entangled state, there will be some plane which separates it from the set of separable states, i.e. there will be some entanglement witness $\mathcal{W}$ for it.

Various entanglement witnesses, and techniques for identifying entanglement which go by way of entanglement witnesses, have been developed [9]. The important point for our purposes is that Bell inequalities are entanglement witnesses [54]—that is, we can write down a self-adjoint operator corresponding to the inequality, and if a state gives a certain expectation value with that operator, we know the state must be entangled.

The operator corresponding to the CHSH inequality, for example, would be of the form: $A_0 \otimes B_0 + A_0 \otimes B_1 + A_1 \otimes B_0 - A_1 \otimes B_1$, with A_x, B_y dichotomous self-adjoint operators for x/y measurements on Alice and Bob's systems, respectively.

Therefore if measurements on a multi-party quantum system violate a Bell inequality, we know that the system is in an entangled state. (The converse does not in general hold: not all entangled states are such that there is some Bell inequality that they will violate.)

In various circumstances it can be important to know whether the states one is concerned with are entangled or not, and it may be an interesting question how highly entangled they are. (Various measures of degree of entanglement exist, typically guided by respecting the ordering given by the principle that amount of entanglement cannot increase under local operations and classical communication. The singlet state is a maximally entangled state for a pair of two dimensional systems, for example.) Bell inequality violation (and degree of Bell inequality violation) can certainly help us with these tasks. Importantly, assessing in detail the degree of entanglement of quantum states produced in high energy experiments can be an important test for beyond-Standard-Model particles and interactions [55, 56].

6. Conclusions

We may care about Bell inequality violation because we want to show something deep and general about the world, not just about some particular theory: that Local Causality (as mathematically formulated) does not obtain. That no satisfactory theory of the world—no theory which can make the predictions we have observed—can also obey that principle.

We may care about Bell inequality violation because, sticking firmly within quantum theory, we want to show something about some particular quantum systems we are dealing with—that they are entangled, or entangled to a certain degree. This fact might have important further consequences, including for new physics, as is the case in these high energy experiments.

In the context of high energy experiments, we have seen that given the lack of independent access to how x and y values are selected, we have difficulty in maintaining the necessary high degree of conviction that λ-independence should hold. Thus we cannot readily infer failure of Local Causality from Bell inequality violation as things stand. But further investigation, starting from a careful analysis of quantum-to-classical measurement models for high energy particle detection, may well allow us to move to a more general understanding of the measurement processes involved, which might in turn furnish lines of support for λ-independence. Either way, an increasingly detailed understanding of how these high energy experiments can approach the ideal for a Bell test for Local Causality is a highly desirable aim.

Thanks

I thank Alan Barr for my introduction to the topic of high energy Bell tests, for discussion, and for the invitation to the workshop to give the talk on which this chapter is based. I thank Harvey Brown for discussions on Bell matters going back many years now.

Author details

Christopher G. Timpson
Faculty of Philosophy, University of Oxford, UK

*Address all correspondence to: christopher.timpson@bnc.ox.ac.uk

References

[1] Barr AJ. Testing Bell inequalities in Higgs boson decays. Physics Letters B. 2022;**825**:136866

[2] Ashby-Pickering R, Barr AJ, Wierzchucka A. Quantum state tomography, entanglement detection and Bell violation prospects in weak decays of massive particles. JHEP. 2023; **05**:020

[3] Fabbrichesi M, Floreanini R, Gabrielli E, Marzola L. Bell inequality is violated in $b^0 \to j/\psi k^\star(892)^0$ decays. arXiv:hep-th 2305.04982. 2023

[4] Hensen B, Bernien H, Dréau A, et al. Loophole-free Bell inequality violation using electron spins separated by 1.3 kilometres. Nature. 2015;**526**:682-686

[5] Bell JS. On the Einstein-Podolsky-Rosen paradox. Physics. 1964;**1**:195-200 Reprinted in [56]

[6] Gao S, Bell M, editors. Quantum Nonlocality and Reality: 50 Years of Bell's Theorem. Cambridge: Cambridge University Press; 2016

[7] Brunner N, Gühne O, Huber M. Special issue on fifty years of Bell's theorem. Journal of Physics A: Mathematical and Theoretical. 2014; **47**(42):420301

[8] Bell JS. The theory of local beables. Epistemological Letters. 1976;**9**:11-24 Reprinted in [56]

[9] Horodecki R, Horodecki P, Horodecki M, Horodecki K. Quantum entanglement. Reviews of Modern Physics. 2009;**81**:865-942

[10] Schrödinger E. Discussion of probability relations between separated systems. Mathematical Proceedings of the Cambridge Philosophical Society. 1935;**31**(4):555-563

[11] Feynman R, Leighton R, Sands M. Lectures on Physics, volume 1, chapter 37. Reading, Mass: Addison-Wesley; 1964

[12] Peres A. Quantum Theory: Concepts and Methods. Dordrecht: Kluwer; 1993

[13] Seevinck M, Uffink J. Sufficient conditions for three-particle entanglement and their tests in recent experiments. Physical Review A. 2001; **65**:012107

[14] Nielsen M, Chuang I. Quantum Computation and Quantum Information. Cambridge: Cambridge University Press; 2010

[15] Bohr N. Discussion with Einstein on epistemological problems in atomic physics. In: Schilpp PA, editor. Albert Einstein: Philosopher Scientist. La Salle, Illinois: Open Court; 1949. pp. 199-242

[16] Bacciagaluppi G, Valentini A. Quantum Theory at the Crossroads: Reconsidering the 1927 Solvay Conference. Cambridge: Cambridge University Press; 2009

[17] Bohr N. The quantum postulate and the recent development of atomic theory. Nature. 1928;**121**:580-590

[18] Fine A. The Shaky Game: Einstein Realism and the Quantum Theory. Chicago, Illinois: Univeristy of Chicago Press; 1988

[19] Einstein A, Podolsky B, Rosen N. Can quantum mechanical description of physical reality be considered complete? Physical Review. 1935;**47**:777-780

[20] Einstein A. Autobiographical notes. In: Schilpp PA, editor. Albert Einstein: Philosopher-Scientist. La Saller, Illinois: Open Court; 1949. pp. 1-95

[21] Bohr N. Can quantum mechanical description of physical reality be considered complete? Physical Review. 1935;**48**:696-702

[22] Bell JS. Bertlemann's socks. Journal de Physique. 1981;**42**(C2):41-61 Reprinted in [56]

[23] Fine A. Bohr's response to EPR: Criticism and defense. Iyyun: The Jerusalem Philosophical Quarterly. 2007; **56**:31-56

[24] Brown HR, Timpson CG. Bell on Bell's theorem: The changing face of nonlocality. 2016. pp. 91-123. Chapter 6 in [6]

[25] Clauser JF, Horne MA, Shimony A, Holt RA. Proposed experiment to test local hidden-variable theories. Physical Review Letters. 1969;**23**:880-884

[26] Brunner N, Cavalcanti D, Pironio S, Scarani V, Wehner S. Bell nonlocality. Reviews of Modern Physics. 2014;**86**: 419-478

[27] Bell JS. La nouvelle cuisine. In: Between Science and Technology. Amsterdam: Elsevier; 1990 Reprinted in [56]

[28] Aspect A, Dalibard J, Roger G. Experimental test of Bell's inequalities using time-varying analyzers. Physical Review Letters. 1982;**49**:1804-1807

[29] Wallace D. The Emergent Multiverse: Quantum Theory According to the Everett Interpretation. Oxford: Oxford University Press; 2012

[30] Wallace D, Timpson CG. Quantum mechanics on spacetime i: Spacetime state realism. The British Journal for the Philosophy of Science. 2010;**61**(4): 697-727

[31] Henson J. Non-separability does not relieve the problem of Bell's theorem. Foundations of Physics. 2013;**43**: 1008-1038

[32] Healey R. Local causality, probability and explanation. 2016. pp. 172-194. Chapter 11 in [6]

[33] Friederich S. Interpreting Quantum Theory: A Therapeutic Approach. Basingstoke: Palgrave; 2015

[34] Fuchs CA, David Mermin N, Schack R. An introduction to QBism with an application to the locality of quantum mechanics. American Journal of Physics. 2014;**82**(8):749-754

[35] Clauser JF, Shimony A. Bell's theorem: Experimental tests and implications. Reports on Progress in Physics. 1978;**41**(12):1881

[36] David N, Mermin. Quantum mechanics vs local realism near the classical limit: A Bell inequality for spin *s*. Physical Review D. 1980;**22**(2):356-361

[37] Norsen T. Against 'Realism'. Foundations of Physics. 2007;**37**:311-340

[38] Maudlin T. What Bell proved: A reply to Blaylock. American Journal of Physics. 2010;**78**:121-125

[39] Laudisa F. How and when did locality become 'local realism'? A historical and critical analysis (1963–1978). Studies in History and Philosophy of Science. 2023;**97**:44-57

[40] Popescu S, Rohrlich D. Quantum nonlocality as an axiom. Foundations of Physics. 1994;**24**:379-385

[41] Brans CH. Bell's theorem does not eliminate fully causal hidden variables. International Journal of Theoretical Physics. 1988;**27**:219-216

[42] Barrett J, Gisin N. How much measurement independence is needed to demonstrate nonlocality? Physical Review Letters. 2011;**106**:100406

[43] Hall MJW. Local deterministic model of singlet state correlations based on relaxing measurement independence. Physical Review Letters. 2010;**105**: 250404

[44] Pütz G, Rosset D, Barnea TJ, Liang Y-C, Gisin N. Arbitrarily small amount of measurement independence is sufficient to manifest quantum nonlocality. Physical Review Letters. 2014;**113**:190402

[45] Bell JS, Shimony A, Hall M, Clauser J. An exchange on local beables. Dialectica. 1985;**39**(2):85-110 Includes [46]

[46] Bell JS. Free variables and local causality. Epistemological Letters. 1977; **15**:79-84 Reprinted in [56]

[47] Price H. Time's Arrow and Archimedes' Point. Oxford: Oxford University Press; 1996

[48] Evans P, Price H, Wharton K. A new slant on the EPR-Bell experiment. The British Society for the Philosophy of Science. 2013;**64**(2):297-324

[49] Palmer T. Superdeterminism without conspiracy. arxiv:quant-ph 2308.11262. 2023

[50] Hooft GT. The Cellular Automaton Interpretation of Quantum Mechanics. Berlin/Heidelberg: Springer; 2016

[51] Afik Y, Muñoz JR, de Nova. Quantum discord and steering in top quarks at the LHC. Physical Review Letters. 2023;**130**(22):221801

[52] ATLAS Collaboration. Observation of Quantum Entanglement in Top-Quark Pair Production using pp Collisions of $\sqrt{s} = 13$ tev with the ATLAS Detector. 2023. Available from: https://atlas.web.cern.ch/Atlas/GROUPS/PHYSICS/CONFNOTES/ATLAS-CONF-2023-069/

[53] Horodecki M, Horodecki P, Horodecki R. Separability of mixed states: Necessary and sufficient conditions. Physics Letters A. 1996;**223** (1–2):1-8

[54] Terhal BM. Bell inequalities and the separability criterion. Physics Letters A. 2000;**271**(5):319-326

[55] Aoude R, Madge E, Maltoni F, Mantani L. Probing new physics through entanglement in diboson production. arxiv:hep-ph/2307.09675. 2023

[56] Bell JS. Speakable and Unspeakable in Quantum Mechanics. 2nd ed. Cambridge: Cambridge University Press; 2004

Chapter 9

Perspective Chapter: Experiments in Entangled Time

Karin Marie Fierke

Abstract

The purpose of this chapter is to revisit the concept, 'To See is to Break an Entanglement,' through an exploration of insights from a three-year project (2020–2023), 'Mapping the Empire: The Contemporary Legacy of Historical Trauma and Forced Displacement.' The project arose from observations that have no explanation in classical physics, and sought to explore the significance of the 'quantum effects' that underpin the dynamics of a particular form of systems therapy and its potential adaptation to the analysis of global entanglements of past, present and future. The chapter develops insights relating to entanglement, language and consciousness that arose from an 'experiment.'

Keywords: time, affective resonance, entanglement, language, consciousness

1. Introduction

The past few years have seen the publication of several books and special issues about the implications of quantum theory for the social sciences [1–8] many of which ask how particular socio-political phenomena look different when lifted from the assumptions of classical physics and placed in an uncertain quantum world. My co-authored contribution with Nicola Mackay, 'To See is to Break an Entanglement' [9], reflected on an 'experiment' that at the time was at a very early stage. Just after the article was published, we received funding to do exploratory research on a project titled 'Mapping the Empire: The Contemporary Legacy of Historical Trauma and Forced Displacement' (MtE).

The concept 'to see is to break an entanglement' is significant against the backdrop of the award of the 2022 Nobel Prize in Physics to John Clauser, Alain Aspect and Aton Zellinger for experiments with entangled photons. Within quantum mechanics, nature is not locally real, that is, particles lack properties such as spin up or spin down prior to measurement, which corresponds with observation. They also communicate and respond to one another regardless of the distance. Entanglement is thus a non-local phenomenon. The prize winners were said to have 'proved' that the Universe is not locally real, on the one hand; on the other hand, as noted by Charles Bennett, a quantum researcher at IBM, 'The experiments...show that this stuff isn't just philosophical, it's real – and like other real things, potentially useful' [10]. The contrast suggests that quantum entanglement is real, but not in the way that 'real' is usually understood. As Niels Bohr, the father of the Copenhagen School, noted, 'The word "reality" is a word, a word which we must learn to use correctly.' [11].

The reality/philosophy question has haunted discussions of quantum physics [12]. The quantum social theory debate has been mired in the binary either/or of whether, when applied to the social sciences, the quantum is real or mere analogy, metaphor or metaphysics ([5], p. 44). The application of insights derived from the analysis of microscopic phenomenon to macroscopic phenomena seems to pose a question of how to scale up from one level of life, that is, the subatomic, which is the pre-occupation of quantum physicists, to another, that is, human social and political life, which is the focus of social scientists. In this respect, the language of quanta, the smallest amount, contributes to the problem, as do claims that quantum effects wash out above the macroscopic level. The MtE project is unique within the social science debate in so far as it grew out of an observation that has no explanation in classical physics.

The 'quantum effects' in this case arose from the dynamics of a particular form of systems therapy, which has been widely used with families and organizations. The MtE experiment, which took place over a two and a half year period, sought to adapt the method to the analysis of more global phenomena. Material entanglements [1] and entanglements in language and social structure [2] have had a central place in the quantum social theory debate. The affective resonances that arose within a mapping process, discussed below, can only be explained in terms of non-local micro-scopic entanglements, although also relevant to both materiality and language. The study thus provides a potential bridge between discussions of quantum entanglement in the natural and social sciences.

The current race to develop quantum computing and artificial intelligence has included attempts to create emotional robots [13]. The potential simulation of human emotions in machines at the very least implies that similar properties of quantum entanglement can be found within human and other life, yet, until recently, there has been resistance to the study of consciousness, including emotions, in both the natural and social sciences. The question raised by 'to see is to break an entanglement' is whether in seeing non-local entanglements between past and present, spaces might be opened for turning toward the loss and 'ungrieved grief' of centuries of empire and war and whether an act of 'seeing,' as a measurement, breaks an entanglement between past and present.'

In what follows, section two brings a problem of seeing and not seeing to an 'experiment,' exploring its broad contours. Section three examines a concept of affective resonance as a diffracted experience within the experiment. Section four illustrates several different cuts into the analysis of the collapse of affective resonances into language within a process of blind mapping. Section five then turns to a discussion of consciousness. The conclusions highlight provisional insights from the exploratory research and further avenues for potential interdisciplinary and practical development.

2. The puzzle and experiment

Several years ago, I stumbled onto a form of family systems therapy, the underlying dynamics of which have no explanation in classical physics. A central principle of systems therapy more generally is that how and where people stand—whether facing each other, or turning their backs, or alone in a corner embodies a relationship between family members. The particular approach I encountered had a further dimension. The family constellations modality, developed by a German, Bert Hellinger in the 1960s,[1] has been used widely in the treatment of transgenerational trauma in Germany, relating in particular to the experience of World War II, and has since spread globally. Rather than working with actual family members, Hellinger brought in complete strangers who knew nothing about one another or the family of the person who was the focus of any particular constellation. The task of these volunteers was to stand in or represent a place within a system [16].

Standing in or representing a place is a form of second party experience that can be distinguished from either first party (subjective) or third party (objective) experience (see [17]). 'Representatives' for different family members are able, through speaking and moving within a created system, to access the feelings and dynamic relationships of the client [18], who is seeking to resolve an issue.[2] Through a facilitated process, the problem is disentangled and family members who were previously excluded are given a place, which 'releases' the problem and restores balance to the system [19].

The title of Hellinger's seminal book, *Acknowledging What Is* [20], suggests that acknowledging 'what is,' even when it is staring us in the face, may be less than straightforward. In this respect, the central dynamic revolves around a question of what can be seen and what is unseen. The potential that individuals could stand within a created system and experience traces of affect relating to someone they do not know and have no knowledge of, not least because they belong to an earlier generation, only begins to make sense in the context of some notion of non-locality, quantum field, and transgenerational entanglement. If family systems are entangled non-locally, the assumed ontological separateness of individuals is an illusion, and a question arises about the existence of entanglements within other domains of experience.

The MtE project sought to explore whether the modality could be adapted to the analysis of entanglements of historical experience and contemporary memory and practice of a more social and political nature, particularly as these relate to a global legacy of historical trauma and forced displacement. What is meant by entanglement as used here is discussed in the next section. Our focus was the legacy of the past in the present, and how entanglements of memory and habit impact on what we are able to look at and what we turn away from. The Covid pandemic provided an unusual opportunity to explore the problem. The pandemic saw an unfolding field of trauma that impacted on populations globally which amplified a range of historical memories, and not least that of the Spanish Flu Pandemic (1918–1920). The run-up to Brexit was haunted by memories of British empire. In May of 2020 George Floyd's murder at the hands of a white policeman in Minneapolis gave rise to a greater ability 'to see' systemic racism, which is a legacy of enslavement, and to challenge it. By early 2022, Western attention shifted away from questions of racial inequality as Russia invaded Ukraine. Different memories of experience during World War II or the Cold War framed this context, which also revealed evidence of a struggle over who would lead a future post-pandemic order, magnified by memories of various historical empires. A

---

[1] Hellinger's method was influenced by other family systems methods, such as that of Virginia Satir; German phenomenology; and observations he made of Zulu healing practices in South Africa where he served as a priest for 16 years before becoming a psychotherapist (see, e.g. [14]). Mackay was trained by Hellinger and has 20 years of experience with the modality, as well as writing about it as a practitioner [15].

[2] For an introduction to family constellations, which includes live sessions of Hellinger working with representatives, see: https://www.youtube.com/watch?v=wZ1oYiR5VpM.

more recent link between a history of colonialism and a future of climate change has highlighted the relationship between past, present and future.

The methodological objective of the exploratory research was to examine the potential to adapt a particular form of systems analysis from its use within family or organizational systems to the analysis of global entanglements of past, present and future. The substantive objective of the project was to explore the legacy of historical trauma and forced displacement in the present.

2.1 A quantum experiment?

The project developed two complementary methods of analysis, both of which evolved over the 3 years as we brought insights from each to the other. A method of *tracking* involved research assistants in the exploration of the relationship between contemporary trending events, including the memories expressed by political agents within them; historical experiences of trauma, particularly in the context of various empires; and the multiplicity of future potentials at work in relation to the events in question. The method evolved over the 3 years from a more linear and local application which was gradually replaced by a non-linear approach to the analysis of global entanglements of empire. The tracking looked at empirical contexts, while developing theoretical insights about how a legacy of the past impacts on events in the present and shapes probabilities for the future.[3]

A complementary method of non-linear *mapping*, which is the focus of this chapter, fed off of the data developed by the trackers.[4] Constructing a map begins with an intentional question relevant to both contemporary events and the existing historical record. An advantage of the more socio-political application of the systems method, over the family application, is the easier access to accounts of the past. The obvious problem, which is increasingly recognized by historians [21], regards the extent to which historical narratives have been impacted by, for instance, dominant national stories, which in the context of empire, facilitated an inability to see the more negative consequences. So, for instance, national histories of empire, such as the British, have highlighted the positive role of Britain's past global presence.

A growing literature has argued that this positive take on empire has made it difficult to see or acknowledge the suffering imposed on colonized populations [21–26]. While existing historical narratives might not provide a complete picture, the unseen elements of a system become more evident in the process of the mapping itself. The historical experiences explored by the trackers revealed material practices of horrific violence. The affective resonances of these experiences were the subject matter of the maps. The conversation within an active map makes it possible to observe how different parts of a system stand in relation to each other, who or what is seen or unseen, the affective dynamics of the system, how the separate parts triangulate (to make certain other parts unseen), as well as obstacles that stand in the way of acknowledging the unseen and giving them a place.

What does this have to do with quantum entanglement and why call it an experiment? First, as already suggested, the initial puzzle arose from an experience of *quantum effects*. Quantum effects are said to wash out at the macroscopic level due to quantum decoherence, by which a system's behavior changes from that which

[3] The Research Assistant trackers hailed from diverse global locations, including India, China, Vietnam, Egypt, South Korea, Nigeria, Germany and Scotland.

[4] We refer to the method as *Dynamic Entangled Memory Mapping* [9].

can be explained by quantum mechanics to that which can be explained by classical mechanics. The question is whether the effects disappear altogether or are simply unobservable in daily life. The experiment provided a modality and an apparatus for observing an otherwise unobservable and non-local phenomena over a sustained period of time.[5] The 'quantum effects' would seem to point to the superposition of entangled states that exist in physically separated locations, past and present, which in the process of the mapping are 'measured' as they collapse into language.

Second, an entangled *complementarity* is evident in several aspects of the experiment, starting with the engagement of two complementary methods, one more oriented to the observable material world, and the other to unobservable entanglements of consciousness. Further, the emphasis on presence and non-presence, seen and unseen, resonates with Bohr's claim that complementary opposites are both mutually exclusive and intertwined, which makes it difficult to observe two parts of an entangled pair simultaneously. Presence and non-presence is an entangled relationship which means any account that deals only with the directly observable will be incomplete.

Third, the mapping process involved analysis of the affect, bodily movements and/or language of the 'mappers,'[6] or those who represented positions within a given map. To insure that any patterns emerging from the process could be understood to arise from a system or field rather than individual interpretation, the experiment was undertaken *blind*. Consistent with the definition of a blind experiment, information that would influence the participants in the mapping was withheld until any one session was completed.

An *experiential* method is difficult to grasp without having had the experience. Readers may be inclined to assume that the words or gestures of the mappers express their subjective consciousness as individuals, but this is *not* the case. Instead, in stepping into a created system, the mappers become temporarily entangled with a category of socio-political experience which they represent and speak from. Recognizing that most people have traumas in their family histories, arising from larger social and political phenomena, the ethical criteria emphasized the importance that the mappers had priori training, including analysis of their own family entanglements, in order to minimize the potential to be overly triggered by stepping into a more global map.[7] In this respect, the individual, as part of a holographic world, does not disappear altogether, but the blindness of the process minimized the role of individual interpretation, thereby maximizing the potential to analyze the extent to which the collapse of affective resonance into language within the maps arose from an holographic field. During the mapping process, the individual participants knew neither the context,

---

[5] The blind group mapping took place once a month over a two and a half year period. An online application of the method made it possible to continue with the group work during the pandemic. Word-for-word transcripts were made from recordings of the sessions. Due to the terms of participants consent, the data underpinning this research cannot be made publicly available.

[6] Use of the term 'mappers,' rather than 'representatives,' as is common with family constellations, was an attempt to avoid a potential misreading through Barad's [1] quantum critique of representation.

[7] Prior experience with and training in the family constellation method was identified as an ethical requirement for participation in the mapping. This background also meant they had some idea of what to expect, given the palpable energy circulating through any one map. In the rare case that a mapper experienced distress, a designated therapist provided an individual post-mapping session to further separate individuals from their experience within the map.

the question, nor the category they represented, which were only introduced to the transcripts after a session was completed.

In setting up a map, the focus of family constellations on individuals within a family system was replaced by categories of memory (e.g. the Siege of Leningrad, the U.S. Civil War, World War II, the Opium Wars), of experience (e.g. forced displacement, hunger and scarcity, war, enslavement, genocide), of emotion (e.g. grief or ungrieved grief, betrayal, guilt, shame, compassion), of collective identity (e.g. particular states, present populations, children), of happenings (e.g. Brexit, climate change, invasions) or belief (e.g. profit as belief, safety, collective agreements not to know), which are derived from an intentional question, the formulation of which drew on the contemporary and historical data compiled by the trackers. At face value, there is no reason to believe that the blind engagement of what on the surface appear to be abstract categories would be anything except a chaotic jumble of words. The surprising and significant observation regards the extent to which, as illustrated in Section 4, patterns that emerged from the maps revealed the broad contours of the topic in question, as well as unseen aspects.

The emergence of patterns from a blind conversation is difficult to grasp and easily conflated with words that evoke negative connotations in Western culture, related, for instance, to parapsychology or talking to the dead. The created maps revolved around categories of experience and memories related to *socio-political phenomena*, rather than individuals, whether alive or dead, and as such express something qualitatively different, i.e. affective resonances or traces of past experience that remain alive, and are threaded through socio-political experience in the present. In this respect, as the physicist/social theorist Karen Barad notes, memory is 'written into the fabric of the world' ([27], p. 260) rather than being purely a property of individual mind. In the next section, I set the stage by thinking about the entanglement of the present with affective resonances of the past.

3. Entanglements of trauma

There is a growing literature on artificial intelligence and the potential that robots, i.e. machines, might be endowed with properties of emotion and consciousness, yet we lack sufficient understanding of how the assumptions of quantum entanglement, which makes such an endeavor conceivable, works in sentient life. Much of the research relating to the latter has focused on the human brain and subjective consciousness. A focus on machines, individuals and the brain would seem to resonate more with the assumptions of classical Newtonian physics, which has sidelined the sentient aspects of life. Quantum mechanics is said to provide a better point of departure for taking emotions seriously. In a discussion of quantum decision theory, Alexander Wendt ([2], p. 167) makes the point that emotions do not follow a classical logic; neuroscientific studies have confirmed that reason and emotion are deeply intertwined in decision making.

The compatibility is not only evident in regard to the reason/emotion relationship. Emotions are *indeterminate*, often changing as frequently as the weather in Scotland. The indeterminism of quantum mechanics highlights randomness and chance, and multiple possible outcomes and thus the importance of probability rather than singular cause. Emotions are *complementary*, often existing in oppositional pairs, which cannot simultaneously be seen and thus express a relationship of presence and non-presence. Hate can mask love. The attribution of blame can mask guilt. Grief may be masked by anger. The manifestation of any one will be subject to a measurement by participant observers. Emotions are also both individual and *relational*—they are experienced or expressed in relation to others, and are heavily shaped by histories, language, memory and habit, particularly as one moves toward their circulation in a global space. Global relations, often forged through war and practices of mutual harm, of betrayal, humiliation, and fear are themselves entanglements, despite the apparent fragmentation. For example, the anthropologist Gregory Bateson has noted that the humiliation of Germany at Versailles (1919) set the stage for all that has happened since. His main point was to highlight that emotional patterns matter [28]. 'Hurt' or trauma, particularly when it is silenced and hidden away, can be radioactive in its persistence. While positive emotions of compassion or love involve 'seeing' an other, the 'ungrieved grief' that accompanies trauma often gives rise to a tendency to turn away and to not see or know, as the emotions, whether guilt, shame or grief, may be too difficult for all parties involved to sit with [29]. The silence that so often surrounds historical traumas is part of the 'radioactivity,' i.e. persistence over generations, of *non-local entanglements* of past and present. The purpose of this section is to briefly set the stage by exploring a notion of non-local affective entanglements with the past, which maintain a resonance in the present, i.e. the 'quantum effects.' The subject matter connects to diverse interdisciplinary literatures, but, as space does not allow for a comprehensive review, the objective is to make sense of the affective dynamics of the mapping process.

3.1 Time and affective resonance

'Seeing' and 'not seeing' may be a spatial relationship but may also be temporal, non-local and non-linear. Linear constructions of time draw lines between past, present and future. Past 'events' may be known factually, but are separated out and relegated to a bygone era where they exist in a linear relationship to the present as something that no longer is. Theoretical mathematicians, physicists and cosmologists have argued that linear notions of time as flowing forward, separated neatly into past, present and future, rest on an illusion [30]. In discussing superposition and entanglement, which she identifies with diffraction, Barad argues that 'different times bleed through one another' [31].

The literature on socio-political memory includes claims that contemporary populations use memory for purposes of the present (e.g [32]) or in the realization of future projects (e.g. [33]). While not denying the importance of the observation, the mapping sought to explore non-local entanglements between historical *experiences* of trauma by segments of populations, and performances of *memory* in contemporary socio-political contexts, a distinction that emerged from the development of the tracking method. While important for analytical purposes, the distinction disguises the extent to which memory arises from an accumulation of past experience, which in its continuous performance, sediments aspects of the past, thereby thickening memory and habits relating to it. The experience/memory distinction, relates to a further distinction between *affect*, which usually refers to an experience of sentience, and the articulation of *emotion* in the political performance of memory and in language, which are often also entangled. The following unpacks these distinctions to make some sense of the 'bleeding' of socio-political trauma across time within the 'experiment.'

'To see is to break an entanglement' rested on a claim that not only individual memories of trauma, but also social memories are entangled with both sensory imprints of past trauma and language [9]. Commenting on a century of study of traumatic memories, the psychiatrist Bessel van der Kolk [34] argued that semantic memories may coexist with sensory imprints, which often remain stable over time, but may return when triggered by reminders. Sensory imprints of trauma appear as fragments, rather than narratives, and are expressed in images, physical sensations and intense emotions [35]. Traumatic memories are also more persistent in their recurring presence than 'normal' everyday memory; they intrude without welcome and often in response to triggers in the environment that resemble an earlier trauma. In cases of complex trauma, woven over extended periods of time, the intrusions become part and parcel of a relational world, structured around types of habitual experience and the use of language [36].

Bateson's point about patterns of hurt suggests that sensory imprints are not exclusive to individuals but infuse experiences of traumatic memory that are more socio-political. Given the association between sensory imprints and the physical body, a concept of *affective resonance* is more useful for understanding a sentient experience that crosses over and through individual bodies, in which past and present bleed through each other. Within the context of the maps, the words, facial expressions or bodily movements of the participants expressed traces or fragments of traumatic historical experience, although, it should be emphasized, without re-experiencing the historical trauma itself. The affective resonance did not arise from personal experience, but rather expressed a diffracted resonance of past social experience in the present. Diffraction in its basic form is the bending of light around an obstacle. The two-slit experiment at the heart of quantum physics used a diffraction grating, to reveal the entanglement of particles and waves, and the importance of the observer and their apparatus for what is seen. Diffraction patterns highlight the presence of waves in matter, which is less about where difference appears than mapping where the effects of difference appear [37].

The 2020 article made a further claim that 'articulations within the context of a created map express a form of wave-function collapse and a pattern of diffracted entanglement by which the attributes of a system become visible or "seen" as the vibrational frequencies surrounding a particular represented space, and the affect that arises from it, are transformed into language and thus become available for analysis' [9]. Within the spacetime of the blind maps, the affective resonances experienced by the mappers might be said to work on the same principle as the forced-resonance tuning fork experiment, in which a vibrating tuning fork forces a stationary tuning fork into resonant vibration when placed beside it [38]. In stepping into a map, the participants come into resonance with the diffracted interference of the category of experience that they represent, where the boundaries of linear time are erased; as they speak from a position of blindness, the affective resonance is 'measured' in the collapse into language.

The experiment also revealed the greater complexity of multiple differently positioned mappers accessing past entangled states blindly. A cloud, which stores the memories of computer systems, might provide an analogy. The entangled mappers resonate with information stored in a cloud-like space—a sort of supercomputer in the cosmos—which is an interesting contrast to the efforts of physicists to use supercomputers to simulate the universe,[8] a contrast that suggests the former may not be that far-fetched. The theoretical physicist David Bohm [39] refers to an implicate order, a seamless quantum field, of all that has been, with all potentials, habits and memories recorded in spacetime. Bohm contrasts the wholeness of this implicate order with the fragmentation of the macroscopic explicate order, with which it is

[8] Examples include NASA's Pleiades supercomputer or Japan's ATERUI II.

entangled. The mapping provides a means to observe the relationship between the potentials within this larger quantum field and the active presence of fragmented entanglements within the world.

In so far as the diffracted affect relates to past experience, whether dispossession, famine, scarcity or enslavement, which is larger than individuals, the resonances arise from a *field* of experience which may or may not be contiguous with the 'collective memory' of a state or culture. The basic assumption of a system or field, while taking diverse forms in the literature, is that the whole coordinates the relationship between parts. In the context of the maps, patterns of affect emerge from the whole rather than the intentions or consciousness of the individuals involved. The biologist Rupert Sheldrake suggests that what he refers to as 'morphic resonance' is the past 'pressed up' against the present, which is potentially present everywhere ([40], p. 161). The resonance arises from past patterns of activity that influence the fields of subsequent and similar systems, which involves a kind of action at a distance in both space and time. Morphogenetic fields are probability structures within which the most common past types, sedimented in memory and habit, combine to increase the probability of a recurrence of that type ([40], p. 157).

At stake is a presence of the past expressed through memory and habit. Epigenetics suggests a potential that traumas from earlier generations may reappear in the bodies of successor generations, often triggered by environmental conditions that resemble an earlier trauma [41–44].[9] Transgenerational trauma has become a widespread concern, particularly in relation to entanglements of race in the U.S. [45]; or as evidenced in a PBS documentary about Pike Island, a sacred Native American space at the meeting point of the Minnesota and Mississippi Rivers, which was transformed into a concentration camp in the aftermath of the U.S.-Dakota War of 1862 [46]. Ruth Davern, the granddaughter of a previous owner of the Island, explores the roots of her family, who fled persecution in Ireland at the hands of the British, only to participate in similar processes of exploitation by European settlers in the U.S.

It is not that work on the more social and political dimensions of transgenerational trauma is new; the problem has been a tendency to turn away from it. Gabriella Schwab discusses the engendering of silence in relation to the transgenerational trauma passed on to German children by those who perpetrated or accommodated the horrors of the Nazi regime [47]. Ngugi wa Thong'o refers to a similar problem of transgenerational memory in post-colonial Africa, in relation to the millions of unmourned victims of a system of enslavement and colonialism [48]. How grief is processed or not has implications for how further relations unfold [49], and whether traumatic entanglements are reinforced or seen and broken.

A concept of affective resonance points to something like a sensory imprint, not in the individual body but within the world as a living organism, bristling with hurt accumulated through repeated practices of harm, also to nature, and expressed in non-local affective resonances, which can be more clearly seen within the created system of a map. The key question raised by 'to see is to break an entanglement' is whether in seeing these entanglements, a space might be opened for turning toward the loss and 'ungrieved grief' of centuries of empire and war and whether an act of 'seeing,' as a measurement, breaks an entanglement between past and present.

[9] Epigenetics shifts attention to the influence of environment on bodies and specifically on the expression or suppression of genes. Over the past decade, the interface between environment and biology has increasingly been used to understand the intergenerational transmission of trauma.

4. Language

The analysis of affect, resonance and field in the previous section raises a difficult question about how we know what we know and why it is important to know. What happens in the collapse of the affective resonances, within the context of a given map, into language? What happens in the shift from a more classical understanding of language as representing, mirroring or corresponding to 'things' in the world to its role in the constitution of non-local relationships between entangled *phenomena*? To examine a phenomenon is to ask 'what it is like' [50, 51] to be in a state of X, in relation to other phenomena, which is to move from an exploration of essence to experience and meaning within a particular entangled moment that is unfolding between past, present and future. The question is what we can know from the process of mapping, or how to make sense of the relationship between an intentional question, as an apparatus, and the words expressed by the blind mappers in the unfolding of a particular map.

There is little agreement among physicists about the nature of 'reality,' and its relationship to language, including mathematical language, and, for instance, whether reality *is* a mathematical structure ([52], italics added), or the world is physical but mathematics is a way of representing it [53]; or that interpretation and storytelling are prior to mathematical formalism ([54], pp. 5, 17). Bohr's claim that we are 'suspended in language' [11] suggests that every aspect of human communication involves the use of language, which in the case of physics extends to mathematical tools, but is not exclusive to them ([55], p. 2). The physicist Stig Stenholme identifies a family resemblance between Bohr's thought on language and that of the philosopher Ludwig Wittgenstein [55]. From this perspective, the scientific apparatus, as the unit of description, functions in much the same way as everyday language, playing a role in the constitution of limited systems rather mirroring universal ones. As such, the relationship between language and reality is both/and rather than either/or. Mathematics captures something real, but, in so far as this is 'useful' [10] also modifies that reality, e.g. in the form of MRI's, AI or quantum computing.

The French philosopher Michael Bitbol explores the relevance of Wittgenstein's philosophy for physics, starting with a distinction between factual propositions and mathematical propositions [56]. He identifies the latter with the performative use of particular rules, norms or signs, the meaning of which is conferred by their relation to phenomena. Rather than mirroring isolated objects or forms, whether empirical or platonic, respectively, the mathematical language invokes a 'system of propositions' that is constitutive of phenomena [57]. The use of language is a measurement that brings into being a realm of possibility, which does not represent *a* reality, but rather that allows one to act or speak *as if* such a correspondence has been established ([56], p. 192). In the physics laboratory, the apparatus produces the conditions that the scientist, who is entangled with it, seeks to explore, including obstacles, which are then read through the semantic grid that is imposed by the project ([56], p. 193). A similar process, I would argue, happens within the experimental conditions of the mapping.

The intentional question of a particular map, like, the semantic grid of the scientist, positions the investigation of a phenomenon. As the apparatus, the question is entangled with the facilitator and other participants, even while the latter neither know what they represent or the context with which they engage. Nor do they control what emerges from the field or system. The dynamic captured within any one map cannot be said to represent an independent 'reality' (the worlds 'as it is') but provides the broad contours of a positional phenomenon in space and time, which reveals unseen aspects (of 'what is') in the world.

The following three illustrations make distinct cuts into the analysis of the non-local entanglements arising from the maps. The first explores the *replication* of an unfolding system of propositions relating to World War II memory, based on the collapse of affective resonances into language within successive maps over the period of a year in the lead up to the Russian invasion of Ukraine in 2022. The second provides a *visualization* of a non-local field expressed within a particular map, and discusses its relationship to the material context at a particular moment in time. The third illustration provides a *narrative construction* of a map from a single day, which emerged from stringing together the fragments of the blind conversation. The memory of World War II is only a single category of experience within a vast amount of data, which due to the terms of participants consent, cannot be made publicly available. The analysis provides a glimpse of the working of memory both within specific maps and as it unfolded across the maps and over time, and what this potentially says about the relationship between the physicality of the non-local affective resonances, language and materiality.

While material and physical are often conflated, a distinction can be made between the former as 'made of matter' and the latter as having a 'real existence' that can be measured and represented mathematically, an example being an electromagnetic field that is physical but not material [58]. In the present analysis, a similar distinction might be made between the present material reality of war and the physical resonances of memory, which in our case are measured through the blind mapping rather than mathematically.

Where do the textual analyses and visualizations below come from? In setting up any one map, a letter was assigned to the participants, which corresponded with the category of experience they represented. After the exercise, and prior to analysis, the letters were replaced with the categories of memory and experience within the map. Each of the transcripts was then processed, line by line, to clarify the relationship between the represented space and the words spoken by the mappers. Visual maps of each session were also created using a Miro tool. The illustrations provide three different cuts into the blind maps, which provide a glimpse of what can be known through the seeing of the otherwise unobservable relationship between the physical and non-local affective resonances and the contours of an unfolding material context.

4.1 The unfolding phenomenon of World War II memory

The following analysis, drawn from a series of maps, structured around different intentional questions at specific points in time over a year, reinforces two insights in particular. First, as suggested in the last section, in stepping into a map, the mappers become entangled with non-local affective resonances surrounding a particular category, of memory in this case, which has been assigned to them in advance, but is unknown to them while in the maps. Second, what collapses into language is less as an essence of the 'event' World War II, than a phenomenon, surrounded by an affective resonance, which is active in the unfolding of a context, prior to, during and in the aftermath of the invasion. Given that the context is not static, but continuously changing, replication does not take the form of seeing the same result, given that any one map is a limited system, revolving around an intentional question at a specific moment in time. The replication is evident in the expression of potentials that can be recognized as 'what it is like' to be World War II. For instance, in the example below, the identification of World War II memory with war (1/27/22), rather than diplomacy, or with revenge and humiliation, rather than compassion, is consistent with knowledge of the phenomenon, which would not be the case if the contrasts appeared in reverse. The memory of World War II has a sense and a use, which is neither fixed nor arbitrary.

The table below illustrates fragments of 'what it is like' to be World War II memory that are replicated *across successive blind maps* in the year preceding and month following the Russian invasion of Ukraine. Each item in this brief sampling of memory fragments is preceded by the date of the map and the intentional question that was its point of departure and context (**Table 1**).

The table illustrates an unfolding system of propositions that is consistent with our knowledge of the phenomenon of World War II memory, the affective resonance

Feb25,21 (*What historical memories are animating the aftermath of the British withdrawal from the EU against the backdrop of pandemic?*): The memory of World War II is wary, particularly of emotions and what they might do. The memory holds hidden agreements [59].
Mar25,21 (*What needs to be acknowledged in order to take the aggression out of the relationship between the EU and the Westminster government following Brexit?*): A dynamic potential for the re-emergence of World War has been unleashed, with the breakdown of old agreements for peace in Europe. The grief surrounding the memory of World War II needs to be acknowledged, but it remains hard to do so. The memory of World War II animates everything in the map [60].
June24,21 (*How is memory shaping the structure of the post-pandemic agreements?*): Russia wants to play a game with the memories of World War II...The memory of World War II is full of potentials but it is not yet clear which potential the post-pandemic agreement will activate [61].
Jan27,22 (*What needs to be seen in order to neutralise the threat of a Russian invasion of Ukraine?*): The memory of World War II doesn't see the usefulness of diplomacy and wants a place for war. There is a strong threat of Russian invasion. The memory of World War II is being dressed up, but it is unclear whether it will play a role in battle or provide a framework for academic analysis of the conflict. The memory is in any case being prepared for an entrance on the stage. Bringing in World War II will start a fire. From the perspective of World War II, multiple memories and potentials inform the context...World War II memories are animated by revenge and humiliation, which are becoming stronger and more dangerous...The memory of World War II invokes a predictable recurrence of the past, which arises from a particular way of thinking in the world that is lacking in compassion [62].
Feb24,22 (*What would support and facilitate a shift into seeing a collective agreement for peace?*): War is happening in Europe again...The past and present political environment has been organized around a heavily militarised aggressive game, and, while it would be desirable to change this, there is a lack of agency and a powerlessness...Seeing a more global safety and a future for children requires looking at memories of war. Past sacrifice and memories of war are being used for political purposes rather than being acknowledged and grieved [63].
Mar24,22 (*Does witness to and grieving memories of WWI and WWII support a humanitarian response in Ukraine?*): The memory of Stalin connects to the memory of the Siege of Leningrad and the Nazi invasion, which have a solid place and resonate with the larger map. NATO is on the periphery, watching but not reacting. It is unclear how the context will unfold, but Putin is like a gamemaster, not necessarily in charge but pulling the cards and making up the story. The memory of the Siege of Leningrad is within him, making him feel very big[*]. The memory of the Nazi Invasion, which impacts on the sovereignty of land, provides the distorted lens through which Putin sees others. For NATO, the memory of the Nazi invasion is only a single part of a larger whole, relating to other memories of atrocity during World War II. A numbness of emotions is pervasive and everything is focused on the Nazi invasion, which seems to provide cohesion but is fragmented and experienced differently from different angles, with no common thread. Unacknowledged guilt underpins the memory of the Nazi invasion. Introducing guilt into the map connects the memory of Versailles to the memory of Nazi invasion [64].
**Putin, who is from Leningrad, hadn't yet been born at the time but his brother died in the siege.*

Table 1.
Memories of World War II: Sampling across successive maps from February 2021 to March 2022.

surrounding it, and its working in the world. In successive maps, the 'hidden agreements' (1/21) reappear as the 'breakdown of old agreements for peace in Europe' (2/21), which, in both maps is linked to Brexit, followed later by the potentials for a new post-pandemic agreement (6/21). Over time, further entailments of World War II memory unfold. The memory 'wants a place for war' rather than diplomacy and is 'being dressed up' for entrance on the stage (1/22). Specific memories (3/22) arise from multiple positions (6/21, 1/22) but also empower Putin (3/22), who initially wants 'to play a game with the memories' (6/21) and after the invasion becomes 'like a gamemaster' (3/22). The memory is infused with emotions, from grief, which needs to be acknowledged (3/21), to revenge and humiliation, which animates the memory (1/22), to numbness (3/22), to unacknowledged guilt and a lack of compassion (3/22). The memory, which invokes a 'predictable recurrence of the past' (1/22), is used strategically for 'political purposes' (2/22) and provides cohesion while remaining fragmented at the level of experience (3/22).

The system of propositions within the maps overlap with the empirical context and as such captures the broad contours of the latter. For instance, Putin did put the memory to use in justifying the defensive nature of his 'special military operation,' pointing to 'Nazi's' in Ukraine. NATO, consistent with its position in the early days of the war, stands at the sidelines without responding to the invasion. The clustering of memories of Stalin, the Siege of Leningrad, and Nazi invasion reflects an historical relationship. The memory of Versailles, which the French President Macron brought in at an early stage of the conflict, received little support and faded from use; as the map from 24 March 2022 (above) suggests, not seeing the memory of Versailles has implications, which I return to below.

4.2 Analysis of single maps

Two types of analysis were brought to the individual maps, as illustrated in the following examples. The first provides a *visualization* of the represented spaces and how the 'blind' mappers engaged within a specific map a month before the Russian invasion. The eyes on the symbols point either toward or away from other parts within the whole, which indicates what each can see or not see. As already stated, prior to the exercise, the mappers were assigned a letter that corresponded with the category of experience they represented. After the exercise, the letters were replaced with the categories presented in the illustration below. In the brief summary that follows, each of the positions is placed in quotation marks to indicate that it is a space within the map. Again, as it is so difficult to comprehend, it is worth reinforcing the point, which is relevant to all three illustrations, that the blindness of the process meant that the mappers had no knowledge of the context or the particular categories that they represented within the map (**Figure 1**).

As the map unfolded, 'Ukraine' was apprehensive, stating it was looking for an ally, but did not think it had one. 'Russia in the present' was aware of the memory of the 'Versailles Treaty 1919,' which suggests the potential influence of the memory on Russia. None of the other categories were looking at the memory of the 'Versailles Treaty,' which is consistent with the larger context of the invasion. Memories of 'World War II' and 'Cold War,' which are clustered at this stage and aware of each other, are the greater influence on the map as a whole. 'NATO' is aware of the 'threat of Russian Invasion,' which is angry, and of the 'threat of nuclear war,' and states that it is both in danger and dangerous. The 'threat of nuclear war,' which is aggravated, could see an affinity between the 'memory of

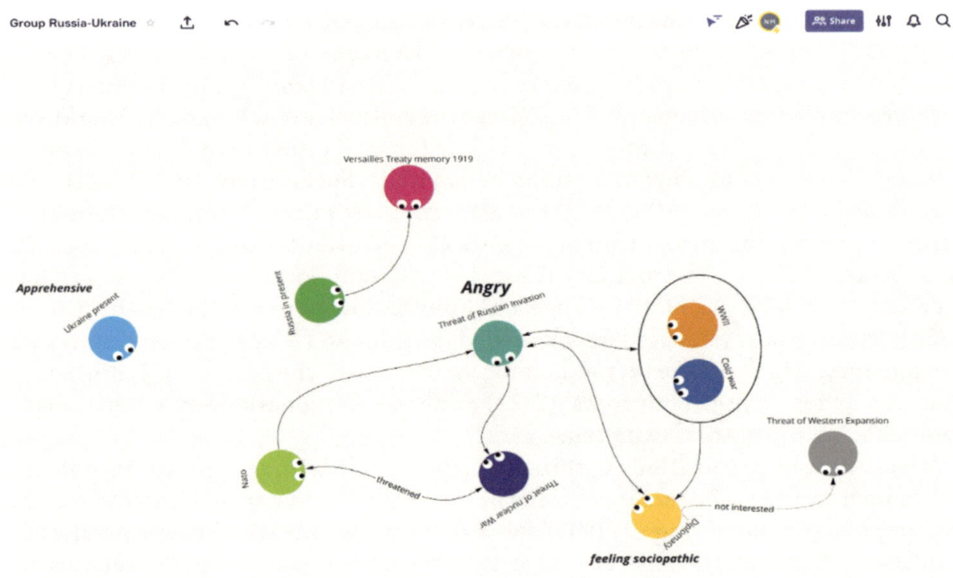

Figure 1.
27Jan 2022: What needs to be seen in order to neutralize the threat of a Russian invasion of Ukraine? [65].

World War II' and the 'threat of Russian invasion,' which are both experiences of invasion. 'Diplomacy' wasn't interested in the 'threat of Western Expansion' but can see it, which is interesting in light of NATO arguments that minimized the significance of its Western expansion [66]. 'Diplomacy' describes itself as feeling *'sociopathic'* (lacking in empathy, anti-social) which is compatible with the failure of diplomacy in the run-up to the invasion.

This is just a brief snippet from an early stage of a session that continued for an hour and a half. As the mapping progressed, and further objects, representing other categories were brought in, the relationship between the different spaces further unfolded. Among these, memory of the Ukrainian famine (Holodomor, 1932–1933) and the Siege of Leningrad (1941–1944) shifted the lens bringing attention to the role of hunger and scarcity as well as profit (during the famine, wheat was Stalin's 'Gold'). The Siege had a solid place within the map and resonated with it. It also made Putin feel 'big,' which is interesting in light of his family history and the resemblance between Hitler's tactics in Leningrad and Putin's in Mariupol and elsewhere. Consistent with the economic sanctions against Russia and Putin's later attempts to obstruct grain shipments from Ukraine, which raised concerns about global food supplies and famine, diplomacy played into food scarcity. The introduction of guilt formed a connection between memories of Versailles and Nazi invasion, which points to the role of punishing sanctions on Germany in paving the ways for Hitler's rise and World War II. NATO wasn't seeing either aspect of the context, but, based on the map, needed to. The larger trans-map analysis of World War II memory above shows a repeated inability to see an earlier legacy of memory, against the dominance of World War II memory. The pattern was repeated in a later map, on 24 March.

The analysis of the unseen memories in the last paragraph highlights several points. First, not only the memories that are directly articulated by leaders in the present were important; others memories woven into a longer past had an unseen but active physical presence. It further highlights the positional experience of the different actors involved in the conflict. Second, further memories are diffracted through the dominant one. The memories together reveal long-standing and criss-crossing patterns of hurt and complicity. Third, the physical reality of past experience expressed in memory infuses a material context of war.

The illustration above provides a visualization of 'seeing' within a particular map at an early stage of its unfolding, which intersects with and replicates positional memories expressed in other maps. The third example illustrates the *narrative* construction of a map from the day Russia invaded Ukraine, a month later, which provides a more cohesive story. The full decoded transcript from that day is too long to include here; the summary answer more directly shapes the narrative to the question. It is interesting that the full decoded transcript began with 'War is happening in Europe again.' (**Table 2**).

It is hard to account for the philosophical nature of the narrative that emerged from stringing together the fragments of a particular blind conversation or where it comes from.[10] Interestingly, it points to a problem that has defined the study of international relations, that is, the desirability of more global security, contrasted with the reality of state action driven by insecurity, commonly referred to as the 'security dilemma.' However, in contrast to more traditional accounts, the narrative highlights

Question: *What would support and facilitate a shift into seeing a shared global collective agreement for peace?* **Answer:** Addressing the absence of trust, which is the heart of the problem. Trust requires turning toward ungrieved grief, with its roots in the past, but grief is presently being used as a political football. There is a need to let go of a sense of safety that arises from fragmentation (*a world divided into states*) in order to give a place to holistic (*global*) safety*. The past and present political environment has been organized around a heavily militarised aggressive game, and, while it would be desirable to change this, there is a lack of agency and a powerlessness. This environment has been a source of grief for past and present populations, and has given rise to an irreversible accumulation of unacknowledged guilt. Because the fragmentation of populations has been seen in the past as a solution, it has left them continuously responding to triggers that repeat old patterns. Seeing a more global safety and a future for children requires looking at memories of war. Unacknowledged guilt is continuously reproducing itself in new forms, and a new post-pandemic agreement is emerging from this. The media, the military-industrial complex, global institutions and scientific observers are watching what is going on but through a lens influenced by beliefs that peace comes through war. Present populations can't yet see what is coming as a result of the Russian invasion of Ukraine. The past is in the process of being repeated and there is a need to gain insight from what has happened before, i.e. looking at the past, rather than seeing the present as unique through a lens of unacknowledged guilt. The emphasis on fragmented safety is energised through war and madness and Russia could destroy us all. This sense of fragmented safety becomes a means to mobilise present populations to escalate war as a way of doing something, but is antagonising Russia into self-destruction and destruction of the world. While Russia appears to be the agent of destruction, belief in the sanctity of fragmentation (*sovereignty*) contributes to the present chaos. Past sacrifice and memories of war are being used for political purposes rather than being acknowledged and grieved. Profit as belief goes hand in hand with the fragmentation, but will eventually bring about its own death. The threat of nuclear war is a potential that arises from the fragmentation of the political environment. The threat of environmental destruction is endangering the future of children, who want the ungrieved grief to be grieved in order to restore a balance. Creating a place for a holistic safety would contribute to seeing the situation more clearly and to compassion. While it seems too late to acknowledge the guilt (*given that it appears irreversible*), from the perspective of ungrieved grieve there is a choice and it is never too late. From the perspective of a global collective agreement for peace a totally new approach is needed; the proposed mobilization of present populations won't help the situation [63].
Italicized points in parentheses have been added by the author for purposes of clarification.

Table 2.
Sample question and answer from blind mapping on the day that Russia invaded Ukraine.

[10] Given the huge amount of detail involved in wading through any one transcript line by line, it would be impossible from within the process of analysis to make this up.

the importance of memories of war and ungrieved grief, which is a function of the apparatus.[11]

The various cuts into the analysis suggest that consciousness is a property of a *system*, which is entangled with the apparatus of measurement, i.e. an intentional question. The system is not static but expresses an unfolding series of propositions which resonate with 'what it is like' to be a particular kind of experience at a moment in time and space. The non-local affective resonances arising from the blind maps provide a window into a material reality that has been woven by political actors in memories of historical experience, which cannot be separated from the drawing of boundaries in a context of war. They further raise a question about the tendency to separate materiality from consciousness. Like the particle and wave, the two exist in a complementary relationship. The final section shifts to an exploration of preliminary insights regarding the nature of consciousness as it emerged from the 'experiment.'

5. Consciousness

The quantum mind/consciousness debate in quantum mechanics, which has been underway for more than a century, began with a question about the role of the observer and consciousness in measurement [67]. Despite these origins, there has been a resistance within physics [68], no less than the social sciences [2], to tackling the hard problem of consciousness. While the simulation of human consciousness in artificial intelligence is a rapidly expanding field, quantum consciousness is often conflated with new age mysticism and thus discredited, although this is beginning to change.[12] Within biology, there is growing evidence of, for instance, the place of quantum effects in photosynthesis, by which plants turn sunlight into fuel (e.g. [71, 72]), or the role of the Earth's magnetic field in providing a 'quantum compass' for migratory birds in navigation ([73], p. 2). The physicist Matthew Fisher [69, 70] has proposed that the nuclear spins of phosphorus atoms may act as rudimentary 'qubits' in the human brain that enable the brain to function like a 'quantum computer.'

Our experiment did not set out to explain consciousness but rather to observe the 'quantum effects' and to identify patterns of non-local entanglement between past and present arising from a blind process over several years. Having said that, the exercise raised significant questions about the nature of consciousness. The first regards the frequent reference in the literature to consciousness as an individual subjective experience (e.g. [6], pp. 323–342; [68, 74]). The analysis above of 'what it is like' [50] to be a particular kind of experience, i.e. of war, does not to deny the importance of subjective consciousness or the role of the brain in processing, but shifts emphasis from the mechanism of the 'container' to experience. 'What it is like to be in a state' is a problem of experience which is more than individual experience, or for that matter, human experience; it was indeed used by the philosopher Thomas Nagel in relation to bats [50] and arguably applies to any category of experience. 'What it is like' to be in a state is a property of field or system, which in the first instance is holographic and non-local rather than subjective.

[11] For an expanded analysis of the dynamics and difference between fragmented and holistic safety, and a thought experiment regarding the 'Safety Paradox,' see ref. [29].

[12] Critics have referred, e,g., to 'hocus pocus' and 'pixie dust in the synapses' [69, 70]. The emphasis on mysticism arguably relies on a misreading of Asian traditions such as Buddhism and Daoism [7].

The 'experiment' highlighted the positional nature of consciousness. In the illustration, the range of propositions belonging to the memory of World War II take multiple forms within an unfolding context, even while remaining a phenomenon that is recognizable as 'what it is like' to be World War II. Further, how World War II was experienced at the time, or its memory in the present, is a function of many different positions, whether as a victim or perpetrator of invasion, famine, rape, etc. where the individual experience was part of a larger social positioning, which was differentiated within and across states in Europe, but also refers to war as an experience across time and space. The maps suggest that affective resonances arise from categories of human experience, within a context of relationships. Consciousness is a product of experience which, through repetition is both shaped by and shapes memory, habit and practice, as well as the meaningful potentials that arise from them. Traces of past experience persist as affective resonances. The collapse of these resonances into language within the blind maps makes it possible to examine the positioning of particular experiences vis-à-vis multiple other experiences at specific moments in space and time.

A second question regards the apparent contradiction that the mappers were 'seeing' or witnessing within a 'blind' map. The consciousness that emerges from a given map does not express a subjective individual state but a non-local entanglement between past experience and present memory. The second party experience of consciousness reveals patterns of diffracted affect, i.e. traces of affect diffracted as the past 'presses up' against the present [40] in the material world, which are also diffracted through the maps [9]. The perplexing problem in regard to our 'experiment' regards a second party experience that is not a first party experience because of the 'blindness,' on the one hand, and in light of the holographic and positional nature of experience within the maps, on the other hand. Consciousness is not a property of separate individuals per se, although expressed in its particularity by them; it is a holographic property of memory, which can be found in life all the way up and all the way down, although specific to species [40].

The mapping is potentially useful, not as a tool of calculation, like the mathematical formulae of physics, but rather as a tool for expanding and changing consciousness. As the project evolved, the investigators noted a change in themselves and how they were viewing the subject matter, often in a more visceral manner, which gave rise to a question of how other participants were impacted by their involvement. In order to better understand the 'witness effect,' we asked both trackers and mappers to regularly reflect on their experience. During the first year, the more empirical research of the trackers was understood to be in a totally different category than the mapping. The one involved reading and writing; the other involved blind engagement within the maps. However, having monitored the 'witness effect' of both, the two experiences do not appear to be qualitatively different. The consciousness that emerges from the maps is more concentrated and a function of a particular intentional question, but is only an intensified experience of everyday consciousness. Both are manifestations of the entangled nature of mind and consciousness with the material world and with language, as distinct from a tendency to identify mind with the mechanism of the individual brain.

Does the holographic entanglement of consciousness and memory suggest that seeing in one place percolates out, expanding the potential for the larger world to see? The question was highlighted by an exploration of memories of slavery and Civil War in 2017, against the backdrop of the U.S. 'politics of hate' and the Trump presidency, several years before the funded research began. Over a series of blind maps, we saw the emergence of a pattern: The mappers could not look at the person who represented the category of slavery, which reflected a more widespread tendency to turn away from this shameful aspect of the U.S. past. The mapping process facilitated a shift into seeing the slavery position within the maps. A month later, the events in Charlottesville, VA, unfolded and have since been referred to as a marker of a new conversation in the U.S. about this history [75]. We questioned whether the correlation was pure coincidence. Charlottesville was a manifestation of violence that has been there all along; the difficulty of facing or reconciling with this past over decades and a century since the Civil War is the core problem. To address a problem requires acknowledgement that it exists, which may be difficult when 'collective agreements not to know' [29] stand in the way of seeing the loss and suffering that often hides behind them. Denial and silence accompany trauma of all kinds and can carry into future generations.

As exploratory research, the MtE project sought to develop tools for analyzing the affective dynamics of seen and unseen within the maps, which expressed non-local quantum entanglements. However, measuring non-local impacts is far more problematic. In any case, neither the blind mapping process nor potential non-local impacts can be understood in classical causal terms, of a scientist manipulating variables. Indeed, given the emphasis on acknowledging 'what is' the point is to achieve a more complete view and, in so far as seeing breaks an entanglement, contribute to opening spaces for reflection and agency that are less burdened by traumatic entanglements with the past.

During the first year of the exploratory research, we became aware, both conceptually and experientially, of the extent to which we were part of the uncertain world surrounding us. From a quantum perspective, the scientist or analyst is a participant observer who is bound up with the apparatus of observation and what is seen is a function of the positioning of both. Those involved in the tracking and the mapping began to see the world in a different way, just as in the midst of the chaos of the pandemic, potentials for acknowledging past harm, some of which have been turned away from for centuries, began to be seen, e.g. in demands for reparations or the return of stolen artifacts to post-colonial spaces or the linkages between colonialism in the past and climate change in the future. Whether and how the two processes intersected non-locally remains an open question.

The affective resonances within the maps, and the palpable experience of them,[13] provided evidence that something interesting was happening, even while it was impossible to say precisely what. The group mapping began with an assumption that the blind nature of the exercise would minimize individual interpretation, and indeed, given that the mappers did not know what they or others represented, it is safe to say that this was the case. In the aftermath of a session, once the question and the positions became known, there was, however, a tendency for individuals to place their own transgenerational legacies within a larger global socio-political context, thereby increasing consciousness of their relationship to land in different parts of the world, and of earlier collective experiences of dispossession, scarcity or war, among others. There was also a greater sense that things in the present are entangled, not only with something older but also with that which has been denied life or is yet to come. Looking back also made it possible to look forward, raising questions about the legacy of the past and present in the future.

A frequent theme of the anonymous survey data was one of hopefulness arising from participation in the project, although the enhanced awareness of suffering in the world was sometimes tinged with futility, particularly for the trackers. The hopefulness, I would suggest, arose in part from a feeling that the act of witnessing is itself

[13] Specific protocols were developed for entering and leaving the maps in order to minimize the possibility that participants would carry resonances into their daily lives.

important. But the sense of empowerment was also due to experiences of seeing and feeling the world from different positions in global space, and thus a perspective that is larger than the narrow positional perspective each of us carry as individuals. The mapping experience was akin to stepping into someone else's shoes; the shoes in this case did not belong to individuals but categories of collective experience across global space and time. The experience of trackers, as already suggested, was not qualitatively different, as captured in one reflection:

> *At an early stage...the choices of tracking events and historical memories were heavily influenced by my own background. It was also easy to get very emotional or to be influenced by historical traumas which are more likely to have resonances with my own experiences. As the tracking unfolded, I started to shift to topics that I am not familiar with...Through the witness of other people's traumas, I found myself become less emotional both in terms of looking at historical traumas and the present situation – I don't know whether this is good or bad. But by looking at traumas from multiple perspectives and traumas of people unlike me, I have a feeling of stretching, which means I feel that my body and my memories themselves become records and carriers of various traumas through 'seeing.' But this process is quite different from the process of traumatization as I am not personally traumatized. For me, it is more like a process of biological recording as if I have become an automatic memory component of a larger social scheme [76].*

Both types of experience potentially gave rise to a more visceral sense of 'seeing' particular kinds of historical and present experience, including the experience of 'what it is like' to be unseen. This 'witness effect' potentially provides a point of departure for developing pedagogical tools that might increase sensitivity among a broader population, and particularly more privileged populations, thereby minimizing a tendency, noted by one of the mappers, to blame victims for their suffering, or to fail to see our own complicity in global practices that are a source of suffering, e.g. the use of child labourers in Cobalt mining in Africa which is integral to the development of green technologies. The cumulative effect of participating in the project over an extended period of time was greater awareness of the entanglement of multiple historical empires, and of practices of appropriating land and what this has meant for occupants of the land, who were dispossessed and/or enslaved, among others. The practices are not purely a thing of the past, but bleed into the present, not least in relation to continuing practices of land appropriation, displacement, deforestation and the extraction of natural resources. As was recognized at the COP27 gathering in 2022, there is a link between colonialism and climate change, and victims of the former are suffering more directly from the latter as a result of this history. The failure to act decisively to address either as we walk blindly to the brink of environmental catastrophe itself suggests an unwillingness to 'see' the problem or the implications for future generations.

6. Conclusions

As exploratory research, the focus was less on testing or falsification than proof of concept. The analysis above suggests three preliminary insights:

- Memories and habits of past trauma are a legacy that we find it hard to look at. The mapping process provides a way to step into affective resonances of past trauma, diffracted in the present, to 'see' their continuing presence.

- The collapse of the non-local affective resonances into language within the blind maps expresses a limited system, or the positioning of phenomena in space and time. The lines of difference within the maps provide insight into what is seen and unseen and the working of memory within the world.

- Consciousness is a property of a holographic field of experience rather than individuals per se.

The conceptual focus on affective resonance, entanglement, language and consciousness would benefit from a more interdisciplinary engagement around questions raised by the experiment.

The continuing analysis of data from the MtE project may bring further insights, including the potential practical significance of the work, and whether the maps are primarily a tool of analysis or might be developed for pedagogical or 'therapeutic' use. In regard to the latter, the modality raises a question of whether 'seeing' in the context of the maps potentially provides 'release' of resonances of past trauma from present experience, making it possible to deal with the present more on its own terms. In the family constellation model, release results from giving a place to the unseen parts, thereby restoring balance to a family system, which is an important component of the therapy. A further example from the individual treatment of trauma is possibly comparable and provides a way to think about the potential importance. EMDR (Eye Movement Desensitization and Reprocessing) can very quickly desensitize and reprocess an individual's bodily experience of trauma. Through the reprocessing of memories of negative events, the individual no longer experiences high levels of distress associated with a particular memory. The reprocessing integrates the traumatic material, separating it from the present, thereby restoring a potential for agency [35]. The question is whether 'seeing' global entanglements of 'ungrieved grief' from centuries of empire, and giving it a place within the maps, similarly provides 'release' from affective resonances of past trauma, thereby reprocessing and taking some of the distress out of engaging with legacies of the past in the present.

Second, given the fragmented nature of traumatic memory, might the reconstruction of the blind conversations in narrative form itself represent a form of reprocessing, similar in intention to truth and reconciliation commissions, but without the accompanying risk of re-traumatization, and/or open spaces for ethical reflection and agency? As the geographer Karen O'Brien notes, telling a different story may be critical to enhancing individual and collective agency to generate transformative change ([77], p. 75). The problem, as suggested by the MtE project, regards the difficulty of opening spaces to 'see' differently, given a tendency to turn away from the suffering that often hides behind assumptions and 'agreements' embedded in everyday language. In so far as traumatic memory intrudes and triggers powerful and defensive reactions, it can be an obstacle to seeing the present on its own terms, which is necessary for exercising autonomy and agency.

The topic of the project is very large in scope but no more so than the magnitude of the problems we presently face as a planet. Further research might develop methods that facilitate the opening of potentials for a change of consciousness and a transition to another way of living together in the world. The experiment was not about the manipulation of variables separate from us but rather the potential to turn toward seeing the self and others, whether individual or collective, from a different angle, with a more complete view such that the unseen might be seen and given a place. Unacknowledged harm in the past remains a toxic influence not only in the present but potentially on the future as well. To acknowledge the presence of the past is itself to break entanglements with 'collective agreements not to know' [29], which stand in the way of restoring balance to a dangerously imbalanced global system. Entanglement is not by definition positive or negative; both potentials exist but are different in kind. The issue is less one of being entangled or not, but rather the ability to acknowledge 'what is,' and the extent to which this recognition opens a space for global conversation [78] and agency in addressing the legacy of human inequality and the destruction of nature, which is blindly propelling us toward the precipice of climate emergency.

Acknowledgements

The Mapping the Empire project was funded by the U.S. Human Family Unity foundation through a donation to the University of St. Andrews. The research was undertaken with Nicola Mackay who has complementary expertise as a systems therapist with training in medical physics. The present analysis grows out of the project but the chapter was written solely by the author as a first step of the project evaluation.

Thanks

I would like to thank the Human Family Unity Foundation, Nicola Mackay and all of the mappers and trackers for their contribution to the project, including Ahmed Abozaid, Andrew Milne, Phuong Anh Nguyen, Adagbo Onoja, Joost Pietschmann, Mary Kay Reinemann, D. Lauren Ross, Aarushi Sharma, Denise Stallcup, Alison Strandberg, Lorraine Tolmie, Shambhawi Tripathi, Sayre Vickers, Kirsty Walsh, Karen Withaus, Chaeyoung Yong, and Yang Yuanfuyi, several of whom also, along with David Sylvan and Nadine Voelkner, provided insightful comments on various drafts of the chapter. Thanks also to Roberta Weber, who, during a acupuncture session in 2015 pulled me off the therapy table and into a family constellation which gave rise to my curiosity and to exploration of that curiosity in the years since.

Author details

Karin Marie Fierke
University of St. Andrews, St. Andrews, Scotland

*Address all correspondence to: kf30@st-andrews.ac.uk

References

[1] Barad K. Meeting the Universe Halfway: Quantum Physics and the Entanglement of Matter and Meaning. Durham, NC: Duke University Press; 2007. DOI: 10.1215/9780822388128

[2] Wendt A. Quantum Mind and Social Science: Unifying Physical and Social Ontology. Cambridge: Cambridge University Press; 2015. DOI: 10.1017/CBO9781316005163

[3] Zanotti L. Ontological Entanglements, Agency and Ethics in International Relations: Exploring the Crossroads. London: Routledge; 2019. DOI: 10.4324/9781315227764

[4] Kurki M. International Relations in a Relational Universe. Oxford: Oxford University Press; 2020. DOI: 10.1093/oso/9780198850885.001.0001

[5] Murphy MPA. Quantum Social Theory for Critical International Relations Theorists: Quantizing Critique. London: Palgrave Macmillan; 2020. DOI: 10.1007/978-3-030-60111-9

[6] DerDerian J, Wendt A. Quantizing international relations: The case for quantum approaches to international theory and security practice. Security Dialogue. 2020;**51**(5):399-413. DOI: 10.1177/0967010620901905

[7] Fierke KM. Snapshots from Home: Mind, Action and Strategy in an Uncertain World. Bristol, UK: Bristol University Press; 2022. DOI: 10.46692/9781529222647

[8] Voelkner N, Zanotti L. Ethics in a quantum world. Special Issue of Global Studies Quarterly. 2022;**3**(2):1-5. DOI: 10.1093/isagsq/ksac044

[9] Fierke KM, Mackay N. Mint: To see is to break an entanglement: Quantum measurement, trauma and security. Security Dialogue. 2020;**51**(5):450-466. DOI: 10.1177/0967010620901909

[10] Garisto D. The universe is not locally real, and the physics Nobel prize winners proved it. Scientific American. 6 October 2022. Available from: https://www.scientificamerican.com/article/the-universe-is-not-locally-real-and-the-physics-nobel-prize-winners-proved-it/

[11] Peterson A. Mint: The philosophy of Niels Bohr. Bulletin of Atomic Scientists. 1963;**19**(7):8-14. DOI: 10.1080/00963402.1963.11454510

[12] Ball P. Beyond Weird: Why Everything you Thought you Knew about Quantum Physics Is ... Different. London: The Bodley Head; 2018. DOI: 10.7208/chicago/9780226594989.001.0001

[13] Stock-Homberg R. Mint: Survey of emotions in human-robot interactions: Perspectives from robotic psychology on 20 years of research. International Journal of Social Robotics. 2022;**14**:389-411. DOI: 10.1007/s12369-021-00778-6

[14] Mayer C, Viviers R. Mint: Constellation work and Zulu culture: Theoretical reflections on therapeutic and cultural concepts. Journal of Sociology and Social Anthropology. 2016;**7**(2):101-110. DOI: 10.1080/09766634.2016.11885706

[15] Mackay N. Your Invisible Inheritance: Releasing the Hidden Legacy of Ancestral Trauma. London: Rebel Magic Books; 2020

[16] Bilger B. When Germans make peace with their dead. The New Yorker. 2016.

Available from: https://www.newyorker.com/magazine/2016/09/12/familienaufstellung-germanys-group-therapy

[17] Scott JW. Mint: The evidence of experience. Critical Inquiry. 1991;**17**(4):773-797

[18] Ulsamer B. The Healing Power of the Past: The Systemic Therapy of Bert Hellinger. Nevada City, NE: Underwood; 2005

[19] Cohen DB. Systemic Family Constellations and the Use with Prisoners Serving Long-Term Sentences for Murder or Rape [Thesis]. San Francisco, CA: Saybrook Graduate School and Research Centre; 2008

[20] Hellinger B, ten Hovel G. Acknowledging What Is: Conversations with Bert Hellinger. Phoenix, AZ: Zieg, Tucker and Co., Inc.; 1999

[21] Satia P. Times Monster: History, Consciousness and Britain's Empire. London: Allen Lane; 2020

[22] Andrews K. The New Age of Empire. London: Allen Lane; 2021. DOI: 10.1080/17449855.2023.2175951

[23] Gildea R. Empires of the Mind: The Colonial Past and the Politics of the Present. New York: Cambridge University Press; 2019

[24] Armitage D. The Ideological Origins of British Empire. Cambridge: Cambridge University Press; 2010. DOI: 10.1017/CBO9780511755965

[25] Bell D. Reordering the World: Essays on Liberalism and Empire. Princeton, NJ: Princeton University Press; 2016. DOI: 10.23943/princeton/9780691138787.001.0001

[26] Inca O. Colonial Capitalism and the Dilemmas of Liberalism. Oxford: Oxford University Press; 2018

[27] Barad K. Mint: Quantum entanglements and hauntological relations of inheritance: Dis/continuities, spacetime enfoldings and justice-to-come. Derrida Today. 2010;**3**(2):240-268. DOI: 10.3366/drt.2010.0206

[28] Bateson G. Steps to an Ecology of Mind. Chicago: University of Chicago Press; [1972]2000

[29] Fierke KM, Mackay N. Mint: The safety paradox: Unknown knowns, ungrieved grief and collective agreements not to know. International Relations. 2023. online first. DOI: 10.1177/04711782231187499

[30] Kuhn RL. The illusion of time: What's real? Spaceflight. 2022. Available from: https://www.space.com/29859-the-illusion-of-time.html [Accessed: August 29, 2023]

[31] Barad K. Mint: Troubling time/s and ecologies of nothingness: Re-turning, re-membering and facing the incalculable. New Informations. 2017;**92**(5). DOI: 10.3898/NEWF

[32] Eyerman R. Memory, Trauma, Identity. London: Palgrave Macmillan; 2019

[33] Kratochwil F. Praxis: On Acting and Knowing. Cambridge: Cambridge University Press; 2018

[34] Kolk Bvd. Mint: Trauma and memory. Psychiatry and Clinical Neurosciences. 1998;**52**(S1):S52-S64. DOI:10.1046/j.1440-1819.1998.0520s5S97.x

[35] Kolk Bvd. The Body Keeps the Score: Mind, Brain and Body in the Transformation of Trauma. London: Penguin; 2014

[36] Fierke KM. Bewitched by the past: Social memory, trauma and international relations. In: Bell D, editor. Memory, Trauma and World Politics. Basingstoke: Palgrave; 2006

[37] Barad K. Mint: Diffracting difference: Cutting-together-apart. Parallax. 2014;**20**:168-187. DOI: 10.1080/13534645.2014.927623

[38] Ooe, H., et.al. Mint: Resonance frequency-retuned quartz tuning fork as a force sensor for noncontact atomic force microscopy. Applied Physics Letters. 2014, 105, 043107 DOI: 10.1063/1.4891882

[39] Bohm D. Wholeness and the Implicate Order. London: Routledge; [1980] 2002. DOI: 10.4324/9780203995150

[40] Sheldrake R. Presence of the Past: Morphic Resonance and the Habits of Nature. London: Icon Books; 2011

[41] Walters KL, Beltran RE, Huh D, Evans-Campbell T. Dis-placement and dis-ease: Land, place and health among American Indians and Alaska natives. In: Burton LM, Kemp SP, Leung MC, Matthews SA, Takeuchi DT, editors. Communities, Neighborhood, and Health: Expanding the Boundaries of Place. Philadelphia, PA: Springer Science+Business Media; 2011. pp. 163-199. DOI: 10.1007/978-1-4419-7482-2

[42] Lehrner A, Yehuda R. Mint: Cultural trauma and epigenetic inheritance. Development and Psychopathology. 2018;**30**(5):1763-1177. DOI: 10.1017/so954579418001153

[43] Yehuda R, Lehrner A. Mint: Intergenerational transmission of trauma effects: Putative role of epigenetic mechanisms. World Psychiatry. 2018;**17**(3):243-257. DOI: /10.1002%2Fwps.20568

[44] Jablonka E. Mint: Cultural epigenetics. The Sociological Review. 2016;**64**(1-suppl):42-60. DOI: 10.1002/2059-7932.12012

[45] Menakem R. My Grandmother's Hands: Racialized Trauma and the Pathway to Mending our Hearts and Bodies. Las Vegas: Central Recovery Press; 2017

[46] PBS. Stories I did not Know. 2021. Available from: https://www.pbs.org/show/stories-i-didnt-know/ [Accessed: May 22, 2023]

[47] Schwab G. Haunting Legacies: Violent Histories and Transgenerational Trauma. New York: Columbia University Press; 2010

[48] Ngugi wa Thong'0. Something Torn and New: An African Renaissance. New York: Basic Books; 2009

[49] Hutchison E, Bleiker R. Grief and the transformation of emotions after war. In: Ahall L, Gregory T, editors. Emotions, Politics and War. London: Routledge; 2015. DOI: 10.4324/9781315765068

[50] Nagel T. Mint: What is it like to be a bat? Philosophical Review. 1974;**83**(4):435-450

[51] Block N. Mint: On the confusion about a state of consciousness. Brain and Behavioral Sciences. 1995;**18**:227-247. DOI: 10.1017/s0140525x00038188

[52] Tegmark M. Our Mathematical Universe: My Quest for the Ultimate Nature of Reality. London: Allen Lane; 2014

[53] Carroll, S. S02E01 Is Consciousness Emergent? Mind Chat. You Tube

[54] Fuchs C. My Struggles with the Block Universe: Selected Correspondence, January 2001–May 2011; 2014-2015.

Available from: https://arxiv.org/pdf/1405.2390.pdf [Accessed: July 10, 2023]

[55] Stenholm S. The Quest for Reality: Bohr and Wittgenstein, Two Complementary Views. Oxford: Oxford University Press; 2011. DOI: 10.1093/acprof:oso/9780199603589.001.0001

[56] Bitbol M. Mint: Mathematical demonstration and experimental activity: A Wittgensteinian philosophy of physics. Philosophical Investigations. 2018;**41**(2):188-203. DOI: 10.1111/phin.12187

[57] Wittgenstein L. On Certainty. Wiley-Blackwell; 1975

[58] Ron Davis, physicist. Available from: https://www.quora.com/Is-there-a-difference-in-the-meaning-of-physical-and-material [Accessed: July 13, 2023]

[59] Based on unpublished preliminary data analysis of transcripts from the Mapping the Empire Project: 7GT240421

[60] Based on unpublished preliminary data analysis of transcripts from the Mapping the Empire Project: 8GT250321

[61] Based on unpublished preliminary data analysis of transcripts from the Mapping the Empire Project: 10GT240621

[62] Based on unpublished preliminary data analysis of transcripts from the Mapping the Empire Project: 14GT270122

[63] Based on unpublished preliminary data analysis of transcripts from the Mapping the Empire Project: 15GT240222

[64] Based on unpublished preliminary data analysis of transcripts from the Mapping the Empire Project: 16GT240322

[65] Based on unpublished preliminary data analysis from the Mapping the Empire Project: 14GM270122

[66] NATO. NATO-Russia Relations: the facts. 2023. Available from: https://www.nato.int/cps/en/natohq/topics_111767.htm [Accessed: July 10, 2023]

[67] Tarlacı S, Pregnolato M. Mint: Quantum neurophysics: From non-living matter to quantum neurobiology and psychopathology. International Journal of Psychophysiology. 2016;**103**:161-173. DOI: 10.1016/j.ijpsycho.2015.02.016

[68] Tegmark M. Why Consciousness is one of the most Divisive Issues in Physics Today, 2022. Available from: https://www.youtube.com/watch?v=Lul6M13F-ao [Accessed: July 14, 2023]

[69] Fisher M. Mint: Are we quantum computers, or merely clever robots? International Journal of Modern Physics B. 2017;**31**. DOI: 10.1142/S0217979217437000019

[70] Fisher M. Mint: Quantum cognition: The possibility of processing with nuclear spins in the brain. Ann. Physics. 2015;**362**:593-602. DOI: 10.1016/j.aop.2015.08.020

[71] Arndt M, Juffmann T, Vedral V. Mint: Quantum physics meets biology. HFSP Journal. 2009;**3**(6):386-400. DOI: 10.2976/1.3244985

[72] Lloyd S. Mint: Quantum coherence in biological systems. Journal of Physics: Conference series. 2011:302. DOI: 10.1088/1742-6596/302/1/012037

[73] Ouellet, J. A New Spin on the Quantum Brain. Quanta Magazine.

2016. Available from: https://www.quantamagazine.org/20161102-quantum-n [Accessed: May 22, 2023]

[74] Bitbol M. Mint: La Conscience a-t-elle une origine? Des neuroscience a' la pleine conscience: une novelle approche de l'esprit. Paris: Flammarion; 2014

[75] PBS. A Year Ago Charlottesville Shined a Light on White Supremacists and Sparked Overdue Conversations. 2018. Available from: https://www.pbs.org/newshour/show/a-year-ago-the-charlottesville-rally-shined-a-light-on-white-supremacists-and-sparked-overdue-conversations [Accessed: May 11, 2023]

[76] Based on unpublished preliminary data analysis from the Mapping the Empire Project: YY/19/07/22

[77] O'Brien K. You Matter More than you Think: Quantum Social Change for a Thriving World. Oslo, Norway: cChange Press; 2021

[78] Fierke KM, Jabri VM. Global conversations: Relationality, embodiment and power in the move toward global IR. Global Constitutionalism. 2019;**8**(3):506-535. DOI: 10.1017/S2045381719000121

Chapter 10

Entanglement in High-Energy Physics: An Overview

Mohammed Nadir

Abstract

This abstract explores the entwined realms of quantum field theory, holography, and the AdS/CFT correspondence, converging upon the enigmatic phenomenon of entanglement within high-energy physics (HEP). At the core of this narrative lies the concept of entanglement entropy—a profound measure of quantum entanglement that threads the connections between quantum information, correlations, and the very architecture of spacetime. As the journey unfolds, the AdS/CFT correspondence illuminates entanglement's holographic nature, decoding its role in deciphering the enigmas of HEP. Bell's inequality emerges as a lighthouse, probing the non-local essence of entanglement and challenging the classical boundaries of reality. Quantum cryptography emerges as a practical extension, harnessing the unique attributes of entanglement for secure communication. The tensor product formalism weaves together the quantum tapestry, while gravity—nature's sculptor of spacetime—molds the dynamics of entanglement within HEP. This abstract paves the path for a chapter that traverses based on original findings, unraveling the secrets of entanglement's significance within the intricate fabric of high-energy physics. The Nobel Prize in Physics 2022, awarded to Alain Aspect, John F. Clauser, and Anton Zeilinger, enriches this narrative. Their experiments solidify entanglement's non-locality, bridging the realms of quantum mechanics and HEP. This abstract encapsulates the entangled narrative and its dialog with gravity.

Keywords: quantum field theory, holography, AdS/CFT (anti-de sitter/conformal field theory) correspondence, entanglement entropy, quantum information, quantum correlations, spacetime, Bell's inequality, quantum cryptography, tensor product, gravity

1. Introduction

Entanglement, a quintessential phenomenon of quantum mechanics, has transcended its role as a conceptual enigma to become a pivotal player in the narrative of high-energy physics (HEP). This chapter embarks on a journey to unravel the deep-seated connections between entanglement and the fundamental aspects of our universe. The principle of superposition, where particles can exist in multiple states simultaneously, and entanglement, where particles become inherently connected across vast distances, are at the core of this mysterious quantum world [1].

1.1 Quantum field theory: Foundations and Frontiers

Quantum field theory (QFT) is the bedrock of modern theoretical physics, providing a framework for describing the interactions of particles and fields. At the heart of QFT lies the peculiar phenomenon of entanglement, where particles become inexorably intertwined, their fates interlinked in a manner that defies classical intuition [1]. This interconnection transcends spatial separation, manifesting non-local correlations that have confounded physicists and spurred debates on the very nature of reality [2].

Quantum field theory intersects with the fabric of spacetime, giving rise to the intricate interplay between entanglement and gravity. The holographic principle, symbolized by the AdS/CFT correspondence, presents a radical perspective that links a gravitational theory in anti-de Sitter space to a quantum field theory (Ads/CFT) on its boundary [3]. This duality evokes profound implications for understanding the emergence of spacetime and the role of entanglement in the underpinnings of gravity [4].

The quest for unraveling entanglement secrets has led to the notion of entanglement entropy, a measure of the quantum entanglement inherent in complex systems [5]. This quantity, intertwined with the fabric of quantum information theory, has offered fresh insights into the architecture of quantum states, phase transitions, and the fabric of spacetime [6].

1.2 Beyond Bell's inequality

Bell's inequality and its experimental violations mark a turning point in the exploration of entanglement. Experiments have confirmed the non-local nature of entanglement, challenging classical notions of causality and realism [7, 8]. The ramifications of these violations span from the foundations of quantum theory to the realm of quantum cryptography, promising secure communication protocols that hinge on the inseparability of entangled particles [9].

1.3 Navigating the quantum terrain

The quantum landscape brims with uncharted territories, offering tantalizing prospects for revolutionizing computation, cryptography, and our understanding of the universe. Quantum computing and quantum cryptography leverage entanglement's unique properties to potentially outperform classical counterparts and ensure communication security in the quantum realm [10]. The tensor product formalism, a powerful mathematical tool, lies at the heart of these endeavors, allowing for the description of composite quantum systems and their interactions [11].

2. Motivation for studying entanglement in high-energy physics

Motivating research in high-energy physics concerning entanglement is essential due to its potential to provide deeper insights into fundamental physical phenomena and their significance in various theoretical frameworks. Here are some key motivations, along with relevant references:

1. *Understanding Quantum Complexity:* High-energy physics deals with complex quantum systems, and entanglement serves as a fundamental tool to understand

and quantify their intricate quantum correlations. Entanglement entropy, in particular, has been extensively studied to characterize the entanglement structure in quantum field theories and many-body systems [5].

2. *Quantum Gravity and Holography:* Investigating the role of entanglement in the holographic AdS/CFT correspondence provides crucial insights into the connection between quantum field theories and gravity in higher-dimensional spacetime. Entanglement entropy in the boundary theory is related to the area of minimal surfaces in the bulk AdS spacetime, offering tantalizing clues about the quantum nature of spacetime [12].

3. *Quantum Computing and Information:* Entanglement plays a central role in quantum information theory and quantum computing. Understanding the role of entanglement in quantum algorithms and quantum communication is crucial for developing efficient computational tools and communication protocols [11].

4. *Quantum Field Theories and Phase Transitions:* The study of entanglement entropy has led to significant advances in understanding phase transitions and critical phenomena in quantum field theories. Entanglement measures have proven to be powerful tools in characterizing quantum phase transitions [13–15].

5. *Black Hole Physics and Information Paradox:* Investigating the entanglement properties of black holes can provide insights into the nature of black hole entropy and the resolution of the information paradox. Entanglement entropy in black hole systems is a fascinating topic in the study of quantum gravity [3].

6. *Quantum Foundations:* Entanglement is at the heart of quantum non-locality and the Einstein-Podolsky-Rosen (EPR) paradox [16]. Understanding the foundational aspects of entanglement and its implications for quantum mechanics can deepen our understanding of quantum reality [1, 14]

7. *Quantum Information in Collider Physics:* In collider experiments, the efficient handling of enormous data sets and the optimization of event selections are critical. Quantum computing and quantum machine learning, harnessing entanglement and quantum correlations, promise to revolutionize data analysis in HEP [17]. It enables us to probe entanglement in extreme conditions. These experiments can provide empirical evidence for the quantum nature of particle interactions [18, 19].

8. *Fundamental Tests of Quantum Mechanics:* HEP provides a unique testing ground for fundamental principles of quantum mechanics. Future experiments may explore aspects like entanglement, non-locality, and quantum correlations to probe the boundaries of our current understanding [1, 20]. These research areas represent just a glimpse of the exciting possibilities that entanglement holds in the realm of High-Energy Physics. As technology advances and our knowledge deepens, these investigations are likely to shape our understanding of the fundamental forces of the universe.

9. *Beyond the Standard Model:* HEP is dedicated to probing the fundamental forces and particles of the universe. Entanglement can be a powerful tool for discover-

ing deviations from the Standard Model of particle physics, including the search for dark matter and understanding neutrino properties [21, 22].

10. *Quantum Gravity and Entanglement:* The search for a theory of quantum gravity remains a fundamental challenge. Entanglement is expected to play a significant role in reconciling quantum mechanics with general relativity. Research in this area aims to uncover the quantum nature of spacetime itself [23, 24].

11. *Quantum Cryptography and Secure Communication:* While not exclusive to HEP, quantum cryptography harnesses entanglement to enable secure communication. Its applications extend to safeguarding sensitive data in HEP experiments, ensuring the integrity of results [25–28].

12. *Quantum Chromodynamics (QCD) and Quark-Gluon Plasma (QGP):* HEP experiments often involve extreme conditions, such as the quark-gluon plasma (QGP) created in heavy-ion collisions. Understanding the entanglement properties of QGP can shed light on its quantum behavior and phase transitions, which are of great interest for studying the early universe and fundamental interactions. Investigating the entanglement properties of particles in the early universe, specifically in the Quark-Gluon Plasma phase, remains a critical area of research. This includes understanding the role of entanglement in QCD and its implications for the early universe [29–31].

In conclusion, entanglement is a versatile and indispensable concept in high-energy physics, offering avenues for understanding fundamental physics, quantum gravity, quantum information, and technological advancements. Research in this area can lead to breakthroughs with wide-ranging implications for our understanding of the universe and our ability to harness quantum phenomena for practical applications.

3. Explanation of the significance of the chosen keywords

The following explanation underscores the significance of the chosen keywords, showcasing how they interconnect to shape the landscape of high-energy physics and our understanding of entanglement's role in fundamental particles, quantum gravity, quantum technologies, and more.

1. *Quantum Field Theory:* Quantum field theory (QFT) serves as the mathematical framework for describing fundamental particles and their interactions. It unifies quantum mechanics with special relativity and provides a foundation for understanding the behavior of particles in high-energy processes [32]. QFT elucidates how entanglement operates in particle interactions, paving the way for unraveling the intricate quantum dynamics that govern these phenomena.

2. *Holography (AdS/CFT Correspondence):* Holography, specifically the AdS/CFT correspondence, establishes a profound connection between quantum field theories and gravity in higher-dimensional spacetimes [2]. This duality suggests that entanglement patterns in a quantum field theory correspond to geometric features in its gravitational counterpart. The AdS/CFT correspondence hints at a deeper unity between these seemingly disparate realms, ushering in novel insights into entanglement's role in both contexts.

3. *Entanglement Entropy:* Entanglement entropy quantifies the amount of entanglement between different components of a quantum system [4]. It provides a powerful tool for probing the entanglement structure within complex systems, such as those found in high-energy physics. Entanglement entropy unveils critical information about the behavior of particles, phase transitions, and emergent phenomena, enriching our understanding of quantum systems.

4. *Quantum Information:* Quantum information theory explores how information is processed and manipulated in the quantum realm [11]. In the context of high-energy physics, understanding how entanglement and quantum correlations can be harnessed for quantum information processing is crucial for the development of quantum technologies such as quantum computing and cryptography. These technologies hold transformative potential for solving complex problems that are beyond the reach of classical computers.

5. *Quantum Correlations:* Quantum correlations extend beyond classical correlations, giving rise to phenomena like entanglement. These correlations defy classical intuitions and hold the key to harnessing the power of quantum mechanics for information processing and fundamental research [33]. Investigating quantum correlations in various platforms, from photons to qubits, is essential for understanding the entanglement landscape in different physical systems.

6. *Quantum Simulation:* Quantum simulation involves using quantum systems to simulate and study complex quantum processes that are challenging to simulate on classical computers [34]. By utilizing entangled states, quantum simulation enables researchers to explore high-energy processes, study particle interactions, and uncover insights into the behavior of particles in extreme conditions.

7. *Gravity and Spacetime:* Gravity, as described by general relativity, governs the curvature of spacetime and the behavior of massive objects [35]. Entanglement's connection to gravity through concepts like the holographic principle highlights its potential role in revealing the underlying nature of spacetime. Exploring entanglement in the context of gravity and spacetime can provide insights into the fundamental fabric of the universe.

8. *Quantum Computing and Quantum Cryptography:* Quantum computing leverages entanglement to perform computations with a quantum advantage [10]. Quantum cryptography exploits the non-local nature of entanglement to establish secure communication protocols [36]. These emerging technologies hold promise for revolutionizing various aspects of high-energy physics, from simulating complex systems to securing communication channels. A more recent entanglement-based secure quantum cryptography has been reported by Yin, Juan et al. [27].

9. *Tensor Products:* The tensor product formalism, a powerful mathematical tool, lies at the heart of the Quantum Field Theory, allowing the representation of composite quantum systems and interactions which are central to the simulation of complex HEP systems [11].

4. Quantum field theory and entanglement

4.1 Entanglement in quantum field theory

In the realm of high-energy physics (HEP), quantum field theory (QFT) takes on a paramount role in modeling the fundamental forces and particles that govern the universe. Entanglement, a quintessential feature of quantum mechanics, plays a pivotal role in this context. As particles are described by quantum fields, their interactions and correlations are inherently entangled [32]. The entanglement entropy, which quantifies the degree of entanglement in a quantum system, has found profound applications in unveiling the complex nature of QFTs, shedding light on quantum phase transitions and critical phenomena [5].

4.2 Entanglement as quantum correlations

Entanglement serves as a manifestation of quantum correlations beyond classical limits, defying our classical intuitions about the separability of particles. In HEP, this non-local interdependence of particles is crucial for understanding phenomena like confinement, where quarks and gluons remain entangled within hadrons [37]. Entanglement entropy not only characterizes the entanglement but also provides a window into the structure of the underlying quantum states, revealing connections between particles and their collective behavior [29].

4.2.1 Furthermore

For example, "von Neumann entropy" quantifies the amount of entanglement between two subsystems in a quantum system. Mathematically, it is described as follows:

$$S(\rho_A) = -Tr(\rho_A \log(\rho_A)) \tag{1}$$

Where:

- $S(\rho_a)$ is the von Neumann entropy of subsystem A.
- ρ_A is the reduced density operator for subsystem A, obtained by taking the partial trace over subsystem B.

To be more explicit, let us express ρ_A and its components:

$$\rho_A = Tr_B(\rho) \tag{2}$$

Here, $Tr_B(\rho)$ represents taking the partial trace over subsystem B, which mathematically means summing over all degrees of freedom in subsystem B. This operation leaves you with a reduced density operator that only contains information about subsystem A.

So, in a more detailed form, the von Neumann entropy for subsystem A is:

$$S(\rho_A) = -Tr(\rho_A \log(\rho_A)) = -Tr(Tr_B(\rho)\log(Tr_B(\rho))) \tag{3}$$

Some recent references related to von Neumann entropy are cited in [38–42], while the original work is in Von Neumann [43].

4.3 Quantum field theory and Bell's inequality (Nobel prize 2022)

The study of Bell's inequality in HEP explores the boundaries of classical correlations and sets the stage for exploring the non-locality inherent in entanglement. In experiments involving entangled particles, such as those in the domain of HEP, violations of Bell's inequality affirm the predictions of quantum mechanics over classical theories [1, 8]. These violations underscore the profound implications of entanglement for our understanding of particle interactions and the non-local nature of reality [16]. The Nobel Prize in Physics 2022, awarded to Alain Aspect, John F. Clauser, and Anton Zeilinger, enriches this narrative [6, 44]. Their experiments solidify entanglement's non-locality, bridging the realms of quantum mechanics and HEP. The entangled narrative encapsulates entanglement's centrality in HEP, its weaving into quantum simulations, and its dialog with gravity.

4.4 Quantum information and HEP

Quantum information theory finds synergy with HEP, particularly in the context of quantum computing and cryptography. Quantum field theory allows for the representation of complex systems using tensor product formalism [11], a crucial tool in quantum information processing. Quantum computers, leveraging entanglement, hold the potential to revolutionize simulations of quantum systems relevant to HEP [10]. Moreover, quantum cryptography exploits the principles of entanglement to establish secure communication protocols, offering insights into particle interactions and information transfer in HEP [45].

5. Holography and the AdS/CFT correspondence

5.1 The holographic principle and emergent gravity

In the pursuit of understanding the elusive connection between quantum mechanics and gravity, the holographic principle emerges as a revolutionary concept. It posits that the information contained in a region of space can be encoded on its boundary, suggesting a deep connection between quantum systems and their gravitational descriptions [46]. The holographic principle not only challenges our traditional understanding of spacetime but also offers insights into the nature of entanglement.

5.2 The AdS/CFT correspondence: Quantum fields and gravity

At the forefront of this intersection lies the AdS/CFT correspondence, a groundbreaking duality that relates certain quantum field theories in "anti-de Sitter" (AdS) spacetime to theories of gravity in one higher dimension [2]. This correspondence provides a remarkable tool to study the interplay between entanglement and gravity. In HEP, the AdS/CFT correspondence allows researchers to explore strongly coupled quantum field theories by mapping them to weakly coupled gravitational theories, enabling investigations into complex phenomena that were previously inaccessible.

Quantum entanglement in the context of gravity is an area of ongoing research that poses significant challenges due to the complex interplay between quantum mechanics and general relativity. While there is ongoing progress in developing rigorous mathematical frameworks, a complete and universally accepted mathematical description of quantum entanglement in gravity is still an active area of investigation. However, several approaches have been proposed that aim to provide more rigorous mathematical tools and derivations in this field:

1. AdS/CFT Correspondence: The Anti-de Sitter/Conformal Field Theory (AdS/CFT) correspondence, a concept derived from string theory, relates certain gravitational theories in anti-de Sitter space (AdS) to conformal field theories (CFTs) in lower dimensions. This duality has been extensively studied and provides a powerful framework to explore the connection between gravity and quantum entanglement. Mathematical techniques from both AdS and CFT sides are employed to understand the entanglement properties of the dual theories.

2. Holography and Entanglement Entropy: Holography, a principle stemming from the AdS/CFT correspondence, suggests that the information within a gravitational system can be encoded on its boundary. In this framework, the entanglement entropy of a quantum field theory living on the boundary is related to the gravitational properties of the corresponding bulk theory. Mathematical tools such as the Ryu-Takayanagi formula and the Hubeny-Rangamani-Takayanagi formula [12] have been derived to calculate the entanglement entropy in holographic setups.

5.3 Entanglement and geometry: Insights from holography

The AdS/CFT correspondence unveils the holographic nature of entanglement, where the entanglement structure of a quantum field theory is mirrored in the geometry of its gravitational counterpart [3]. The "holographic entanglement entropy" captures this relationship, showing that the entanglement entropy of a boundary region corresponds to the area of a minimal surface in the bulk spacetime. This deep connection between entanglement and geometry enriches our understanding of both quantum mechanics and gravity.

5.4 Spacetime as an emergent phenomenon

Emergence lies at the heart of the AdS/CFT correspondence, where gravity and spacetime emerge from the collective behavior of quantum fields. This hints at a deeper understanding of gravity as a consequence of entanglement and quantum correlations at a more fundamental level [3]. This perspective has profound implications for black hole physics, information theory, and the nature of spacetime itself.

The exploration of holography and the AdS/CFT correspondence uncovers the intricate connection between quantum field theories, entanglement, and the emergence of gravity, reshaping our understanding of the fabric of the universe.

6. Entanglement entropy: unveiling quantum information

6.1 Entanglement entropy in quantum field theories

In the realm of high-energy physics (HEP), the concept of entanglement entropy serves as a powerful tool to probe the intricate nature of quantum field theories. Entanglement entropy quantifies the amount of quantum entanglement between different regions of a quantum system, revealing crucial insights into its underlying structure [5]. In HEP, it provides a means to explore the fundamental nature of particle interactions and their entangled correlations.

6.2 Renormalization and scale dependence

The study of entanglement entropy within HEP has led to a deep connection between the behavior of entanglement and the renormalization group flow of quantum field theories [47]. The entanglement entropy exhibits intriguing scale dependence, reflecting the way that quantum fluctuations and particle interactions manifest at different energy scales. This perspective sheds light on the behavior of particles as they undergo energy changes, providing an intricate understanding of quantum field dynamics.

6.3 Holographic insights into entanglement entropy

The AdS/CFT correspondence also plays a pivotal role in understanding entanglement entropy in HEP. The holographic entanglement entropy prescription relates the entanglement structure of a boundary theory to the geometry of its corresponding bulk spacetime [12]. This deepens our comprehension of how quantum correlations and entanglement are intertwined with the geometry of the universe. Moreover, it offers a novel perspective on the emergence of spacetime itself from the entanglement of underlying degrees of freedom.

6.4 Applications in critical phenomena

Entanglement entropy finds significant applications in the study of critical phenomena, where phase transitions occur in quantum systems. In HEP, this concept provides a unique lens to explore the behavior of quantum fields as they undergo dramatic changes. By characterizing the scaling properties of entanglement entropy near critical points, researchers gain valuable insights into the underlying quantum phase transitions [4]. This has the potential to illuminate the behavior of particles in extreme conditions, such as the early universe.

> *Exploring entanglement entropy within HEP provides a unique lens to probe the fundamental interactions of particles, revealing insights into renormalization, holography, and critical phenomena, and unveiling the deep connections between quantum mechanics and spacetime.*

7. Quantum correlations and spacetime

7.1 Quantum correlations in high-energy physics

In the realm of high-energy physics (HEP), the exploration of quantum correlations takes the central stage. Quantum correlations, often manifested as entanglement, play a fundamental role in describing the behavior of particles and their interactions. These correlations go beyond classical correlations, enabling particles to exhibit non-local connections that challenge our intuition about the nature of reality [1].

7.2 Entanglement and spacetime

One of the most intriguing aspects of quantum correlations in HEP is their connection to spacetime geometry. The AdS/CFT correspondence illuminates how entanglement in a boundary quantum field theory corresponds to geometric features in the dual higher-dimensional spacetime [2]. This suggests that the fabric of spacetime itself might emerge from the intricate web of quantum correlations. Such a perspective provides a new paradigm for understanding gravity as an emergent phenomenon arising from quantum entanglement [3].

7.3 Quantum correlations in extreme conditions

In HEP, the study of quantum correlations becomes particularly fascinating in the context of extreme conditions, such as those found in the early universe or near black holes. Quantum entanglement may play a crucial role in unraveling the mysteries of cosmic inflation, the Big Bang, and the behavior of matter in ultra-high-energy collisions [48]. By harnessing the power of quantum correlations, researchers aim to gain deeper insights into the fundamental nature of particles and their interactions.

7.4 Spacetime as an entanglement structure

Recent research within HEP has unveiled tantalizing connections between the structure of spacetime and the entanglement of quantum degrees of freedom. Entanglement wedges and causal wedges, emerging from the AdS/CFT correspondence, provide new insights into how spacetime is intricately woven from quantum correlations [49]. This suggests that spacetime may be an emergent consequence of quantum entanglement, raising profound questions about the nature of reality at the smallest scales.

Exploring the quantum correlations within HEP offers insights into the nature of spacetime, its connection to entanglement, and the role of quantum correlations in extreme conditions, ultimately leading to a deeper understanding of the fundamental interactions of particles.

8. Bell's inequality and quantum cryptography

8.1 Bell's inequality: Testing the limits of classical correlations

In the realm of high-energy physics (HEP), the violation of Bell's inequality stands as a crucial experiment that challenges classical notions of correlations [16].

Bell's theorem demonstrates that no local realistic theory can reproduce the quantum mechanical predictions of entangled systems, thereby highlighting the non-classical nature of quantum correlations [1].

8.2 Experimental verification and violation of Bell's inequality

Experiments involving entangled particles, such as photons or ions, have validated the violation of Bell's inequality and confirmed the fundamentally non-local nature of entanglement [20]. These experiments not only provide compelling evidence against classical realism but also open avenues for harnessing quantum entanglement for practical applications.

8.3 Quantum cryptography: Leveraging quantum correlations for security

The violation of Bell's inequality has paved the way for revolutionary advancements in quantum cryptography. Quantum key distribution protocols, such as the famous BB84 protocol, utilize entanglement to enable secure communication channels that are inherently immune to eavesdropping [36]. This has immense implications for secure communication, with the potential to revolutionize data encryption and transmission. However, entanglement-based secure quantum cryptography over 1120 kilometers has been reported by Yin, Juan et al. [27].

The exploration of Bell's inequality and its violation not only challenges classical theories but also finds applications in quantum cryptography, offering a secure foundation for quantum communication protocols.

9. Quantum information processing and tensor product

9.1 Harnessing quantum information for simulation

In the context of high-energy physics (HEP), quantum information processing has gained significant traction due to its potential to simulate complex quantum systems that are computationally intractable with classical methods [34]. Quantum simulators, often implemented using platforms like trapped ions or superconducting qubits, can mimic the behavior of fundamental particles and interactions, offering insights into the behavior of HEP phenomena [10].

9.2 Tensor product formalism: Enhancing computational power

The tensor product, a mathematical construct representing the combined state of multiple quantum systems, plays a pivotal role in quantum information processing. By encoding the entanglement structure between particles, the tensor product enables the manipulation of composite quantum systems, which is central to the simulation of complex HEP systems [11].

9.3 Quantum parallelism and exponential speedup

Quantum computers leverage the inherent parallelism of quantum states to perform certain calculations exponentially faster than classical counterparts. This capability has the potential to revolutionize fields like lattice gauge theory, aiding in the understanding of strong interactions in HEP [50].

> *Quantum information processing, guided by the tensor product formalism, holds immense promise in simulating complex HEP systems and unlocking unprecedented computational power for addressing fundamental interactions.*

10. Entanglement in gravity

10.1 Quantum Entanglement's role in gravity

Entanglement, a cornerstone of quantum mechanics, has intriguing implications for gravity in the context of HEP. Quantum entanglement's role in gravitational theories can offer insights into the nature of spacetime itself and the unification of fundamental forces [51]. Entanglement entropy, a measure of quantum entanglement, has revealed a deep connection between gravity and quantum information. The AdS/CFT correspondence suggests that the entanglement entropy of a boundary quantum field theory is related to the area of a bulk black hole horizon, indicating a profound link between quantum information and spacetime geometry [12].

10.2 Entanglement entropy and black holes

In the study of black holes, entanglement entropy has emerged as a crucial concept. The famous "holographic principle" suggests that the information content of a black hole is encoded in its event horizon's surface area, highlighting the deep connection between entanglement entropy and the geometry of spacetime [52]. Quantum entanglement seems to imply that information is never truly lost, posing a challenge to classical notions of black hole evaporation [52]. The study of entanglement entropy in the context of black holes has opened new avenues for understanding the quantum nature of spacetime [53].

10.3 Emergent spacetime from entanglement

The idea that spacetime may emerge from entanglement has gained traction through the AdS/CFT correspondence. This concept implies that gravity itself could be a consequence of entanglement, reshaping our understanding of the fundamental nature of spacetime in the realm of HEP [3].

> *The interplay between entanglement and gravity in HEP opens the door to novel perspectives on spacetime's nature and the fundamental forces governing the universe.*

11. Breakthroughs and their awaited applications

Breakthroughs in entanglement within High-Energy Physics (HEP) are eagerly awaited and could have profound implications for our understanding of the fundamental forces of the universe. It could provide new insights into how spacetime emerges from entangled quantum states, potentially leading to a theory of quantum gravity as exploring *entanglement's implications* for quantum gravity could reveal deeper insights into the fabric of spacetime Susskind [54]. The potential entanglement structure of the quantum vacuum remains a tantalizing enigma, with implications for phenomena like vacuum energy and the cosmological constant.

1. *Entanglement and the Quantum Nature of Spacetime:* Understanding how entanglement is related to the fundamental structure of spacetime is a key goal. Breakthroughs in this area might involve new insights into the nature of spacetime itself and how it emerges from entangled quantum states [2].

2. *Quantum Gravity Unification:* One of the most significant open questions lies in reconciling General Relativity with Quantum Mechanics, giving rise to a theory of quantum gravity. Entanglement's role here is pivotal. The holographic principle, stemming from the AdS/CFT correspondence, suggests that the universe's fabric is encoded in lower-dimensional entanglement patterns [2, 54]. Bridging this principle with real-world observations could unlock the elusive theory of everything.

3. *Quantum Computing for HEP Simulations:* Leveraging entanglement for quantum simulations of high-energy physics processes may lead to breakthroughs in our ability to understand complex particle interactions more efficiently than classical computers [10, 55].

4. *Quantum Simulations Revolutionized:* Harnessing entanglement for quantum simulations stands at the forefront. Simulating complex quantum systems, inaccessible through classical computation, could revolutionize material science, drug discovery, and even cosmology. In HEP, simulating quantum chromodynamics to understand confinement and strong interactions is a compelling avenue. Entanglement-based quantum simulations Preskill [10], Cirac & Zoller [55] have the potential to revolutionize our understanding of fundamental interactions. Quantum computers, exploiting entanglement, could efficiently simulate quantum field theories, yielding insights into phenomena that are computationally intractable using classical methods [11].

5. *Fundamental Symmetries Under Scrutiny:* Entanglement provides a unique lens for probing fundamental symmetries in particle interactions. Research into the violation of Bell's inequality Aspect [1] and entangled neutrino oscillations Bilenky et al. [56] could uncover new facets of the Standard Model and point the way to physics beyond it.

6. *Testing Bell's Inequality Violation in HEP:* Further experimental tests of Bell's inequality violation in particle physics could provide insights into the non-local nature of entanglement and quantum correlations [1, 8].

7. *Black Hole Information Paradox:* The information paradox, which arises when quantum information seems to be lost in black holes, is a major challenge surrounding black holes and continues to baffle physicists. Discoveries related to entanglement could help resolve this paradox by shedding light on how information is encoded and preserved in black holes [57, 58]. Discovering how entanglement properties of particles near black holes contribute to this paradox could lead to a breakthrough, resolving the information paradox [59].

8. *Quantum Gravity and Black Holes:* Entanglement plays a role in addressing the black hole information paradox. The ER = EPR conjecture [60] suggests that entangled particles might be connected by a wormhole, shedding light on the connection between quantum entanglement and gravity [61].

9. *Entanglement and Black Hole Information Paradox:* Resolving the information paradox, which arises when quantum information seems to be lost in black holes, is a major challenge. Discovering how entanglement properties of particles near black holes contribute to this paradox could lead to a breakthrough [58].

10. *Entanglement in Particle Collisions:* In HEP experiments like those at the Large Hadron Collider (LHC), understanding how entanglement plays a role in particle collisions could lead to new discoveries about the fundamental forces and particles of the universe [18].

11. *Entanglement and Quantum Field Theory:* Advancements in our understanding of how entanglement is described within the framework of quantum field theory can lead to deeper insights into the behavior of particles and fields [5].

12. *Quantum Information Protocols in HEP:* Developing practical quantum information protocols, such as quantum cryptography, tailored to the needs of high-energy physics experiments, could enhance data security and communication [62].

13. *Entanglement Entropy in Quantum Field Theory (QFT):* In condensed matter physics, entanglement entropy has been used to study topological phases of matter. It helps classify and understand different quantum states [5].

14. *AdS/CFT Correspondence:* The AdS/CFT correspondence has transformative significance in high-energy physics. It provides a remarkable bridge between two distinct theoretical frameworks—gravity and quantum field theory—enabling researchers to address complex problems in a simpler setting [63, 64]. It has facilitated the study of strongly coupled systems, such as quark-gluon plasmas (QGP) produced in heavy-ion collision [45, 65].

15. *Dark Matter and Dark Energy:* These systems are notoriously difficult to analyze using traditional methods. Researchers have explored quantum correlations and entanglement to understand the nature of dark matter and dark energy, critical components of the universe [45, 66].

Breakthroughs in these areas have the potential to revolutionize our understanding of the universe, the behavior of particles, and the nature of spacetime, and they may also have practical applications in quantum technologies.

12. Conclusion

In this chapter, we have delved into the intricate interplay between entanglement and high-energy physics (HEP), showcasing how this fundamental quantum phenomenon has profound implications for our understanding of the universe. From the vantage point of quantum field theory (QFT), entanglement emerges as a key player in elucidating the behavior of particles and their interactions [67]. The holographic nature of entanglement, illuminated by the AdS/CFT correspondence, demonstrates its potential as a bridge between quantum field theories and gravitational theories [2].

Entanglement entropy serves as a powerful tool to quantify and explore the intricate connections within HEP systems. Its applications in deciphering critical phenomena have led to a deeper comprehension of quantum phase transitions [14, 15]. Moreover, its role in black hole physics through the holographic principle hints at the profound relationship between information, spacetime, and gravity [54].

We have also explored how entanglement plays a pivotal role in testing the foundations of quantum mechanics, as exemplified by Bell's inequality and its implications for quantum correlations [1]. Furthermore, the deployment of entanglement for quantum simulations promises to revolutionize computational capabilities for HEP [34]. The potential of entangled states in quantum cryptography underscores its importance in securing sensitive information [26].

As we look to the future, the prospects for harnessing entanglement to unlock the mysteries of gravity, advance quantum computing, and deepen our understanding of quantum information remain tantalizingly promising. Yet, open questions persist, beckoning researchers to explore uncharted territories at the nexus of entanglement and HEP. How can we leverage entanglement to enhance our computational and communicative capacities? These questions and more constitute the frontier of entanglement in HEP, inviting future explorations that promise to reshape our understanding of the fundamental fabric of the cosmos [68]. The Nobel Prize in Physics for 2022 highlights the fundamental role of quantum entanglement nexus and HEP [44]. Can entanglement guide us toward a unified theory encompassing quantum and Bell violations in our understanding of quantum mechanics and its relevance to HEP? These experiments have opened new avenues for exploring the quantum behavior of particles, testing fundamental theories, and shedding light on the mysteries of the universe. In the context of HEP, these developments have significance for understanding the behavior of particles and forces in the universe, especially under extreme conditions.

> *In conclusion, the entanglement in HEP serves as a guiding thread, weaving through the intricate fabric of the cosmos, offering glimpses into the deepest mysteries of nature, and paving the way for new frontiers in research and understanding.*

Acknowledgements

The author thanks the publisher IntechOpen and Tampere University.

Author details

Mohammed Nadir
Tampere University, Finland

*Address all correspondence to: mohammed.nadir.fi@ieee.org

References

[1] Aspect A, Dalibard J, Roger G. Experimental test of Bell's inequalities, using time-varying analyzers (time-resolved photon counting). Physical Review Letters. 1982;**49**(25):1804-1807. DOI: 10.1103/PhysRevLett.49.1804

[2] Maldacena JM. The large N limit of superconformal field theories and supergravity. Advances in Theoretical and Mathematical Physics. 1999;**2**(2):231-252

[3] Van Raamsdonk M. Building up spacetime with quantum entanglement. General Relativity and Gravitation. 2010;**42**(10):2323-2329. DOI: 10.1007/s10714-010-1034-0

[4] Amico L, Fazio R, Osterloh A, Vedral V. Entanglement in many-body systems. Review of Modern Physics. 2008;**80**(2):517-576

[5] Calabrese P, Cardy J. Entanglement entropy and quantum field theory. Journal Statistical Mechanics: Theory and Experiment. 2004;**2004**(06):P06002. DOI: 10.1088/17425468/2004/06/P06002

[6] Clauser JF, Horne MA, Shimony A, Holt RA. Proposed experiment to test local hidden-variable theories. Physical Review Letters. 1970;**24**:549. DOI: 10.1103/PhysRevLett.24.549

[7] Fox R, Rosner B. Proposed experiment to test local hidden-variable theories. Physical Review D. 1971;**4**(4):1243-1244

[8] Hensen B et al. Loophole-free Bell inequality violation using electron spin separated by 1.3 kilometres. Nature. 21 Oct 2015;**526**:682-686. DOI: 10.1038/nature15759

[9] Ekert AK. Quantum cryptography is based on Bell's theorem. Physical Review Letters. 1991;**67**:661-663

[10] Preskill J. Quantum computing in the Noisy intermediate-scale quantum (NISQ) era and beyond. Quantum. 2018;**2**:79. DOI: 10.22331/q-2018-08-06-79.ni

[11] Nielsen MA, Chuang IL. Quantum Computation and Quantum Information. UK: Cambridge University Press; 2010 (One of the most cited books in physics of all time - Google)

[12] Ryu S, Takayanagi T. Holographic derivation of entanglement entropy from AdS/CFT. Physical Review Letters. 2006;**96**(18):181602. DOI: 10.1103/Phys Rev Lett.96.181602

[13] Vidal G, Latorre JI, Rico E, Kitaev A. Entanglement in quantum critical phenomena. Physical Review Letters. 2003;**90**(22):227902. DOI: 10.1103/PhysRevLett.90.227902

[14] Aspect A, Grangier P, Roger G. Experimental tests of realistic local theories via Bell's theorem. Physical Review Letters. 1982;**47**(7):460-463. DOI: 10.1103/Phys

[15] Shi X, Chen L. An extension of entanglement measures for pure states. Annalen der Physik. 26 Jul 2020;**533**(4):2000452. Available from: https://api.semanticscholar.org/CorpusID:220794033. DOI: 10.1002/andp.202000462. Available from: arXiv:2007.13164v1 [quant-ph]

[16] Bell JS. On the Einstein Podolsky Rosen paradox. Physics. 1964;**1**(3):195-200

[17] Benedetti M, Faccin M. Quantum-enhanced machine learning. Physical Review X. 2019;**9**(4):041011

[18] Chatrchyan S et al. Observation of a new boson at a mass of 125 GeV with

the CMS experiment at the LHC. Physics Letters B. 2012;**716**(1):30-61

[19] The ATLAS Collaboration. Observation of a new particle in the search for the standard model Higgs boson with the ATLAS detector at the LHC. Physics Letters B. 2012;**716**(1):1-29

[20] Aspect A. Bell's theorem: The naïve view of an experimentalist. Text of a Talk in Memory of John Bell. 2004. DOI: 10.48550/arXiv.quantph/0402001

[21] Bertone G et al. Physics Reports. 2005;**405**(5–6):279-390

[22] Giunti C, Kim CW. Fundamentals of Neutrino Physics and Astrophysics. New York, US: Oxford University Press; 2007

[23] Almheiri A et al. Journal of High Energy Physics. 2013;**2013**(2):82

[24] Harlow D. Reviews of Modern Physics. 2016;**88**(1):015002

[25] Wilde MM. Quantum Information Theory. Cambridge: Cambridge University Press; 2013

[26] Gisin N, Thew R. Quantum communication. Nature Photonics. 2007;**1**(3):165-171. DOI: 10.1038/nphoton.2007.22

[27] Yin J et al. Entanglement-based secure quantum cryptography over 1,120 kilometres, Nature. 20 Jun 2020;**582**. DOI: 10.1038/s41586-020-2401-y

[28] Gisin N, Ribordy G, Tittel W, Zbinden H. Quantum cryptography. Reviews of Modern Physics. 2002;**74**(1):145

[29] Casini H, Huerta M. Entanglement entropy in free quantum field theory. Journal of Physics A. 2009;**42**(50:)504007. p. 41. arXiv:0905.2562arXiv:0905.2562. DOI: 10.1088/1751-8113/42/50/504007

[30] Shuryak EV. What RHIC experiments and theory tell us about properties of quark–gluon plasma? Nuclear Physics A. 2004;**750**(1-2):64-83. arXiv:hep-ph/0405066 (hep-ph)

[31] Abdallah J et al. Simplified models for dark matter and missing energy searches at the LHC. Physics of the Dark Universe, Elsevier. Sep–Dec 2015;(**9-10**):8-23. Available from: arXiv:1808.10036 arXiv:1808.10036. DOI: 10.1016/j.dark.2015.08.001

[32] Peskin ME, Schroeder DV. An Introduction to Quantum Field Theory. US: CRC Press; 2018

[33] Horodecki R, Horodecki P, Horodecki M, Horodecki K. Quantum entanglement. Reviews of Modern Physics. 2009;**81**(2):865-942

[34] Georgescu IM, Ashhab S, Nori F. Quantum simulation. Reviews of Modern Physics. 2014;**86**(1):153-185

[35] Carroll SM. Spacetime and Geometry: An Introduction to General Relativity. UK: Cambridge University Press; 2019

[36] Bennett CH, Brassard G. Quantum cryptography: Public key distribution and coin tossing. In: Proceedings of IEEE International Conference on Computers, Systems and Signal Processing. Banglore, India. 10-12 Dec 1984. pp. 175-177. arXiv:2003.06557 [quant-ph]

[37] Wilczek F. Quantum Chromodynamics. The Modern Theory of the Strong Interaction. US AIP; 1982;**32**:177-209

[38] Rajibu I. Entanglement entropy in quantum many-body systems. Nature. 2020;**528**:77-83. DOI: 10.1038/nature15750

[39] Nishioka T, Takayanagi T. Entanglement entropy: Holography

and renormalization group. Reviews of Modern Physics. Jul-Sep 2018;**90**:035007-1. DOI: 10.1103/RevModPhys.90.03500

[40] Laflorencie N. Quantum entanglement in condensed matter systems. Physics Reports. Elsevier; 2016;**646**:1-59

[41] Casini H, Huerta M, Myers RC. Towards a derivation of holographic entanglement entropy. Journal of High Energy Physics. 2011;**036**:1105 DOI: 10.1007/JHEP05%282011%29036

[42] Eisert J, Cramer M, Plenio MB. Area laws for the entanglement entropy – A review. Reviews of Modern Physics. 2010. DOI: 10.48550/arXiv.0808.3773

[43] Von Neumann J. Mathematical Foundations of Quantum Mechanics. US, London, UK: Princeton University Press, Geoffrey Cumberlege, Oxford University Press; 1995

[44] Nobel Prize. Awarded to Alain Aspect, Clauser and Anton Zeilinger that entanglement is real as described on webpages. 2022. Available from: https://www.nobelprize.org/uploads/2022/10/popular-physicsprize2022.pdf, https://www.nobelprize.org/educational-nobel-prize-lessons-physics-2022

[45] Casalderrey-Solana J, Mateos D. Gauge/string duality, hot QCD and heavy ion collisions. Journal of Physics A: Mathematical and Theoretical. 2014;**47**(25):253001. DOI: 10.1017/CBO9781139136747

[46] 't Hooft G. Dimensional reduction in quantum gravity. 1993. arXiv preprint gr-qc/9310026

[47] Swingle B. Entanglement renormalization and holography. Physical Review D. 2012;**86**:065007 DOI: 10.1103/PhysRevD.86.065007

[48] Hirvonen H, Eskola KJ, Niemi H. Deep learning for flow observables in ultrarelativistic heavy-ion collisions. 8 Mar 2023. arXiv:2303.04517v1 [hep-ph]. DOI: 10.48550/arXiv.2303.04517

[49] Geoffrey P. Entanglement wedge reconstruction and the information paradox. Aug 2020:1-75. arXiv:1905.08255 [hep-th] DOI: 10.48550/arXiv.1905.08255

[50] Jordan SP, Lee KSM, Preskill J. Quantum algorithms for quantum field theories. Science. 2012. DOI: 10.48550/arXiv.1111.3633

[51] Maldacena SJ, Alexey MSYK wormhole formation in real-time, Published for SISSA by Springer, JHEP04 2021:258. DOI: 10.1007/JHEP04(2021)258

[52] Bekenstein JD. Information holographic universe. Scientific American. Aug 2003:45-52

[53] Harlow D. Quantum gravity at a distance. Journal of High Energy Physics. 2014;**2014**(11):1-39

[54] Susskind L. The world as a hologram. Journal of Mathematical Physics. Nov 1995;**36**(11). DOI: 10.1063/1.531249

[55] Cirac JI, Zoller P. Goals and opportunities in quantum simulation. Nature Physics. 2012;**8**(4):264-266

[56] Bilenky SM, Giunti C, Grimus W. Phenomenology of neutrino oscillations. Progress in Particle and Nuclear Physics. 1999;**43**:1-86.

[57] Hawking SW. Particle creation by black holes. Communications in Mathematical Physics. 1975;**43**(3):199-220

[58] Stefano A et al. The black hole information paradox. 2018. Available from: https://www.cs.umd.edu/class/fall2018/cmsc657/projects/group_2.pdf

[59] Hawking SW. Breakdown of predictability in gravitational collapse. Physical Review D. 1976;**14**:2460

[60] Baez JC, Vicary J. Wormholes and Entanglement. 4 Feb, 2014. arXiv:1401.3416v2 [gr-qc] 3 Feb 2014. Available from: https://arxiv.org/abs/1401.3416

[61] Baez JC, Vicary J. Wormholes and entanglement. Classical and Quantum Gravity. 2014;**31**(21):214007

[62] Scarani V et al. The security of practical quantum key distribution. Reviews of Modern Physics. 29 Sep 2009;**81**:1301

[63] Sadeghi D et al. Application of AdS/CFT in quark-gluon plasma, (HEP_theory). 2013. DOI: 10.1155/2013/759804

[64] Swingle B. Entanglement renormalization and holography. 2012. arXiv:1209.3304v1 [hep-th] [Accessed: Sep 14, 2012]

[65] Casalderrey-Solana J, Liu H, Mateos D, Rajagopal K, Wiedemann UA. Gauge/string duality, hot QCD and heavy ion collisions. CUP. 2014. arXiv:1101.0618 (HEP-TH)

[66] Verlinde EP. Emergent Gravity and the Dark Universe. 2016. arXiv:1611.02269 [hep-th]. DOI: 10.21468/SciPostPhys.2.3

[67] Weinberg S. The Quantum Theory of Fields. Vol. 1. UK: Cambridge University Press; 1995

[68] Arkani-Hamed N et al. The universe as a hologram. Journal of High Energy Physics. 2001;**2001**, JHEP08. DOI: 10.1088/1126-6708/2001/08/017

Section 4

Experiments

Chapter 11

Translational Symmetry of Intermediate Nodes and Antinodes of Entangled Particles

Kisalaya Chakrabarti

Abstract

In a 1935 publication, Albert Einstein, Boris Podolsky, and Nathan Rosen present a thought experiment meant to demonstrate the apparent absurdity of quantum entanglement, which appeared to be in direct opposition to a fundamental law of the world. It was obvious to the men who created quantum mechanics in the decade of 1920, that quantum entanglement is unsettling because of the fact that quantum superposition is a real phenomenon. Thomas Vidick, a professor of physics, says it may be alluring to believe that the particles are somehow interacting with each other across such vast distances. A key resource for quantum networks is entanglement, a fundamental aspect of quantum theory that enables the sharing of unbreakable quantum connections between distant participants. Distributing entanglement across distant items that can act as quantum memory is very important. This has been accomplished in the past employing systems like warm and cold atomic vapours, single atoms and ions, and flaws in solid-state system

Keywords: K-state, entangled pair, standing waves, distributed locality, POVM

1. Introduction

Einstein splendidly criticised the phenomenon of quantum physics, particularly quantum entanglement, which shows how particles can share a combined state while separated arbitrarily far apart. Struggling with the oddity of this phenomenon, he referred to it as "spooky action at a distance" [1, 2]. This uncertainty has recently become more pronounced in various non-locality forms. The truth of this phenomenon has recently been confirmed by scientists over ever-greater distances, even from Earth to an orbiting spacecraft. Many times, it defies the special theory of relativity's fundamental tenet that nothing can move faster than the speed of light, which has baffled scientists for years. In this study, the author suggests a potential quantum entanglement mechanism that upholds the special theory's basic premise. It is important to note that the spin directions of the entangled particles such as vector bosons in Higgs boson decays are comparable to the corresponding phases of the standing wave's intermediate field particles. Furthermore, a crucial phenomenon known as "phase transitions" can be observed in one-dimensional overlaid mixed wave functions. A crucial finding is that the application of a magnetic field to any entangled particle, no matter how minute, results in the same spin orientations in the intermediate antinodes regardless of temperature changes.

By effectively splitting a system into two parts, where the sum of the portions is known, entangled particles are produced. For instance, it is possible to divide a particle with spin zero into two particles whose spins will inevitably be opposite one another, resulting in a sum of zero.

David Bohm is credited with coming up with a condensed version of this thought experiment that takes the decay of the pi meson into account. The electron and positron that are created when this particle decays have opposite spins and are travelling away from one another. So, if the measured spin of the electron is up, the measured spin of the positron can only be down, and the opposite is true. Even if the particles are billions of miles distant, this is still true.

If the measured positron spin were always down and the measured electron spin were always up, everything would be OK. Each particle's spin, however, has both an up and a down component until it is detected due to quantum physics. The quantum state of the spin does not "collapse" into either up or down until the measurement takes place, at which point the other particle instantly assumes the opposite spin. This appears to indicate that the particles are in communication with one another via a method that is moving faster than light. However, nothing can move more quickly than the speed of light, according to the laws of physics. Undoubtedly, it is impossible to instantly predict the state of one particle from its measured condition.

A formula created by Bell is now known as Bell's inequality, and it is only and always true for hidden variable theories—never for quantum mechanics. This means that local hidden variable theories cannot account for quantum entanglement if Bell's equation is proven to be unsatisfied in a real-world experiment. The idea that entangled particles are communicating with one another faster than the speed of light contradicts Einstein's special theory of relativity, and is a popular fallacy about entanglement. This is not the case, and quantum physics cannot be employed to deliver messages that travel at a faster-than-light rate. The origin of the entanglement phenomena, which at first glance appears strange, is still a topic of controversy among scientists, but they are aware that the theory behind it holds up to repeated testing. Despite the fact that Einstein famously called entanglement "spooky action at a distance," modern quantum scientists assert that it is not spooky in any way. According to Schrödinger, with a certain probability, a particle in another location could be guided into one of a set of states by an entangled state. In reality, the 'remote guiding' potential is considerably more spectacular than Schrödinger's experiment. Assuming, two experimenter Bob and Alice both possess a pair of entangled photons that are in an entangled pure state similar to those Bell considered, with Alice possessing one of the entangled photons and Bob is holding the other. Assume that Alice also receives an additional photon with unknown polarisation $|a\rangle$ where "$|\rangle$" represents a quantum state. If Alice performs an operation on the two photons in her possession, Bob's photon may change into one of the four states. The four possible (random) outcomes of Alice's operation are either the state $|a\rangle$ or a state that is clearly related to $|a\rangle$. The two photons Alice is holding become entangled as a result of Alice's operation, which also manages to separate Bob's photon and direct it into a state of $|a^*\rangle$. Bob is either aware that $|a^*\rangle = |a\rangle$ or understands how to do a local operation to change $|a^*\rangle$ to $|a\rangle$ after Alice tells him the results of her operation. Quantum teleportation is the term used to describe this phenomenon. Both Alice and Bob are unable to determine the state $|a\rangle$ following the teleportation technique [3, 4]. Two conventional bits of information can be used to specify Alice's operation's result, which has four potential outcomes and an equal probability of 0.25. Surprisingly, Bob can recreate the state $|a\rangle$ from just two bits of classical information that Alice has communicated. It appears that Bob did this by using the entangled state as a quantum communication channel. The noteworthy aspect of these phenomena is that Alice and Bob were able to use their shared entangled state as a quantum communication channel to alter the state $|a\rangle$ of a photon in Alice's part of the universe and reproduce it in Bob's portion of the universe.

In 1989, Daniel Greenberger, Michael Horne, and Anton Zeilinger conducted the initial research on it. A Greenberger-Horne-Zeilinger state (GHZ state) is a specific kind of entangled quantum state in the field of quantum information theory that involves three subsystems (particle states, qubits, or qudits). In advancement of this aforesaid theory on the year 2020, Kisalaya Chakrabarti proposed an entangled quantum state in the field of quantum information theory that involves four subsystems called Kisalaya state (K state) [4].

In this regard, it is worth mentioning, aground-breaking workshop that was held at the Brookhaven National Laboratory in 2018, from onwards the ideas of using quantum information methods and tools in high-energy physics is gaining more and more attention. We know that, the density matrix of the hadrons – the bound states of quarks and gluons, must be averaged over the phase with the associated Haar integration measure because it is impossible to detect the phase of the wave function in high energy interactions. This results in the density matrix of the Parton model, which is a mixed state density matrix that is diagonal in Parton number. With the help of this method, one can better grasp the limits of the Parton model, such as how it cannot accurately describe spin-dependent processes [5].

2. Probability transfer matrix of an entangled particle at a distance

Mixed state states are possible for quantum mechanical systems. The term "pure state" refers to the states that wave functions describe. In this case, the probability states are superimposed. Starting from the Pauli's matrix of quantum physics, which accounts for the interaction between a particle's spin and an external electromagnetic field and additionally they depict the interaction states of two polarisation filters for 45° right or left, circular or horizontal or vertical polarisation, given by:

$$\sigma_x = \begin{bmatrix} 0 & 1 \\ 1 & 0 \end{bmatrix}, \sigma_y = \begin{bmatrix} 0 & -i \\ i & 0 \end{bmatrix} \text{ and } \sigma_z = \begin{bmatrix} 1 & 0 \\ 0 & -1 \end{bmatrix}$$

Now, taking the tensor products of two parallel operations at superimposed similar states given as follows;

$$\sigma_x \otimes \sigma_x = [\mathrm{A}] = \begin{bmatrix} 0 & 1 \\ 1 & 0 \end{bmatrix}\begin{bmatrix} 0 & 1 \\ 1 & 0 \end{bmatrix} = \begin{bmatrix} 0 & 0 & 0 & 1 \\ 0 & 0 & 1 & 0 \\ 0 & 1 & 0 & 0 \\ 1 & 0 & 0 & 0 \end{bmatrix} = \begin{bmatrix} \rho_i & \rho_i & \rho_i & 1-\rho_i \\ \rho_i & \rho_i & 1-\rho_i & \rho_i \\ \rho_i & 1-\rho_i & \rho_i & \rho_i \\ 1-\rho_i & \rho_i & \rho_i & \rho_i \end{bmatrix} \tag{1}$$

Instead of using wave functions in this situation, we must employ the idea of a probability transfer matrix. A mixed state can be represented as an incoherent summation of Orthonormal bases $|\psi_i\rangle$'s as $\rho = \sum_i \rho_i |\psi_i\rangle\langle\psi_i|$ where ρ_i is the probability P for the system in the state of ψ_i, and ψ_i's are the diagonal basis for ρ. ρ_i are referred to as the density matrix's Eigen values ρ.

Similarly,

$$\sigma_y \otimes \sigma_y = [\mathrm{B}] = \begin{bmatrix} 0 & -i \\ i & 0 \end{bmatrix} \begin{bmatrix} 0 & -i \\ i & 0 \end{bmatrix} = \begin{bmatrix} 0 & 0 & 0 & -1 \\ 0 & 0 & +1 & 0 \\ 0 & +1 & 0 & 0 \\ -1 & 0 & 0 & 0 \end{bmatrix}$$
$$= \begin{bmatrix} \rho_i & \rho_i & \rho_i & |\rho_i - 1| \\ \rho_i & \rho_i & 1-\rho_i & \rho_i \\ \rho_i & 1-\rho_i & \rho_i & \rho_i \\ |\rho_i - 1| & \rho_i & \rho_i & \rho_i \end{bmatrix} \tag{2}$$

and

$$\sigma_z \otimes \sigma_z = [\mathrm{C}] = \begin{bmatrix} 1 & 0 \\ 0 & -1 \end{bmatrix} \begin{bmatrix} 1 & 0 \\ 0 & -1 \end{bmatrix} = \begin{bmatrix} 1 & 0 & 0 & 0 \\ 0 & -1 & 0 & 0 \\ 0 & 0 & -1 & 0 \\ 0 & 0 & 0 & 1 \end{bmatrix}$$
$$= \begin{bmatrix} 1-\rho_i & \rho_i & \rho_i & \rho_i \\ \rho_i & |\rho_i - 1| & \rho_i & \rho_i \\ \rho_i & \rho_i & |\rho_i - 1| & \rho_i \\ \rho_i & \rho_i & \rho_i & 1-\rho_i \end{bmatrix} \tag{3}$$

Tensor Products or Tensor Matrices A, B and C are a set of three 4×4 complex matrices which are Hermitian, involutory and unitary. Here worth mentioning that, if and only if; $\mathrm{M}^2 = \mathrm{I}$, where I is the 4×4 identity matrix, then multiplying by the matrix M results in an involution. Each and every involutory matrix is the square root of the identity matrix. This is merely a result of the fact that the identity is obtained by multiplying any non-singular matrix by its inverse.

It is an important observation that none of the Tensor matrices (Tensor products) A, B and C have elements consisting of imaginary quantity. The two Eigen values of the three Hermitian Tensor products are +1 and -1.

In the same fashion one can achieve:

$$\sigma_x \otimes \sigma_y = [\mathrm{D}] = \begin{bmatrix} 0 & 1 \\ 1 & 0 \end{bmatrix} \begin{bmatrix} 0 & -i \\ i & 0 \end{bmatrix} = \begin{bmatrix} 0 & 0 & 0 & -i \\ 0 & 0 & i & 0 \\ 0 & -i & 0 & 0 \\ i & 0 & 0 & 0 \end{bmatrix}$$
$$= i \begin{bmatrix} \rho_i & \rho_i & \rho_i & |\rho_i - 1| \\ \rho_i & \rho_i & 1-\rho_i & \rho_i \\ \rho_i & |\rho_i - 1| & \rho_i & \rho_i \\ 1-\rho_i & \rho_i & \rho_i & \rho_i \end{bmatrix} \tag{4}$$

$$\sigma_y \otimes \sigma_z = [\mathrm{E}] = \begin{bmatrix} 0 & -i \\ i & 0 \end{bmatrix} \begin{bmatrix} 1 & 0 \\ 0 & -1 \end{bmatrix} = \begin{bmatrix} 0 & 0 & -i & 0 \\ 0 & 0 & 0 & i \\ i & 0 & 0 & 0 \\ 0 & -i & 0 & 0 \end{bmatrix}$$
$$= i \begin{bmatrix} \rho_i & \rho_i & |\rho_i - 1| & \rho_i \\ \rho_i & \rho_i & \rho_i & 1-\rho_i \\ 1-\rho_i & \rho_i & \rho_i & \rho_i \\ \rho_i & |\rho_i - 1| & \rho_i & \rho_i \end{bmatrix} \tag{5}$$

$$\sigma_z \otimes \sigma_x = [\mathrm{F}] = \begin{bmatrix} 1 & 0 \\ 0 & -1 \end{bmatrix} \begin{bmatrix} 0 & 1 \\ 1 & 0 \end{bmatrix} = \begin{bmatrix} 0 & 1 & 0 & 0 \\ 1 & 0 & 0 & 0 \\ 0 & 0 & 0 & -1 \\ 0 & 0 & -1 & 0 \end{bmatrix} = \begin{bmatrix} \rho_i & 1-\rho_i & \rho_i & \rho_i \\ 1-\rho_i & \rho_i & \rho_i & \rho_i \\ \rho_i & \rho_i & \rho_i & |\rho_i - 1| \\ \rho_i & \rho_i & |\rho_i - 1| & \rho_i \end{bmatrix} \tag{6}$$

In the reverse order the Tensor products are given as follows:

$$\sigma_y \otimes \sigma_x = [\mathrm{G}] = \begin{bmatrix} 0 & -i \\ i & 0 \end{bmatrix} \begin{bmatrix} 0 & 1 \\ 1 & 0 \end{bmatrix} = \begin{bmatrix} 0 & 0 & 0 & -i \\ 0 & 0 & -i & 0 \\ 0 & i & 0 & 0 \\ i & 0 & 0 & 0 \end{bmatrix}$$
$$= i \begin{bmatrix} \rho_i & \rho_i & \rho_i & |\rho_i - 1| \\ \rho_i & \rho_i & |\rho_i - 1| & \rho_i \\ \rho_i & 1-\rho_i & \rho_i & \rho_i \\ 1-\rho_i & \rho_i & \rho_i & \rho_i \end{bmatrix} \tag{7}$$

$$\sigma_x \otimes \sigma_z = [\mathrm{H}] = \begin{bmatrix} 0 & 1 \\ 1 & 0 \end{bmatrix} \begin{bmatrix} 1 & 0 \\ 0 & -1 \end{bmatrix} = \begin{bmatrix} 0 & 0 & 1 & 0 \\ 0 & 0 & 0 & -1 \\ 1 & 0 & 0 & 0 \\ 0 & -1 & 0 & 0 \end{bmatrix} = \begin{bmatrix} \rho_i & \rho_i & 1-\rho_i & \rho_i \\ \rho_i & \rho_i & \rho_i & |\rho_i - 1| \\ 1-\rho_i & \rho_i & \rho_i & \rho_i \\ \rho_i & |\rho_i - 1| & \rho_i & \rho_i \end{bmatrix} \tag{8}$$

$$\sigma_z \otimes \sigma_y = [\mathrm{I}] = \begin{bmatrix} 1 & 0 \\ 0 & -1 \end{bmatrix} \begin{bmatrix} 0 & -i \\ i & 0 \end{bmatrix} = \begin{bmatrix} 0 & -i & 0 & 0 \\ i & 0 & 0 & 0 \\ 0 & 0 & 0 & i \\ 0 & 0 & -i & 0 \end{bmatrix}$$
$$= i \begin{bmatrix} \rho_i & |\rho_i - 1| & \rho_i & \rho_i \\ 1-\rho_i & \rho_i & \rho_i & \rho_i \\ \rho_i & \rho_i & \rho_i & 1-\rho_i \\ \rho_i & \rho_i & |\rho_i - 1| & \rho_i \end{bmatrix} \tag{9}$$

Measurements of this type are referred to as Positive operator valued measure (POVM). The most general type of POVM in which sums are substituted by integrals; so they are termed measures, which has a continuous family of outcomes. In this case there are a finite number of potential outcomes, over a finite dimensional quantum space. Hermitian matrices having finite dimensional positive operators are used here as the elements. We use these measures implicitly when we measure some qubits of a quantum computer while leaving other qubits unaffected, described in a proper mathematical definition given below:

$$\prod_i \prod_j = 0 \quad \text{if } i \neq j \text{ and } \sum_{i-1}^{k} = 1 \tag{10}$$

Considering the case when we have a collection of projectors onto the orthogonal subspaces $\prod_1, \prod_2, \ldots.., \prod_k$ which implies that the tensor products to be $[D] = [E] = [F] = [G] = [H] = [I] = 0$, from Ortho-normality condition depicted in Eq. (10) where the projective measurement in these subspaces that evolves a quantum state $|\psi\rangle$ in the state given by

$$|\psi'\rangle = \frac{\prod_{i\neq j\neq k}|\psi\rangle}{\langle\psi|\prod_{i\neq j\neq k}|\psi\rangle^{1/2}} \text{ where Probablity } P = \langle\psi| \prod_{i\neq j\neq k} |\psi\rangle \tag{11}$$

A minimum rate of unresolved events or error probability P_e, is imposed by the state's impossible to differentiate detection and is given by the following equation:

$$P_e = 1 - \mu \text{ where, } \mu = \left\langle \psi_i | \psi_j \right\rangle \tag{12}$$

From Eq. (12) it is clearly seen that when $i \neq j$, $P_e = 1$.

Therefore, the equations of importance are (1), (2) and (3) only corresponds to [A], [B} and [C] tensor product matrices where

$$\prod_i \prod_i = \prod_j \prod_j = \prod_k \prod_k = 1 \tag{13}$$

Solving for Eigenvalues and Eigenvectors for [A], we find that the Eigenvalues $\lambda = \pm 1$ and the Eigenvectors are as follows:

$$\text{for } \lambda = 1 \qquad \text{Eigenvectors } X = \begin{pmatrix} \psi_1 \\ \psi_2 \\ \psi_2 \\ \psi_1 \end{pmatrix} \tag{14}$$

Eq. (14) can be written as:

$$X = \psi_1 \begin{pmatrix} 1 \\ 0 \\ 0 \\ 1 \end{pmatrix} + \psi_2 \begin{pmatrix} 0 \\ 1 \\ 1 \\ 0 \end{pmatrix} \tag{15}$$

and

$$\text{for } \lambda = -1 \qquad \text{Eigenvectors } X = \begin{pmatrix} \psi_1 \\ \psi_2 \\ -\psi_2 \\ -\psi_1 \end{pmatrix} \tag{16}$$

Eq. (16) can be written as:

$$X = \psi_1 \begin{pmatrix} 1 \\ 0 \\ 0 \\ -1 \end{pmatrix} + \psi_2 \begin{pmatrix} 0 \\ 1 \\ -1 \\ 0 \end{pmatrix} \tag{17}$$

In the same fashion Eigenvalues and Eigenvectors for [B] are as follows:

$$\text{for } \lambda = 1 \qquad \text{Eigenvectors } X = \begin{pmatrix} \psi_1 \\ \psi_2 \\ \psi_2 \\ -\psi_1 \end{pmatrix} \tag{18}$$

Eq. (18) can be written as:

$$X = \psi_1 \begin{pmatrix} 1 \\ 0 \\ 0 \\ -1 \end{pmatrix} + \psi_2 \begin{pmatrix} 0 \\ 1 \\ 1 \\ 0 \end{pmatrix} \tag{19}$$

and;

$$\text{for } \lambda = -1 \qquad \text{Eigenvectors } X = \begin{pmatrix} \psi_1 \\ \psi_2 \\ -\psi_2 \\ \psi_1 \end{pmatrix} \tag{20}$$

Eq. (20) can be written as:

$$X = \psi_1 \begin{pmatrix} 1 \\ 0 \\ 0 \\ 1 \end{pmatrix} + \psi_2 \begin{pmatrix} 0 \\ 1 \\ -1 \\ 0 \end{pmatrix} \tag{21}$$

Similarly, Eigenvalues and Eigenvectors for [C] are as follows:

$$\text{for } \lambda = 1 \qquad \text{Eigenvectors } X = \begin{pmatrix} \psi_1 \\ 0 \\ 0 \\ \psi_4 \end{pmatrix} \tag{22}$$

Eq. (22) can be written as:

$$X = \psi_1 \begin{pmatrix} 1 \\ 0 \\ 0 \\ 0 \end{pmatrix} + \psi_4 \begin{pmatrix} 0 \\ 0 \\ 0 \\ 1 \end{pmatrix} \tag{23}$$

and;

$$\text{for } \lambda = -1 \qquad \text{Eigenvectors } X = \begin{pmatrix} 0 \\ \psi_2 \\ \psi_3 \\ 0 \end{pmatrix} \tag{24}$$

Eq. (24) can be written as:

$$X = \psi_2 \begin{pmatrix} 0 \\ 1 \\ 0 \\ 0 \end{pmatrix} + \psi_3 \begin{pmatrix} 0 \\ 0 \\ 1 \\ 0 \end{pmatrix} \tag{25}$$

Now, from the set of two Eqs. (15) and (17); and (19) and (21) it is noticeable that only two vector states $|\psi_1\rangle, |\psi_2\rangle$ with two different orientations $|+,-\rangle$ are originated from the interference of forward and reverse moving waves at the antinodes sections resulting in Standing waves, whereas Eqs. (21) and (23) describe that presence of all the four states $|\psi_1\rangle, |\psi_2\rangle, |\psi_3\rangle, |\psi_4\rangle$ have been noticed with a single orientation $|+\rangle$. As there are only two Eigen vector states $|\psi_1\rangle, |\psi_2\rangle$ in three dimensional vector space and with two different orientations $|+,-\rangle$, it is obvious that $|\psi_4\rangle = -|\psi_1\rangle$ and $|\psi_3\rangle = -|\psi_2\rangle$.

So we can modify Eq. (23) as follows:

$$\text{for } \lambda = 1 \qquad X = \psi_1 \begin{pmatrix} 1 \\ 0 \\ 0 \\ -1 \end{pmatrix} \tag{26}$$

Similarly, Eq. (25) as follows and

$$\text{for } \lambda = -1 \qquad X = \psi_2 \begin{pmatrix} 0 \\ 1 \\ -1 \\ 0 \end{pmatrix} \tag{27}$$

From Eqs. (15) and (17); (19) and (21) and (26) and (27) we find that only two pure states or the combination of two states are manifested with two different orientations $|\psi_{1\pm}\rangle, |\psi_{2\pm}\rangle$ at the antinodes positions due to interference of two waves originated from two entangled particles lying far apart. It is important and observed from

DOI: http://dx.doi.org/10.5772/intechopen.1002292

(26) and (27) that only two pure states $|\psi_{1\pm}\rangle$, $|\psi_{2\pm}\rangle$ exist at antinodes at the standing wave perpendicular to the direction of propagations.

3. Entanglement theory with stationary phase approximation

Consider two entangled particles with superimposed states between them at positions "A" and "B" separated by an arbitrarily large distance "d" and having superimposed states between them ψ_s. Since there is no net energy transfers in either the forward or backward direction because the two particles are entangled, this composite wavefunction ψ_S is composed of two ψ states, ψ_1 and ψ_2, where, ψ_1 denotes the wave moving forward from "A" to "B" and a backward travelling wave ψ_2 moving from "B" to "A". The propagation of forward or backward travelling waves does not yet take the spin directions of either particle into consideration; therefore we can express the superimposed wavefunction in the context of the Schrodinger equation [4],

$$\psi_S = \psi_1 + \psi_2 = \left(\psi_{01}e^{\frac{iPx}{\hbar}} + \psi_{02}e^{\frac{-iPx}{\hbar}}\right) \tag{28}$$

Without taking spin effect on directions into account, Eq. (28) can be written as

$$\psi_S = \left(\psi_{01}e^{\frac{iPx}{\hbar}} - \psi_{01}e^{\frac{-iPx}{\hbar}}\right) \tag{29}$$

Now, taking into account the spin direction of the particle at location "A" during the forward wave propagation ψ_1 and we must select an axis to line up with the measuring device's axis in order to measure the spin direction. Finally, we denote this observed spin of the particle at "A" by unit vector by moving the axis of the measurement device located at point "A" in the direction of the forward travelling wave ψ_1 in the forward "+x" direction and we will denote this observed spin of the particle at point "A" by unit vector $\hat{x}$, and bringing in this factor to "ψ_1" we obtained the modified wavefunction expressed as:

$$\psi_1 = \psi_{01}e^{\frac{iPx}{\hbar}}\hat{x} \tag{30}$$

Taking into account the time evolution of the combined wavefunction, Eq. (29) is customised and can be written as follows:

$$\psi_S(t) = \text{Re}\,\left(\psi_S e^{i\omega t}\right) = 2\psi_{01}\cos\left(\frac{Px}{\hbar}\right)\,\cos(\omega t)\,\hat{x} \tag{31}$$

where, ω is the angular velocity of $\psi_S(t)$, given by $E/\hbar$.

It is now necessary to comprehend the stationary phase approximation of the combined wavefunction of all entangled particles, which is given by combined wave function $\psi_S(t)$, as shown in equation, in order to comprehend how quantum entanglement changes with time (31).

$$J = \lim_{\Omega\to 0}\int_{-\infty}^{\infty} e^{i\psi_S(t)/\Omega}dt \tag{32}$$

Eq. (32) can be stretched out as follows:

$$J \approx \sum_{n} \int_{-\infty}^{\infty} e^{i\left\{\frac{\psi_S(t_0)}{\Omega} + \frac{\frac{d\,\psi_S(t_0)}{dt}(t-t_0)}{\Omega} + \frac{\frac{d^2\,\psi_S(t_0)}{dt^2}(t-t_0)^2}{2\Omega}\right\}} dt \approx e^{i\frac{\psi_S(t_0)}{\Omega}} \sqrt{\frac{2\Omega\pi}{i\frac{d^2\,\psi_S(t_0)}{dt^2}}} \tag{33}$$

In order to formally follow the convention of Gaussian integration, one should replace "i" to "-i" inside the square root portion of Eq. (33). The imaginary unit "i" is assumed to be a true positive constant even though the integral mathematically speaking does not exist. We obtain nonzero contributions precisely where, just like in the Path Integral case;

$$\delta\, S[\,\psi_S(t), p(t), t] = 0 \tag{34}$$

which is the prerequisite for the functional of the classical trajectory "S" and is expressed in units of $\hbar$; $p(t)$ is the momentum of the entangled system; as a result, in any physical domain where "S" is much larger than, even the slightest deviation from the classical trajectory causes $\psi_S(t)$ to wiggle out so quickly that it has no impact on the value of "J". The conventional boundary is $\frac{S}{\hbar} \gg 1$.

As is evident, that $J = \psi_S(t)$, therefore squaring and comparing Eq. (31) with Eq. (33) and allowing for the real part, we obtain

$$\frac{\Omega\pi\, \sin\left(\frac{2\psi_S(t_0)}{\Omega}\right)}{2{\psi_{01}}^2 \cos^2(\omega t)} = \frac{d^2\,\psi_S(t_0)}{dt^2} \cos^2\left(\frac{Px}{\hbar}\right) \tag{35}$$

Putting the average value of $\cos^2\left(\frac{Px}{\hbar}\right) = \frac{1}{2}$ at x direction, Eq. (35) can be written as

$$\frac{\Omega\pi\, \sin\left(\frac{2\psi_S(t_0)}{\Omega}\right)}{{\psi_{01}}^2 \cos^2(\omega t)} = \frac{d^2\,\psi_S(t_0)}{dt^2} \tag{36}$$

Now, Eq. (36) can be rewritten as;

$$\frac{d^2\psi_S(t_0)}{dt^2} = \kappa \frac{\sin(\psi_S(t_0))}{\cos^2(\omega t)} \tag{37}$$

where, $\kappa = \frac{\Omega\,\pi}{{\psi_{01}}^2} =$ constant and equating it to "1" with book-keeping constant "Ω" for the sake of arithmetical straightforwardnesswhich serves only as a scaling factor.

Now solving the differential Eq. (37) with angular frequency $\omega = 1$, we can derive the solution of $\psi_S(t_0)$ by double integration at any point of time t_0, is in the form:

$$\psi_S(t_0) = \ln(|\,tan\,(t_0) + sec\,(t_0)|) + \mathbf{C} \tag{38}$$

Now, assuming any value of Constant "**C** = 7" and plotting the graph we get **Figure 1**.

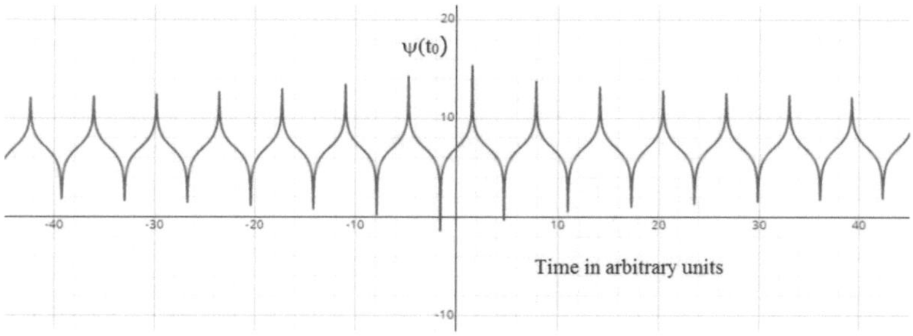

Figure 1.
Superimposed wave function $\psi_S(t_o)$ field distribution with time.

4. Results and discussions

An interesting fact is observed from the **Figure 1** that the Standing wave pattern formed and has a Translational symmetry. This pattern is periodic and is repeated after a time interval "T". This phenomenon is called "Distributed locality" which implies the same pattern is repeated throughout the space after regular time interval. Therefore once such pattern is formed, the information of spin is translated immediately between two entangled particles kept in far apart [2]. This forms a "String". For an "N" dimensional observations there are "N" number of such strings which are superimposed to each other in different phases and amplitudes. Now, when we measure the spin at any end of the entangled pair particle at any particular direction or orientation, immediately the string correlated with the other directions or orientations collapses. This is called String lowering Operation.

When the measurement has taken place to calculate the Spin Orientation of an entangled particle for any particular direction; supposedly in the $\hat{x}$ direction, we get:

$$|\psi\rangle = \frac{1}{\sqrt{2}}(|\psi_1\rangle + |\psi_2\rangle) \tag{39}$$

where, $\hat{x} = \sqrt{\frac{\hbar}{2m\omega}}(a + a^\dagger)$

Now,

$$\begin{aligned}\langle\psi|\hat{x}|\psi\rangle &= \frac{1}{2}\langle(\langle\psi_1| + \langle\psi_2|)|\sqrt{\frac{\hbar}{2m\omega}}(a + a^\dagger)|(|\psi_1\rangle + |\psi_2\rangle)\rangle \\ &= \frac{1}{2}\sqrt{\frac{\hbar}{2m\omega}}\langle(\langle\psi_1| + \langle\psi_2|)\Big|\Big(|\psi_1\rangle + |\psi_2\rangle + \sqrt{2}\Big)|\psi_3\rangle\Big\rangle \\ &= \frac{1}{2}\sqrt{\frac{\hbar}{2m\omega}}(1+1) = \sqrt{\frac{\hbar}{2m\omega}}\end{aligned} \tag{40}$$

We find a finite expectation value in the $\hat{x}$ direction which is quiet obvious and all the others strings are immediately readjusted to "0" as follows:

We know, due to Ortho-normality conditions, $|\psi\rangle = \frac{1}{\sqrt{2}}(|\psi_1\rangle + i|\psi_2\rangle) = 0$.

Therefore,

$$\begin{aligned}\langle\psi|\hat{x}|\psi\rangle &= \frac{1}{2}\langle(\langle\psi_1| + i\langle\psi_2|)|\sqrt{\frac{\hbar}{2m\omega}}(a + a^\dagger)|(|\psi_1\rangle + i|\psi_2\rangle)\rangle \\ &== \frac{1}{2}\sqrt{\frac{\hbar}{2m\omega}}(1-1) = 0\end{aligned} \tag{41}$$

This Operation is "String Lowering Operation".

5. Conclusion

A particle with spin zero can be split into two particles whose spins are invariably opposite one another, giving rise to a sum of zero. Until the measurement is made, the quantum state of the spin does not "collapse" into either up or down; nevertheless, when the measurement is made, the other particle immediately assumes the opposing spin.

Author details

Kisalaya Chakrabarti
Department of Electronics and Communication Engineering, Haldia Institute of Technology, Haldia, W.B, India

*Address all correspondence to: kisalayac@gmail.com

References

[1] Einstein A et al. Can quantum-mechanical description of physical reality be considered complete? Physical Review. 1935;**47**:777-780

[2] Chakrabarti K. Is there any spooky action at a distance? In: Maji AK, Saha G, Das S, Basu S, Tavares JMRS, editors. Proceedings of the International Conference on Computing and Communication Systems. Lecture Notes in Networks and Systems. Vol. 170. Singapore: Springer; 2021. DOI: 10.1007/978-981-33-4084-8_65

[3] Bennett C, Brassard G, Crepeau C, Jozsa R, Peres A, Wootters W. Teleporting an unknown quantum state via dual classical and Einstein-Podolsky-Rosen channels. Physical Review Letters. 1993;**70**:1895-1899

[4] Chakrabarti K. Realization of original quantum entanglement state from mixing of four entangled quantum states. In: Castillo O, Jana D, Giri D, Ahmed A, editors. Recent Advances in Intelligent Information Systems and Applied Mathematics: ICITAM 2019. Studies in Computational Intelligence. Vol. 863. Cham: Springer; 2020. DOI: 10.1007/978-3-030-34152-7_12

[5] Kharzeev DE. Quantum information approach to high energy interactions. Philosophical Transactions on Royal Society A. 2022;**380**:20210063. DOI: 10.1098/rsta.2021.0063

Chapter 12

Perspective Chapter: EPR Paradox – Experimental and Quantum Field Theoretical Status of Light Meson Resonances

Alexander Machavariani

Abstract

The inclusive reaction of the V-meson resonance production is studied in order to check consistency of the field-theoretical approach with formulation based on the Einstein-Podolsky-Rosen (EPR) concepts. For this aim we have constructed the relativistic field-theoretical amplitudes of the V-meson resonance decay $1 + 2 \leftarrow V$ with structureless (pointlike) and composite (non-pointlike) resonance. Nonlocal composite states of a resonance are obtained using quark-gluon degrees of freedom in accordance with quantum chromodinamics (QCD). Particle 1 and 2 have opposite momentums $\mathbf{p}_1 = -\mathbf{p}_2$ and strongly correlated spin states in the rest frame of V-meson decay. Moreover, for electron-positron decay of the V-meson resonance, electron and positron have the opposite helicity if the electron mass is neglected. Therefore, the decay of V-meson into two particles can be considered as the formation of the EPR-pair according to EPR gedanken experiment. In addition, the color quark-gluon states can be interpreted as hidden (non-observed) states within EPR formulation. As an example of such an approaches, we have considered papers with high-energy experimental data where for description of observable was used quark-antiquark states and concepts of EPR pairs and entanglement.

Keywords: paradox, entanglement, quantum, field, theory, meson, resonance

1. Introduction

Paradox of Einstein, Podolsky and Rosen [1] has motivated numerous experimental and theoretical investigations of basic concepts in quantum mechanics. In particular, in review papers [1–7] were studied the highly correlated so called EPR-pairs of particles with following possibility of determination of the location of one of these particles through the location of another particle. Schrödinger [8] showed that the EPR paradox lies in the possibility of separabelization of the two-particle wave function $|\psi_{A+B}>$ as product of the single-particle wave functions $|\psi_{A+B}> \ \Leftrightarrow \ |\psi_A> \ |\psi_B>$. The next principal result was obtained by Bell [9], where it is proved that EPR paradox in quantum mechanics should be supplemented by

IntechOpen

additional local hidden variables which enables to restore causality and locality. Subsequently, particles which satisfy the conditions of Schrödinger and Bell were called EPR pairs and interaction (correlation) of these particles in final states are defined as entanglement [2]. The notable exception of the Bell theory are given in [5] using nonlocal hidden variables and in [10–12] using the concept of superdeterminism. Superdeterministic hidden-variable theories can be local Superdetermine yet be compatible with observations. Theoretical and experimental evidences for the violation of Bell's inequalities for n-particle systems of with spin are considered in [13].

On the other hand, Bell's inequalities and modified Bell's inequalities are based on simple, model-independent mathematical proofs [14]. Therefore, the mechanism and mathematical the justification for the violation of Bell's inequalities is of particular interest.

At present, interest in the study of the EPR paradox and EPR pairs has grown due to the possibility of their practical application in different technologies and in the quantum computing. Review papers in this direction [2–7] are constantly updated with new investigations (see e.g. [15, 16]).

Comprehensive study of the EPR paradox and its modifications involves analysis of locations of particles from EPR-pair in coordinate space. Measurements of cross sections and other characteristics of particle interactions in the high-energy region are usually carried out in momentum space since the time and radius of interaction of hadrons are very small and are not available for the direct measurement. The nonrelativistic quantum mechanical measurement of trajectory of particle is also strongly restricted via Heisenberg's uncertainty principle. An alternative point of view on the theoretcal possibility of measuring the trajectory of quantum particles is discussed in the review article [17] using formulations of Einstein, De Broglie and Bohm. In relativistic quantum mechanics and relativistic quantum field theory interaction (entanglement) of particles and their trajectories are restricted also by strong interconnection between the temporal and spatial components of coordinate of each particle. This interconnection follows from the principles of relativity and causality. However, entanglement between two hadrons in coordinate space can be established on the basis of a theoretical model, which allows one to unambiguously determine the dependence of the observed characteristics from a two-hadron system. For example, entanglement of two atoms at a stand of 1 cm was found in [18] using one-photon exchange. Similarly, one can establish entanglement between two hadrons at microscopic distance using the theoretical model dependence of measurements of the two-hadron production.

Another aspect of the EPR paradox is problem of hidden variables to determine particle states and particle interactions. The color-charged quark-gluon states one can consider as hidden degrees of freedom because they can not be observed or isolated due to confinement in quantum chromodinamics (QCD). Hadronic and nuclear states are formed from colorless states. confinement was analytically reproduced using two-dimensional Schwinger model or Abelian gauge theories in 2 and 3 space-time dimensions. Moreover QCD as well as confinement obtained in the framework of the multi-dimensional String theory which is not yet justified experimentally. Note that the introduction of unobservable generalizations of particle states for convenient description of their interactions or reproduction of experimental data, is long-known and effective method in theoretical physic. For example, in the review article [19] are analyzed the five-dimensional theoretical models of many outstanding physicists from 1922 to 1991 years, where the fifth dimension has an auxiliary, mathematical meaning.

The above considerations can be applied to the analysis of high-energy experimental data using quark interaction models and with the emergence of the EPR-pairs.

The production of the EPR pair of the neutral B^o-mesons mesons at the $\Upsilon(4S)$-resonance was measured in [20] in order to study the non-local entanglement between B^o and $\overline{B}^o$ mesons. For this aim was measured the time-dependent asymmetry between two B^o states with different flavor, which is compared with prediction of quantum mechanical model and predictions of two realistic models. The flavor asymmetry of quarks for weak interacted particles as well as the mass difference and mixing (oscillating) of B^o and $\overline{B}^o$ was predicted by standard model and was supported by many experiments (see e.g. [21]).

Another application of the concept of entanglement between quark states in the high energy particle scattering experiments is done in [22, 23] using LHC CERN data set collected by ATLAS detector for $\sqrt{s} = 13$ TeV between 2013 and 2018 years. From this data were extracted differential cross section of the reaction $e^+ + e^- \leftarrow t + \bar{t}$ with top diquark masses $m_{t\bar{t}} = 350GeV$ and and $m_{t\bar{t}} \geq 1TeV$. It was shown that the separability of finite states e^+ and e^- are violated due to interaction in the reaction $e^+ + e^- \leftarrow t + \bar{t}$, which significantly depends on the scattering angle in this subprocess. Besides in these papers is considered application of suggested entanglement in quantum information theory and quantum technologies. Thus the entanglement of the electron-positron pair in [22, 23] is determined by inelastic scattering of a quark-antiquark pair $e^+ + e^- \leftarrow t + \bar{t}$.

Most detailed and complete measurements of the exclusive ρ^o-meson resonance with the polarized virtual photons were fulfilled in numerous experiments during many years at CERN (see ref. [24] and papers quoted there), where the applications of various kinematic, quark-parton models were implemented.

The purpose of this paper is to consider the the implementation of the EPR-criteria for resonances in the framework of the relativistic quantum field theory. In particular, it will be considered the spin 1 meson (V-meson) resonance decay into hadronic or leptonic pairs in the high-energy experiments, where the measurement results are analyzed using concepts of EPR-pair and entanglement of particles in this pair.

Measuring the mass and width of the resonance M_V and $\Gamma_V 2$ is usually performed using selection of the resonating pair of particles 1 and 2 by their four momentums $p_1 = \sqrt{m_1^2 + \mathbf{p}_1^2}, \mathbf{p}_1))$ and $p_2 = \sqrt{m_2^2 + \mathbf{p}_2^2}, \mathbf{p}_2)$ which are located in the resonant region $\left(M_V - \frac{\Gamma_V}{2}\right)^2 \leq (p_1 + p_2)^2 \leq \left(M_V + \frac{\Gamma_V}{2}\right)^2$. This condition determines the relationship or entanglement of particles 1 and 2 after the decay of the resonance. In the rest frame of the V-meson resonance, $\mathbf{p}_1$ and $\mathbf{p}_2$ have opposite direction and $\mathbf{p}_1 = -\mathbf{p}_2$. Moreover, spin of these particles is strongly correlated. For instance, neglecting mass of electrons in vector meson decay one obtains that helicity of the electron and positron are also opposite. Therefore the V-meson resonance decay particle 1 and 2 form the EPR pair-type system.

For the relativistic description of the resonances we will use the field-theoretical formulation of Huang and Weldon [25–28], where hadrons are bound states of quarks. In this formulation quarks and gluons are the point-like (structureless) particle which are described by local field operator. The field operators of quarks and gluons satisfy the QCD equation of motion and causality condition. Hadron states are constructed via colorless time-ordered product of the quark fields. In this approach is constructed also the asymptotic one hadron creation or annihilation operator and corresponding $\mathcal{S}$-matrix of the hadron-hadron interaction, which satisfies unitarity condition in hadronic sector. Therefore propagation of quarks in the intermediate states does not destroy the unitarity of the hadron-hadron $\mathcal{S}$-matrix. Therefore one can obtain the equation of motion for amplitudes of the hadron-hadron and hadron-nuclear interaction, where color states appear only between colorless hadronic or nuclear states.

This enables to avoid problems with confinement in this approach. The 3D field-theoretical equations for the scattering amplitude as an alternate approach for the 4D Bethe-Salpeter equations is considered in Appendix A.

In contrast to the non-relativistic formulation, relativistic quantum mechanics requires strong connection between the temporal and spatial components of coordinates according to invariance under Lorentz transformation. Besides in the relativistic approach arise additional mechanisms of entanglement which are caused by off-energy and off mass shell effects and contributions from creation and annihilation of particles in intermediate states.

The organization of this papers is as follows. In Section 1 the general properties of the relativistic field-theoretical $\mathcal{S}$-matrix of reactions $1+2+X \leftarrow V+X \leftarrow A+B$ and corresponding construction of the resonance decay amplitude $1+2 \leftarrow V$ are outlined. Section 2 is devoted to the V-meson resonance decay amplitude with intermediate quark-gluon degrees of freedom. In Section 3 angular distribution of the reaction $1+2+X \leftarrow V+X \leftarrow A+B$ is given in terms of the decay angles of the V-meson resonance and is considered the alignment of the particles 1 and 2 in this reaction. The last fourth part of the article provides a brief formulation of the results and examines the formation of an EPR pair and their entanglement based on relativistic field-theoretical description of the resonance. In particular are compared amplitudes with the local point-like resonance and with the nonlocal resonance consisting of quarks.

2. Amplitude of the structureless (point-like) resonance

In relativistic quantum field theory [29–31] $\mathcal{S}$-matrix of inclusive collision of two particles A and B which generated particles 1, 2 and subsystem X is

$$\mathcal{S}_{1+2+X\leftarrow A+B} \equiv \, < out; \mathbf{p}_1, \mathbf{p}_2, X|\mathcal{S}|\mathbf{p}_A, \mathbf{p}_B; in >$$

$$= \, < 0|a_{\mathbf{p}_1}(out)a_{\mathbf{p}_2}(out), b_{\mathbf{p}_1}(out), \ldots, b_{\mathbf{p}_X}(out)a^+_{\mathbf{p}_A}(in), a^+_{\mathbf{p}_B}(in)|0> \qquad (1)$$

where $a_{\mathbf{p}_1}(out)$, $a_{\mathbf{p}_2}(out)$ and $a^+_{\mathbf{p}_A}(in)$ and $a^+_{\mathbf{p}_B}(in)$ denotes corresponding particle creation operators in *out* and *in* states with momentum $\mathbf{p}_1, \mathbf{p}_2, \mathbf{p}_A, \mathbf{p}_B$ correspondingly. $b_{\mathbf{p}_1}(out), \ldots, b_{\mathbf{p}_X}(out)$ stands for particle from subsystem X. We shall suppose that particles 1 and 2 compose intermediate vector meson V with spin 1 and we shall consider V-meson decay into electron-positron e^-e^+ and two pion $\pi^+\pi^-$ pairs. Therefore in the V-resonance energy region one gets following chain of subreactions $A+B \rightarrow V+X \rightarrow 1+2+X$. It is convenient to express $\mathcal{S}$-matrix (1) via corresponding amplitude

$$\mathcal{S}_{1+2+X\leftarrow A+B} = i(2\pi)^4\delta^4(p_A + p_B - p_1 - p_2 - P_X)\mathcal{A}_{1+2+X\leftarrow A+B} \qquad (2)$$

where p_A, p_B, p_1, p_2 and P_X are four momentums of on shell asymptotic particles $p_{A,B} = \left(\sqrt{m^2_{A,B} + \mathbf{p}^2_{A,B}}, \mathbf{p}_{A,B}\right), p_{1,2} = \left(\sqrt{m^2_{1,2} + \mathbf{p}^2_{1,2}}, \mathbf{p}_{1,2}\right)$ with masses $m_{A,B}, m_{1,2}$ and three momentums $\mathbf{p}_{A,B}, \mathbf{p}_{1,2}$. Besides we shall use four-vector of V-meson resonance $P_V \equiv (P^o_V, \mathbf{P}_V) = \left(\sqrt{M^2_V + \mathbf{P}^2_V}, \mathbf{P}_V\right)$.

Spin of relativistic particle or resonance is defined in the rest frame of this particle or resonance. Therefore we shall consider the rest frame of system of particles 1 and 2, where .

Resonance in quantum mechanics is defined as pole of the amplitude in the complex momentum plane at $\sqrt{s_{12}} \equiv \sqrt{(p_1 + p_2)^2} \to M_V - i\Gamma_V/2$, where M_V and Γ_V denote the pole position and the V-resonance decay width in the particle system $1 + 2$. Therefore $\mathcal{A}_{1+2+X\leftarrow A+B}$ must have the following form

$$lim_{s_{1,2}\to M_{12}-i\Gamma_{12}/2}\left(\sqrt{s_{1,2}} - M_{12} + i\Gamma_{12}/2\right)\mathcal{A}_{1+2+X\leftarrow A+B} = \mathcal{A}_{1+2\leftarrow V}\mathcal{A}_{V+X\leftarrow A+B} \tag{3}$$

or

$$\mathcal{A}_{1+2+X\leftarrow A+B} = \mathcal{A}_{1+2\leftarrow V}\frac{1}{\sqrt{s_{1,2}} - M_V + i\frac{\Gamma_V}{2}}\mathcal{A}_{V+X\leftarrow A+B} + nonresonant\ \ part \tag{4}$$

where the pole part of the amplitude is represented by the V-meson exchange term. It is convenient to express $\mathcal{A}_{1+2\leftarrow V}$ via the product of the wave function $|\Psi_V>$ and two-body potential $<1,2|V|1',2'>$

$$\mathcal{A}_{1+2\leftarrow V}\left(\sqrt{s_{1,2}}\right) = <1,2|V\Psi_V> \tag{5}$$

where $|\Psi_V>$ satisfies the equation

$$<1,2|\Psi_V> = \frac{1}{\sqrt{s_{1,2}} - M_V + i\frac{\Gamma_V}{2}} <1,2|V\Psi_V>. \tag{6}$$

Unlike to the bound state, $<1,2|\Psi_V>$ is not the eigen-function of the Hamiltonian of the system $1 + 2 - 1' + 2'$.

Eqs. (4) and (6) enables to describe the resonant and non-resonant parts of $\mathcal{A}_{1+2+X\leftarrow A+B}$ using the same Lagrangian and same phenomenogical potential for all partial waves of $1 + 2$ subsystem. In this approach were reproduced low energy $\pi N - \pi N$, $NN - NN$ and $\gamma N - \pi N$ reactions [26, 32, 33]. In particular, description of low-energy πN scattering amplitude in the Δ-resonance region was performed within the Born approximation in [30] (see eq. (10.67) and (10.79)).

In other relativistic approach resonance is treated as particle-type degrees of freedom with imaginer mass $M_V - i\Gamma_V/2$, spin s_V and with corresponding Lagrangian. Corresponding S,P,D,F pion-nucleon S,P,D,F phase shifts are calculated within third order chiral effective field theory in [34].

Cross section of the reaction $1 + 2 + X \leftarrow A + B$ according to (4) is [31].

$$\frac{d\sigma_{1+2+X\leftarrow A+B}}{d\Omega_{1+2+X\leftarrow A+B}} = \sum_{MM'}\frac{d\tilde{\sigma}^{MM'}_{1+2\leftarrow V}}{d\Omega_{1+2\leftarrow V}}\frac{1}{\left(m_V - \sqrt{s_{12}}\right)^2 + \Gamma_V^2/4}\frac{d\sigma^{MM'}_{V+X\leftarrow A+B}}{d\Omega_{V+X\leftarrow A+B}} \tag{7}$$

where M, M' are magnetic quantum number of V resonance

$$\frac{d\sigma^{MM'}_{V+X\leftarrow A+B}}{d\Omega_{V+X\leftarrow A+B}} = \mathcal{N}_{A+B}\prod_i^X \mathcal{A}^M_{V+X\leftarrow A+B}\left(\mathcal{A}^*_{V+X\leftarrow A+B}\right)^{M'} \tag{8}$$

$$d\Omega_{V+X\leftarrow A+B} = (2\pi)^3 \frac{d\mathbf{P}_{12}}{(2\pi)^3} \prod_{i=1}^{X} d\tilde{\mathbf{p}}_i (2\pi)^4 \delta^3(\mathbf{P}_A + \mathbf{P}_B - \mathbf{P}_{12} - \mathbf{P}_X)\delta(P_A^o + P_B^o - P_{12}^o - P_X^o)$$

where $\mathbf{P}_{12} = \mathbf{p}_1 + \mathbf{p}_2$ and $P_{12}^o = p_1^o + p_2^o$,

$$\frac{d\tilde{\sigma}_{1+2\leftarrow V}^{MM'}}{d\Omega_{1+2\leftarrow V}} = \mathcal{A}_{1+2\leftarrow V}^{M} \left(\mathcal{A}_{1+2\leftarrow V}^{*}\right)^{M'} \tag{9}$$

$$d\Omega_{1+2\leftarrow V} = d\tilde{\mathbf{p}}_1 d\tilde{\mathbf{p}}_2 \delta^3(\mathbf{p}_1{}^* + \mathbf{p}_2^*)$$

where $\mathrm{N}_A = 1/(2p_A^o)$ for bosons and $\mathrm{N}_A = m_A/p_A^o$ for fermions

$$\mathcal{N}_{A+B} = \frac{1}{(2S_A+1)(2S_B+1)} \frac{\mathrm{N}_A \mathrm{N}_B}{\sqrt{(P_A P_B)^2 - m_A^2 m_B^2}}; \qquad d\tilde{\mathbf{p}}_a = \mathrm{N}_a \frac{d\mathbf{p}_a}{(2\pi)^3} \tag{10}$$

$d\tilde{\sigma}_{1+2\leftarrow V}^{MM'}/d\Omega_{1+2\leftarrow}$ does not contain $\delta(P_V^o - p_1^o - p_2^o)$ and $\mathcal{A}_{1+2\leftarrow V}^{M}$ depends on the decay width Γ_V of resonance according to (5) and (6). That's why amplitude $\mathcal{A}_{1+2\leftarrow V}^{M}$ is nonlocal and off shell even for the point-like, structureless resonance. General field theoretical construction of the nonlocal V-resonance with quark degrees of freedom is given in next section.

According to the vector meson dominance model the main contribution into vector meson decay is made by diagram of the V-meson transformation into off shell photon $\gamma^* \leftarrow V$. Therefore in phenomenogical $\mathcal{A}_{1+2\leftarrow V}^{M}$ is used the form factor of electromagnetic-type (see e.g. [29] ch. 10.6), where current operator of the V-meson resonance is replaced by the current operator of the elementary particle with the mass M_V, spin 1 and the coupling constant g_V. Then $\mathcal{A}_{1+2\leftarrow V}^{M}$ for the leptonic decay takes the form

$$\mathcal{A}_{\ell^+ + \ell^- \leftarrow V}^{M} = g_{\ell^+ + \ell^- \leftarrow V} \xi_\mu{}^* (\mathbf{P}_V, M) \bar{u}(\mathbf{p}_\ell) \gamma_\mu v(\mathbf{p}_{\bar{\ell}}) \tag{11}$$

where $\xi^{\mu *}(\mathbf{P}_V, M)$ is the polarization vector of the spin 1 particle and coupling constant $g_V \equiv g_{\ell^+ + \ell^- \leftarrow V}$ is defined according to vector meson dominance model.

In this approximation the hadronic decay amplitude is

$$\mathcal{A}_{h+\bar{h}\leftarrow V}^{M} = g_{h+\bar{h}-V}\left((p_h + p_{\bar{h}})^2\right)(\xi_\mu{}^* (\mathbf{P}_V, M) p_h^{*\mu}) \tag{12}$$

where for simplicity only spineless hadrons are considered. $g_{h+\bar{h}-V}\left((p_h + p_{\bar{h}})^2\right)$ is form factor of $h + \bar{h} \leftarrow V$.

3. V-meson decay amplitude with the quark-gluon degrees of freedom

Amplitude $\mathcal{A}_{1+2\leftarrow V}$ (5) can be constructed with intermediate quark-antiquark-gluon degrees of freedom $q - \bar{q} -$ using the field-theoretical approach of the composite particle states [25, 26]. In this formulation the Heisenberg field operator of the composite particle is defined via the Bethe-Salpeter wave functions (see [31] ch. 10.2.1)

$$\chi_{m_V,\mathbf{P}_V}\left(Y,y_{1,2}\right) = <0|\mathrm{T}(\psi_1(y_1)\psi_2(y_2))|m_V,\mathbf{P}_V,M;in>, \tag{13}$$

where

$$Y = y_1 + y_2, \qquad y_{1,2} = \eta_2 y_1 - \eta_1 y_2; \qquad \eta_1 + \eta_2 = 1 \tag{14}$$

are Jacobi coordinates. For the sake of simplicity in (1) we have introduced the complex mass $m_V = M - V - i\Gamma_V/2$ for the resonance state. In alternate formulation the dependence on Γ_V of wave function implicitly through solution of the Bethe-Salpeter equation. In alternate formulation the dependence on Γ_V of wave function occurs implicitly through solution of Bethe-Salpeter equation.

One can introduce nonlocal operator of the resonance in analogy with bound-state field

$$\mathcal{B}_{m_V,\mathbf{P}_V}(Y_o) = i\int d^3\mathbf{Y}d^4y_{12}\exp(iP_VY)\left[\frac{\partial}{\partial Y_o} - iP^o_V\right]\chi_{m_V\mathbf{P}_V}(Y,y_{12})\mathrm{T}(\psi_1^+(x_1)\psi_2^+(x_2)). \tag{15}$$

which transforms into single composite particle creation or annihilation operator in asymptotic regions $\pm\infty \Leftarrow Y_O$

$$<\Psi|\mathcal{B}_{m_V,\mathbf{P}_V}(out(in))|\Phi> = lim_{\pm\infty\Leftarrow Y_o} <\Psi|\mathcal{B}_{m_V,\mathbf{P}_V}(Y_o)|\Phi>, \tag{16}$$

where $|\Phi>$ and $|\Psi>$ denote an arbitrary wave functions.

The asymptotic operators satisfy the same commutation relation as elementary (point-like) particle

$$\left[\mathcal{B}^+_{m_V,\mathbf{P}_V}(out(in)), \mathcal{B}_{m_V,\mathbf{P}_V}(out(in))\right] = (2\pi)^3 2P^o_V\delta(\mathbf{P}'_\mathbf{V} - \mathbf{P}_\mathbf{V});$$
$$[\mathcal{B}_{m_V,\mathbf{P}_V}(out(in)), \mathcal{B}_{m_V,\mathbf{P}_V}(out(in))] = 0; \tag{17}$$

But unlike elementary (point-like, structureless) particle, the same commutation relations for $\mathcal{B}_{m_V\mathbf{P}_V}(Y_o)$ (15) are violated

$$\begin{gathered}\left[\mathcal{B}^+_{m_V\mathbf{P}_V}(Y_o), \mathcal{B}_{m_V\mathbf{P}_V}(Y_o)\right] \neq (2\pi)^3 2P^o_V\delta(\mathbf{P}_{\mathbf{V}'} - \mathbf{P}_\mathbf{V}); \\ [\mathcal{B}_{m_V\mathbf{P}_V}(Y_o), \mathcal{B}_{m_V\mathbf{P}_V}(Y_o)] \neq 0.\end{gathered} \tag{18}$$

Inequalities (18) are caused by nonlocality of $\mathcal{B}_{m_V\mathbf{P}_V}(Y_o)$ (15) which depends exactly on interacted fields $\psi_1(y_1)$, $\psi_2(y_2)$. and on the resonance decay width $\Gamma_V/2$.

Using $\mathcal{B}_{m_V\mathbf{P}_V}(Y_o)$ (15) one can built the V meson field $\Upsilon_{m_V\mathbf{P}_V}(Y)$ and corresponding source operator $\mathcal{J}_{m_V\mathbf{P}_V}(Y)$

$$\Upsilon_{m_V\mathbf{P}_V}(Y) = \int\frac{d^3\mathbf{P}_V}{(2\pi)^3 2P^o_V}\left[e^{-iP_VY}\mathcal{B}_{m_V\mathbf{P}_V}(Y_o) + e^{iP_VY}\mathcal{B}^*_{m_V\mathbf{P}_V}(Y_o)\right], \tag{19}$$

$$\mathcal{J}_{m_V\mathbf{P}_V}(Y) = \left(W_Y + M_V^2\right)\Upsilon_{m_V\mathbf{P}_V}(Y) \tag{20}$$

which enables to obtain the general expression for amplitude of V meson resonance

$$\mathcal{A}_{1+2\leftarrow V}(nonlocal) = <out; 1, 2|\mathcal{J}_{m_V\mathbf{P}_V}(0)|0> \tag{21}$$

where $m_V = M_V - i\Gamma_V/2$.

4. Spin density matrix and alignment for reaction $1+2+X \leftarrow V+X \leftarrow A+B$

The cross sections (7), (8) and (9) make it possible to obtain a convenient formula for the expansion of the angular distribution $W_{1+2+X\leftarrow A+B}$ in terms of the sum spin density matrices (see e.g. ch. 8 [35]) $\rho^{MM'}_{V+X\leftarrow A+B}$ and $\rho^{MM'}_{1+2\leftarrow V}$.

$$W_{1+2+X\leftarrow A+B} = \sum_{M,M'} \rho^{MM'}_{1+2\leftarrow V}\rho^{MM'}_{V+X\leftarrow A+B}, \tag{22}$$

where

$$\rho^{MM'}_{V+X\leftarrow A+B} = \frac{d\sigma^{MM'}_{V+X\leftarrow A+B}/d\Omega^*_{V+X\leftarrow A+B}}{\int d\Omega^*_{V+X\leftarrow A+B}\sum_M d\sigma^{MM}_{V+X\leftarrow A+B}/d\Omega^*_{V+X\leftarrow A+B}}; \tag{23}$$

where

$$\rho^{MM'}_{1+2\leftarrow V} = \frac{d\tilde{\sigma}^{MM'}_{1+2\leftarrow V}/d\Omega_{1+2\leftarrow V}}{\int d\Omega^*_{1+2\leftarrow V}\sum_M d\tilde{\sigma}^{MM}_{1+2\leftarrow V}/d\Omega^*_{1+2\leftarrow V}} \tag{24}$$

Upper index * in (22)-(24) indicate the rest frame of V-meson, where $\mathbf{p}^* \equiv \mathbf{p}^*_1 = -\mathbf{p}^*_2$ and decay angles θ^* Đ¸ ϕ^* are defined by momentum of the final particle

$$\mathbf{p}^\star \equiv \mathbf{p}^\star_1 = |\mathbf{p}^\star|(sin\,\theta^\star\,cos\,\phi^\star,\, sin\,\theta^\star\,\sin\phi^\star,\, \cos\theta^\star), \tag{25}$$

$d\Omega^\star_{V+X\leftarrow A+B}$ and $d\Omega_{1+2\leftarrow V}$ stands for Lorentz invariant phase space elements

$$d\Omega^\star_{V+X\leftarrow A+B} = \frac{d^3\mathbf{P}^*_V}{(2\pi)^3}\prod_{i=1}^{X} \mathrm{N}_i \frac{d^3\mathbf{p}^*_i}{(2\pi)^3} \tag{26}$$

$$d\Omega_{1+2\leftarrow V} = \mathrm{N}_1\mathrm{N}_2\frac{d^3\mathbf{p}^*}{(2\pi)^3} \tag{27}$$

Cross sections (7) and decomposition of the corresponding angular distribution (22) through products of spin density matrices are often used to describe experimental data of exclusive and inclusive reactions using the helicity basis. Polarization effects in the corresponding generalizations of the cross sections (7) and corresponding spin density matrices are taking into account in [36, 37]. Beam polarization effects, decomposition of the spin density matrix by the transverse and longitudinal components of the virtual photon are considered in [24] for the exclusive reactions $\mu' + p' + \rho^o \leftarrow \mu + p$ and $p' + \rho^o \leftarrow \gamma^* + p$.

Using vertex functions (11) and (12) and relationship between $\xi^{\mu}(0,M)$ and Wigner D-matrices $D^1_{mn}(\theta^*,\phi^*)$

$$\left(\xi^{\mu}(0,M)p^*_{\mu}\right) = |\mathbf{p}^*|D^1_{M0}(\theta^*,\phi^*) \tag{28}$$

we obtain

$$W_{1+2+X\leftarrow A+B} = C + \sum_{MM'} {D^*}^1_{M0}(\theta^*,\phi^*)\rho^{MM'}_{V+X\leftarrow A+B}D^1_{M'0}(\theta^*,\phi^*) \tag{29}$$

where $C = 0$ if $1,2$ are pions and $C = 1$ if $1,2$ are electron-positron pair.

Angular distributions $W_{1+2+X\leftarrow A+B}$ (29) describe decay of the point-like (structureless) resonance V on two particles 1 and 2, where dependence of $W_{1+2+X\leftarrow A+B}$ on the resonance decay angles is contained in $D^1_{M'0}(\theta^*,\phi^*)$. Therefore anisotropy regarding angles θ^*,ϕ^* in $W_{1+2+X\leftarrow A+B}$ (29) appear due to spin density matrix $\rho^{MM'}_{V+X\leftarrow A+B}$ of the sub-reaction $V+X\leftarrow A+B$. This anisotropy is called alignment of the decay particles. In particular, if $\rho^{M=M'}_{V+X\leftarrow A+B}\simeq 1/3$ and $\rho^{M\neq M'}_{V+X\leftarrow A+B}\simeq 0$, then V-meson decay is isotropic. In this case the decay particles 1 and 2 move independently from the interactions in the initial sub-reaction $V+X\leftarrow A+B$. Thus for isotropic decay entanglement between particles 1 and 2 caused only by location of effective mass of subsystem $1+2$ in resonance region

$$\left(M_V - \frac{\Gamma_V}{2}\right)^2 \leq (p_1+p_2)^2 \leq \left(M_V + \frac{\Gamma_V}{2}\right)^2. \tag{30}$$

Vertex (21) for composite V-meson resonances can be presented as

$$\begin{aligned}\mathcal{A}_{1+2\leftarrow V} &= <out;\mathbf{p}^*,S^*_1M^*_1;-\mathbf{p}^*,S^*_2M^*_2|{\mathcal{J}^*}^{1M}_{m_V\mathbf{0}_V}(0)|0> = \\ &= \delta_{J^*,1}\sum_{L^*,S^*}\mathcal{Y}^{JM}_{L^*,S^*}(\hat{\mathbf{p}}^*)\mathrm{F}^{J^*}_{S^*L^*}(\sqrt{s_{12}}),\end{aligned} \tag{31}$$

where $\mathcal{Y}^{JM}_{L^*,S^*}(\hat{\mathbf{p}}^*)$ and $Y^*_{L^*M}(\hat{\mathbf{p}}^*)$ are tensor spherical harmonics and spherical harmonics

$$\mathcal{Y}^{JM}_{L^*,S^*}(\hat{\mathbf{p}}^*) = \sum_{M^*_L} <L^*M^*_LS^*M^*_S|1M> Y^*_{L^*M}(\hat{\mathbf{p}}^*)\chi_{S^*M^*_S}(S^*_1M^*_1,S^*_2M^*_2), \tag{32}$$

and

$$\begin{aligned}\mathrm{F}^1_{S^*L^*}(\sqrt{s_{12}}) &= \left(\xi_{\mu}(0,M)\mathbf{p}^{*\mu}\right)\int d^3\mathbf{q}^*\,\tilde{\mathrm{F}^{(1)}}_{S^*L^*}(\mathbf{p}^*,\mathbf{q}^*,\sqrt{s_{12}}) \\ &+ \int d^3\mathbf{q}^*\left(\xi_{\mu}(0,M)\mathbf{q}^{*\mu}\right)\tilde{\mathrm{F}^{(2)}}_{S^*L^*}(\mathbf{p}^*,\mathbf{q}^*,\sqrt{s_{12}})\end{aligned} \tag{33}$$

is the radial part of $\mathcal{A}_{1+2\leftarrow V}$ with spin S^*, orbital momentum L^* and their projections M^*_S and M^*_L.

Using (31)–(33) one can represent the cross section of reaction $1+2+X\leftarrow A+B$ in the form

$$\frac{d\sigma_{1+2+X\leftarrow A+B}}{d\Omega^*_{1+2+X\leftarrow A+B}} = \sum_{MM'}\left(\sum_{L^*S^*}\left[\mathcal{Y}^{L^*S^*}_{JM}(\hat{\mathbf{p}}^*_{12})\mathrm{F}^1_{S^*L^*}(\sqrt{s_{12}})\right]^+ \frac{d\sigma^{MM'}_{A+B\Rightarrow V+}}{d\Omega^*_{A+B\Rightarrow V+X}} \frac{\sum_{L'S'}\left[\mathcal{Y}^{L'S'}_{JM'}(\mathbf{p}^*_{12})\mathrm{F}^1_{S'L'}(\sqrt{s_{12}})\right]}{(M_V^2 - s_{12})^2 + M_V^2\Gamma_V^2}\right), \tag{34}$$

Thus cross sections (34) and (7) with the amplitude of the composite and structureless V-meson resonance decay (31), (11) and (12) have different angular momentum dependence and different meson decay form factors.

5. Conclusion and outlook

In this paper we have discussed the entanglement of two hadrons and two leptons, which occur in high energy collision $1+2+X \leftarrow V+X \leftarrow A+B$ after V meson resonance decay $1+2 \leftarrow V$. For the most general reproduction of entanglement mechanisms, we used a field-theoretic construction of the resonance decay amplitude with and without quark-gluon degrees of freedom.

In the both cases the two-body decay of any resonance automatically forms EPR-pair-type system in their rest frame, where these particles have opposite momentums and the strongly correlating spin states.

Big difference in the leptonic and hadronic resonance decay of V-meson produce corresponding difference between form factors (11) and (12). In particular, the hadronic and leptonic decay of the point-like (structureless) V-meson contains form factors $g_{h\bar{h}-V}\left((p_h+p_{\bar{h}})^2\right)$ (12) and $g_{\ell\bar{\ell}-V} \simeq const$ (11). There are even more differences between the hadron decay form factors (33) and (12) with and without quark-gluon degrees of freedom for the non-point-like (non-local) and point-like (local) resonances correspondingly. Even more difference appear between the hadronic decay form factors (12) and (33) point-like (local) and non-point-like (nonlocal) resonances.

If decay of the V-meson resonance is approximately isotropic $\rho_{11} = \rho_{00} = \rho_{-1-1} \simeq 1/3$, then decay of it is identical with the decay of the isolated resonance. In this case decay of the V-meson is independent on the influence of the initial subprocess $V+X \leftarrow A+B$. After that, decay $1+2 \leftarrow V$ can be considered as the decay of a quasi-free V-meson resonance and the entanglement between particles 1 and 2 can be determined only by the resonant decay mechanism.

Nonlocality of hadrons or nuclears in quantum field theory is a consequence of the composite (nonlocal) structure of hadrons or nuclei. The composite hadrons are built through the interactions of local (point-like) fields of quarks and gluons, and the composite (nonlocal) structure of nuclei in nuclear physics is determined through the interaction of local (point-like) nucleons. The equal-time commutators of the composite field operators composite particles (15) or (19) and (20) are not equal to zero even in a space-like region.

Cross sections (7) are valid for both local and non-local resonant decay of the VmesonÑŽ. One can rewrite this expression as

$$\frac{d\tilde{\sigma}^{MM'}_{1+2\leftarrow V}}{d\Omega_{1+2\leftarrow V}} = \mathcal{A}^M_{1+2\leftarrow V}\left(\mathcal{A}^*_{1+2\leftarrow V}\right)^{M'} \tag{35}$$

Representation of (35) in the form of a product of cross sections of the reactions $1+2 \leftarrow V$ and $V+X \leftarrow A+B$ is a consequence of the separability of scattering amplitudes with a single-particle s-channel exchange in quantum field theory or in nonrelativistic scattering theory [38].

Thus the nonlocality of the hadrons in quantum field theory [25] differs from Bell nonlocality (see [5] eq. (3)), where nonlocality arises due to integration over a certain variables λ, "having a joint causal influence on both outcomes, and which fully account for the dependence between" 1 and 2.

This discrepancy with Bell nonlocality is a consequence of the two reasons. First, nonlocality in quantum field theory and nonrelativistic nuclear and collision theories is introduced via construction of the composite particle states through degrees of freedom of local, structureless particles. And Bell's nonlocality is determined through integration over λ of the local probability in the Bell formulation.

Secondly, EPR thought experiments and similar subsequent formulations imply independent measurements of the location of a single particle from an EPR pair in coordinate space based on the uncertainty principle Heisenberg. But in real experiments in elementary particle physics and nuclear physics, the momentum of particles and the states of these particles in the momentum space are measured. Determining the position of particles in coordinate space requires not only a relativistic generalization of the Heisenberg uncertainty principle, but also model-independent calculations of the trajectory of quantum particles and other tasks beyond the scope of this article.

Appendix A: relativistic 4*D* and 3*D* field-theoretical equations

Field-theoretical Bethe-Salpeter equations for 4D amplitude

$$\mathcal{A}^{4D} \equiv \mathcal{A}^{4D}\left(p^{o}{}_{A}\mathbf{p}_{A}, p^{o}{}_{B}\mathbf{p}_{B}; p^{o}_{C}\mathbf{p}_{C}, p^{o}_{D}\mathbf{p}_{D}\right) \tag{36}$$

and relativistic Lippmann-Schwinger-type equations for 3D amplitude

$$\mathcal{A}^{3D} \equiv \mathcal{A}^{3D}\left(\mathbf{p}_{A}, \mathbf{p}_{B}; \mathbf{p}_{C}, \mathbf{p}_{D};\right) \tag{37}$$

describe reactions $C + D \leftarrow A + B$ in framework of the relativistic field theory taking into account the absorption and appearance of particles. In (A1) $\mathcal{A}^{4D}$ depends on the off shell four-momentum of each particle , p_B, p_C and p_D. In (A2) $\mathcal{A}^{3D}$ is function of on mass shell momentums $p_A^{on} = \left(\sqrt{m_A^2 + \mathbf{p}_A^2}, \mathbf{p}_A\right)$, $p_B^{on} = \left(\sqrt{m_B^2 + \mathbf{p}_B^2}, \mathbf{p}_B\right)$, $p_C^{on} = \left(\sqrt{m_C^2 + \mathbf{p}_C^2}, \mathbf{p}_C\right)$ and $p_D^{on} = \left(\sqrt{m_D^2 + \mathbf{p}_D^2}, \mathbf{p}_D\right)$. Observed particles of the reaction $C + D \leftarrow A + B$ are on shell, where $p_k = p_k^{on}$ $(k = A, B, C, D)$ and $p_C^{on} + p_D^{on} = p_A^{on} + p_B^{on}$. On shell 4D and 3D amplitudes coincide

$$\mathcal{A}^{4D}_{on} = \mathcal{A}^{3D}_{on} \tag{38}$$

In case of spin less particles Bethe-Salpeter equation for $\mathcal{A}^{4D}$ (A1) in coordinate space have the form (see e.g. [31] ch. 10)

$$\mathcal{A}^{4D}(x_C x_D; x_A x_B) = \mathcal{V}^{4D}(x_C x_D; x_A x_B)$$
$$+ \int d^4x_\alpha d^4x_\beta \mathcal{V}^{4D}(x_C x_D; x_\alpha x_\beta)\left(W_{x_\alpha} + m_\alpha^2 + i\epsilon\right)^{-1}\left(W_{x_\beta} + m_\beta^2 + i\epsilon\right)^{-1} A^{4D}(x_\alpha x_\beta; x_A, x_B), \tag{39}$$

where $\mathcal{V}^{4D}(x_C x_D; x_A x_B)$ consist of sum of all Feynman diagrams with all possible intermediate states excluding asymptotic two-body states $\alpha + \beta \equiv C + D, A + B$.

General form of Bethe-Salpeter equations are derived in framework of the method of functional integrals [31] and graphical summation method for Feynman diagrams [39, 40]. These equations depend on non-physical relative time variables $x_A^o - x_B^o$, $x_C^o - x_D^o$ which greatly complicates their application and requires additional approximations. Therefore, were proposed so called 3D quasi-potential reductions of Bethe-Salpeter equations, where they are replaced by a set of four-dimensional equations for the relation between the 4D and 3D potentials and 3D reduced equation for the sought amplitude, 3D potential (so called quasipotential of Bethe-Salpeter equation) and the 3D Green's functions [41–43]. However, there are an infinite number of three-dimensional reductions of same Bethe-Salpeter equation which produce different 3D equations with different potentials, different 3D off shell amplitudes and different Green function. Another version of the 3D reduction of the Bethe-Salpeter equations is given in [44].

From the beginning 3D field-theoretical equations $\mathcal{A}^{3D}$ (A2) are derived within modified "old perturbation theory" [45–47] where all particles are being on shell. In this formulation arose the additional quasi-particle degrees of freedom, which enables to take into account off-shellness of the intermediate states.

Another type of initially three-dimensional field-theoretic equations [26, 33, 48] was obtained within the S-matrix reduction method [29–31]. In this case

$$\mathcal{A}^{3D} = < out; \mathbf{p}_A | j_B(0) | \mathbf{p}_c \mathbf{p}_D; in > \qquad (40)$$

where $j_B(x) = \left(W_x + m_B^2\right)\phi_B(x)$ is the source operator of particle B. In this approach final equations have the form of the relativistic Lippmann Schwinger-type equations

$$\mathcal{T}^{3D}(\mathbf{p}', \mathbf{p}; P_o) = \mathcal{U}^{3D}(\mathbf{p}', \mathbf{p}; P_o)$$
$$+ \int \mathcal{U}^{3D}(\mathbf{p}', \mathbf{p}''; P_o) \frac{d^3\mathbf{p}_1'' d^3\mathbf{p}_2'' \delta(\mathbf{P} - \mathbf{P}'')}{2E_{\mathbf{p}_1''} 2E_{\mathbf{p}_2''}\left(P_o - E_{\mathbf{p}_1''} - E_{\mathbf{p}_2''} + i\epsilon\right)} \mathcal{T}^{3D}(\mathbf{p}'', \mathbf{p}; P_o), \qquad (41)$$

where $P_o = E_{\mathbf{p}_1} + E_{\mathbf{p}_2}$, $\mathbf{P} = \mathbf{p}_1 + \mathbf{p}_2$ denote total energy and total momentum of the system $1 + 2$, $E_{\mathbf{p}_{1,2}} = \sqrt{\mathbf{p}_{1,2}^2 + m_{1,2}^2}$, and $\mathbf{p} = 1/2(\mathbf{p}_1 - \mathbf{p}_2)$ are the relative momentum of particles 1, 2 in the rest frame $\mathbf{P} = \mathbf{p}_1 + \mathbf{p}_2 = 0$.

Potential $\mathcal{U}^{3D}(\mathbf{p}', \mathbf{p}; P_o)$ of (A6) consists of all three-dimensional time ordered diagrams with on mass shell particle exchanges in u, $\overline{u}$ and $\overline{s}$. Besides $\mathcal{U}^{3D}$ contains so called seagull term which produces off mass shell one-particle exchange term and overlapping (contact) terms.

On shell $\mathcal{T}_{on}^{3D}$ and $\mathcal{A}_{on}^{3D}$ coincides.

Unlike the Bethe-Salpeter equations and their 3D reductions, the potential of three-dimensional field-theoretic Eq. (A6) is determined by vertex functions and amplitudes $< out; n | j_a(0) | m; in >$, which contain only one particle a with off shell source operator $j_a(0)$. Therefore, in these three-dimensional equations, it is not required to make approximations which simplify off shell behavior of the vertex functions and amplitudes.

Note that taking into account the quark-gluon degrees of freedom does not change the form of the Eq. (41) for the amplitudes of hadron interactions.

Author details

Alexander Machavariani
High Energy Physics Institute of Tbilisi State University, Tbilisi, Georgia

*Address all correspondence to: aimachavariani@gmail.com

References

[1] Einstein A, Podolsky B, Rosen N. Can quantum-mechanical description of physical reality Be considered complete? Physics Review. 1935;**47**:777

[2] Reid MD et al. Colloquium: The Einstein-Podolsky-Rosen paradox: From concepts to applications. Reviews of Modern Physics. 2009;**81**:1727. arXhiv: 0806.0270v2[quant-ph] 08Dec 2008

[3] Cavalcanti D, Skrzypczyk P. Quantum steering: A. Reports on Progress in Physics. 2017;**80**:024001

[4] Peise J et al. Satisfying Einstein-Podolsky-Rosen criterion with massive particle. Nature Communications. 2015. Available from: www.nature.com/naturecommunications

[5] Brunner N et al. Bell nonlocality. Reviews of Modern Physics. 2014;**86**: 419. arXhiv:1303.2849v3 [quant-ph] 10 Apr 2014

[6] Fine A, Ryckman TA. In: Zalta EN, editor. The Einstein-Podolsky-Rosen Argument in Quantum Theory. Stanford, USA: The Stanford Encyclopedia of Philosophy; 2020. Available from: https://plato.stanford.edu/archives/sum2020/entries/qt-epr

[7] Klyshko DN. A simple method of preparing pure states of an optical field, of implementing the Einstein-Podolsky-Rosen experiment and of demonstrating the complementarity principle. Soviet Physics Uspekhi. 1988;**31**:74. Available from: http://iopscience.iop.org/0038-5670/31/1/A05

[8] Schrödinger E. Discussion of probability relations between separated systems. Proceedings–Cambridge Philosophical Society. 1935;**31**:555 and Probability relations between separated systems. Proceedings–Cambridge Philosophical Society. 1936;**32**:446

[9] Bell JS. On Einstein Podolsky Rosen paradox. Physics. 1964;**1**:195

[10] G. ‘t Hooft. The cellular automamaton interpretation of quantum mechanics. In: Fundamental Theories of Physics. Vol. 185. Heidelberg, New York, Dordrecht, London: SpringerOpen; 2016. Open Access: https://library.oapen.org/bitstream/id/470b7d7e-768c-4b5f-936e-d9cb5fb6fa13/1002003.pdf

[11] Andreoletti G, Vervoort L. Superdeterminism: A reappraisal. Synthese. 2022;**200**:361. Available from: https://philarchive.org/rec/ANDSAR-9

[12] Hossenfelder S, Palmer T. Rethinking superdeterminism. 2020;**8**:1. Available from: https://www.frontiersin.org/articles/10.3389/fphy.2020.00139/full

[13] Cabello A. Bell’s inequality for n spin-s particles. Physical Review A. 2002;**65**:062105

[14] Maccone L. A simple proof of Bell’s inequality. American Journal of Physics. 2013;**81**:854

[15] Colgiachi P et al. Einstein-Podolsky-Rosen experiment with two Bose-Einstein condensates. Physical Review X. 2023;**13**:021031

[16] Schnabel R. The solution to the “Einstein-Podolsky-Rosen” paradox. arXiv: 2208.13831v4 [quant-ph] 2023

[17] Aristarhov S. Heisenberg’s uncertainty principle and particle trajectories. Foundations of Physics. 2023;**53**:7

[18] Hagley E et al. Generation of Einstein-Podolski-Rosen pairs of atoma. Physical Review Letters. 1997;**79**:1

[19] Fanchi JR. Review of invariant time formulations of relativistic quantum theories. Foundations of Physics. 1992; **23**:487

[20] Go A, Bay A. Measurement od EPR-type flavor entanglement in $\Upsilon(4S) - > B^o \to \overline{B}^o$ decays. Physical Review Letters. 2007;**99**

[21] Nestle U. Three lectures on meson mixing and CKM phenomenology. Karlsruhe. 2009. Available from: https://www.ttp.kit.edu/media/preprints/2009/ttp09-07.pdf

[22] Afik Y, de Nova JRM. Entanglement and quantum tomography with top quarks at the LHC. European Physical Journal - Plus. 2021;**136**:907; Quantum discord and steering in top quarks at the LHC. Physical Review Letters. 2023;**130**: 221801

[23] Xinari T. Quantum Entanglement with Top Quarks at the LHC.CERN-Students-Students-Note-2022-062. CERN 2022

[24] Alexeev GD et al. Spin Density Matrix Elements in Exclusive o Meson Muon-production CERN-EP-2022-23 and 1arXiv: 2210.16932v2 [hep-ex] 2023. COMPASS Collaboration. To be publ [Preprint]

[25] Huang K, Weldon HA. Bound state wave functions and bound state scattering in relativistic field theory. Physical Review D. 1975;**11**:257

[26] Machavariani A. Three-dimensional field-theoretical equations for a two-particle system and the problem of πN, NN and $q\overline{q}$ scattering. Physics of Particles and Nuclei. 1993;**24**:317. American Institute of Physics

[27] Machavariani AI, Faessler A. Current conservation and analytic determination of the magnetic moment of the Δ resonance in the πN bremsstrahlung: II. Formulation with quark degrees of freedom. III. Magnetic moment of the Δ^o and Δ^- resonances. Journal of Physics G: Nuclear and Particle Physics. 2011;**38**: 035002

[28] Machavariani A. TMD PDF model of the ρ^o-meson production in the inclusive proton-proton collision. Available from: https://arxiv.org/pdf/1712.06395.pdf

[29] Weinberg S. The Quantum Theory of Fields – Volume I: Foundations. Cambridge: Cambridge University Press; 1995

[30] Bjorken JD, Drell SD. Relativistic Quantum Fields. New York, USA: McGraw Hill Book Company; 1965

[31] Itzykson C, Zuber C. Quantum Field Theory. New York: McGraw-Hill; 1980

[32] Banerjee MK, Cammarata JB. Theory of low-energy pion0pion-nucleon interaction. Physical Review C. 1978;**17**: 11251

[33] Machavariani AI, Rusetsky AG. On the field-theoretical formulation of the low-energy pion-nucleon scattering problem. Nuclear Physics A. 1990;**515**: 671

[34] Yao D-L et al. Pion-nucleon scattering in covariant baryon chiral perturbation theory with explicit Delta resonances. Journal of High Energy Physics. 2016;**38**:05. ArXiv: 1603.03638

[35] Pilkuhn HM. Relativistic Quantum Mechanics. Berlin Heidelberg: Springer-Verlag; 2003

[36] Abdulameer NJ et al. Improving constraints on gluon spin-momentum

correlations in transversely polarized protons via midrapidity open-heavy-flavor electrons in $p\uparrow + p$ collisions at $\sqrt{s} = 200GeV$; (PHENIX collaboration). Physical Review D. 2023;**107**:052012

[37] S. Adhikari et al. Measurement of spin-density matrix elements in $\rho(770)$ production with a linearly polarized photon beam at $E_\gamma = 8.2 \cdot 8.8GeV$. ArXiv: 2305.09047v1 [nucl-exp] 2023. GLUEX collaboration.

[38] Goldberger MI, Watson M. Collision Theory. New York-London-Sydney: Wiley; 1965

[39] Taylor JG. Graphical methods for Bethe-Salpeter equations. Supplimento of Nuovo Cimento. 1963;**1**:13

[40] Nakanishi N. A general survey of the theory of the Bethe-Salpeter equation. Progress of Theoretical Physics Supplement. 1969;**43**:1

[41] Logunov AA, Tavkhelidze AN. Quasi-optical potential in quantum field theory. Nuovo Cimento. 1963;**29**:380

[42] Blankenbecler R, Sugar R. Linear integral equations for relativistic multichannel scattering. Physics Review. 1966;**142**:1051

[43] Gross F. Three dimensional covariant integral equations for low energy systems. Physics Review. 1969; **186**:1448

[44] Yamasaki T et al. Relation between scattering amplitude and Bethe-Salpeter wave functions in quantum field theory. UTHEP-704 UTTCS-P104. ArXiv: 1709.09779v2 [hep-lat] 2017

[45] Kadyshevsky VG. Quasipotential type equation for the relativistic scattering amplitude. Nuclear Physics B. 1968;**6**:125

[46] Kadyshevsky VG. Three dimensional formulation of the relativistic two-body problem. Fizika Elementarnikh Chastits at Yadra. 1972;**2**: 635. Soviet Journal of Nuclear Physics 1972;**2**:69

[47] Carbonell J et al. Explicitly covariant light-front dynamics and relativistic few-body systems. Physics Reports. 1998;**300**:215

[48] Machavariani AI, Dj Chelidze A. On the field-theoretical approach to nucleon-nucleon scattering problem in the low energy region. Journal of Physics G. 1993;**24**:317

Chapter 13

Perspective Chapter: Squeezing and Entanglement of Two-Modes Quantum X Waves

Ali Saif M. Hassan, Waleed S.A. Hasan and Mohamed A. Shukri

Abstract

Quantum theory of generalized X waves with orbital angular momentum in dispersive media, and the interaction of quantized X waves in quadratic nonlinear media were studied in (J. opt,20,065201 (2018)). We present a kind of phase matching, which is called velocity phase matching, and this phase matching can be used for determining the length of the nonlinear crystal or the interaction time in the experiment setup, to produce X waves with particular velocity v. Moreover, we introduce more analysis for the dependence of squeezing of X waves on its spectral order, and for spectral orders $j > 0$, we predict the existence of a characteristic axicon aperture for maximal squeezing. Then, we find the quantum squeezed state of the down-converted state generated by the χ^2-nonlinear process. Finally, we detect their entanglement using a criterion of separability.

Keywords: entanglement, X waves, squeezed state, nonlinear quantum optics, phase matching, scalar field

1. Introduction

Based on the work of Tesla, German physicist Konstantin Meyl created a new unified field and particle theory. In Meyl's theory [1], vortices are used to explain quantum and classical physics, mass, gravitation, the constant speed of light, neutrinos, waves, and particles. With the help of this model, the properties of subatomic particles may be precisely estimated. The unified equation can also be used to derive well-known equations. He offers resources for carrying out an experiment by Tesla that proves the existence of scalar waves. Simple energy vortices in the form of particles are what scalar waves are. The electromagnetic, eddy current, potential vortex, and special distributions are all covered by the unified field theory. According to Meyl, the field—which creates particles through decay or conversion—always comes first. Energy particles (i.e., potential vortices) were not included in the theory because they are not recognized by classical physics. Quantum physics attempted to explain everything in terms of vortices which is why it is incomplete. Gravitation is from the speed of light difference caused by proximity that, proportional to field strength, decreases the distance of everything for the field strength [2]. The spatially localized vortex structures include pseudo-nondiffracting (P-N) vortex beams and their superpositions and provide additional effects interesting for theoretical research, experiments, and applications [3]. As we see, this kind of vortex in its quantum form can connect to Meyl's vortex to explain unified field and particle theory of Meyl. Also, it can connect to the localizability of Busch [4] and others for elementary particle physics to solve the theory of measurement in relativistic quantum mechanics problems as stated in Ref. [5]. Vortex solutions produced in some particle physics models like extended Abelian Higgs model [6], Galaxies and their dark-matter halos [7].

The primary characteristic that unifies numerous advancements in quantum physics is entanglement [8]. Entanglement is required, as a prime example, to distinguish between classical and quantum physics using Bell inequalities [9, 10]. Another way to think of entanglement is as the tool that permits real quantum protocols like teleportation [11] and Bell inequalities-based cryptography [12]. When a quantum method, like Shor's algorithm, yields a meaningful performance advantage over a classical computer, large entanglement is anticipated to be present in quantum registers [13]. It is obvious that entanglement is essential to comprehending and using quantum physics. Therefore, it makes sense to examine the creation of entanglement at its most fundamental source, which are the particle physics theories of fundamental interactions. A Bell inequality would never be violated in nature if the quantum theory of electromagnetic, or QED, never produced electron entanglement. This suggests that at a fundamental level, quantum unitary development must produce entanglement. The gauge principle, which states that the physical laws are invariant despite internal local rotations for particular symmetry groups, has become widely accepted as a way to explain the fundamental interactions seen in nature. With regard to electroweak and strong interactions, the Standard Model uses a Lagrangian that is essentially constrained by gauge symmetry requirements, except from quantum gravity. It is only logical to continue looking for a principle that is even more basic. Previous research has looked at entanglement's function in particle physics. Orthopositronium has been demonstrated in Ref. [14] to decay into 3-photon states that can be used in Bell-like experiments that reject classical physics more quickly than the conventional 2-particle Bell inequality. Bell inequalities have also been considered in regard to neutrino oscillations [15], kaon physics [16, 17], and their relationship with the characterization of T-symmetry violation [18]. The S-matrix formalism has been used to examine how entanglement varies in an elastic scattering process in Ref. [19]. A recent study on entanglement in deep elastic scattering is also worth mentioning [20]. Bell inequalities have also been discussed in regard to quantum correlations in CMB radiation [21].

Squeezed states, and two-mode-squeezed states, are foundational for describing the basis nature of quantum mechanical phenomena. Their quantum entanglement plays a central role in quantum physics, experiments of quantum optics, and quantum information science [22–27]. Nonlinear quantum optics, in particular, parametric down-conversion (PDC), gives us ways to study issues in the foundations of quantum mechanics that are not readily available using other technologies [27, 28]. The quantum communication encode information into the polarization degrees of freedom of photons [12] or the orbital angular momentum (OAM) [29–35]. The important problem related to multi-modes quantum communications is the diffraction and dispersion of the electromagnetic waves [30]. Electromagnetic waves packet is usually subjected to diffraction and dispersion. Diffraction create a broadening in space during propagation of the wave and dispersion create a broadening in the time during the propagation. Ultimately, these effects are connected with the bounded nature of the wave spectrum and, therefore, to its finite energy content [36]. A great effort has been done in studying the effect of atmosphere turbulence in free-space communication [31, 37–40]. Maxwell's equations admit diffraction-free and dispersion-free solutions, the so-called localized waves [3]. In particular, such solutions in the monochromatic domain are the Bessel beams [41], and in the pulsed domain, the most renowned localized waves are the X waves, first introduced in acoustics in 1992 by Lu and GreenLeef [42]. Despite the great work in the literature concerning X waves, the investigations of their quantum properties are very few [43, 44]. Generalization of the traditional X waves to the case of OAM-carrying X wave and the coupling between angular momentum and the temporal degrees of freedom of ultrashort pulses have been investigated [45]. Very recently, quantum and squeezing of X waves with OAM in nonlinear dispersive media have been proposed [46, 47], and they can open a new direction for free-space quantum communication and different areas of physics. In this work, we state the propagation of a scalar electromagnetic field in a linear dispersive medium. We present a kind of phase matching called velocity phase matching and introduce a relation between the interaction time and the velocity of X wave as well as a relation between the length of the $\chi^{(2)}$-nonlinear crystal and the velocity of X wave generated by the spontaneous parametric down-conversion (SPDC) process in Section 2. We study the SPDC process in a quadratic medium, in particular the dependence of squeezing of the down-converted state generated by the $\chi^{(2)}$-nonlinear crystal on the spectral order of quantum X waves in Section 3. We introduce the quantum squeezed form of the state generated by the SPDC process and its entanglement in Section 4. Finally, the results are summarized in Section 5.

2. Propagation of a generalized X wave in dispersive media and its quantization

We consider the propagation of a scalar electromagnetic field in a linear dispersive medium with refractive index $n = n(\omega)$. Applying the paraxial and slowly varying envelope approximation to the envelope function $A(\mathbf{r}, t)$ of an electric field

$$E(\mathbf{r}, t) = \sqrt{\frac{2}{\varepsilon_0 n^2}} A(\mathbf{r}, t) e^{i(kz - wt)}, \tag{1}$$

where $A(\mathbf{r}, t)$ varies slowly with $z$and $k = n(\omega)\omega/c$, which are the propagation coordinate and the wave number, respectively.

The field envelope $A(\mathbf{r}, t)$ satisfies the following eq. [48],

$$i\frac{\partial A}{\partial t} + i\omega'\frac{\partial A}{\partial z} - \frac{\omega''}{2}\frac{\partial^2 A}{\partial z^2} + \frac{\omega'}{2k}\Delta_\perp^2 A = 0, \tag{2}$$

where $\omega' = c^2 dk/d\omega$ and $\omega'' = c^2 d^2k/d\omega^2$ are the first and the second order dispersion, respectively.

The solution of Eq. (2) can be written as a superposition of the generalized X waves as follows [47]:

$$A(\mathbf{r}, t) = \sum_{m,p} \int dv \; C_{mp}(v) \; e^{-i\frac{v^2}{2\omega''}t} \; \psi_{m,p}^{(v)}(R, \zeta), \tag{3}$$

where $\psi_{m,p}^{(v)}(R, \zeta)$ is the generalized X wave of OAM number m, spectral order p and velocity v defined as,

$$\psi^{(v)}_{m,p}(R,\zeta)=\int_0^\infty d\alpha f_p(\alpha)J_m\left(\sqrt{\omega''k/\omega'}\alpha R\right)\ e^{i(\alpha-v/\omega'')\zeta}\ e^{im\theta},\tag{4}$$

with $R=\sqrt{x^2+y^2}$, $\zeta=z-(\omega'+v)t$ which is the co-moving reference frame associated to the X waves, and $f_p(\alpha)=\sqrt{k/\pi^2\omega'(1+p)}(\Delta\alpha)L^{(1)}_p(2\alpha\Delta)e^{-\alpha\Delta}$ which is the spectrum function, where $L^{(1)}_p(x)$ is the generalized Laguerre polynomials of the first kind of order p [48], and Δ is the reference length related to the spatial extension of the spectrum [47]. The quantized field of Eq. (3) can be written as [47],

$$\hat{A}(\mathbf{r},t)=\sum_{m,p}^{\infty}\int dv\ e^{\frac{-i}{\hbar}\left(\frac{Mv^2}{2}\right)t}\sqrt{\hbar\omega_{m,p}(v)}\ \psi^{(v)}_{m,p}(R,\zeta)\ \hat{a}_{m,p}(v)+h.c.,\tag{5}$$

where $h.c.$ denotes to the Hermitian conjugate, $\omega_{m,p}(v)=v^2/(2\omega'')$ and $M=\hbar/\omega''$ is the mass of X wave. The Hamiltonian operator can be written as follows [47]:

$$\hat{H}=\sum_{m,p}\int dv\ \hbar\omega_{m,p}(v)\left(\hat{a}^\dagger_{m.p}(v)\ \hat{a}_{m,p}(v)+\frac{1}{2}\right),\tag{6}$$

with the following usual bosonic commutation relations of creation and annihilation operators of the field,

$$\begin{aligned}\left[\hat{a}_{m,p}(v),\hat{a}^\dagger_{n,q}(u)\right]&=\delta_{m,n}\ \delta_{p,q}\ \delta(u-v),\\ \left[\hat{a}_{m,p}(v),\hat{a}_{n,q}(u)\right]&=0=\left[\hat{a}^\dagger_{m,p}(v),\hat{a}^\dagger_{n,q}(u)\right].\end{aligned}\tag{7}$$

Now, we turn our attention to the quadratic nonlinear process involving X waves, particularly to the spontaneous parametric down-conversion (SPDC) [28]. The interaction Hamiltonian of this process, with a practical case $\rho=\sqrt{k_1\omega_2'/k_2\omega_1'}\simeq 1$, can be written in two different forms [47]:

The first form is for the time-dependent interaction Hamiltonian which can be written as follows [47]:

$$\mathscr{H}_I=\chi\langle A_1|A_2^*\rangle+\chi^*\langle A_2|A_1^*\rangle,\tag{8}$$

$$\chi\langle A_1|A_2^*\rangle=\chi\int d^3r\hat{A}_1^\dagger(\mathbf{r},t)\hat{A}_2^\dagger(\mathbf{r},t)\tag{9}$$

$$\begin{aligned}\chi\langle A_1|A_2^*\rangle=\chi&\sum_{m,p,n,q}\int du\ dv\ e^{\frac{i(u^2+v^2)t}{2\omega''}}\sqrt{\hbar^2\omega_{m,p}(u)\omega_{n,q}(v)}\\ &\times\int d^3r\psi^{(u)*}_{m,p}(R,\zeta_1)\ \psi^{(v)*}_{n,q}(R,\zeta_2)\hat{a}^\dagger_{m,p}(u)\hat{b}^\dagger_{n,q}(v),\end{aligned}\tag{10}$$

$$\hat{H}_I(t)=\hbar\sum_{m,p,q}\int du\ dv\ \chi_{m,p,q}(u+v)\sqrt{\omega_{m,p}(u)\omega_{-m,p}(v)}\ e^{iF(u,v)t}\ \hat{a}^\dagger_{m,p}(u)\ \hat{b}^\dagger_{-m,q}(v)+h.c.,\tag{11}$$

with

$$F(u,v) = \frac{1}{2\omega''}\left[\ 2uv + (v-u)\left(\omega_1' - \omega_2'\right)\right], \tag{12}$$

and the interaction function is

$$\chi_{m,p,q}(u+v) = \frac{(-1)^m\ \chi\ \Delta^2}{\omega''^2\sqrt{(1+p)(1+q)}}(u+v)L_p^{(1)}\left(\frac{(u+v)\Delta}{2\omega''}\right)L_q^{(1)}\left(\frac{(u+v)\Delta}{2\omega''}\right)e^{\frac{-(u+v)\Delta}{2\omega''}}\ \theta(u+v), \tag{13}$$

where $\theta(u+v)$, is the Heaviside step function [48]. The second form is for the finite length interaction Hamiltonian which can be given by Ref. [47]

$$\hat{H}_I(z) = \int dt\ \left[\ \chi\ \langle A_1|A_2^*\rangle + \chi^*\langle A_2|A_1^*\rangle\right]. \tag{14}$$

The final form of $\hat{H}_I(z)$ can be written as follows:

$$\hat{H}(z) = \hbar\sum_{m,p,q}\int du\ dv\ \Xi_{m,p,q}(u,v)\ \sqrt{\omega_{m,p}(u)\ \omega_{-m,q}(v)}\ e^{i\Lambda(u,v)z}\ \hat{a}^\dagger_{m,p}(u)\ \hat{b}^\dagger_{-m,q}(v) + h.c., \tag{15}$$

where $\Lambda(u,v) = 2uv/\omega''(u+v)$, and $\Xi_{m,p,q}(u,v)$, which is the modified vertex function can be given by Ref. [47],

$$\Xi(u,v) = \frac{(-1)^m\ 2\ \chi\ \Delta^2}{\omega''^2\sqrt{(1+p)(1+q)}}\ \frac{u^2+v^2}{u+v}\ L_p^{(1)}\left(\frac{(u^2+v^2)\Delta}{(u+v)\omega''}\right)\ L_q^{(1)}\left(\frac{(u^2+v^2)\Delta}{(u+v)\omega''}\right)\ e^{-\frac{(u^2+v^2)\Delta}{(u+v)\omega''}}. \tag{16}$$

To calculate the state of the system after the interaction, we apply Schwing-Dyson expansion of the propagator $\exp\left[-i\hat{H}_I(t)/\hbar\right]$, and consider only the first term of the expansion [49]. For the time-dependent form of the interaction Hamiltonian stated in Eq. (11), the final state after the $\chi^{(2)}$-nonlinear interaction can be expressed as follows [47]:

$$|\psi^{(1)}(t)\rangle = \frac{-i}{\hbar}\int_0^t d\tau\ \hat{H}_I(\tau)|0\rangle. \tag{17}$$

Substituting Eq. (11) into the above equation and after some calculations, the final state can be written in the following form:

$$|\psi^{(1)}(t)\rangle = \sum_{m,p,q}\int du\ dv\ \mathcal{G}_{m,p,q}(u,v)\ \mathscr{F}(u,v,t)\ |m,p,u;\ -m,q,v\rangle, \tag{18}$$

where $|m,p,u;\ -m,q,v\rangle \equiv \hat{a}^\dagger_{m,p}(u)\ \hat{b}^\dagger_{-m,q}(v)|0,0\rangle$, the amplitude $\mathcal{G}_{m,p,q}(u,v)$ is defined as

$$\mathcal{G}_{m,p,q}(u,v) = -i\sqrt{\omega_{m,p}(u)\ \omega_{-m,q}(v)}\ \frac{\chi_{m,p,q}(u+v)}{F(u,v)}, \tag{19}$$

and $\mathscr{F}(u,v,t) = \left[e^{iF(u,v)t} - 1\right]$. The above state represents the superposition of two particles corresponding to the modes ω_1 and ω_2 (generated by the nonlinear interaction) traveling with velocities u and v, respectively. For the second form of the interaction Hamiltonian stated in Eq. (15), the final state of the system can be written as

$$|\psi^{(1)}(L)\rangle = \frac{-i}{\hbar}\int_0^L dz\ \hat{H}_I(z)|0\rangle. \tag{20}$$

Substituting Eq. (15) into the above equation, the state of the field at the output face of the crystal can be written as follows:

$$|\psi^{(1)}(L)\rangle = \sum_{m,p,q}\int du\ dv\ \mathscr{L}_{m,p,q}(u,v)\ \mathscr{F}(u,v,L)|m,p,u;\ -m,q,v\rangle, \tag{21}$$

where $\mathscr{F}(u,v,L) = \left[e^{i\Lambda(u,v)L} - 1\right]$, and

$$\mathscr{L}_{m,p,q}(u,v) = \frac{-i}{2}\sqrt{\omega_{m,p}(u)\omega_{-m,q}(v)}\ \left[\frac{\omega''(u+v)}{uv}\right]\ \Xi_{m,p,q}(u,v). \tag{22}$$

The transition probability for the field to be in such a state after the time-dependent interaction with the $\chi^{(2)}$-nonlinear crystal is proportional with $|\mathscr{F}(u,v,t)|^2$ as follows:

$$P(t) \propto |\mathscr{F}(u,v,t)|^2, \tag{23}$$

similarly, the transition probability for the finite length interaction of the $\chi^{(2)}$-nonlinear crystal is proportional with $|\mathscr{F}(u,v,L)|^2$ as follows [47]:

$$P(L) \propto |\mathscr{F}(u,v,L)|^2. \tag{24}$$

Explicitly, $|\mathscr{F}(u,v,t)|^2$ can be expressed as follows:

$$|\mathscr{F}(u,v,t)|^2 = 2[1 - \cos(F(u,v)t)]. \tag{25}$$

From the above equation, we notice that the transition probability reaches its maximum, when $|\mathscr{F}(u,v,t)|^2$ reaches its maximum, and that occurs when the condition $F(u,v)t = (2m+1)\pi$, with $m = 0,1,2,3,\ \ldots$, is achieved. The above condition is considered as a new phase matching condition, called velocity matching condition over the velocities of the two-modes X waves generated from SPDC process. This condition requires that

$$u = \frac{2k'_m + v\ \Delta\omega'}{2v + \Delta\omega'}, \tag{26}$$

where $\Delta\omega' = \omega'_1 - \omega'_2$, and $k'_m = (2m+1)\pi\omega''/t$. For the case, when $\Delta\omega' = 0$, we get $uv = k'_m = (2m+1)\pi\omega''/t$, and the interaction time can be given by $t = (2m+1)\pi\omega''/uv$. When $u = v$, regardless of that $\Delta\omega' = 0$, we get $v^2 = k'_m = (2m+1)\pi\omega''/t$.

Thus, the interaction time can be written in terms of the velocity of X waves modes as follows:

$$t = \frac{(2m+1)\pi\omega''}{v^2}. \tag{27}$$

For the finite length of the nonlinear crystal, the velocity matching condition is given by Ref. [47],

$$u = \frac{k_n \ v}{v - k_n}, \tag{28}$$

where $k_n = \left(n + \frac{1}{2}\right)\pi\omega''/L$, with $n = 0,1,2,\cdots$. Therefore, the velocity matching condition fixes either the length of the crystal L or the relative velocity of the two modes X wave involved in the SPDC process as, $uv/(u+v) = k_n = (n+1/2)\pi\omega''/L$, and the length of the crystal L can be given by,

$$L = \frac{\left(n+\frac{1}{2}\right)\pi\omega'' \ (u+v)}{uv}. \tag{29}$$

For $u = v$, the above equation becomes

$$L = \frac{(2n+1)\pi\omega''}{v}. \tag{30}$$

From Eqs. (27) and (30), we note that the ordinary relation between the length and the time is achieved ($L = vt$). This result can be used for determining the length of the nonlinear crystal in the experimental setup, to produce X waves with velocity v. This is the first result of our work.

3. Squeezing of quantum X waves

Now, we can study the general case of the squeezing effect in the SPDC process, in particular, the dependence of squeezing of the down-converted state, generated by $\chi^{(2)}$-nonlinear media, on the spectral orders of quantum X waves, for the case, when the velocities of the two-modes X waves are equal $u = v$, and their spectral orders are equal as well $p = q = j$. The interaction time Hamiltonian in Eq. (11) becomes

$$\hat{H}_I(t) = \hbar \sum_{mj} \int dv \ \chi_{mj}(2v) \ \omega(v) \ \hat{a}^\dagger_{mj}(v,t) \ \hat{b}^\dagger_{-mj}(v,t) + h.c., \tag{31}$$

where $\hat{a}^\dagger_{mj}(v,t) = e^{iF(v)t/2} \ \hat{a}^\dagger_{mj}(v)$, and $\hat{b}^\dagger_{-mj}(v,t) = e^{iF(v)t/2} \ \hat{b}^\dagger_{-mj}(v)$. Also, the interaction function in Eq. (13) becomes

$$\chi_{mj}(2v) = \frac{2(-1)^m \ \chi\Delta^2 v}{\omega''^2(1+j)} \ L_j^{(1)}\left(\frac{v\Delta}{\omega''}\right) \ L_j^{(1)}\left(\frac{v\Delta}{\omega''}\right) \ e^{\frac{-v\Delta}{\omega''}}. \tag{32}$$

And the modified vertex function in Eq. (16) can be rewritten as follows:

$$\Xi_{mj}(v) = \frac{2(-1)^m \ \chi \Delta^2 \ v}{\omega''^2 (1+j)} \ L_j^{(1)}\left(\frac{v\Delta}{\omega''}\right) \ L_j^{(1)}\left(\frac{v\Delta}{\omega''}\right) \ e^{\frac{-v\Delta}{\omega''}}. \tag{33}$$

From Eqs. (32) and (33) we notice that the interaction function and the modified vertex function have the same form. In the interaction picture, we consider the time evolution controlled by $\hat{H}_I(t)$. Thus, the equations of motion for $\hat{a}_{mj}(v,t)$ and $\hat{b}_{-mj}(v,t)$ are:

$$\frac{d\hat{a}_{mj}(v,t)}{dt} = \frac{1}{i\hbar} \left[\hat{a}_{mj}(v,t), \hat{H}_I(t)\right], \tag{34}$$

$$\frac{d\hat{b}_{-mj}(v,t)}{dt} = \frac{1}{i\hbar} \left[\hat{b}_{-mj}(v,t), \hat{H}_I(t)\right]. \tag{35}$$

Substituting Eq. (31) into the above equations, we obtain

$$\frac{d\hat{a}_{mj}(v,t)}{dt} = -2i \ \omega(v) \ \chi_{mj}(2v) \ \hat{b}^{\dagger}_{-mj}(v,t), \tag{36}$$

$$\frac{d\hat{b}_{-mj}(v,t)}{dt} = -2i \ \omega(v) \ \chi_{mj}(2v) \ \hat{a}^{\dagger}_{mj}(v,t). \tag{37}$$

The general solutions of these equations are [50],

$$\hat{a}_{mj}(v,t) = \cosh\left(\beta_{mj}(v) \ t\right) \ \hat{a}_{mj}(v) + \sinh\left(\beta_{mj}(v) \ t\right) \ \frac{\alpha_{mj}}{\beta_{mj}(v)} \ \hat{b}^{\dagger}_{-mj}(v), \tag{38}$$

$$\hat{b}_{-mj}(v,t) = \cosh\left(\beta_{mj}(v)t\right)\hat{b}_{-mj}(v) + \sinh\left(\beta_{mj}(v) \ t\right) \ \frac{\alpha_{mj}}{\beta_{mj}(v)} \ \hat{a}^{\dagger}_{mj}(v). \tag{39}$$

Here, we consider the expression $\alpha_{mj} = -2i \ \omega(v) \ \chi_{mj}(2v)$, with $\omega(v) = v^2/2\omega''$. To find the squeezing parameter $\xi_{mj}(v) = \beta_{mj}(v)t$, we use the relation $\beta_{mj}(v) = \sqrt{\alpha_{mj}\alpha^*_{mj}}$, then we can write the squeezing parameter as follows:

$$\xi_{mj}(v) = \frac{2(-1)^m \chi \ t}{(1+j)\Delta} \ \frac{\tilde{v}^3}{\omega''^3} \left(L_j^{(1)}\left(\frac{\tilde{v}}{\omega''}\right)\right)^2 e^{-\frac{\tilde{v}}{\omega''}}, \tag{40}$$

with $\tilde{v} = v\Delta$. The above equation represents the squeezing parameter which depends on the OAM number m, the spectral order j, and the velocity of the two-modes X waves v. Now, we can show the normalized squeezing parameter modulus $|\bar{\xi}_{mj}| = \frac{\Delta|\xi_{mj}|}{4\chi t}$ as a function of the normalized velocity $\frac{\tilde{v}}{\omega''}$ for different values of j, ($j = 0,1,2,3,4,5$), and fixing the time t so that $\frac{4\chi t}{\Delta} = 1$, when $m = 0$, as shown in **Figure 1**.

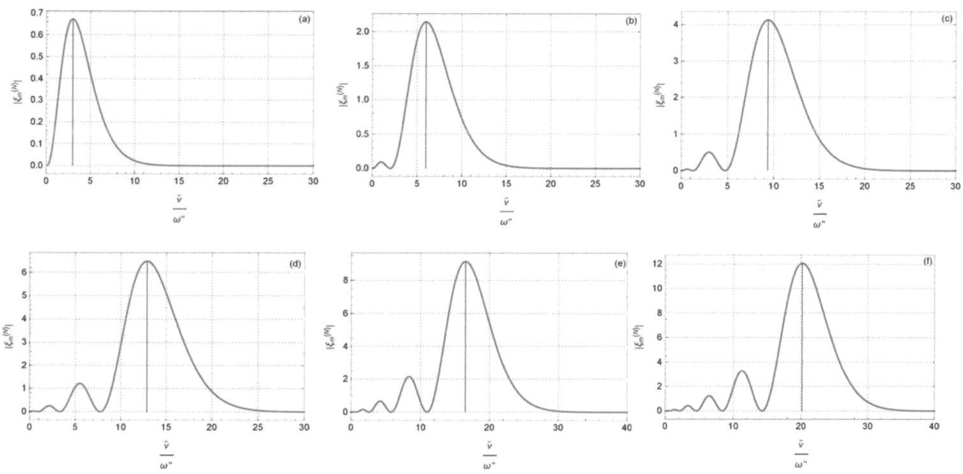

Figure 1.
Plot of the normalized squeezing parameter modulus $|\bar{\xi}_{mj}| = \frac{\Delta|\xi_{mj}|}{\chi}$ in function of the normalized velocity $\frac{\bar{v}}{\omega''}$, for different values of the spectral order j where, (a) for $j = 0$, (b) for $j = 1$, (c) for $j = 2$, (d) for $j = 3$, (e) for $j = 4$, (f) for $j = 5$, with fixing the time t, so that $\frac{4\chi t}{\Delta} = 1$, and $m = 0$.

Figure 1 shows the dependence of the squeezing on the spectral order j of X waves. This dependence can be explained in the following points:

1. The maximum value of the squeezing parameter $|\bar{\xi}_{mj}|$ and the corresponding normalized velocity $\frac{\bar{v}}{\omega''}$ increases as the spectral order j increases.

2. The number of the peak values of squeezing depends on the spectral order j and this number can be given by $j + 1$, where the smallest peak is the first and the biggest one is the last.

3. The squeezing parameter depends on the velocity of X wave as shown in Eq. (40) where the amount of squeezing parameter produced by SPDC process will be maximized at an optimal velocity for every value of the spectral order j, and this velocity can be given by $v_{opt} = n_j\omega''/\Delta$, where $n_0 = 3, n_1 = 6, n_2 = 9.36$, $n_3 = 12.88, n_4 = 16.49, n_5 = 20.156$ for $j = 0,1,2,3,4,5$, respectively as shown in **Figure 1. Figure 1(a)** shows that for $j = 0$ as in Ref. [46], and the corresponding optimal axicon angles for different values of j can be given by $\cos\theta_j^{opt} = \Delta/n_j\lambda$. For example, given a nondiffracting pulse with a duration of $\Delta t = 8$ fs, a carrier wavelength of $\lambda = 850$ nm and assuming that $\chi^2 \simeq 10^{-12}$ m/V, the optimal axicon angles θ_j^{opt} that maximize the squeezing for $j = 0,1,2,3,4,5$ will take the following values respectively, $\theta_0^{opt} \simeq 20°$, $\theta_1^{opt} \simeq 62°$, $\theta_2^{opt} \simeq 72°$, $\theta_3^{opt} \simeq 77°$, $\theta_4^{opt} \simeq 80°$, $\theta_5^{opt} \simeq 82°$. Then, we can evaluate the corresponded maximal squeezing parameters to get the following values respectively, ($\xi_m^{(0)} \simeq 100s^{-1}$, $\xi_m^{(1)} \simeq 280s^{-1}$, $\xi_m^{(2)} \simeq 520s^{-1}$, $\xi_m^{(3)} \simeq 820s^{-1}$, $\xi_m^{(4)} \simeq 1160s^{-1}$, $\xi_m^{(5)} \simeq 1550s^{-1}$).

To illustrate the effect of the OAM on the squeezing more, we can introduce the quadrature operators $\hat{X}_{mj}(v,t) = \hat{a}_{mj}(v,t) + \hat{a}^{\dagger}_{mj}(v,t)$, and $\hat{Y}_{mj}(v,t) =$

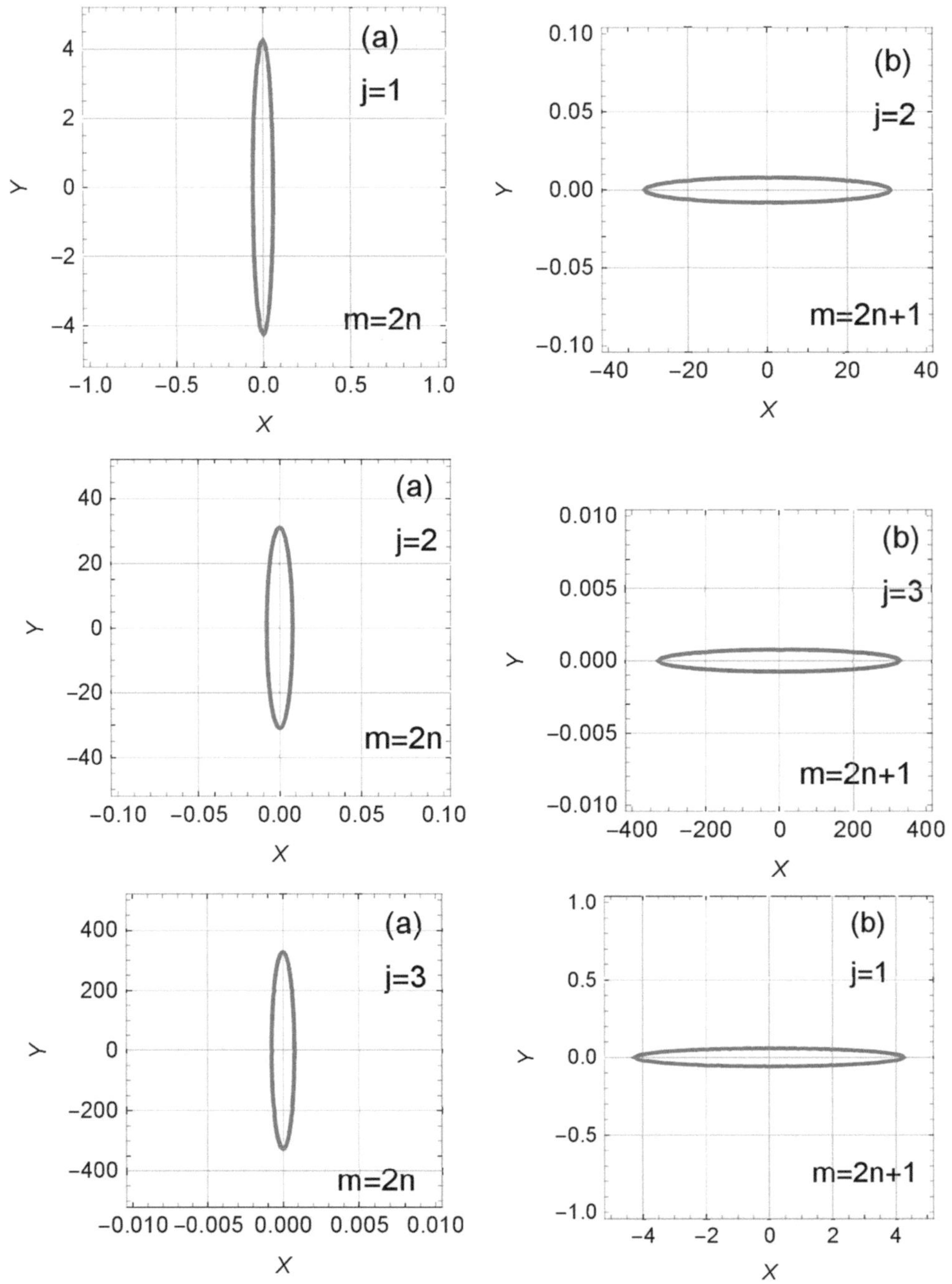

Figure 2.
Quadrature space representation of the squeezed down-converted state in the case of even [panel (a)] and odd [panel (b)] values of the OAM number m with a normalized velocity $\frac{\bar{v}}{\omega''} = 6{,}9.36{,}12.88.$ *for a spectral order* $j = 1{,}2{,}3$*, respectively.*

$i\left[\hat{a}^{\dagger}_{mj}(v,t) - \hat{a}_{mj}(v,t)\right]$. For $e^{i\phi} = \alpha_{mj}/\beta_{mj} = 1$ or $\phi = 0$, we get $\hat{X}_{mj}(v,t) = e^{\xi_{mj}(v)}\hat{X}_{mj}(v,0)$, and $\hat{Y}_{mj}(v,t) = e^{-\xi_{mj}(v)}\hat{Y}_{mj}(v,0)$, and the variance of them is given by $\Delta\hat{X}_{mj}(v,t) = e^{\xi_{mj}(v)}\hat{Y}_{mj}(v,0)$. This shows that the SPDC interaction Hamiltonian for the generalized X waves acts as a two-modes squeeze operator which will be illustrated more in the next section. Noticeably, the effect of the OAM number m changes only the sign of the squeezing parameter $\xi_{mj}(v)$, that is, the squeezed quadrature changes depending on the parity of the orbital angular momentum number m. In particular, if m is even number, $\xi_{mj}(v) > 0$, the squeezing accrues on the Y quadrature. On the other hand, if m is an odd number, $\xi_{mj}(v) < 0$ and the squeezing occurs on X quadrature as in Ref. [46] for $j = 0$, and for $j = 1,2,3$. this effect can be shown as in **Figure 2**.

4. The two-modes squeezed state of X wave and its entanglement

In analogy to traditional case [51, 52], we look for the eigenvalues of Hamiltonian operator, represented by Eq. (6). Let us introduce the N-particle states $|m,j,v;N\rangle$ as states with N field excitation with velocity v in the traveling mode (i.e., the X wave) $\psi_{m,p}^{(v)}(\mathbf{r},t)$ with energy $Mv^2/2$ [47]. The eigenstates $|m,p,v,N\rangle$ of the photon (particle) number operator $\hat{N}(v) = \hat{a}_{mp}^{\dagger}(v)\ \hat{a}_{mp}(v)$ are called Fock states, where $\hat{a}_{mp}(v)$ is the annihilation operator of the single mode in Eq. (7). These states are orthogonal

$$\langle n,q,u,N|m,p,v,N\rangle \equiv \delta_{nm}\ \ \delta_{pq}\ \ \delta(u-v), \tag{41}$$

and complete. The number state $|m,p,v,N\rangle$ can be generated from the vacuum state (i.e., the eigenstate of $\hat{N}(v)$ with the eigenvalue equal to zero) as

$$|m,p,v,N\rangle = \frac{\left[\hat{a}_{mp}^{\dagger}(v)\right]^N}{\sqrt{N!}}|0\rangle. \tag{42}$$

The expectation value of the electric field operator is zero on the particle number operator eigenstates [47],

$$\left\langle m,p,v,N|\hat{A}(r,t)|m,p,v,N\right\rangle = 0. \tag{43}$$

However, the Fock states $|m,p,v,N\rangle$ are not normalizable [47], since their representation in configuration state $\langle \mathbf{r},t|m,p,v,N\rangle = \psi_{mp}^{(v)}(r,t)$ carries infinite energy. The particle number states of the two modes (A,B) with particle numbers (N_A,N_B) are defined as

$$|m,j,v;N_A\rangle_A = \frac{\left[\hat{a}_{mj}^{\dagger}(v)\right]^{N_A}|0\rangle_A}{\sqrt{N_A!}}, \tag{44}$$

and

$$|-m,j,v,N_B\rangle_B = \frac{\left[\hat{b}_{-mj}^{\dagger}(v)\right]^{N_B}|0\rangle_B}{\sqrt{N_B!}} \tag{45}$$

Together with the conditions for the vacuums, $\hat{a}_{mj}(v)|0\rangle_A = 0$ and $\hat{b}_{-mj}(v)|0\rangle_B = 0$.

The operators $\hat{a}_{mj}(v)$ and $\hat{b}_{-mj}(v)$ obey the bosonic commutation relations in Eq. (7). Now, we can define the two-mode squeeze operator, which is a unitary operator, using the normalized squeezing parameter $\overline{\xi}_{mj}(v) = \frac{(-1)^m}{2(1+j)} \frac{\tilde{v}^3}{\omega''^3} \left(L_j^{(1)}\left(\frac{\tilde{v}}{\omega''}\right)\right)^2 e^{-\frac{\tilde{v}}{\omega''}}$, as

$$V_{\overline{\xi}_{mj}} = \exp\left(e^{-i\phi}\ \overline{\xi}_{mj}\ \hat{a}_{mj}(v)\ \hat{b}_{-mj}(v) - e^{i\phi}\ \overline{\xi}_{mj}\ \hat{a}^\dagger_{mj}(v)\ \hat{b}^\dagger_{-mj}(v)\right). \tag{46}$$

The two-mode squeezed vacuum TMSV state can be obtained from the vacuum state as [52],

$$|0,0,\overline{\xi}_{mj}\rangle_{AB} = V_{\overline{\xi}_{mj}}|0,0\rangle = \left[\cosh\left(\overline{\xi}_{mj}\right)\right]^{-1} \sum_{N=0}^{\infty} \left(e^{i\phi}\tanh\left(\overline{\xi}_{mj}\right)\right)^N |m,j,v,N;\ -m,j,v,N\rangle. \tag{47}$$

Also, we can define the annihilation operators crossing over the separation of part A and part B as [53],

$$\begin{aligned} \hat{A}_{\overline{\xi}_{mj}} &= V_{\overline{\xi}_{mj}} \hat{a}_{mj}(v) V^\dagger_{\overline{\xi}_{mj}} &= \cosh\left(\overline{\xi}_{mj}\right)\left[\hat{a}_{mj}(v) + e^{i\phi}\tanh\left(\overline{\xi}_{mj}\right)\hat{b}^\dagger_{-mj}(v)\right] \\ \hat{B}_{\overline{\xi}_{mj}} &= V_{\overline{\xi}_{mj}} \hat{b}_{-mj}(v) V^\dagger_{\overline{\xi}_{mj}} &= \cosh\left(\overline{\xi}_{mj}\right)\left[\hat{b}_{-mj}(v) + e^{i\phi}\tanh\left(\overline{\xi}_{mj}\right)\hat{a}^\dagger_{mj}(v)\right]. \end{aligned} \tag{48}$$

The new field operators satisfy the bosonic commutation relations in Eq. (7)

$$\left[\hat{A}_{\overline{\xi}_{mj}}, \hat{A}^\dagger_{\overline{\xi}_{m'j'}}\right] = \left[\hat{B}_{\overline{\xi}_{mj}}, \hat{B}^\dagger_{\overline{\xi}_{m'j'}}\right] = \delta_{m,m'}\delta_{j,j'}\delta(v-v'), \tag{49}$$

and

$$\left[\hat{A}_{\overline{\xi}_{mj}}, \hat{B}_{\overline{\xi}_{m'j'}}\right] = \left[\hat{A}_{\overline{\xi}_{mj}}, \hat{B}^\dagger_{\overline{\xi}_{m'j'}}\right] = 0. \tag{50}$$

The TMSV state corresponds to the vacuum of the new field operators, as we have used $\hat{a}_{mj}(v)|0\rangle = 0$, $\hat{b}_{-mj}(v)|0\rangle = 0$, Eq. (47), Eq. (48), and the unitary property of the squeeze operator, we get $\hat{A}_{\overline{\xi}_{mj}}|0,0,\overline{\xi}_{mj}\rangle = 0$, and $\hat{B}_{\overline{\xi}_{mj}}|0,0,\overline{\xi}_{mj}\rangle = 0$.

The two-mode number state is defined by the number state of nonlocal modes as [53],

$$|N_A, N_B, \overline{\xi}_{mj}\rangle = \frac{\left[\hat{A}^\dagger_{\overline{\xi}_{mj}}\right]^{N_A}\left[\hat{B}^\dagger_{\overline{\xi}_{mj}}\right]^{N_B}}{\sqrt{N_A!}\sqrt{N_B!}}|0,0,\xi_{\overline{\xi}_{mj}}\rangle. \tag{51}$$

In our case, we have $N_A = N_B = 1$, so we obtain

$$|1,1,\overline{\xi}_{mj}\rangle = \hat{A}^\dagger_{\overline{\xi}_{mj}}\hat{B}^\dagger_{\overline{\xi}_{mj}}|0,0,\overline{\xi}_{mj}\rangle. \tag{52}$$

The state in the above equation represents a two-particles squeezed state of the two modes X wave. The superposition of the two particles corresponding to the two modes ω_1 and ω_2 see Eq. (18) (when $u = v$) can be written as

$$|\psi^{(1)}(t)\rangle = \sum_{mj}\int dv\ \mathcal{G}_{mj}(v)\ \mathcal{F}(v,t)\ \hat{a}^\dagger_{mj}(v)\hat{b}^\dagger_{-mj}(v)|0,0\rangle, \tag{53}$$

where $\mathcal{F}(v,t) = \left[e^{iF(u)t} - 1\right]$, and $\mathcal{G}_{mj}(v) = -i\ \omega(v)\chi_{mj}(2v)/F(v)$, with $\chi_{mj}(2v)$ as in Eq. (32). The superposition of the two particles squeezed state corresponds to the state in Eq. (53) is

$$|\psi^{(1)}_{\overline{\xi}_{mj}}(t)\rangle = V_{\overline{\xi}_{mj}}|\psi^{(1)}(t)\rangle = \sum_{mj}\int dv\ \mathcal{G}_{mj}(v)\ \mathcal{F}(v,t)V_{\overline{\xi}_{mj}}\hat{a}^{\dagger}_{mj}(v)\hat{b}^{\dagger}_{-mj}(v)|0,0\rangle. \tag{54}$$

Using Eq. (48), the above state can be written in the form

$$|\psi^{(1)}_{\overline{\xi}_{mj}}(t)\rangle = \sum_{mj}\int dv\ \mathcal{G}_{mj}(v)\ \mathcal{F}(v,t)\hat{A}^{\dagger}_{\overline{\xi}_{mj}}\hat{B}^{\dagger}_{\overline{\xi}_{mj}}|0,0,\overline{\xi}_{mj}\rangle, \tag{55}$$

which is a superposition of the squeezed states $|1,1,\overline{\xi}_{mj}\rangle$ in Eq. (52).

Now, for calculating the entanglement of the state in Eq. (55), we consider the separable conditions based on the operators that form $SU(2)$ algebra and $SU(1,1)$ algebra [53]. The measurements of quadrature moments of the fourth order are sufficient to verify the entanglement of this state. Let us define the operators,

$$\begin{aligned} J_z &= \frac{1}{2}(N_A - N_B), \\ K_x &= \frac{1}{2}\left(\hat{a}^{\dagger}_{mj}(v)\ \hat{b}^{\dagger}_{-mj}(v) + \hat{a}_{mj}(v)\ \hat{b}_{-mj}(v)\right), \\ K_y &= \frac{1}{2i}\left(\hat{a}^{\dagger}_{mj}(v)\hat{b}^{\dagger}_{-mj}(v) - \hat{a}_{mj}(v)\hat{b}_{-mj}(v)\right) \end{aligned} \tag{56}$$

where $\hat{N}_A = \hat{a}^{\dagger}_{mj}(v)\hat{a}_{mj}(v)$ and $\hat{N}_B = \hat{b}^{\dagger}_{-mj}(v)\hat{b}_{-mj}(v)$ are number operators of the local modes.

The relations of the annihilation and creation operators $\hat{a}_{mj}(v)$, $\hat{a}^{\dagger}_{mj}(v)$, $\hat{b}_{-mj}(v)$, and $\hat{b}^{\dagger}_{-mj}(v)$ with the new field operators in Eq. (48) can be represented by

$$\begin{aligned} \hat{a}_{mj}(v) &= \cosh\left(\overline{\xi}_{mj}\right)\left(\hat{A}_{\overline{\xi}_{mj}} - e^{i\phi}\tanh\left(\overline{\xi}_{mj}\right)\hat{B}^{\dagger}_{\overline{\xi}_{mj}}\right) \\ \hat{a}^{\dagger}_{mj}(v) &= \cosh\left(\overline{\xi}_{mj}\right)\left(\hat{A}^{\dagger}_{\overline{\xi}_{mj}} - e^{-i\phi}\tanh\left(\overline{\xi}_{mj}\right)\hat{B}_{\overline{\xi}_{mj}}\right) \\ \hat{b}_{-mj}(v) &= \cosh\left(\overline{\xi}_{mj}\right)\left(\hat{B}_{\overline{\xi}_{mj}} - e^{i\phi}\tanh\left(\overline{\xi}_{mj}\right)\hat{A}^{\dagger}_{\overline{\xi}_{mj}}\right) \\ \hat{b}^{\dagger}_{-mj}(v) &= \cosh\left(\overline{\xi}_{mj}\right)\left(\hat{B}^{\dagger}_{\overline{\xi}_{mj}} - e^{-i\phi}\tanh\left(\overline{\xi}_{mj}\right)\hat{A}_{\overline{\xi}_{mj}}\right) \end{aligned} \tag{57}$$

The separable condition is [53],

$$\left[\langle\Delta^2 K_y\rangle - \frac{1}{4}\right]\langle\Delta^2 J_z\rangle \geq \frac{1}{4}|\langle K_x\rangle|^2 \tag{58}$$

where $\left\langle\hat{O}\right\rangle = tr\left(\hat{O}\rho\right)$ denotes the expectation value of the observable $\hat{O}$ with respect to state ρ, $\Delta\hat{O} = \hat{O} - \left\langle\hat{O}\right\rangle$, and $\left\langle\Delta^2\hat{O}\right\rangle = \left\langle\hat{O}^2\right\rangle - \left\langle\hat{O}\right\rangle^2$. The state we use in Eq. (58) is the superposition of simultaneous eigenstates of the number operators $\hat{A}^{\dagger}_{\overline{\xi}_{mj}}\hat{A}_{\overline{\xi}_{mj}}$ and $\hat{B}^{\dagger}_{\overline{\xi}_{mj}}\hat{B}_{\overline{\xi}_{mj}}$. From the definition of J_z in Eq. (56), we have

$$\hat{A}^{\dagger}_{\bar{\xi}_{mj}}\hat{A}_{\bar{\xi}_{mj}} - \hat{B}^{\dagger}_{\bar{\xi}_{mj}}\hat{B}_{\bar{\xi}_{mj}} = N_A - N_B = 2J_z \tag{59}$$

This implies that the two-mode squeezed (TMS) number states are eigenstates of J_z and $\langle \Delta^2 J_z \rangle = 0$.

Also, we can write

$$K_x = \frac{1}{2}\cosh^2\left(\bar{\xi}_{mj}\right)\left[\left(\hat{A}^{\dagger}_{\bar{\xi}_{mj}} - \eta^*\hat{B}_{\bar{\xi}_{mj}}\right)\left(\hat{B}^{\dagger}_{\bar{\xi}_{mj}} - \eta^*\hat{A}_{\bar{\xi}_{mj}}\right) + \left(\hat{A}_{\bar{\xi}_{mj}} - \eta\hat{B}^{\dagger}_{\bar{\xi}_{mj}}\right)\left(\hat{B}_{\bar{\xi}_{mj}} - \eta\hat{A}^{\dagger}_{\bar{\xi}_{mj}}\right)\right], \tag{60}$$

with $\eta = e^{i\phi}\tanh\left(\bar{\xi}_{mj}\right)$.

Then, the expectation value of K_x for the TMS number state of Eq. (55) can be given by

$$\langle K_x \rangle = -\sum_{mj}\int \frac{dv}{2}\cosh^2\left(\bar{\xi}_{mj}\right)\left|\mathcal{G}_{mj}(v)\right|^2 \ |\mathcal{F}(v,t)|^2(\eta^* + \eta)\left\langle \hat{A}^{\dagger}_{\bar{\xi}_{mj}}\hat{A}_{\bar{\xi}_{mj}} + \hat{B}^{\dagger}_{\bar{\xi}_{mj}}\hat{B}_{\bar{\xi}_{mj}} + 1\right\rangle \tag{61}$$

with $\left\langle \hat{A}^{\dagger}_{\bar{\xi}_{mj}}\hat{A}_{\bar{\xi}_{mj}} + \hat{B}^{\dagger}_{\bar{\xi}_{mj}}\hat{B}_{\bar{\xi}_{mj}} + 1\right\rangle = 3$. From Eq. (25) we get the maximum value $|\mathcal{F}(v,t)|^2 = 4$, thus

$$\langle K_x \rangle = -6\sum_{mj}\int dv\cosh^2\left(\bar{\xi}_{mj}\right)\left|\mathcal{G}_{mj}(v)\right|^2(\eta^* + \eta) \tag{62}$$

where, $(\eta^* + \eta) = 2\cos(\phi)\tanh\left(\bar{\xi}_{mj}\right)$, which implies that $\langle K_x \rangle = 0$, if $\phi = \frac{\pi}{2}$, this problem can be avoided by applying a local phase shift [53], we obtain

$$\begin{aligned}\langle K_x \rangle &= -12\cos(\phi)\sum_{mj}\int dv\left|\mathcal{G}_{mj}(v)\right|^2\cosh\left(\bar{\xi}_{mj}\right)\sinh\left(\bar{\xi}_{mj}\right) \\ &= -6\cos(\phi)\sum_{mj}\int dv\left|\mathcal{G}_{mj}(v)\right|^2\sinh\left(2\bar{\xi}_{mj}\right).\end{aligned} \tag{63}$$

If we take $\phi = (2n+1)\pi,\ n = 0,1,2,3,\cdots$, then $\langle K_x \rangle$ will be maximized. This implies that $|\langle K_x \rangle| > 0$, for $\phi = n\pi,\ n = 0,1,2,3,\cdots$. Therefore, for the state in Eq. (55), the L.H.S. of Eq. (58) is equal to zero and its R.H.S. is positive. Hence, we can confirm the inseparability of the state in Eq. (55) from the violation of the condition in Eq. (58), means that we can take it as an indicator of quantifying the amount of the entanglement of squeezed states which can be detected by increasing $|\langle K_x \rangle|$. In other words, we have found that $|\langle K_x \rangle|$ increases as the spectral order j increases, which indicates that the amount of entanglement of the state in Eq. (55) increases.

5. Summary and comments

To summarize, we obtain the velocity phase matching, which can be used for determining the length of nonlinear crystal in the experiment setup to produce X a wave with velocity v. We have introduced the relation between the squeezing parameter and the spectral order of X waves which shows that the maximal squeezing increases as the spectral order increases, and there exists an optimal velocity (i.e., axicon angle) that maximizes the amount of squeezing generated for each value of the spectral order. Moreover, we have constructed the squeeze operator of the two modes X wave, then we act by it on the two-particle state generated by SPDC process to obtain the squeezed form of it. Finally, we have used the criterion of the quadrature moment of the fourth order to verify the entanglement of the two modes squeezed state of X wave. We have observed that the entanglement of the state increases as the spectral order increases.

Author details

Ali Saif M. Hassan[1*], Waleed S.A. Hasan[1,2] and Mohamed A. Shukri[2]

1 Department of Physics, University of Amran, Amran, Yemen

2 Department of Physics, University of Sana'a, Sana'a, Yemen

*Address all correspondence to: alisaif@amu.edu.ye; alisaif73@gmail.com

References

[1] Ing Konstantin Meyl, Scalar Waves, Theory and Experiments. Available from: http://www.k-meyl.de/go/Primaerliteratur/Scalar-Waves.pdf

[2] Zohuri B. Scalar Wave Driven Energy Applications. Cham, Switzerland: Springer Nature Switzerland AG; 2019

[3] Hernandez-Figueroa HE, Zamboni-Rached M, Recami E. Localized Waves. New York: Wiley; 2008

[4] Busch P. Journal of Physics A: Mathematical and General. 1999;**32**: 6535-6546

[5] Jaeger G. Journal of Physics: Conference Series. 2020;**1638**:012010

[6] Adhikari P, Choi J. Acta Physica Polonica B. 2017;**48**:145-161

[7] Noah Glennon, Anthony E. Mirasola, Nathan Musoke, Mark C. Neyrinck, Chanda Prescod-Weinstein 2023, arXiv: 2301.13220v1

[8] Schrödinger E. Discussion of probability relations between separated systems. Mathematical Proceedings of the Cambridge Philosophical Society. 1935;**31**:555

[9] Bell JS. On the Einstein-Podolsky-Rosen paradox. Physics 1. 1964:195-200

[10] Aspect A, Grangier P, Roger G. Experimental tests of realistic local theories via Bell's theorem. Physical Review Letters. 1981;**47**:460-463

[11] Bennett CH, Brassard G, Crépeau C, Jozsa R, Peres A, Wootters WK. Teleporting an unknown quantum state via dual classical and Einstein-Podolsky-Rosen channels. Physical Review Letters. 1993;**70**:1895

[12] Ekert AK. Quantum cryptography based on Bell's theorem. Physical Review Letters. 1991;**67**:661

[13] Shor PW. Polynomial-time algorithms for prime factorization and discrete logarithms on a quantum computer. SIAM Journal on Computing. 1997;**26**:1484

[14] Acín A, Latorre JI, Pascual P. Three-party entanglement from positronium. Physical Review A. 2001;**63**:042107

[15] Banerjee S, Kumar Alok A, Srikanth R, Hiesmayr BC. A quantum-information theoretic analysis of three-flavor neutrino oscillations. European Physical Journal C: Particles and Fields. 2015;**75**:487

[16] Bertlmann RA, Hiesmayr BC. Kaonic qubits. Quantum Information Processing. 2006;**5**:421

[17] Hiesmayr BC. Limits of quantum information In weak interaction processes of hyperons. Scientific Reports. 2015;**5**:11591

[18] Bernabeu J. T and CPT symmetries in entangled neutral meson systems. Journal of Physics: Conference Series. 2011;**335**:012011

[19] Peschanski R, Seki S. Entanglement entropy of scattering particles. Physics Letters B. 2016;**758**:89

[20] Kharzeev DE, Levin EM. Deep inelastic scattering as a probe of entanglement. Physical Review D. 2017; **95**:114008

[21] Maldacena J. A model with cosmological Bell inequalities. Fortschritte der Physik. 2015;**64**:10-23

[22] Nielsen MA, Chuang IL. Quantum Computation and Quantum Information. Cambridge: Cambridge University Press; 2000

[23] Bennett CH, DiVincenzo DP. Nature (London). 2000;**404**:247

[24] Cerf NJ, Leuchs G, Polzik ES, editors. Quantum Information with Continuous Variables of Atoms and Light. London: Imperial College Press; 2007

[25] Braunstein SL, van Loock P. Reviews of Modern Physics. 2005;**77**:513

[26] Adesso G, Illuminati F. Journal of Physics A: Mathematical and Theoretical. 2007;**40**:7821

[27] Reid MD et al. Reviews of Modern Physics. 2009;**81**:1727

[28] Drummond P, Hiliry M. The Quantum Theory of Nonlinear Optics. New York, USA: Cambridge University Press; 2014

[29] Boyd RW, Jha A, Malik M, O'Sullivan C, Rodenburg B, Gauthier DJ. Proceedings of SPIE. 2011;**7948**: 79480L

[30] Mirhosseini M, Magaña-Loaiza OS, O'Sullivan MN, Rodenburg B, Malik M, Lavery MPJ, et al. New Journal of Physics. 2015;**17**:033033

[31] Mirhosseini M, Rodenburg B, Malik M, Boyd R. Journal of Modern Optics. 2013;**61**:43

[32] Hamadou Ibrahim A, Roux FS, McLaren M, Konrad T, Forbes A. Physical Review A. 2013;**88**:012312

[33] Gröblacher S, Jennewein T, Vaziri A, Weihs G, Zeilinger A. New Journal of Physics. 2006;**8**:75

[34] Marrucci L, Karimi E, Slussarenko S, Piccirillo B, Santamato E, Nagali E, et al. Journal of Optics. 2011;**13**:064001

[35] Cardano F, Massa F, Qassim H, Karimi E, Slussarenko S, Paparo D, et al. Science Advances. 2015;**1**:1500087

[36] Jackson JD. Classical Electrodynamics. 3rd ed. New York: Wiley; 1998

[37] Gopaul C, Andrews R. New Journal of Physics. 2007;**9**:94

[38] Tyler GA, Boyd RW. Optics Letters. 2009;**34**:142

[39] Anguita JA, Neifeld MA, Vasic BV. Applied Optics. 2008;**47**:2414

[40] Krenn M, Fickler R, Fink M, Handsteiner J, Malik M, Scheidl T, et al. New Journal of Physics. 2014;**16**:113028

[41] Durnin J, Miceli JJ, Eberly JH. Physical Review Letters. 1987;**58**:1499

[42] Lu J, Greenleaf JF. IEEE Transactions on Ultrasonics, Ferroelectrics, and Frequency Control. 1992;**39**:19

[43] Conti C. Generation of entangled 3D localized quantum wave-packets via optical parametric amplification. 2004. arXiv:quant-ph/0309069

[44] Ciattoni A, Conti C. Journal of the Optical Society of America B: Optical Physics. 2007;**24**:2195

[45] Ornigotti M, Conti C, Szameit A. Physical Review Letters. 2015;**115**: 100401

[46] Ornigotti M, Di Mauro Villari L, Szameit A, Conti C. Physical Review A. 2017;**95**:011802(R)

[47] Ornigotti M, Conti C, Szameit A. Journal of Optics. 2018;**20**:065201

[48] Olver FWJ, Lozier DW, Boisvert RF, Clark CW. NIST Handbook of Mathematical Functions. Cambridge: Cambridge University Press; 2010

[49] Messiah A. Quantum Mechanics. New York: Dover; 2014

[50] Mandel L, Wolf E. Optical Coherence and Quantum Optics. Cambridge, UK: Cambridge University Press; 1995

[51] Loudon R. The Quantum Theory of Light. 3rd ed. Oxford: Oxford University Press; 1997

[52] Gerry C, Knight P. Introductory Quantum Optics. Cambridge, UK: Cambridge University Press; 2005

[53] Namiki R. Physical Review A. 2012; **85**:062307

Chapter 14

Universality of Koba-Nielsen-Olesen Scaling in QCD at High Energy and Entanglement

Yizhuang Liu, Maciej A. Nowak and Ismail Zahed

Abstract

Using Mueller's dipole formalism for deep inelastic scattering in Quantum Chromodynamics (QCD), we formulate and solve the evolution for the generating function for the multiplicities of the produced particles in hadronic processes at high energy. The solution for the multiplicities satisfies Koba-Nielsen-Olesen (KNO) scaling, with good agreement with the recently re-analyzed data from the H1 experiment at HERA (DESY) and the old ALEPH detector data for hadronic Z decay at LEP (CERN). The same scaling function with KNO scaling carries to the hadronic multiplicities from jets in electron-positron annihilation. This agreement is *a priori* puzzling, since in Mueller's dipole evolution, one accounts for virtual dipoles in a wave function, whereas in electron-positron annihilation, one describes cross-sections of real particles. We explain the origin of this similarity, pointing at a particular duality between the two processes. Finally, we interpret our results from the point of view of quantum entanglement between slow and fast degrees of freedom in QCD and derive the entanglement entropy pertinent to electron-positron annihilation into hadronic jets.

Keywords: QCD, DIS, entanglement, multiplicity, KNO scaling, perturbation theory

1. Introduction

Using Mueller's dipole formalism for deep inelastic scattering in QCD, we formulate and solve the evolution for the generating function for the multiplicities of the produced particles in hadronic processes at high energy. The solution for the multiplicities satisfies Koba-Nielsen-Olesen (KNO) scaling, with good agreement with the recently re-analyzed data from the H1 experiment at HERA (DESY) and the old ALEPH data for hadronic Z decay at LEP (CERN). The same scaling function with KNO scaling carries to the hadronic multiplicities from jets in electron-positron annihilation. This agreement is *a priori* puzzling, since in Mueller's dipole evolution, one accounts for virtual dipoles in a wave function, whereas in electron-positron annihilation, one describes cross-sections of real particles. We explain the origin of this similarity, pointing at a particular duality between the two processes. Finally, we interpret our results from the point of view of quantum entanglement between slow and fast degrees of freedom in QCD and derive the entanglement entropy pertinent to electron-positron annihilation into hadronic jets.

Universality is a powerful concept permeating several branches of physics, whereby different physical systems can exhibit similar behavior. This is usually captured by universal exponents, given general assumptions. Perhaps, the best example is the universality of the critical exponents in scaling laws in the vicinity of phase changes. Scaling laws, *per se*, form an important theoretical corpus in physics. In general, they describe the functional relationship between two physical quantities, that scale with each other over a significant interval.

In the context of high-energy particle physics, the so-called Koba-Nielsen-Olesen scaling (named KNO scaling hereafter), formulated half a century ago, is of paramount importance in the empirical analysis of many high-energy hadronic multiplicities. Yet, it is usually challenging to derive from the first principles in QCD. Historically, KNO scaling was first formulated in two independent theoretical works [1, 2], which suggested that at high energies with large Mandelstam s (squared invariant mass), the probability distribution of producing n particles in a specific collision process, scales as

$$\overline{n}(s)p_n(s) = f(z), \tag{1}$$

where $\overline{n}(s)$ is the average multiplicity at large s, and $z \equiv \frac{n}{\overline{n}(s)}$ is the argument of the *scaling function* $f(z)$. Remarkably, the KNO scaling hypothesis precedes the emergence of Quantum Chromodynamics (QCD) and the advent of high energy and luminosity data currently available at colliders.

This contribution is motivated by the recent work in [3], where the deep inelastic scattering (DIS) data from the H1 experiment at DESY were re-analyzed, with interest in an assessment of the quantum entanglement in high energy particle physics. Clearly, the data analyzed, especially for the highest energy range, shows KNO scaling (see **Figure 1**). Also, the Shannon entropy of the multiplicities presented bears some similarity to the entanglement entropy. However, the explicit form of the scaling

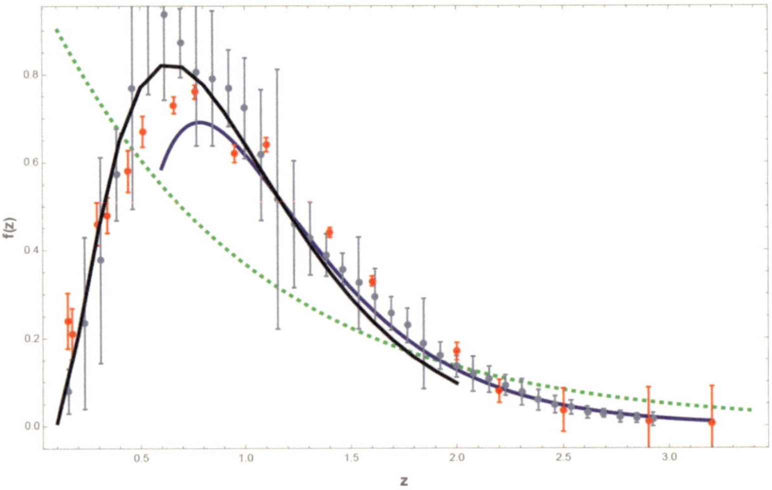

Figure 1.
*The numerical (black-solid curve) up to $z = 2.0$ and asymptotic (blue-solid curve) scaling of the KNO particle multiplicity $f(z)$ based on **Table 2**, compared to the recent data for DIS [3] (red) and the Z-decay data at $\sqrt{s} = M_Z$ by ALEPH [4] (gray).*

function was unknown, and the QCD understanding of the hypothetical entanglement was not specified.

In the first part of this chapter, we discuss the unexpected *a priori* fact, that identical differential equations, such as the ones we derived recently in [5], yield a scaling function that applies to different settings at high energy, e.g., deep inelastic scattering (DIS) and jets in electron-positron or e^+e^- annihilation. Exploiting the duality [6, 7] between the Banfi-Marchesini-Smye (BMS) construction [8] and the Mueller's dipole [9] in the conformal limit, we show that the pertinent generating functions for the multiplicity probabilities in the case of DIS and jets, respectively, are mathematically *identical*. Furthermore, in the double-logarithm approximation (DLA), the duality holds beyond conformal limit by introducing one-loop running in the "minimal" way. Given these, we arrive at the final differential equation for the KNO function, which we then solve using methods based on analyticity. The resulting, parameter-free curve, does not only agree well with DIS data from the H1 experiment at DESY, and the Z-decay data from ALEPH at CERN, but also represents a valuable prediction for future DIS experiments, alike electron-ion collider (EIC) or electron-ion-collider in China (EicC).

In sum, this paper consists of three new results: (1) the derivation of the KNO scaling function in QCD for both DIS and jets in the DLA; (2) the use of the KNO scaling function in the DLA, to show the universality of the hadronic multiplicities from current colliders, for both DIS and jets; (3) the explicit derivation of the entanglement entropy for e^+e^- annihilation into hadronic jets, to be measured at collider energies.

2. The BMS equation for infrared (IR) logarithms in $e^+e^- \to$ multi-gluon cross-section

The BMS equation [8, 10, 11], describes the "nonglobal" logarithms in the e^+e^- annihilation process, and is based on the universal features of the soft divergences in the n-gluon contribution to the total cross-section σ_n. To leading logarithm accuracy, the "most singular" part of σ_n, can be effectively generated through a Markov process. Defining the directions of the quark-antiquark pair as p and n, soft gluons (k) are emitted from harder ones (p) through a universal eikonal current $\frac{gp^\mu}{p\cdot k}$, and the emissions are strongly ordered in time and energy. As a result, the emission depends only on the color charges that are already present in the final state but not on the history of how they are emitted. In the large number of colors N_c, the generating functional for σ_n,

$$Z\left(\frac{E}{E_0};n,p;\lambda\right) = \sum_{n=0}^{\infty} \lambda^n \sigma_n, \tag{2}$$

satisfies a closed integral equation (4)

$$\begin{aligned} Z\left(\frac{E}{E_0};n,p;\lambda\right) &= S\left(\frac{E}{E_0};\hat{p},\hat{n}\right) \\ &+\bar{\alpha}_s\lambda\int_{E_0}^{E}\frac{d\omega}{\omega}S\left(\frac{E}{E_0};\hat{p},\hat{n}\right)\int d\Omega_k K\left(\hat{k};\hat{p},\hat{n}\right)Z\left(\frac{\omega}{E_0};n,\hat{k};\lambda\right)Z\left(\frac{\omega}{E_0};p,\hat{k};\lambda\right), \end{aligned} \tag{3}$$

with the eikonalized gluonic emission kernel $K\left(\hat{k};\hat{p},\hat{n}\right)=\frac{1}{4\pi}\frac{p\cdot n}{k\cdot pk\cdot n}$ and $\overline{\alpha}_s=N_c\alpha_s/\pi$. The Sudakov factor reads

$$S\left(\frac{E}{E_0};\hat{p},\hat{n}\right)=e^{-\overline{\alpha}_s\ln\frac{E}{E_0}\int d\Omega_k K\left(\hat{k};\hat{p},\hat{n}\right)}. \tag{4}$$

More precisely, the first term is the Sudakov contribution, where all the soft gluons are virtual, and the second term is the contribution where at least one soft gluon is real (3) is the integral form of the BMS equation, which can brought to the standard form discussed in [8, 10, 11], by taking a derivative with respect to $\ln\frac{E}{E_0}$ and some re-arrangements.

3. The Mueller's dipole for small-x logarithms and the BMS-Balitskii-Kovchegov (BK) correspondence

Mueller [9] has shown that the small-x evolution equations such as Balitskii-Fadin-Kuraev-Lipatov (BFKL) [12–14], and BK [15, 16], are also based on a very similar branching process, where small-x virtual gluons are released into the light-front wave functions (LFWFs). The same reasoning yields an evolution equation, this time for the generating function for the distribution of the virtual dipoles or $Z\left(b,\frac{x}{x_{\min}},\lambda\right)$, which is exactly of the form (3), with the substitution $\int d\Omega_k\rightarrow\int d^2b_2$:

$$\begin{aligned}Z\left(b_{10},\frac{x_0}{x_{\min}},\lambda\right)&=S\left(b_{10},\frac{x_0}{x_{\min}}\right)\\&+\lambda\frac{\alpha_s N_c}{2\pi^2}\int_{x_{\min}}^{x_0}\frac{dx_1}{x_1}S\left(b_{10},\frac{x_0}{x_1}\right)\int db_2^2\frac{b_{10}^2}{b_{12}^2b_{20}^2}Z\left(b_{12},\frac{x_1}{x_{\min}},\lambda\right)Z\left(b_{20},\frac{x_1}{x_{\min}},\lambda\right).\end{aligned} \tag{5}$$

Here, the corresponding Sudakov or "soft-factor" for "virtual" emissions is

$$S\left(b_{10},\frac{x_0}{x_1}\right)=\exp\left[-\frac{\alpha_s N_c}{\pi}\ln b_{10}^2\mu^2\ln\frac{x_0}{x_1}\right]. \tag{6}$$

In fact, one can show that in the leading order, the BMS and BK equations map onto each through a pertinent conformal transformation [6, 7, 17–19], where the asymptotic real soft gluons at $t=\infty$, map onto the virtual gluons present at $x^+=0$ [5]. In this sense, the mapping is a "virtual-real" duality, in addition to the standard

	Dipole	Cusp
Distribution in	***Virtual* gluon**	***Real* gluon**
Large N_c	Yes	Yes
Kernel	$\frac{b_{10}^2}{b_{12}^2b_{20}^2}$	$\frac{n\cdot p}{k\cdot nk\cdot p}$
Virtual part	TMD soft-factor	Sudakov form factor

	Dipole	**Cusp**
Distribution in	***Virtual* gluon**	***Real* gluon**
Time ordering	In LF time	In center of mass (CM) time
Momentum ordering	Decreasing k^+	Decreasing energy ω
Virtuality ordering	Increasing	Decreasing
Markov process	Yes	Yes
DLA	$b_{10} \gg b_{12} \gg \ldots$	$\theta_{01} \gg \theta_{12} \gg . \ldots$

Table 1.
Mueller hierarchy and BMS hierarchy.

interpretation that it maps rapidity divergence to ultraviolet (UV) divergence [6].[1] In **Table 1**, we have highlighted this duality through a parallel between the two constructions.

4. The DLA limit and the universal DLA KNO scaling function

The BK (BMS) equation resums *single* logarithms in rapidity (energy). However, in both cases, there are two types of divergences instead of one: in the Mueller's dipole construction for the LFWFs [9], there are UV divergences when $k_\perp$ becomes large, while in the BMS construction with Wilson-line cusps, there are rapidity-divergences when the emissions become collinear to the Wilson-lines. It is natural to resum the double logarithms for both. One can then show that the double-logarithm approximation (DLA) for BK (BMS) is generated from the same branching process, with one more strong ordering in dipole sizes (emitting angles) (see **Table 1**). The strong ordering in virtuality is preserved by the DLA limit. Clearly, the two DLA and the underlying size (angle) orderings, simply map onto each other under a conformal transformation. The similarity of the branching process follows from the non-Abelian three-gluon coupling and large N_c in QCD, giving rise to a binary tree branching.

A unique feature of the DLA is that the distribution of dipoles (soft gluons) has a nontrivial KNO scaling function $f(z)$, which coincides with that suggested many years ago for jets in [20–22]. This is due to the same strong ordering in emitting angles in both cases, with **Table 1** making this plausible.

In the DLA, the equation for the generating function (3) simplifies

$$Z(\rho) = e^{-\rho} + \rho\int_0^1 dx\int_0^1 dy e^{-\rho(1-y)} Z(\rho x y) Z(\rho y), \tag{7}$$

where

$$\rho = \frac{2C_F}{\pi\beta_0} y \ln\ln\frac{Q^2}{M^2}. \tag{8}$$

[1] In nonconformal theory, the exact mapping breaks at two-loop already. But for the virtual part, it can be generalized to all orders.

with $2C_F \sim N_c$ and $\beta_0 = \frac{11N_c}{12\pi}$ for large N_c. Note that we have switched from fixed α_s in (3) to one-loop running in (9), with the scale fixed by the emitted small dipoles, hence the emergence of the beta function β_0. A more thorough discussion of the choice of α_s at low-x can be found for example in [23], which reduces to the size of the smallest dipole in the presence of strong ordering. In case of e^+e^-, the same DLA limit holds when the running was introduced through the energy of the soft gluons. Defining $Z = \exp W$, and introducing $u = 2\sqrt{\rho}$, we arrive at the final equation $\Delta_2 W = e^W - 1$, where Δ_2 is a radial part of the 2-dimensional Laplacian. This equation is reminiscent of the Poisson-Boltzmann equation, if Δ_2 was the full Laplacian. For large energies, it reduces to

$$\frac{d^2W}{du^2} = e^W - 1. \tag{9}$$

The solution of this equation, encodes the shape of the KNO scaling function $f(z)$, through a Fourier-Laplace transform

$$Z(t = -e^u) = \int_0^\infty dz e^{-tz} f(z). \tag{10}$$

A detailed investigation of the above equations can be found in our recent analysis [5]. With the help of complex-analytic method, we are able to unravel the scaling function $f(z)$ both in the asymptotic region analytically, and throughout numerically. For $z < 0.1$, the scaling function can be made explicit

$$f(z) \sim \frac{\alpha}{z^2} \ln\frac{\alpha}{z} \exp\left(-\frac{1}{2}\ln^2\frac{\alpha}{z} \ln\frac{\alpha}{z} + \mathcal{O}(1)\right), \tag{11}$$

where $\alpha = 1.50972$. For $z > 2.0$, the scaling function behaves asymptotically as

$$f(z) = 2r\left(rz - 1 + \mathcal{O}\left(\frac{\ln z}{z}\right)\right)e^{-rz}, \quad z \to \infty, \tag{12}$$

where $r = 2.55297$. The explicit expression for r can be found in [20–22], and that for α in [5].

5. Scaling function vs. data

Since this scaling function represents a parameter-free QCD prediction for future experiments, including the EIC and EicC, we record our numerical solution in **Table 2**,

z	**0.1**	**0.2**	**0.3**	**0.4**	**0.5**	**0.6**	**0.7**	**0.8**	**0.9**	**1.0**
f(z)	0.01	0.21	0.45	0.65	0.77	0.82	0.82	0.78	0.72	0.64
z	1.1	1.2	1.3	1.4	1.5	1.6	1.7	1.8	1.9	2.0
f(z)	0.56	0.49	0.42	0.35	0.29	0.24	0.20	0.15	0.12	0.1

Table 2.
Table of the scaling function $f(z)$.

following our analysis in [5]. For completeness, we note that a moment reconstruction of $f(z)$ using different arguments was used in the context of jets in [21, 22, 24].

In **Figure 1**, we compare our results recorded in **Table 2** (black-solid curve) with the H1 data for DIS [3] (red data) and the old ALEPH data for Z decay [4] (gray data). The agreement is very good for both data sets, supporting the universality of our results. For comparison, we also show the exact asymptotic (11) (blue-solid curve), and the KNO particle multiplicity e^{-z} [9], following from the dimensional reduction (diffusive approximation) of Mueller's dipole wave function evolution (dashed-green curve).

6. KNO scaling and entanglement

The knowledge of the effective reduced density matrix for the virtual dipoles in the LFWF [25], and the shape of the KNO function, allow for the evaluation of the entanglement entropy between fast and slow degrees of freedom [25–28]. Remarkably, the reduced density matrix for large rapidities is diagonal [25]

$$\rho = \sum_n p_n \rho_n, \tag{13}$$

with p_n the probability for the emission of n dipoles (gluons), and ρ_n an effective reduced density matrix with n soft dipoles in-out. Hence, the entanglement entropy

$$S = -\sum_n p_n \ln p_n + \sum_n p_n s_n, \tag{14}$$

with $s_n = -\mathrm{tr}\rho_n \ln \rho_n$. Since the wave function peaks at $n = \overline{n}$ for large n, we assume the same to hold for s_n, with the scaling $s\left(\frac{n}{\overline{n}}\right)$, so that

$$S = \ln \overline{n} + \int dz f(z)(-\ln f(z) + s(z)). \tag{15}$$

For DIS in the DLA with KNO scaling, the result is [5]

$$S_{DIS}(y, Q^2) \rightarrow \ln \overline{n} \equiv 2\left(\frac{2C_F}{\pi\beta_0} y \ln \ln \frac{Q^2}{M^2}\right)^{\frac{1}{2}}. \tag{16}$$

a measure of the Sudakov contribution (17) is measurable in the Dokshitzer-Gribov-Lipatov-Altarelli-Parisi (DGLAP) regime of DIS. We note that the KNO scaling function in the diffusive (BFKL) regime with $f(z) = e^{-z}$ [9], leads to maximal decoherence in the entanglement entropy $S_{BFKL} \sim y$ [25–28]. In contrast, the unimodal character of the scaling function in the DLA leads to a smaller entanglement entropy $S_{DIS} \sim \sqrt{y}$.

Using the BMS-BK duality, we can readily formulate the entanglement entropy between soft and hard degrees of freedom in the final state of e^+e^- annihilation into hadronic jets

$$S_{e^+e^-}(Q^2) \equiv 2\left(\frac{C_F}{\pi\beta_0} \ln \frac{Q^2}{M^2} \ln \ln \frac{Q^2}{M^2}\right)^{\frac{1}{2}}, \tag{17}$$

also a measure of the Sudakov contribution (with no extra 2 in the bracket). The rapidity gap between the quark-antiquark pair $y \rightarrow \ln \frac{Q^2}{M^2}$, produces *another logarithm* in Q^2. Note that for e^+e^-, this contribution dominates the single logarithm resummation in the BFKL contribution at large Q^2, which is

$$\ln \overline{n}_{\mathrm{BFKL}} \sim \frac{2C_F \ln 2}{\pi\beta_0} \ln \ln \frac{Q^2}{M^2}. \tag{18}$$

The prediction (17) is amenable to experimental verification in high-energy hadronic jet physics.

7. Conclusion

To summarize, in this chapter, we presented the universal KNO scaling function underlying the DLA limit for two different systems: the $e^+e^- \to$ multi-gluon process and the Mueller's dipole wave function. In both systems, the multiplicity generating function satisfies nonlinear BK type evolution equation with respect to primary evolution variables: energy for the $e^+e^- \to$ multi-gluon process and rapidity for the Mueller's dipole. They map to each other through the BMS-BK correspondence.

On the other hand, in both systems there also exists another large logarithm: in rapidity for $e^+e^- \to$ multi-gluon process and in energy for the Mueller's dipole. The second logarithm appears naturally in the multiplicity distribution. When projected to the double-logarithm limit (DLA limit), which resumes both of the two large-logarithms, the evolution equation simplifies considerably, and the corresponding multiplicity distribution exhibits a nontrivial universal KNO scaling function. This scaling function compares well with the measured hadronic multiplicities in DIS and e^+e^- annihilation, especially in the exponential tail. Moreover, the KNO scaling also constraints, in a universal way, in the DLA limit, the asymptotic behavior of the rapidity space or energy-space entanglement entropies of the underlying process.

Acknowledgements

We are grateful to Jacek Wosiek for bringing [20] to our attention and to Yoshitaka Hatta for informing us about [19]. This work is supported by the Office of Science, U.S. Department of Energy under Contract No. DE-FG-88ER40388, and by the Priority Research Areas SciMat and DigiWorld under the program Excellence Initiative—Research University at the Jagiellonian University.

Author details

Yizhuang Liu[1], Maciej A. Nowak[1,2]* and Ismail Zahed[3]

1 Institute of Theoretical Physics, Jagiellonian University, Kraków, Poland

2 Mark Kac Center for Complex Systems Research, Jagiellonian University, Kraków, Poland

3 Center for Nuclear Theory, Department of Physics and Astronomy, Stony Brook University, Stony Brook, New York, USA

*Address all correspondence to: maciej.a.nowak@uj.edu.pl

References

[1] Polyakov AM. Zhurnal Éksperimental'noĭ i Teoreticheskoĭ Fiziki. 1970;**59**:542-552

[2] Koba Z, Nielsen HB, Olesen P. Nuclear Physics B. 1972;**40**:317-334. DOI: 10.1016/0550-3213(72)90551-2

[3] Andreev V et al. European Physical Journal. 2021;**81**(3):212. DOI: 10.1140/epjc/s10052-021-08896-1

[4] Buskulic D et al. Zeitschrift für Physik. 1995;**69**:15-26. DOI: 10.1007/BF02907382

[5] Liu Y, Nowak MA, Zahed I. Physics Review D. 2023;**2023**

[6] Vladimirov AA. Physical Review Letters. 2017;**118**(6):062001. DOI: 10.1103/PhysRevLett.118.062001

[7] Hatta Y, Iancu E, Mueller AH, Triantafyllopoulos DN. JHEP. 2018; **02**:075. DOI: 10.1007/JHEP02(2018)075

[8] Banfi A, Marchesini G, Smye G. JHEP. 2002;**08**:006. DOI: 10.1088/1126-6708/2002/08/006

[9] Mueller AH. Nuclear Physics B. 1994; **415**:373-385. DOI: 10.1016/0550-3213(94)90116-3

[10] Caron-Huot S. JHEP. 2018;**03**:036. DOI: 10.1007/JHEP03(2018)036

[11] Marchesini G, Mueller AH. Physics Letters B. 2003;**575**:37-44. DOI: 10.1016/j.physletb.2003.09.041

[12] Fadin VS, Kuraev EA, Lipatov LN. Physics Letters B. 1975;**60**:50-52. DOI: 10.1016/0370-2693(75)90524-9

[13] Kuraev EA, Lipatov LN, Fadin VS. Soviet Physics JETP. 1976;**44**: 443-450

[14] Kuraev EA, Lipatov LN, Fadin VS. Soviet Physics. JETP. 1977;**45**:199-204

[15] Balitsky I. Nuclear Physics B. 1996; **463**:99-160. DOI: 10.1016/0550-3213(95)00638-9

[16] Kovchegov YV. Physical Review D. 2000;**61**:074018. DOI: 10.1103/PhysRevD.61.074018

[17] Weigert H. Nuclear Physics B. 2004; **685**:321-350. DOI: 10.1016/j.nuclphysb.2004.03.002

[18] Cornalba L. Eikonal Methods in AdS/CFT: Regge Theory and Multi-Reggeon Exchange. arXiv. 2008. DOI: 10.48550/arXiv.0710.5480

[19] Hatta Y, Ueda T. Nuclear Physics B. 2013;**874**:808-820. DOI: 10.1016/j.nuclphysb.2013.06.021

[20] Dokshitzer YL, Fadin VS, Khoze VA. Z Physics C. 1983;**18**:37. DOI: 10.1007/BF01571703

[21] Dokshitzer YL, Fadin VS, Khoze VA. Z Physics C. 1982;**15**:325. DOI: 10.1007/BF01614423

[22] Bassetto A, Ciafaloni M, Marchesini G. Physical Report. 1983;**100**:201-272. DOI: 10.1016/0370-1573(83)90083-2

[23] Kovchegov YV, Weigert H. Nuclear Physics A. 2007;**784**:188-226. DOI: 10.1016/j.nuclphysa.2006.10.075

[24] Bassetto A. Nuclear Physics B. 1988; **303**:703-712. DOI: 10.1016/0550-3213(88)90426-9

[25] Liu Y, Nowak MA, Zahed I. Physical Review D. 2022;**105**(11):114028. DOI: 10.1103/PhysRevD.105.114028

[26] Stoffers A, Zahed I. Physical Review D. 2013;**88**(2):025038. DOI: 10.1103/PhysRevD.88.025038

[27] Kharzeev DE, Levin E. Physical Review D. 2021;**104**(3):L031503. DOI: 10.1103/PhysRevD.104.L031503

[28] Kharzeev DE, Levin EM. Physical Review D. 2017;**95**(11):114008. DOI: 10.1103/PhysRevD.95.114008

Artificial Intelligence and Education - Shaping the Future of Learning

Edited by Seifedine Kadry

Published in London, United Kingdom

Artificial Intelligence and Education - Shaping the Future of Learning
http://dx.doi.org/10.5772/intechopen.1002135
Edited by Seifedine Kadry

Contributors
Abraham Prieto, Agatha Aballa Nkechi, Akintayo O. Ojo, Ana Paula S. Loures-Elias, António Costa, António Pedro Silva, Bahago S. Benedict, Belhassen Guettat, Carla Abreu, Christian Oscar Orlando Aparicio-Escalante, Cláudia Domingues, Cristina Renda, Dalila Durães, Esmaeil Zaraii Zavaraki, Georg Winder, Gregor Hrastnik, Hacer Türkoğlu, Halil Ersoy, Horst Biedermann, Hwang Eunkyung, Johnbenetic Gnanaprakasam, José María Bey, Llaudett Natividad Escalona-Márquez, Luis Alberto Camputaro, María Meizoso-García, Matthew L. Bernacki, Nataša Meh Peer, Nghia Duong-Trung, Niclas Schaper, Niels Pinkwart, Obinna A. Eneh, Ogunlade B. Olusola, Olivia Rütti-Joy, Paulo Novais, Peter Peer, Ramzi Farhat, Ravi Lourdusamy, Serdar Engin Koc, Serpil Yalcinalp, Sheikh Faisal Rashid, Shotayo E. Olusola, Stefanía Johanna Cedeño-Tapia, Syrine Karoui, Žiga Emeršič

First published in London, United Kingdom, 2024 by IntechOpen
IntechOpen is the global imprint of INTECHOPEN LIMITED, registered in England and Wales, registration number: 11086078, 167-169 Great Portland Street, London, W1W 5PF, United Kingdom

British Library Cataloguing-in-Publication Data
A catalogue record for this book is available from the British Library

Additional hard and PDF copies can be obtained from orders@intechopen.com

Artificial Intelligence and Education - Shaping the Future of Learning
Edited by Seifedine Kadry
p. cm.

This title is part of the Artificial Intelligence Book Series, Volume 28
Topic: Applied Intelligence
Series Editor: Andries Engelbrecht
Topic Editor: Carlos M. Travieso-Gonzalez

Print ISBN 978-0-85466-782-6
Online ISBN 978-0-85466-781-9
eBook (PDF) ISBN 978-0-85466-783-3
ISSN 2633-1403

For EU product safety concerns:
IN TECH d.o.o., Prolaz Marije Krucifikse Kozulić 3, 51000 Rijeka, Croatia, info@intechopen.com or visit our website at intechopen.com.

IntechOpen Book Series
Artificial Intelligence
Volume 28

Aims and Scope of the Series

Artificial Intelligence (AI) is a rapidly developing multidisciplinary research area that aims to solve increasingly complex problems. In today's highly integrated world, AI promises to become a robust and powerful means for obtaining solutions to previously unsolvable problems. This Series is intended for researchers and students alike interested in this fascinating field and its many applications.

Meet the Series Editor

Andries Engelbrecht received the Masters and Ph.D. degrees in Computer Science from the University of Stellenbosch, South Africa, in 1994 and 1999 respectively. He is currently appointed as the Voigt Chair in Data Science in the Department of Industrial Engineering, with a joint appointment as Professor in the Computer Science Division, Stellenbosch University. Prior to his appointment at Stellenbosch University, he has been at the University of Pretoria, Department of Computer Science (1998-2018), where he was appointed as South Africa Research Chair in Artifical Intelligence (2007-2018), the head of the Department of Computer Science (2008-2017), and Director of the Institute for Big Data and Data Science (2017-2018). In addition to a number of research articles, he has written two books, Computational Intelligence: An Introduction and Fundamentals of Computational Swarm Intelligence.

Meet the Volume Editor

Dr. Seifedine Kadry has a bachelor's degree in 1999 from Lebanese University, an MS degree in 2002 from Reims University (France) and EPFL (Lausanne), a Ph.D. in 2007 from Blaise Pascal University (France), and an HDR degree in 2017 from Rouen University (France). At present his research focuses on data science, education using technology, system prognostics, stochastic systems, and applied mathematics. He is an ABET program evaluator for computing and an ABET program evaluator for engineering tech. He is a professor of data science at Noroff University College.

Contents

Preface

In today's rapidly evolving 21st-century landscape, artificial intelligence (AI) has become a transformative force across various sectors, including education. Our book *Artificial Intelligence and Education – Shaping the Future of Learning* explores the significant impact of AI on educational paradigms, discussing both the opportunities and challenges it brings.

This book aims to provide educators, policymakers, students, and technology enthusiasts with a comprehensive understanding of how AI is reshaping the educational environment. From personalized learning experiences and intelligent tutoring systems to administrative efficiencies and new ethical considerations, the integration of AI in education is paving the way for a future where learning is more accessible, engaging, and tailored to individual needs.

As we navigate this transformation, it's crucial to address the implications of AI on traditional educational models and the broader societal context. This book seeks to spark a thoughtful dialogue about the role of AI in education, encouraging readers to reflect on its potential for enhancing learning outcomes and preparing students for an AI-driven world.

By combining theoretical insights, practical examples, and visionary perspectives, the book aims to inspire innovation and critical thinking. We hope that this book will serve as a valuable resource for those dedicated to advancing education in an era defined by technological progress and human ingenuity. Welcome to a journey into the future of learning.

Seifedine Kadry
Department of Computer Science and Mathematics,
Lebanese American University,
Beirut, Lebanon

Department of Applied Data Science,
Noroff University College,
Kristiansand, Norway

Chapter 1

Utilization of Artificial Intelligence in Education: A Perspective on Learning Strategies

Serpil Yalcinalp, Hacer Türkoğlu, Serdar Engin Koc and Halil Ersoy

Abstract

When considered in scope of our digitally surrounded world, new approaches unavoidably find their places in education as well as other fields in society. In such a world, AI is one of the recent and most sound technical developments with its fast widespread at an unprecedented pace in all educational levels. Students, who did not even know the name of a specific generative AI, are starting to use it as soon as they hear about it. A vast variety of rich resources for learning exist and approaches in learning like individual learning, personalized learning, and connected learning are gaining considerable importance. While there is a strong emphasis on learning based on learners' own needs and their freedom in the selection of their own strategies, we need to consider the place of AI as the facilitator of such trends so that learning could be implemented effectively. With this in mind, this section aims to present and discuss how AI tools and applications could be used in accordance with learning strategies to give chances to learners to learn efficiently in their way. Within that perspective connected learning and cognitive-affective strategies in AI environments were also elaborated with sample scenarios of cases provided.

Keywords: AI in education, AIEd, learning strategies, connectivism, personalized learning

1. Introduction

This chapter aims to introduce and categorize Artificial intelligence (AI) technologies both in general and within the educational context and explore their use in education with a focus on implementing appropriate learning strategies. First, definitions of AI from the literature were given, reflecting various considerations in categorizing AI from technological and educational perspectives. After providing a brief overview of learning strategy definitions, this chapter highlights connected learning as one of the most crucial strategies in today's digitally immersive educational environments. The discussion delves into the advantages and disadvantages of its implementation. Specifically, personalized and deep learning emerge as central themes, with a focus on these approaches dominating the discourse.

Section 4 of this chapter shows examples of popular AI tools, as categorized by Zawacki et al. [1], with a primary focus on their educational utility. These tools include ChatBots, visual and design aids, lesson design and content creation tools, teaching aids, quiz and assessment generators, collaboration and communication platforms, and learning platforms. Furthermore, appropriate learning strategies tailored to each category are proposed with the aim of maximizing learning efficacy. In Section 5, scenarios and case studies are developed to illustrate the effective utilization of AI tools within exemplary domains and to show the implementation of effective learning strategies.

2. What is the scope of artificial intelligence?

2.1 Main definitions of artificial intelligence

Artificial intelligence (AI) spans numerous research and application domains. Among these, ChatGPT and other popular ChatBots have emerged since 2022, finding application in various fields such as economics, business, and, notably, education. To understand those applications, we need to pay attention to the following definitions and descriptions.

ChatGTP: is an AI tool, developed by OpenAI company, that uses a generative pre-trained transformer model (GPT) as a large language model (LLM) to understand and generate text in human languages. Users enter their questions, called prompts, into a text box on the ChatGPT website. ChatGPT performs input processing by tokenization and segmentation, and unnecessary words are removed based on the generative pre-trained transformer (GPT) model. Next, ChatGPT encodes these meaningful pieces of data into a vector format to see what the input means [2]. Then, based on its trained GPT model and a huge dataset collected over the Internet, a response in human language is given to the user. The user may provide extra prompts to reflect on the responses or to change the direction of the conversation.

One of the important features of ChatGPT is its dataset and GPT model. OpenAI provides different versions of ChatGPT based on different datasets and models covering recent information on the Internet and is capable of more diverse analysis of input [3].

Natural Language Processing (NLP) is a research area in computer science that aims to create computers or applications that understand and generate responses for human language in textual, auditory or speech forms [4]. NLP is being used in many diverse areas such as speech-to-text or text-to-speech engines, translators among languages, etc. [4].

Large Language Models (LLM) are a kind of NLP models that are trained with a rich dataset of a specific language to understand the input and generate proper output in that language. The generation process is aimed at getting the most expected or proper token (smallest unit of meaningful words or phrases) in the context [5].

Generative Pre-Trained Transformer (GPT) is a specific type of LLM created by OpenAI in 2018 (version 1). The main purpose was to generate output by predicting the next word in a sentence. In 2019, the second version, GTP-2, and in 2020, the third version, GTP-3, were introduced with larger datasets and parameters. Open AI continued enlarging the size and diversity of the training dataset and released GTP-3.5 in 2022 and GTP-4 in 2023 so that it is almost impossible to distinguish if the output is generated by a ChatBot or a human [6].

Every definition of artificial intelligence makes some assumptions. The first assumption is that it is modeled by a set of programmes, the second assumption is that it has input and output as a set of information, and the third assumption is that it exists in a world of some type which is also influenced by the output of the information. From a different perspective, Ng et al. ([7], p. 1) state AI as "intelligent machines and algorithms that can reason and adapt based on sets of rules and environment which mimic human intelligence." Furthermore, AI is termed as machines emulating human behaviour [8]. Dobrev ([9], p. 2) defines AI as "...such a program which in an arbitrary world will cope not worse than a human" [10] sees no correct definition of AI due to the wide research field that AI covers and AI still remains a label for historical reasons, whereas Baker and Smith ([11], p. 10) define AI as "Computers which perform cognitive tasks, usually associated with human minds, particularly learning and problem-solving."

2.2 Criteria for categorizing AI systems in literature

The above definitions lead us to believe that there are two types of AI; one being the learning machine that is open to evolve, and the other is trained on sets of knowledge and statistics with limited memory and limited to a single medium or field lacking the cognitive abilities of a human. The latter is categorized as narrow or weak AI that is designed for specific tasks and effective only in a certain domain. The former is only in theory, and the closest thing to a strong AI is Elon Musk's self-driving car. The reason is that while it is possible to train AI using datasets and all sorts of information, programmes do not learn, but they try to perfect their operations using repetition over finely grained data and tasks.

Based on functionality, AI is divided into being reactive and having limited memory with a theory of mind. Reactive machines cannot learn from past experiences but only respond to a certain set of rules. Some AI such as self-driving cars use past experiences to make better decisions and improve performance. On the other hand, AI that understands human emotions, intentions, beliefs, and mental states sets the theoretical mind of AI, which only exists in the form of understanding facial expressions with the help of input devices till today. In the category of AI learning approach, that is, the way AI is supposed to be trained, there are four types of approaches: supervised, unsupervised, semi-supervised, and reinforcement learning. Supervised learning refers to approximating the behaviour of the function or imitation by analyzing labeled examples in the data. Supervised learning leverages datasets that contain images and labels. According to data, generally, supervised learning requires first training using large and unrelated data and then fine-tuning with data of interest. Unsupervised learning is for training without labels, and AI trying to find patterns and relationships that exist within the data provided. In semi-supervised learning, small portions of labelled data are used as a primer for the large unlabeled data. Reinforcement learning occurs when AI learns a policy of action depending on observation and feedback from its environment, leading to rewards or punishments and, hence, reinforcement. Based on the way the applications of AI operate, we see categories such as natural language processing, computer vision, robotics, and expert systems. Natural language processing models lead to large language models (LLMs), which are capable of self-learning with annotated data and accordingly imitate texts that are written by humans. They can learn more efficiently and generate output specified by users [12]. LLMs use relevant predictions to generate long texts. While LLMs can summarize, and expand texts, it is not clear how to produce reliable outputs from them [13]. GPT, BERT, and RoBERTa are some of the models mentioned that made it possible for further research in natural language processing [14]. Computer vision is the ability of AI to perceive visual stimuli meaningfully. Objects are classified and localized to identify their types through image processing. Recently, medical imagery has advanced in detection with the help of deep learning and visual pattern recognition up to an accuracy that matches or surpasses the level of expert physicians [15]. AI takes forms that are used in robots (Robotics) that can offer enhancement in precision and less error in all kinds of operations. AI helps robots in decision-making and learning algorithms with the ultimate purpose of autonomous robotic systems. Lately, AI techniques have made it possible for robotics to connect wirelessly with applications, increasing efficiency and optimization [16]. Expert systems were built to provide aid in decision-making processes. To conclude, expert systems have to be able to reason and generate explanations. The introduction of machine learning enhanced the capturing process of data patterns for knowledge acquisition. The explanation is generally a compromise between precision and understandability, depending on the audience. Based on the time horizon, batch learning refers to learning from a fixed dataset without further interactions with the world, whereas online learning indicates continuous learning with data updates. Batch learning is advantageous in that it is cheaper, faster, has better sample efficiency and has algorithms that are model-free [17].

3. Utilization of AI in education and learning strategies

3.1 AI in education

Based on the aforementioned categorizations of AI systems such as machine learning, personalized learning, deep learning, learning analytics, and more, Artificial Intelligence in Education (AIEd) has also been categorized in various ways according to different criteria (AIEd) were categorized into three paradigms by Ouyang and Jiao ([18], p. 2) as (i) "AI-directed (learner-as-recipient and AI is used to represent knowledge models and direct cognitive learning). (ii) AI-supported (learner-as-collaborator, AI is used to support learning) while learners work as collaborators with AI, and (iii) AI-empowered (learner-as-leader, AI is used to empower learning) while learners take agency to learn."

Chaudhry and Kazim [19] categorized the work of AIEd as (i) reducing teachers' workload, (ii) contextualized learning for students (proving customized and/or personalized learning), (iii) revolutionizing assessments (knowing how learners learn and pedagogies apply to them individually), and (iv) intelligent tutoring systems (ITS).

In general, the trajectory of AIEd's development has been aimed at empowering learner autonomy and personalization, facilitating learners' reflective processes, and guiding AI systems to adapt accordingly in light of such reflections. This progression has fostered iterative enhancements towards learner-centred and personalized learning approaches.

The educational value of AI is continuously being watched and scrutinized, and educators remain doubtful about the adoption of AI, which is lagging except in fields such as medicine and finance.

Adnan et al. [20] carried out an experiment on 425 students and found that using deep learning techniques and giving learners adaptive, motivational triggers and adaptive learning content had a significant effect on the experimental group. Fryer et al. [21] found that students benefited from the ChatBot system by monitoring their learning behaviour and assigning speaking tasks via ChatBot. Baillifard et al. [22] used an AI Tutor to assess and understand concepts understood by students for further personalization of practice to each learner. The results showed significantly increased achievement compared to the control group and 15% more active engagement.

We cannot deny the disconnect between vision and the present status of AI in educational systems as they are not being practised on large scales and use simple methods for testing student mastery. However, the use of intelligent tutoring systems includes the integrated form of AI, so the spread of ITS in time will further disseminate these systems that are being used in education. These systems are used for adaptive feedback [23–28], presenting and recommending content [24, 27, 29, 30].

3.2 Learning strategies

Duffy ([31], p. 368) defines learning strategies as algorithmic "secrets." In the student-centred approach, it is up to the learner to choose which strategy to learn effectively. Mariani [32] explains learning strategies as "the processes (or sequences of processes) that, when matched to the requirements of tasks, facilitate performance."

Focusing on learning strategies and their place in educational phenomena, it is important to keep in mind that;

- There are many ways to classify and categorize learning strategies. This varies from the point of view of the educators, instructors, and even the learner himself.

- Selection of learning strategies is up to the learner, so it is not convenient to label strategies as best or worse.

- Learning and instructional strategies overlap generally. It is important to note that the use of learner strategies varies in situations in which (i) the learner has full autonomy on his learning (learning strategies), (ii) the learner has half autonomy but is guided by instructor-tutor (learning and instructional strategies), and (iii) the learners have no autonomy; instructional strategies are dominant (instructional strategies).

It must be kept in mind that (i) learner-centred education, learner autonomy, and increasing emphasis on learning strategies, (ii) technological developments such as in AI, and (iii) new approaches and theories like connected learning are the key points for consideration in today's educational phenomena and technology-driven world.

Learning activities, which were the first indications for categorizing learning strategies, were identified as (i) cognitive, (ii) affective, and (iii) metacognitive-regulation by educators in the field [33–35]. Boekaerts' [36] explanation of those strategies with examples fits well with learning activities as cognitive strategies like elaboration to deepen the understanding of the domain, metacognitive strategies like planning to regulate the learning process, and motivational strategies like self-efficacy to motivate oneself to engage in learning.

New categories of learning strategies also appeared in parallel to developments in technology, especially with the wide use of the Internet and increased autonomy of learners in that environment. For example, Dignath et al. [37] added a fourth category—management strategies such as finding, navigating, and evaluating resources.

Dunlosky et al. [38] provided an overview of the most common effective cognitive learning strategies as practice testing (or retrieval practice), distributed practice,

	Theoretical underpinning	**Implementations**	**AI techniques**
Paradigm One: *AI-directed, learner-as-recipient*	Behaviorism	Earlier work of Intelligent Tutoring Systems (ITSs)	AI based on statistical relational techniques
Paradigm Two: *AI-supported, learner-as-collaborator*	Cognitive, social constructivism	Dialogue-based Tutoring Systems (DTSs); Exploratory Learning Environments (ELEs)	Bayesian network, natural language processing, Markov decision trees
Paradigm Three: *AI-empowered, learner-as-leader*	Connectivism, Complex adaptive system	The human-computer cooperation; Personalized/ adaptive learning	The brain-computer interface, machine learning, deep learning

Table 1.
Three paradigms of artificial intelligence in education ([18], p. 3).

elaborative interrogation, self-explanation, interleaved practice, summaries, mental imagery, keyword mnemonics, highlighting or re-reading.

In their study, Hattie and Donoghue ([39], p. 2) identified the most effective learning strategies based on the synthesis of 228 meta-analysis studies among more than 400 learning strategies based on a model. Their model described those strategies as "three inputs and outcomes (skill, will and thrill), success criteria, three phases of learning (surface, deep and transfer) and an acquiring and consolidation phase within each of the surface and deep phases." In this section in **Table A1**, learning strategies were mainly suggested by selecting from Hattie and Donoghue's [39] list of such strategies.

Educational technology can hardly receive the adequacy and robustness for integration with educational and learning theories and needs good examples of applications rather than just using AI for the prospect of transforming education [40, 41]. Blayone [41] thinks that theories that depend on technology such as technology-mediated learning theory and activity theory can shed light on AI education.

According to Vygotsky's zone of proximal development, children learn by initiating and interacting with others. As the school's grip over technology loosens, children learn from technology and resources, problem-solving under guidance or in collaboration with peers. Cognitive development is in conjunction with other kinds of development in children, and the learning internalization process shows itself when children solve problems. To realize their potential, children need assistance, consolidation, and cooperation. Given that interaction primarily occurs through language, it is imperative for intellectual development, beginning from early childhood education and extending to all levels of education. This is where AI-related technologies come in handy in aiding students with new concepts exemplified in different contextualized situations. Both children and adults need a learning companion to gain critical thinking skills. They need a safe environment to evaluate their peers and they can deepen their understanding of concepts by conversing with agents of AI.

Generally, AI applications were used to enrich students' online learning experience through teaching robots, learning recommendations, and detecting progress, all under the name of natural language processing. Ouyang and Jiao [18] represented three AI Education paradigms such as AI-directed, AI-supported, and AI-empowered, where learners take the role of recipients, collaborators, and empowered agents that in ways adapt, personalize, and reflect learning. To achieve this goal, ChatBots have made significant progress, particularly in providing assignments, resources, and individualized interactions. Wollny et al. [42] examined the pedagogical role of ChatBots and identified four roles: supporting learning, assisting, and mentoring roles. It can be seen that ChatBots are also in line with self-directed learning, metacognitive theories, and motivational theories, which cover a large part of the educational framework. According to [43], almost all studies related to the use of ChatBots in education adopted guided learning, so they decided that research in this area and the use of artificial intelligence are still in their early stages.

So, what does AI-supported learning entail? Integration of AI technologies must consider learning strategies that include pedagogy. At this juncture, it is worth questioning how effectively AI can incorporate human-like elements to address pedagogical issues ([44], p. 108). Yang [44] suggests that "AI may be able to imitate human emotions; however, imitating human feelings is difficult for AI."

Some models intensively explain the attributes of integrating AI into education. According to Zhai et al. [45], AI-integrated systems are developed for the classification of content, matching of content to learners, recommendations for learners, and for deep learning purposes. In their literature review, they found studies of pedagogical applications under feedback, AI-supported reasoning, and adaptive learning. They also found that applications of AI carried the labels of immersive learning, affection computing, gamification, and role-playing. As technology develops, AI can also take roles that are finely grained in teaching situations that are necessary for visualization and task-specific procedures such as highlighting concept map nodes or reasoning [46, 47].

As seen in **Table 1**, Ouyang and Jiao's [18] separation of AI in Education is historically grounded and goes from behaviourism into cognitive and constructive theoretical foundations and lastly to connectivism where the learner is seen as the leader and has the capability to monitor and evaluate his own learning making personalized content possible. With its possible connections and large application field of choice, AI is seen as a novelty in education that can take part in differing roles to supplement learners.

Interactive learning, feedback and assessment, and personalized learning are essential features to be considered in the scope of AI applications in education.

Interactive Learning: Bhuthoria [48] linked the capability of AI to transform education to both being the judge and the helper in learning with optimization and algorithms on both ends. The role of AI in active teaching manifests itself through ChatBots and intelligent agents, which help learners receive feedback [49] and hints [46], as well as general classroom purposes [50] through natural language processing, machine learning, and deep learning.

Kong et al. [51] developed the Virtual Integrated Patient (VIP) that can mimic realistic conversations that take place between a doctor and patient via a natural language processor. With AI technologies, it is guaranteed that no two patients are the same, and the students are surrounded by a safe environment where they can make mistakes and learn from them. Yang and Shulruf [52] used artificial skin that was connected to an AI system for suture and ligature training of medical students.

This AI-assisted and expert-led tutoring increased the students' performance as well as confidence due to real-time feedback. Kim et al. [53] used different interaction modes such as AI text, AI voice chatting, and face-to-face chatting to test the speaking performance of students. Voice chatting pairs outperformed the two other modes of interaction in speaking tasks. Vasquez-Cano et al. [54] compared two teaching methods, one being exercises written on paper and the other interaction with a ChatBot, and the results showed that the ChatBot group improved their punctuation skills significantly. Vahabzadeh et al. [55] used the Empowered Brain, which is a conveyor of socioemotional and behavioural skills using augmented reality and artificial intelligence to decrease the irritability, hyperactivity, and social withdrawal of subjects. The results were decreased symptoms in all aspects.

Feedback and Assessment: AI can help create feedback and scaffold content according to the continuous assessment from adaptive learning systems and, in a way, reduce teacher burnout to compensate for problems in learning. The AI uses machine learning and deep learning-based models or techniques to balance learner content as it is most effective in outputting the most effective mode of learning for the specific learner at his level. We can mostly see the combined form of AI, personalized, and adaptive learning systems in intelligent tutoring systems, which derive these technologies to provide support in monitoring, deliverance, feedback and interface to learners [56].

Personalized learning: Personalization refers to individual assessment and teaching, so it needs to be handled with learning content and assessment, which AI today is fit to provide with proper integration. After that repetition or introduction of the next level of content is provided depending on the progress of the learner, adaptive learning is a way to personalize learning, and it can be different for each learner to focus on problematic areas in a customized form. Personalized learning centres around the task of connecting the learner's previous knowledge, experience, and abilities with training materials that link that understanding to new information. Edula et al. [57] see the key to personalized learning as levelling prior knowledge and skills of the learner to link to new knowledge.

Murtaza et al. [58] commented on how a personalized e-model should state that a trained machine learning model and adaptive learning model create the past and future behavior of learners through sequential assessment. They added that continuous e-learning data provides information for the recommendation module, and other modules can help to create personalized learning. Peng et al. [59] thought that to achieve personalized adaptive learning, one must rely on teaching strategies, whether based on differences in individual characteristics, differences in individual performance, or differences in personal development vision. These strategies should also rely on data-driven decision-making.

Educause Learning Initiative ([60], p. 1) explains the relationship between adaptive learning and personalized learning as "*Adaptive learning is one technique for providing personalized learning, which aims to provide efficient, effective, and customized learning paths to engage each student. Adaptive learning systems use a data-driven and, in some cases, nonlinear approach to instruction and remediation.*" In addition to that, adaptive learning works by creating paths of learning for students according to their knowledge, skills, and learning speed. Adaptive learning environments also depend on monitoring and drawing results of student activities to supplement their learning [61]. There are examples of using personalized learning that could positively affect learning and reduce anxiety and stress among learners [21, 62, 63] with promising results, but these studies need more proof and a larger sample size to be generalized.

3.3 Use of AI to learn: in scope of learning strategies

Using technology to effectively teach and learn has become one of the most critical issues since the first time people needed to integrate them into educational settings. Today, we grapple more with the dilemma of effectively integrating educational technology, even as intelligent systems continue to evolve, supporting and enhancing educational progress across all levels. While teachers are irreplaceable by computers and AI systems from various perspectives, including pedagogical considerations, advancements in AI systems' capabilities have reached a point where they can assist teachers in sequencing, adjusting, and organizing instruction. This support extends to creating learning activities, measurement tools, and more, shifting this dilemma in a completely new direction to the real roles of teachers. It is evident that teachers cannot be replaced by AI when it comes to applying pedagogy in real-life situations, which underscores the increasing importance of recent human-AI studies. On the flip side, AI systems are revolutionizing learning by providing substantial support for personalized learning in many aspects. These new insights necessitate new roles for learners, not only in effectively utilizing supportive tools and systems but also in engaging with them through constructive and meaningful dialogs.

As we, as instructors and learners, faced with AI more intensively in our educational activities, we get familiar that such technology is assisting and guiding us instead of displaying information and is planning and adapting learning activities instead of sequencing them [64].

In situations where learners have complete autonomy over their learning, the skills and knowledge to select and apply appropriate learning strategies have become increasingly critical, particularly following the emergence of generative AI technologies (see Appendix 1).

One of the reasons why the connected learning paradigm has become such a hot topic today is the increasing autonomy of learners in determining and directing their learning processes. This departure from traditional educational models allows for a greater focus on learners' interests, needs, and learning styles. Additionally, with the emergence of generative artificial intelligence technologies, learners have more choice and control over their learning processes and can personalize their learning experiences. This makes the skills of selecting and applying their own learning strategies increasingly critical. In this context, the connected learning paradigm has become a significant topic of discussion in the field of education today.

In the upcoming section, we will delve into connected learning and its critical role in contemporary education within the context of AI, highlighting its significance as one of the foremost learning paradigms in our digitally interconnected world.

3.4 Learning paradigm in today's world of technology: connected learning

Surrounded by a huge amount of educational resources, easy access to those resources, easy access to educators and specialists in the area, increased interaction with people and the content, and challenges to tracking learners digitally all opened a new area for learners to learn. Such learning allows learners to select their way of learning in line with a lot of choices in their way of learning, or in other words, in controlling their learning.

Moreover, the rapid development and progress in AI technologies in recent years have significantly accelerated such learning phenomena. As a result, the future is now more unpredictable than at any other time in history.

The theory of connectivism was first proposed by Siemens [65] and Downes [66]. It was almost the first theory of Internet-related learning. They identified connectivism as a set of principles. Those principles still have strong ties with generative AI today, and now we discuss them with insights in examples in light of today's AI systems ([65], p. 5).

3.4.1 Learning and knowledge rest in a diversity of opinions

Today's capabilities in generative AI, such as ChatGPT, could enhance learning by swiftly gathering and presenting a multitude of alternative opinions, saving learners the time and effort of searching for sources themselves. However, there are some issues to consider.

On one hand, this convenience might lead students to passively accept information without critically engaging with it, potentially fostering a sense of intellectual laziness. On the other hand, proper training in critical thinking and effective inquiry techniques is crucial for students to engage in meaningful dialog with these systems and extract valuable insights from the wealth of information provided. Thus, while generative AI technologies offer great potential for deepening learning, it is essential to balance convenience with the development of critical thinking skills in learners.

3.4.2 Learning is a process of connecting specialized nodes or information sources

Before the advent of generative AI and personalized tutorials, it was crucial for learners to possess the ability to connect specialized nodes or information sources. This required the application of a series of learning strategies, including self-acquisition, metacognition (such as reflecting on their learning processes), decision-making, problem-solving, questioning, and more. These skills empowered learners to navigate through vast amounts of information, critically assess its relevance and reliability, and synthesize knowledge from various sources. This active engagement in the learning process not only deepened their understanding of the subject matter but also honed their cognitive abilities essential for lifelong learning and success in various endeavours.

Again, generative AI, instead of requiring students to track and connect specialized nodes, may inadvertently foster a sense of laziness. When students can rely on AI to gather and present information, they might become less inclined to actively engage in the process of seeking out diverse perspectives and making connections themselves. This passivity could hinder their development of critical thinking skills and their ability to deeply understand complex topics. These issues could be addressed through careful instructional design, which includes projects requiring students to engage in self-reflection and compare results from AI systems.

3.4.3 Learning may reside in non-human appliances

Day by day, non-human appliances are increasingly entering the scene through generative AI systems that continuously learn from previous interactions and data, as such systems are encountering our daily lives immensely, including educational settings. Here, all issues, including ethical, replacement, social equity and more, must be considered.

3.4.4 Capacity to know more is more critical than what is currently known

For sure, the primary role of generative AI is continuous learning. How do we encourage students to see learning as an ongoing journey where mastering the skill of learning itself is paramount?

3.4.5 Nurturing and maintaining connections is needed to facilitate continual learning

While strong connections between nodes are critical, even more important is the presence of activities and tasks within each node that necessitate actions in other nodes. Designing such connections and maintaining them can be achieved through careful instructional design, emphasizing instructional control. While learners have full autonomy in selecting and following these actions, developing skills and knowledge to manage their learning is crucial. On the other hand, personalized AI systems that provide support and manage the entire learning process for the learner represent significant progress. The question arises: will there be no need for learners to possess such skills in the future, given that AI handles all tasks? Alternatively, could AI potentially undermine the development of these skills, thereby impacting the profile of the next generation of learners?

3.4.6 Ability to see connections between fields, ideas, and concepts is a core skill

Similar questions arise here: how can such skills be enhanced when various AI systems are performing tasks instead of learners? What strategies could be employed to leverage AI systems at this juncture without impeding the development of these skills in learners?

3.4.7 Currency (accurate, up-to-date knowledge) is the intent of all connectivist learning activities

While the primary responsibility for this lies with instructional designers creating learning environments and learners utilizing these environments for learning, contemporary AI systems like ChatGPT may base decisions on a limited timeframe, excluding recent years. Even more so, the occurrence of mistakes and errors is inevitable in such systems, underscoring the need for students' care and inquiry now more than ever. It is inevitable to develop learners' skills and abilities by comparing results from AI systems and reflecting on them.

3.4.8 Decision-making is itself a learning process

In our digitally immersed and AI-enhanced educational landscape, learners are not only equipped with decision-making skills but also have questioning, problem-solving, self-assessment, self-regulation, and decision-making skills that remain essential as cornerstones for healthy societies and individuals. Therefore, in all educational systems, the appropriate integration of AI into learning approaches and tasks, as well as the promotion of effective application of learning strategies, must be the focal point for educators in the field.

4. Exemplary AI tools in education with recommended strategies for implementation

This section introduces several popular AI tools, providing explanations of their general features (in Appendix 1), and recommends learning strategies for their application. The table categorizes AI tools based on the literature review by [1, 63] and for educational use into (i) ChatBots, (ii) visual and design tools, (iii) lesson design and content creators, (iv) teaching aids, (v) quiz/assessment generators, (vi) collaboration-communication, and (vii) learning platforms. In the last column, we suggest learning strategies from [39], which are deemed most suitable for each respective tool.

It is important to note that the selection and implementation of learning strategies for each tool may vary depending on the specific domain and learning objectives. Our suggestions are intended to provide a general framework for effectively integrating AIEd into practice.

5. Exemplary case scenarios for implementing AI tools in education

In this section, exemplary case scenarios were created for implementing AI tools across various domains.

5.1 Case 1: the category of ChatBots - ChatGPT

Anna aims to teach sixth-grade students how to create formulas in Microsoft Excel and utilize the natural language processing capabilities of ChatGPT for *interactive learning*. The goal is to empower students with the ability to use ChatGPT to calculate sales tax, seller's commission, and total, as well as the remaining inventory, in an imaginary Excel table containing January 2023 inventory data. Students will use ChatGPT to write formulas for sales tax and commission, then apply these formulas correctly in Excel by pasting them. Additionally, students will be encouraged to explore creating new formulas with different values using both ChatGPT and external sources. This learning experience not only equips students with the skills to effectively use Excel formulas but also provides them with hands-on experience in natural language processing with ChatGPT.

5.2 Case 2: the category of visual and design - Adobe express with firefly

Graham, contemplating the integration of technology into K-12 education to *enhance students' creativity and storytelling skills, suggests various methods by* combining Generative AI with Adobe Firefly. Initially, students are tasked with using descriptive words to create opening scenes for their stories before even starting to write. Subsequently, the teacher recommends using generative AI for formative assessment, where students input descriptive text to swiftly generate corresponding visuals, saving time and offering a visual representation for assessment and improvement. Finally, students utilize additional keywords obtained from the created visuals to review and enhance their stories, fostering continuous reflection and improvement in storytelling skills through the effective use of technology (**Figure 1**).

Figure 1.
Created the cute little giant by adobe firefly.

5.3 Case 3: the category of lesson design and content creator - Curipod AI

David wants to design a lesson on historical figures. He needs more interactive materials to engage students actively and give them instant feedback. He first creates the classroom with a Curipod. He enters the topic to be taught, and the AI generator creates the interactive slides. The teacher can edit these slides if he wishes. The teacher sends a team code to his students so they can access the presentation. By entering this code at curipod.live, students can access the interactive presentation. The teacher can ask the students questions during the slide transitions and have them log in with the team code to answer the question. The teacher can set the time and duration while the students are replying to the questions. AI gives instant feedback to the students' answers to the question. The teacher can review the feedback, accept it after checking its accuracy, and let the student see it. At the end of the last slide, the student is asked questions to self-evaluate what they learned at the end of the lesson, and all the answers are displayed on the teacher's screen (**Figure 2**).

5.4 Case 4: the category of teaching aid - MagicToDo

Susan, a pre-service teacher, wants to create a lesson plan but does not know how to break down the process. So she writes to MagicToDo about the task she wants to do, but she cannot. MagicToDo explains step by step how to create the lesson plan. If Susan wants, she can ask MagicToDo to explain each sub-task again and decrease or increase by adjusting the spicy ordinal level. So she can plan her time more effectively and find a guide to help her complete the task (**Figure 3**).

5.5 Case 5: the category of quiz/assessment generators - Quizizz

Ivana wants to evaluate her students at the end of the Addition and Subtraction of Fractions lesson. She decides to use Quizizz for this. She types the topic of the test she wants to create in Quizizz, and AI creates the test. The teacher, Ivana, can edit the questions, correct grammatical errors in the wording of the questions, replace the

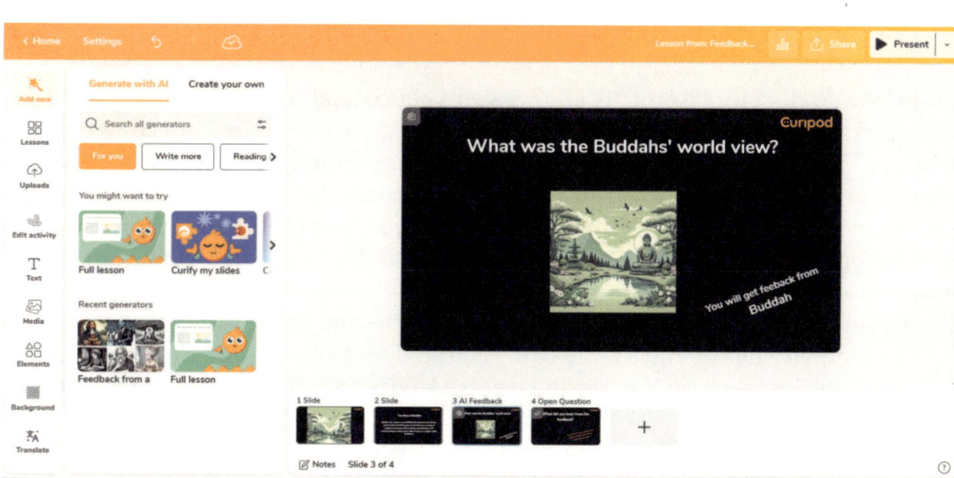

Figure 2.
Created a lesson tutorial about historical figures by Curipod.

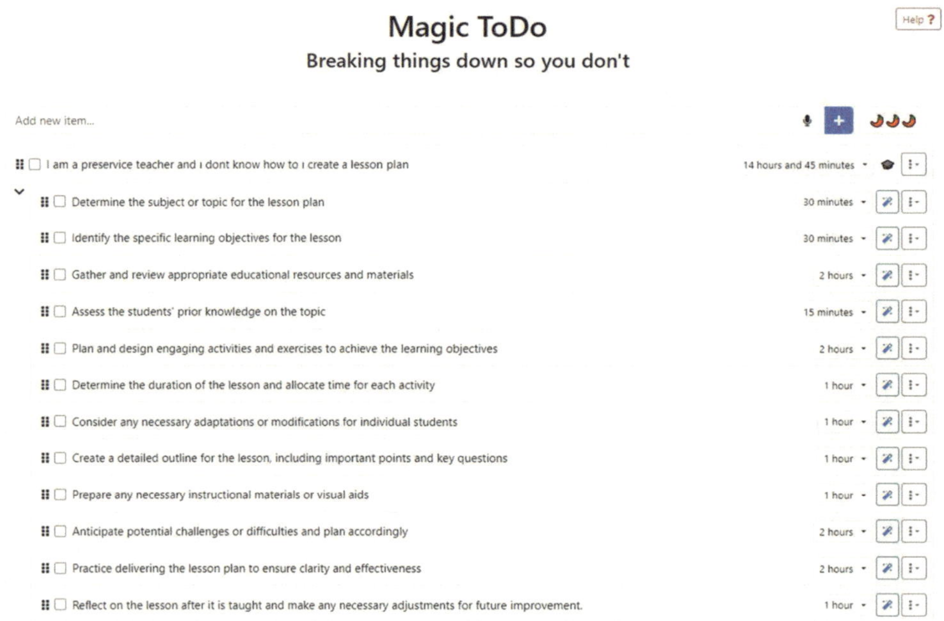

Figure 3.
Created the stages of lesson plan creation with MagicToDo.

questions with a similar question, adapt them to a daily life problem, or add feedback to the options in a way that the student can see. If the teacher wishes, he/she can start and solve the test live with his/her students or give homework to be assigned at a certain time. It is also possible for students to take the test without a mobile device. The teacher distributes Q cards and students can mark the answers on the Q cards. The teacher can scan the cards with the mobile app and see the results.

5.6 Case 6: the category of collaboration and communication - Parlay

Sophie planned to discuss the safe use of the Internet by asking her students the question, "*Should parents control your phone*?" She decided to use the Parlay app to monitor the discussion. First of all, she created a course and invited her students. Then, she created a verbal discussion activity. She deepened the discussion by asking the questions: "*Do you think that it is fair for your parents to control your phone*?" and "*Do parents have the right to read your text messages*?" Discussion rules and evaluation criteria were added with Parlay. In Parlay, students are allowed to express their opinions verbally around a round table by turning on the camera and audio to the student whose turn it is. Parlay also allows students to construct their ideas by posing discussion questions to the students and asking them to respond in writing. At the end of the discussion, the teacher can see a summary report and a table with each student's opinion and the time spent in the discussion. Students can also see the feedback written by the teacher (**Figure 4**).

5.7 Case 7: the category of learning platform - RiPPLE

The teacher gives an assignment using RiPPLE.
This tool helps you study more effectively, actively, and socially.

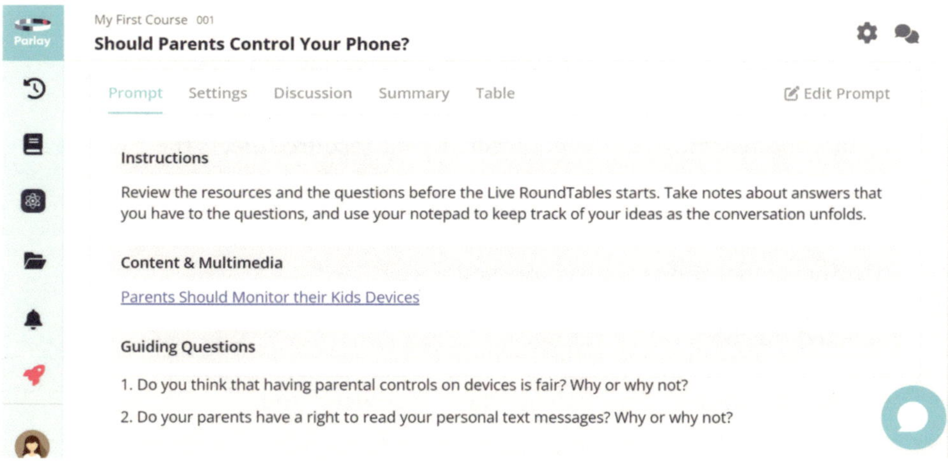

Figure 4.
Created a discussion activity with parlay.

RiPPLE will help you learn by involving you in:

Creating study resources. Creating study resources can improve your mastery and metacognition and encourage higher-order thinking as it requires you to process, organize, integrate, and reconstruct knowledge.

1. Moderation of your peers' resources. Moderating (or checking) your peers' study resources and providing critical feedback encourages self-reflection, communication, and problem-solving skills.

2. Adaptive practice. RiPPLE will provide you with study resources to view and practice based on your current level of knowledge on each course topic. Over 100 years of educational research demonstrates practicing is one of the most effective study techniques – far more effective than re-reading lecture notes or re-watching lectures.

3. Live formative quizzes. Participating in quizzes during lectures or while studying can improve your memory, attention, and engagement while also reducing mind-wandering.

5.7.1 Assessment

There will be a total of four rounds of RiPPLE assessment throughout the semester. In each round, you can earn up to 2.5 marks for correctly creating, moderating, and practicing study resources.

5.7.2 Task details

Creating one or more effective resources.

For a learning resource to be considered effective, it must:

- align with the course content and objective,

- be correct, clear, and easy to understand,

- be appropriately difficult, and
- encourage critical thinking and reasoning.

Your peers and teaching team will decide if your resource meets these criteria.

5.7.3 Moderating four or more resources effectively

An effective moderation is one where you have:

- completed the moderation rubric,
- provided a detailed justification for your judgement, and
- provided constructive feedback on how the resource can be improved.

Simply writing feedback such as "good job" does not qualify. Again, teaching staff will be spot-checking the quality of moderation and change the outcome when necessary.

5.7.4 Answering 10 or more questions correctly

To answer a resource correctly, your first response must be correct. You can attempt as many questions as you want. Only answers from the practice page are counted. Answering questions in an activity or moderation does not count. https://itali.uq.edu.au/digital-learning/learning-analytics/ripple/ripple-guides.

6. Ethical considerations in AI

There are also issues about plagiarism, ethics, and misuse of ChatGPT that are unrelated to normal usage since it is in ChatGPT's natural form to fabricate results in certain situations in the form of annotations, resources, and assumptions.

Li et al. [67] found that ChatGPT gave correct answers to questions but made wrong explanations. In their work, Belletini et al. [13] looked for the problem-solving ability of ChatGPT using the Bebras International Challenge on Informatics and Computational Thinking as background, and it was found that ChatGPT had poor performance and had no confidence in its answers. While it had better answers for procedural knowledge, it had difficulty with tasks requiring knowledge synthesis. Furthermore, the reliability of ChatGPT was in question as it did not provide sources of information but gave accurate answers [68]. Likewise, ChatGPT provides non-academic resources and annotations depending completely on what is available on the web [69]. Jalil et al. [70] showed that for software testing education and software processes, ChatGPT was not well-calibrated, lacked knowledge, and made wrong assumptions. Bordt and Luxburg found that the success of ChatGPT depends on quality entries on the web as the concepts they were investigating were common knowledge, and they warned that if there were not quality information on the web, the results would have been incorrect or poorer [71].

Overuse of ChatGPT would prevent learning, and learning how to use prompts must be a high priority for users [72]. Furthermore, Castillo et al. [73] declared that with such detailed and long answers, ChatGPT can promote laziness among students and can cause decreased learning. The same results can be seen in another study [74], where they thought ChatGPT overuse interferes with critical thinking and problem-solving skills, which are essential and prevent them from seeking alternative solutions to problems, reducing their vision. Using ChatGPT to write term papers and code is plagiarism, and it reduces the experience and knowledge of students [75]. They warned that the students' lack of previous information would undermine their ability for upcoming challenges and declared the best usage of ChatGPT would be as an assistant role as it would lead to personal learning and negate differences between students.

Perceptions of math, science, art, school, and teachers by ChatGPT mirroring the students' perceptions [76]. The used LLMs were GPT-3 and ChatGPT-4, and it was found that they perceived math much more negatively than science, and GPT-3 perceived schools and teachers more negatively than ChatGPT-4. Also, math was viewed as detached from scientific advancements and real-world understanding. The authors suggest a cognitive psychology approach to these perceptions and to understand the behavioural patterns shown by LLMs.

It was very easy to get a university degree using ChatGPT, and the integrity of the academic area was under threat, and that is why changes have to be made in the education system for prevention [75]. Similarly, the study [74] stated that without proper prevention, ChatGPT can weaken the validity of assessments, and it is effortless to create considerable work without attribution and showing resources. Supporting these findings, plagiarism detection is unable to detect ChatGPT-generated content [77].

These results all show that ChatGPT depends more on the user than on itself and can be misunderstood easily. That is why there should be precautions for using ChatGPT or for better results; before experiments of any sort, ChatGPT should be trained with the data that the experiment will be about.

7. Conclusion

In summary, while the integration of AI technologies into education is inevitable, given recent advancements and contemporary approaches to learning and teaching, attention must be paid to several critical issues:

i. AI technologies are not yet sufficiently advanced to address all pedagogical concerns. Therefore, it is essential to recognize that these technologies cannot replace human educators entirely. Research into human-AI systems should be prioritized.

ii. The difference between the success of AI on its own and as a teaching/learning aid should also be realized. AI systems that provide feedback and assessment should be used with larger samples, whereas AI that is used with more specialized purposes that help the learner at every step of learning should continue to determine effectiveness for generalization.

iii. The development of critical skills in students, such as inquiry, problem-solving, metacognition, cooperation, group work, and emotional intelligence, should take precedence over-reliance on AI systems. Over-reliance on AI could lead to a decline in student initiative and engagement.

iv. Ethical considerations, including the prevention of plagiarism, must be carefully addressed with appropriate safeguards.

v. In-service training on AI features and usage should be provided to all stakeholders, including instructors, students, educational professionals, and academics. This training should include guidance on interacting with large language models, which calls for prompt engineering features.

Acknowledgements

ChatGPT is used in language editing in cooperation with an English language specialist.

Conflict of interest

The authors declare no conflict of interest.

Appendix 1

AI categories for education	AI tools	How it works	Pricing
ChatBots	BingChat	It can be accessed through the Bing website or Microsoft Edge browser and can be used to ask both simple and complex questions, and request summaries of articles, books, events, news, sports results, and more. Bing Chat can also be used as a creative tool, helping you write poems, and stories, or even share ideas for a project.	Free
	ChatGPT	ChatGPT is a generative AI ChatBot that interacts with you conversationally, the way a human would. It is trained with information from all over the Internet and tonnes of other sources – and it is been trained by humans to interact with you authentically.	Freemium (free+ premium plan)
	Perplexity	Perplexity is another generative AI ChatBot that lets users ask questions and get responses conversationally. What seems to set Perplexity apart is that it cites sources, which are easily clickable to get more context and more information.	Free
	Google Bard	Google Bard is an experimental conversational AI chat service developed by Google. Google Bard is designed to engage in text-based conversations with users, offering responses and generating human-like dialog	Free
	Pi	To use Pi, you can interact with it through text or voice commands. You can ask questions, seek advice, or simply engage in conversation with the AI.	Free
Which learning strategies could be supported by using ChatBots?			
Skill: Prior achievement *Will:* Using trial and error with computer support can increase self-confidence and competence by coping with mistakes. *Environment:* The students control their overlearning at any time and anywhere *Acquiring surface learning:* Activate prior knowledge; summarization; organization; note taking; highlighting. *Acquiring deeper learning:* A synthesis of what they have learned at the surface phase			
Visual and Design	Adobe Express with Firefly (free)	With generative AI, you can create custom images and text effects using text prompts. As an educator you can use this feature to generate unique and engaging visuals for your lessons and activities.	Free
	Bing Image Creator (limited)	Type in a description of an image you would like to see. Take the resulting images and download or save them – or ask for an adjustment to them.	Free
	Canva Classroom Magic	Canva Classroom Magic is a suite of AI tools designed for educators. It includes features such as Magic Design, a free AI design tool for creating visual content, and Magic Write, an AI text generator and writer that assists in generating written content.	Free
Which learning strategies could be supported by using visual and design tools?			

AI categories for education	AI tools	How it works	Pricing
Skill: Prior achievement *Will:* Increasing self-efficacy, reducing anxiety and self-blame by increasing social support *Thrill-motivation:* Achieving approach *Management of the environment:* The students control their overlearning at any time, anywhere *Acquiring surface learning:* Activate prior knowledge; summarization; organization; note taking; highlighting *Consolidating surface learning* Giving-receiving feedback, help-seeking, effort. *Acquiring deeper learning:* A synthesis of what they have learned at the surface phase *Consolidating deep learning: critical thinking, evaluation and reflection, seeking help from peers* Transfer *adapted learned skills and knowledge to the different tasks.*			
Lesson Design and Content Creator	Curipod	Give Curipod a topic, and it will generate an interactive slide deck for you in seconds, including: polls, word clouds, open-ended questions, a drawing tool, etc.	Freemium
	Diffit	Diffit allows you to get levelled resources for literally anything. You can adapt existing materials for any reader, generate "just right" resources for any topic and then edit and share resources with students.	Freemium
	MagicSchool.ai	MagicSchool.ai is an AI-powered educational platform designed to help teachers save time by automating tasks such as lesson planning, grading, and generating educational content. It offers over 40 AI tools that are searchable by keyword and categorized for planning, student support, community, productivity, and community tools.	Free
	Education Copilot	Education Copilot provides you with AI-generated templates for lesson plans, writing prompts, educational handouts, student reports, project outlines and lots more.	30-day free trial.
	Nolej	Provide content (like text, videos, audio, YouTube, websites, etc.). AI analyzes it and generates educational content. After that, you can embed, export, or integrate with your learning management system (LMS).	Freemium
	Eduaide.ai	Eduaide.ai is an AI-powered teaching assistant designed to help teachers with lesson planning, instructional design, and generating educational content. It offers a resource generator, teaching assistant, feedback bot, and AI chat.	Freemium
	Khanmigo	Khanmigo is an AI-powered learning tool developed by Khan Academy that can be used as a virtual tutor and debate partner. It can also be used by teachers to generate lesson plans and assist with other administrative tasks.	Monthly donation
	Copy.ai	Copy.ai is an AI-powered writing tool that uses machine learning to generate various types of content, including blog headlines, emails, social media content, web copy, and more. Educators can use it to generate lesson plans, substitute teaching plans, newsletter copy, email, writing prompts, and more.	Freemium
	PopAI	A powerful AI tool that boosts productivity. Besides instant answers, explore search engine integration, PDF reading, PowerPoint generation, and more	Freemium
	Tutor AI	TutorAI. Author: Sign In. Learn anything. What do you want to learn about? Start learning. Try topics like physics, history, or your own name!	Freemium

AI categories for education	AI tools	How it works	Pricing
Which learning strategies could be supported by using lesson design and content creator tools?			
Skill: Prior achievement *Will:* Increasing self-efficacy, reducing anxiety and self-blame by increasing social support *Thrill-motivation:* Achieving approach, goals, *Management of the environment:* The students control their overlearning at any time and anywhere, social support *Acquiring surface learning* Activate prior knowledge; summarization; organization; note taking; highlighting. *Consolidating surface learning:* Giving-receiving feedback, help-seeking, effort *Acquiring deeper learning:* A synthesis of what they have learned at the surface phase *Consolidating deep learning:* Critical thinking, evaluation and reflection, seeking help from peers *Transfer* Adapted learned skills and knowledge to the different tasks			
Teaching Aid	gotFeedback by got learning	Choose what you want gotFeedback to provide feedback on. Upload a Word document or PDF file – or paste it into text. It provides an evaluation that you can use as a teacher – or copy/paste to give to a student.	Freemium
	Grammarly	Grammarly is your online writing assistant. Using artificial intelligence helps you compose bold, clear, mistake-free writing. It can work inside your email client, your productivity suites, and even your social media. Grammarly's AI checks grammar, spelling, style, tone, and more	Freemium
	Goblin tools	Goblin Tools is a collection of small, simple, single-task tools designed to help neurodivergent people with tasks they find overwhelming or difficult. The tools include MagicToDo, Formalizer, Judge, Estimator, and Compiler.	Free
	Hello History	Hello History is an app that lets you have life-like conversations with historical figures. Get a personal perspective on life, history, and the world from some of the most influential figures of all time – Cleopatra, Einstein, Buddha, and more.	Freemium
	Chat PDF	ChatPDF is an AI-powered tool that allows users to interact with PDF documents in a conversational format. From the website "ChatPDF in a Nutshell Your PDF AI-like ChatGPT but for PDFs. Summarize and answer questions for free."	Freemium
Which learning strategies could be supported by using Teaching Aid AI Tools?			

AI categories for education	AI tools	How it works	Pricing
Skill: Prior achievement *Will:* Increasing self-efficacy, reducing anxiety and self-blame by increasing social support *Thrill-motivation:* Achieving approach, goals, *Management of the environment:* The students control their learning any time, anywhere; social support *Success criteria* Planning and prediction; setting standards for self-judgment *Acquiring surface learning* Activate prior knowledge; summarization; organization; note taking; highlighting. *Consolidating surface learning* Giving-receiving feedback, help-seeking, effort *Acquiring deeper learning:* A synthesis of what they have learned at the surface phase *Consolidating deep learning* Critical thinking, evaluation and reflection, seeking help from peers *Transfer* Adapting learned skills and knowledge to the different tasks			
Quiz/Assessment Generators	QuestionWell	QuestionWell generates an endless supply of questions so teachers can focus on what matters. Just input some reading, and the AI will write essential questions, learning objectives, and aligned multiple-choice questions.	Free
	Formative AI	Formative has been an assignment and quiz platform for a long time, allowing teachers to ask students a variety of question types. Now, it has incorporated the power of ChatGPT to generate new standard-aligned questions and hints for learners as well as student feedback.	Freemium
	Quizizz AI	Quizizz AI is an artificial intelligence-powered tool that can generate multiple-choice questions with four answer options, of which one is the correct answer. Quizizz AI can automatically determine the appropriate number of questions to generate from the content provided. Quizizz AI can also modify existing quizzes, both manually created and AI-generated, with Enhance to customize activities according to students' needs.	Freemium
	Conker	Conker lets you create multiple-choice, read-and-respond, and fill-in-the-blank quizzes for students at a variety of levels on specific topics. It will also let you copy/paste reading material to generate quizzes.	Free
	Twee	Twee is an AI-powered tool designed to simplify lesson planning for English teachers by generating questions, dialogs, stories, letters, articles, multiple-choice questions, true/false statements, and more.	Free
	Gradescope	It allows students to submit their assignments online and automatically grades multiple-choice, fill-in-the-blank, and coding questions. With Gradescope, students can receive instant feedback on their work and track their progress throughout the semester.	Freemium
Which learning strategies could be supported by using Quiz/Assessment Generators AI Tools?			

AI categories for education	AI tools	How it works	Pricing
Skill: Prior achievement *Will:* Increasing self-efficacy, reducing anxiety and self-blame by increasing social support *Thrill-motivation:* Achieving approach, goals, *Management of the environment:* The students control their learning at any time and anywhere; social support *Success criteria* Planning and prediction; setting standards for self-judgment *Acquiring surface learning* Activate prior knowledge; summarization; organization; note taking; highlighting. *Consolidating surface learning* Giving-receiving feedback, help-seeking, effort *Acquiring deeper learning:* A synthesis of what they have learned at the surface phase *Consolidating deep learning* Critical thinking, evaluation, and reflection, seeking help from peers *Transfer* Adapting learned skills and knowledge to the different tasks			
Collaboration and Communication	FigJam Jambot	Jambot is a FigJam widget that brings the power of ChatGPT's AI to your boards. To find and run the Jambot widget: 1. Hover over the pile of recently used icons in the toolbar, then click More. 2. From the Widgets tab, search Jambot. 3. Click or drag to add the widget to the board. 4. Add text to the input sticky. 5. Select a function.	Freemium
	Parlay Genie	Parlay Genie (part of Parlay) is a discussion prompt generator. It generates higher-order thinking questions for your class based on a topic, a YouTube video, or an article. It uses ChatGPT to generate prompts for your students.	Freemium
	Notion	To manage and organize tasks, projects, and notes. • Create to-do lists • Take class notes • Collaborate with classmates • Stay organized throughout their studies	Freemium
Which learning strategies could be supported by using Collaboration and communication AI Tools?			

AI categories for education	AI tools	How it works	Pricing
Skill: Prior achievement *Will:* Increasing self-efficacy, reducing anxiety and self-blame by increasing social support *Thrill-motivation:* Achieving approach, goals, *Management of the environment:* The students control their over learning any time and anywhere; social support *Success criteria* Planning and prediction; setting standards for self-judgment *Acquiring surface learning* Activate prior knowledge; summarization; organization; note taking; highlighting. *Consolidating surface learning* Giving-receiving feedback, help-seeking, effort.. *Acquiring deeper learning:* A synthesis of what they have learned at the surface phase *Consolidating deep learning* Critical thinking, evaluation and reflection, seeking help from peers *Transfer* Adapting learned skills and knowledge to the different tasks			
Learning Platforms	Ripple	RiPPLE is predominantly a course-level platform that instructors can incorporate into their teaching. to help students with their learning. Includes metacognition, higher-order thinking, evaluative judgment, and retrieval practice. It also has a classroom response system, peer learning, learning analytics, assessment tools and gamification elements.	Paid platform
	Fuma	FUMA uses a range of machine learning models to discover classes of exploratory learning behaviors. These behaviors can then be used to classify users based on their learning needs and to generate personalized hints in real-time that guide the student towards a more effective usage of the tools and affordances available in the OLE (open learning environments).	
Which learning strategies could be supported by using Learning Platforms that use AI?			
Skill: Prior achievement *Will:* Increasing self-efficacy, reducing anxiety and self-blame by increasing social support *Thrill-motivation:* Achieving approach, goals, *Management of the environment:* The students control their learning anytime, anywhere; social support *Success criteria* Planning and prediction; setting standards for self-judgment *Acquiring surface learning* Activate prior knowledge; summarization; organization; note taking; highlighting. *Consolidating surface learning* Giving-receiving feedback, help-seeking, effort *Acquiring deeper learning* A synthesis of what they have learned at the surface phase *Consolidating deep learning* Critical thinking, evaluation, and reflection, seeking help from peers *Transfer* Adapting learned skills and knowledge to the different tasks			

Table A1.
Exemplary AI tools in education with recommended strategies for implementation.

Author details

Serpil Yalcinalp*, Hacer Türkoğlu, Serdar Engin Koc and Halil Ersoy
Baskent University, Ankara, Turkiye

*Address all correspondence to: serpily@baskent.edu.tr

References

[1] Zawacki-Richter O, Marín VI, Bond M, Gouverneur F. Systematic review of research on artificial intelligence applications in higher education – Where are the educators? International Journal of Educational Technology in Higher Education. 2019;**16**(1):39. DOI: 10.1186/s41239-019-0171-0

[2] Bernstein O. How to use ChatGPT & its benefits [Internet]. Sydney: Minsdesigns. 2023. Available from: How to Use ChatGPT & The Benefits (Complete Guide 2023) (mindesigns.com.au) [Accessed: May 6, 2024, cited May 8, 2024]

[3] OpenAI. ChatGPT. Available from: https://openai.com/chatgpt [Accessed: January 2, 2024]

[4] Aydın Ö, Karaarslan E. Is ChatGPT leading generative AI? What is beyond expectations? Academic Platform Journal of Engineering and Smart Systems. 2023;**11**(3):118-134. DOI: 10.21541/apjess.1293702

[5] Ozdemir S. Quick Start Guide to Large Language Models: Strategies and Best Practices for Using ChatGPT and Other LLMs. New Jersey: Addison-Wesley Professional; 2023. ISBN: 9780138199197

[6] VanBuskirk A. A Brief History of the Generative Pre-Trained Transformer (GPT) Language Models [Internet]. 2023. Available from: https://blog.wordbot.io/ai-artificial-intelligence/a-brief-history-of-the-generative-pre-trained-transformer-gpt-language-models/ [Accessed: January 20, 2024, cited January 20, 2024]

[7] Ng DTK, Leung JKL, Chu SKW, Qiao MS. Conceptualizing AI literacy: An exploratory review. Computers and Education: Artificial Intelligence. 2021;**2**:100041. DOI: 10.1016/j.caeai.2021.100041

[8] Dodigovic M. Artificial intelligence and second language learning: An efficient approach to error remediation. Language Awareness. 2007;**16**(2):99-113. DOI: 10.2167/la416.0

[9] Dobrev D. A definition of artificial intelligence. Mathematica Balkanica, New Series. 2005;**19**(Fasc. 1-2):67-74. DOI: 10.48550/arXiv.1210.1568

[10] Wang P. On defining artificial intelligence. Journal of Artificial General Intelligence. 2019;**10**(2):1-37. DOI: 10.2478/jagi-2019-0002

[11] Baker T, Smith L. Educ-AI-tion Rebooted? Exploring the Future of Artificial Intelligence in Schools and Colleges. London: Nesta Foundation; 2019. Available from: Educ-AI-tion Rebooted? Exploring the future of artificial intelligence in schools and colleges (nesta.org.uk) [Accessed: January 30, 2024]

[12] Johnson D, Goodman R, Patrinely J, et al. Assessing the Accuracy and Reliability of AI-Generated Medical Responses: An Evaluation of the Chat-GPT Model. 28 Feb 2023. DOI: 10.21203/rs.3.rs-2566942/v1

[13] Bellettini C, Lodi M, Lonati V, Monga M, Morpurgo A. DaVinci goes to Bebras: A study on the problem solving ability of GPT-3. In: Proceedings of the 15th International Conference on Computer Supported Education, April 2023. Prague, Czech Republic: SciTePress; 2023. pp. 59-69

[14] Küchemann S, Steinert S, Revenga N, Schweinberger M, Dinc Y, Avila KE, et al. Can ChatGPT support

prospective teachers in physics task development? Physical Review Physics Education Research. 11 Sep 2023;**19**(2):020128-020214. DOI: 10.1103/PhysRevPhysEducRes.19.020128

[15] Esteva A, Chou K, Yeung S, Naik N, Madani A, Mottaghi A, et al. Deep learning-enabled medical computer vision. npj Digital Medicine. 2021;**4**(1):5. DOI: 10.1038/s41746-020-00376-2

[16] Vermesan O, Bröring A, Tragos E, Serrano M, Bacciu D, Chessa S, et al. Internet of robotic things–converging sensing/actuating, hyperconnectivity, artificial intelligence and IoT platforms. In: Vermesan O, Bacquet J, editors. Cognitive Hyperconnected Digital Transformation. New York: River Publishers; 2022. pp. 97-155

[17] Fu Y, Di W, Boulet B. Batch Reinforcement Learning in the Real World: A Survey [Internet]. In: Offline RL Workshop. 2020. Available from: https://offline-rl-neurips.github.io/program/offrl_42.html [cited February 3, 2024]

[18] Ouyang F, Jiao P. Artificial intelligence in education: The three paradigms. Computers and Education: Artificial Intelligence. 2021;**2**:100020. DOI: 10.1016/j.caeai.2021.100020

[19] Chaudhry MA, Kazim E. Artificial intelligence in education (AIEd): A high-level academic and industry note 2021. AI and Ethics. 2022;**2**:157-165. DOI: 10.1007/s43681-021-00074-z

[20] Adnan M, AlSaeed DH, Al-Baity HH, Rehman A. Leveraging the power of deep learning technique for creating an intelligent, context-aware, and adaptive m-learning model. Complexity. 2021;**2021**:1-21. DOI: 10.1155/2021/5519769

[21] Fryer LK, Ainley M, Thompson A, Gibson A, Sherlock Z. Stimulating and sustaining interest in a language course: An experimental comparison of chatbot and human task partners. Computers in Human Behavior. 2017;**75**:461-468. DOI: 10.1016/j.chb.2017.05.045

[22] Baillifard A, Gabella M, Lavenex PB, Martarelli CS. Implementing learning principles with a personal AI tutor: A case study [Internet]. 2023. Available from: https://arxiv.org/pdf/2309.13060.pdf [Updated September 10, 2023, cited February 23, 2024]

[23] Grawemeyer B, Mavrikis M, Holmes W, Gutierrez-Santos S, Wiedmann M, Rummel N. Affecting off-task behaviour: How affect-aware feedback can improve student learning. In: Proceedings of the 6th International Conference on Learning Analytics & Knowledge. Edinburgh: ACM Digital Library; 25-29 Apr 2016. pp. 104-113. DOI: 10.1145/2883851.2883936

[24] Harley JM, Bouchet F, Hussain MS, Azevedo R, Calvo R. A multi-componential analysis of emotions during complex learning with an intelligent multi-agent system. Computers in Human Behavior. 2015;**48**:615-625. DOI: 10.1016/j.chb.2015.02.013

[25] Echeverría V, Guamán B, Chiluiza K. Mirroring teachers' assessment of novice students' presentations through an intelligent tutor system. In: 2015 Asia-Pacific Conference on Computer Aided System Engineering, 2015 Jul 14-16. Quito, Ecuador: IEEE; 2015. pp. 264-269

[26] Taele P, Hammond T. Boponoto: An intelligent sketch education application for learning zhuyin phonetic script. In: Guercio A, editor. 21st DMS 2015. The 21st International Conference on Distributed Multimedia Systems, 2015 Aug 31-Sep 2. Vancouver, Canada: Knowledge Systems Institute; 2015. pp. 101-107

[27] Mohammed P, Mohan P. Dynamic cultural contextualisation of educational content in intelligent learning environments using ICON. International Journal of Artificial Intelligence in Education. 2015;**25**(2):249-270. DOI: 10.1007/s40593-014-0033-9

[28] Dolenc K, Aberšek B. TECH8 intelligent and adaptive e-learning system: Integration into technology and science classrooms in lower secondary schools. Computers and Education. 2015;**82**:354-365. DOI: 10.1016/j.compedu.2014.12.010

[29] Grivokostopoulou F, Perikos I, Hatzilygeroudis I. An intelligent tutoring system for teaching FOL equivalence. Processings of AIED. 2013;**2013**:20-29. Available from: https://publications.informatik.hu-berlin.de/archive/cses/publications/aied2013ws_volume9.pdf

[30] Kose U, Arslan A. Optimization of self-learning in computer engineering courses: An intelligent software system supported by artificial neural network and vortex optimization algorithm. Computer Applications in Engineering Education. 2017;**25**(1):142-156. DOI: 10.1002/cae.21787

[31] Duffy GG. Fighting off the alligators: What research in real classrooms has to say about Reading instruction. Journal of Reading Behaviour. 1982;**XIV**(4):357-373. DOI: 10.1080/10862968209547464

[32] Mariani L. Learning strategies, teaching strategies and new curricular demands: A critical view. Perspectives, A Journal of TESOL-Italy. 2002;**29**(2):45-56. Avaliable from: https://www.researchgate.net/publication/263854179_Learning_strategies_teaching_strategies_and_new_curricular_demands_a_critical_view [Accessed: February 3, 2024]

[33] Vermunt JD. The interplay between internal and external regulation of learning, and the design of process-oriented instruction. In: Paper Presented at the 3rd Conference of the European Association for Research on Learning and Instruction, Madrid, Spain. 1989

[34] Vermunt JD. Leerstijlen en Sturen Van Leerprocessen in het Hoger Onderwijs — Naar Procesgerichte Instructie in Zelfstandig Denken [Learning Styles and Regulation of Learning in Higher Education — Toward Process-Oriented Instruction in Autonomous Thinking]. Amsterdam/Lisse: Swets & Zeitlinger; 1992

[35] Short EJ, Weisberg-Benchell JA. The triple alliance for learning: Cognition, metacognition, and motivation. In: McCormick CB, Miller GE, Pressley M, editors. Cognitive Strategy Research: From Basic Research to Educational Applications. New York: Springer Verlag; 1989. pp. 33-63

[36] Boekaerts M. Self-regulated learning: A new concept embraced by researchers, policy makers, educators, teachers, and students. Learning and Instruction. 1997;7:161-186. DOI: 10.1016/S0959-4752(96)00015-1

[37] Dignath C, Buettner G, Langfeldt HP. How can primary school students learn self-regulated learning strategies most effectively? A meta-analysis on self-a meta-analysis on self-regulation training programmes. Educational Research Review. 2008;**3**(2):101-129. DOI: 10.1016/j.edurev.2008.02.003

[38] Dunlosky J, Rawson KA, Marsh EJ, Nathan MJ, Willingham DT. Improving students' learning with effective learning techniques: Promising directions from cognitive and educational psychology. Psychological Science in the Public Interest. 2013;**14**:4-58. DOI: 10.1177/1529100612453266

[39] Hattie JA, Donoghue GM. Learning strategies: A synthesis

and conceptual model. npj Science of Learning. 2016;**1**(1):1-13. DOI: 10.1177/1529100612453266

[40] Bower M. Technology-mediated learning theory. British Journal of Educational Technology. 2019;**50**(3):1035-1048

[41] Blayone TJ. Theorising effective uses of digital technology with activity theory. Technology, Pedagogy and Education. 2019;**28**(4):447-462. DOI: 10.1080/1475939X.2019.1645728

[42] Wollny S, Schneider J, Di Mitri D, Weidlich J, Rittberger M, Drachsler H. Are we there yet?-a systematic literature review on chatbots in education. Frontiers in Artificial Intelligence. 2021;**4**:654924. DOI: 10.3389/frai.2021.654924

[43] Hwang GJ, Chang CY. A review of opportunities and challenges of chatbots in education. Interactive Learning Environments. 2023;**31**(7):4099-4112

[44] Yang SJH. Guest editorial: Precision education - A new challenge for AI in education. Educational Technology & Society. 2021;**24**(1):105-108

[45] Zhai X, Chu X, Chai CS, Jong MSY, Istenic A, Spector M, et al. A review of artificial intelligence (AI) in education from 2010 to 2020. Complexity. 2021;**2021**:1-18. DOI: 10.1155/2021/8812542

[46] Chin DB, Dohmen IM, Cheng BH, Oppezzo MA, Chase CC, Schwartz DL. Preparing students for future learning with teachable agents. Educational Technology Research and Development. 2010;**58**:649-669. DOI: 10.1007/s11423-010-9154-5

[47] Jain GP, Gurupur VP, Schroeder JL, Faulkenberry ED. Artificial intelligence-based student learning evaluation: A concept map-based approach for analyzing a student's understanding of a topic. IEEE Transactions on Learning Technologies. 2014;**7**(3):267-279. DOI: 10.1109/TLT.2014.2330297

[48] Bhutoria A. Personalized education and artificial intelligence in the United States, China, and India: A systematic review using a human-in-the-loop model. Computers and Education: Artificial Intelligence. 2022;**3**:100068. DOI: 10.1016/j.caeai.2022.100068

[49] Munawar S, Toor SK, Aslam M, Hamid M. Move to smart learning environment: Exploratory research of challenges in computer laboratory and design intelligent virtual laboratory for eLearning technology. EURASIA Journal of Mathematics, Science and Technology Education. 2018;**14**(5):1645-1662. DOI: 10.29333/ejmste/85036

[50] Kelly S, Olney AM, Donnelly P, Nystrand M, D'Mello SK. Automatically measuring question authenticity in real-world classrooms. Educational Researcher. 2018;**47**(7):451-464. DOI: 10.3102/0013189X18785613

[51] Kong JS, Teo BS, Lee YJ, Pabba AB, Lee EJ, Sng JC. Virtual integrated patient: An AI supplementary tool for second-year medical students. The Asia Pacific Scholar. 2021;**6**(3):87. DOI: 10.29060/TAPS.2021-6-3/SC2394

[52] Yang YY, Shulruf B. An expert-led and artificial intelligence system-assisted tutoring course to improve the confidence of Chinese medical interns in suturing and ligature skills: A prospective pilot study. Journal of Educational Evaluation for Health Professions. 2019;**16**

[53] Kim HS, Kim NY, Chat Y. Is it beneficial to use AI chatbots to improve learners' speaking performance? Journal of Asia TEFL. 2021;**18**(1):161-178

[54] Vázquez-Cano E, Mengual-Andrés S, López-Meneses E. Chatbot to improve learning punctuation in Spanish and to enhance open and flexible learning environments. International Journal of Educational Technology in Higher Education. 2021;**18**(1):1-20. DOI: 10.1186/s41239-021-00269-8

[55] Vahabzadeh A, Keshav NU, Abdus-Sabur R, Huey K, Liu R, Sahin NT. Improved socio-emotional and behavioral functioning in students with autism following school-based smartglasses intervention: Multi-stage feasibility and controlled efficacy study. Behavioral Sciences. 2018;**8**(10):85. DOI: 10.3390/bs8100085

[56] Seldon A, Abidoye O. The Fourth Education Revolution. London: University of Buckingham Press; 2018. 370 p

[57] Edula R, Shengcheng L, Sasaki H, Viladiro R, Roy D. Personalized learning systems in smart education: A review of the literature & pilot feasibility discussion. In: AIP Conference Proceedings, 28 November 2023. Vol. 2909(1). New York: AIP Publishing; 2023

[58] Murtaza M, Ahmed Y, Shamsi JA, Sherwani F, Usman M. AI-based personalized E-learning systems: Issues, challenges, and solutions. IEEE Access. 2022;**10**:81323-81342. DOI:10.1109/ACCESS.2022.3193938

[59] Peng H, Ma S, Spector JM. Personalized adaptive learning: An emerging pedagogical approach enabled by a smart learning environment. Smart Learning Environments. 2019;**6**(1):114. DOI: 10.1186/s40561-019-0089-y

[60] Moskal P, Carter D, Johnson D. 7 things you should know about adaptive learning [Internet]. Educause Publications; Available from: https://library.educause.edu/~/media/fles/library/2017/1/eli7140.pdf [Updated 2017, cited February 12, 2024]

[61] Paramythis A, Loidl-Reisinger S. Adaptive learning environments and e-learning standards. Electronic Journal on e-Learning [Internet]. 2003;**1**:369-379. Available from: https://files.eric.ed.gov/fulltext/EJ1099144.pdf

[62] Walkington C, Bernacki ML. Personalizing algebra to students' individual interests in an intelligent tutoring system: Moderators of impact. International Journal of Artificial Intelligence in Education. 2019;**29**:58-88. DOI: 10.1007/s40593-018-0168-1

[63] Hwang GJ, Xie H, Wah BW, Gašević D. Vision, challenges, roles and research issues of artificial intelligence in education. Computers and Education: Artificial Intelligence. 2020;**1**:100001. DOI: 10.1016/j.caeai.2020.100001

[64] Cardona MA, Rodríguez RJ, Ishmael K. Artificial intelligence and the future of teaching and learning: Insights and recommendations. Available from: https://policycommons.net/artifacts/3854312/ai-report/4660267/ [Accessed: December 5, 2023]

[65] Siemens G. Connectivism: A learning theory for the digital age. International Journal of Instructional Technology and Distance Education [Internet]. 2004;**2**(1):3-11. Available from: https://www.itdl.org/Journal/Jan_05/Jan_05.pdf [cited January 20, 2024]

[66] Downes S. An Introduction to Connective Knowledge. In: Hug T, editor. Media, Knowledge & Education - Exploring new Spaces, Relations and Dynamics in Digital Media Ecologies. Innsbruck: Innsbruck University Press; 2008. pp. 77-102

[67] Li D, Yu J, Hu B, Xu Z, Zhang M. ExplainCPE: A free-text explanation benchmark of Chinese pharmacist examination. In: The 2023 Conference on Empirical Methods in Natural Language Processing, 6-10 December 2023. Singapore: Association for Computational Linguistics; 2023. pp. 1922-1940

[68] Talan T, Kalinkara Y. The role of artificial intelligence in higher education: ChatGPT assessment for anatomy course. Uluslararası Yönetim Bilişim Sistemleri ve Bilgisayar Bilimleri Dergisi. 2023;7(1):33-40. DOI: 10.33461/uybisbbd.1244777

[69] Rudolph J, Tan S, Tan S. ChatGPT: Bullshit spewer or the end of traditional assessments in higher education? Journal of Applied Learning and Teaching. 2023;**6**(1):342-363. DOI: 10.37074/jalt.2023.6.1.9

[70] Jalil S, Rafi S, LaToza TD, Moran K, Lam W. ChatGPT and software testing education: Promises & perils. In: 2023 IEEE International Conference on Software Testing, Verification and Validation Workshops (ICSTW), 2023 April 16-20. Dublin, Ireland: IEEE; 2023. pp. 4130-4137

[71] Bordt S, von Luxburg U. ChatGPT participates in a computer science exam. ArXiv. 2023. DOI: 10.48550/arXiv.2303.09461

[72] Rajabi P, Taghipour P, Cukierman D, Doleck T. Exploring ChatGPT's impact on post-secondary education: A qualitative study. In: 25th Western Canadian Conference on Computing Education (WCCCE'23). New York: Association for Computing Machinery; 2023. pp. 1-6

[73] Castillo AGR, Silva GJS, Arocutipa JPF, Berrios HQ, Rodriguez MAM, Reyes GY, et al. Effect of Chat GPT on the digitized learning process of university students. Journal of Namibian Studies: History Politics Culture. 2023;**33**:1-15. DOI: 10.59670/jns.v33i.411

[74] Geng C, Yihan Z, Pientka B, Si X. Can ChatGPT pass an introductory level functional language programming course? arXiv:2305.02230. 2023. Available from: https://arxiv.org/abs/2305.02230 [Accessed: March 7, 2024]

[75] Malinka K, Peresíni M, Firc A, Hujnák O, Janus F. On the educational impact of ChatGPT: Is artificial intelligence ready to obtain a university degree? In: Proceedings of the 2023 Conference on Innovation and Technology in Computer Science Education, 2023 July 7-12, Turku, Finland. New York: United States; 2023. pp. 47-53

[76] Abramski K, Citraro S, Lombardi L, Rossetti G, Stella M. Cognitive network science reveals bias in GPT-3, ChatGPT, and GPT-4 mirroring math anxiety in high-school students. Big Data and Cognitive Computing. 2023;7(3):124-148. DOI: 10.3390/bdcc7030124

[77] Neumann M, Rauschenberger M, Schön EM. "We need to talk about ChatGPT": The future of AI and higher education. In: Proceedings of the 2023 IEEE/ACM 5th International Workshop on Software Engineering Education for the Next Generation (SEENG), 2023. Melbourne, Australia: IEEE; 2023. pp. 29-32. DOI: 10.1109/SEENG59157.2023.00010

Chapter 2

Generative AI in Education: Technical Foundations, Applications, and Challenges

Sheikh Faisal Rashid, Nghia Duong-Trung and Niels Pinkwart

Abstract

Generative artificial intelligence (AI) (GenAI) has emerged as a transformative force in various fields, and its potential impact on education is particularly profound. This chapter presents the development trends of "GenAI in Education" by exploring the technical background, diverse applications, and multifaceted challenges associated with its adoption in education. The chapter briefly introduces the technical background of GenAI, particularly the development of large language models (LLMs) such as ChatGPT & Co. It provides key concepts, models, and recent technological advances. The chapter then navigates through the various applications of GenAI or LLMs in education, examining their impact on different levels of education, including school, university, and vocational training. The chapter will highlight how GenAI is reshaping the educational landscape through real-world examples and case studies, from personalized learning experiences to content creation and assessment. It also discusses various technical, ethical, and organizational/educational challenges to using technology in education.

Keywords: generative AI, large language models, ChatGPT, generative AI in education, educational technologies

1. Introduction

In recent years, artificial intelligence (AI) has made remarkable progress in all areas of life, including education. Particularly with advances in generative AI (GenAI), its applications in teaching and learning have attracted stakeholders like students, educators, researchers, and educational institutes worldwide [1]. The development of tools, such as ChatGPT and Dall-E, has further enriched the field by engaging students and teachers in real-time conversations with large language models (LLMs) to create expressive artwork and digital images. Generative artificial intelligence (GenAI) is at the forefront of artificial intelligence and machine learning, focusing on generating new content, such as text, images, music, video, code, or other data types. Unlike traditional AI methods that categorize data into predefined classes, GenAI models learn the underlying patterns and relationships within data to generate entirely new content. This ability of GenAI models to create new content offers immense potential for revolutionizing the educational landscape. Some examples include applications that facilitate personalized learning experiences tailored to the unique needs of each learner; increased accessibility for students facing challenges, such as learning disabilities, anxiety, or language barriers; and support for a variety of tasks, such as coding, writing, art, music, etc. In addition, GenAI enables teachers to provide constructive feedback at scale, fostering iterative learning and improving writing skills. Furthermore, GenAI helps educational institutions provide customized support and information to students, automate tasks such as scheduling events, and generate promotional content, etc.

Despite the potential innovations and opportunities, concerns and risks are also associated with using GenAI in education. For example, the ability of ChatGPT to correctly answer many practice and exam questions has raised concerns among many educational stakeholders about the use of artificially generated solutions [2]. Teachers' initial concerns are that students may use ChatGPT to cheat on their assignments. Teachers may struggle to determine the difference between artificially generated and human-generated content. This scenario leads to undermining the whole assessment and grading process, and a natural demand has arisen to detect the content of GenAI via some other AI tools, such as TurnItIn, GPTZero, or OpenAI's classifier. However, in the absence of other evidence, technical methods currently need to be more helpful in regulating the use of AI in the classroom. The Education Technology Lab (EDTec Lab), German Research Center for Artificial Intelligence (DFKI) Berlin, advocates an objective assessment of the potential and limitations of the technology and warns against software that claims to be able to automatically recognize text generated by ChatGPT [3]. Other concerns include personal data privacy and security, academic dishonesty's consequences, students' overestimation of GenAI's capabilities and trustworthiness, and the inadvertent reinforcement of biases through system output and user interaction. It has led to a debate, in which some educational institutions have banned the use of ChatGPT or similar GenAI tools for students. In contrast, others have welcomed the ethical and transparent use of GenAI tools in education. Policymakers and educators have initiated disseminating guidelines for students, teachers, and educational institutions aiming to promote academic integrity, ensure accessibility, and encourage ethical applications of this technology in educational settings [4–6].

Generative AI (GenAI) has the potential to formulate knowledge models and guide cognitive activities, aid learning while learners actively engage as collaborators, and empower learners to control their own learning experiences. These tools reflect complex thought processes fundamental to human understanding and are potent resources for students and educators. Innovations like AI-generated content and adaptive learning platforms are reshaping how educational content is delivered and consumed, making learning more accessible and tailored to individual learner's needs [1]. This chapter aims to provide a comprehensive overview of GenAI in education, briefly describe technical foundations, and highlight its emergent opportunities while acknowledging the challenges that must be addressed to ensure its responsible and equitable implementation.

2. Technical foundations of generative AI

2.1 What is generative AI?

Before ChatGPT was launched in late 2022, the public domain was primarily influenced by what's known as discriminative AI. This form of AI specializes in sorting and categorizing information, serving as a foundational tool in many applications. However, the narrative began to shift with the rise of generative AI, particularly marked by the public availability of models like ChatGPT. Despite sharing common technological underpinnings, generative and discriminative AI diverge significantly in their objectives. Discriminative AI is adept at discerning and differentiating between various data categories. In contrast, generative AI aims to synthesize new content or outputs, drawing upon the inputs it receives and the extensive data understanding it has developed through its training [7].

Generative AI is a type of AI technology that autonomously creates content in response to natural language prompts through conversational interfaces. Unlike traditional methods that mainly curate content from existing web sources, generative AI actively generates new material. This content spans various formats embodying different aspects of human cognition, including natural language texts, images, videos, music, and even software codes. Training generative AI models frequently involve unsupervised learning methods, which allow these models to process and comprehend large volumes of data autonomously. The goal is to discern patterns in the structure or creation of these data, enabling the AI to emulate and reproduce these patterns.

2.2 Text generative AI

Text generative AI employs a specific type of artificial neural network (ANN) called a general-purpose transformer. Within this category, a particular form called a large language model (LLM) is prevalent. These systems are often called LLMs due to their extensive linguistic capabilities. The type of LLM utilized in text generation AI, such as ChatGPT, is known as a generative pre-trained transformer, or GPT. The 'GPT' in 'ChatGPT' reflects this underlying technology.

ChatGPT is built on GPT-4, a version developed by OpenAI, which represents the latest evolution in their GPT series (as of the time the book chapter was written). Each new iteration of OpenAI's GPT has shown significant improvements over its predecessors, driven by advancements in AI architectures, training methodologies, and optimization techniques. A notable aspect of its continuous development is the increasing scale of data used for training and the exponential growth in the number of parameters.

These parameters are akin to metaphorical knobs, fine-tuning the performance of the GPT. They include the model's 'weights', and numerical values dictating how it processes input data and generates output. As of GPT-4, the model substantially increases these parameters, enabling more nuanced and sophisticated responses than earlier versions. This evolution reflects OpenAI's commitment to enhancing the capabilities and accuracy of its language models, ensuring they remain at the forefront of AI technology in text generation.

2.3 Image generative AI

Image generative AI, such as Dall-E and Stable Diffusion, primarily leverage Diffusion Models, a type of ANN that differs from traditional generative adversarial networks (GANs). These Diffusion Models operate by gradually learning to reverse a process that adds random noise to an image. Initially, the model introduces noise to a clear image, transforming it into a completely random pattern. It then learns to reverse this noise addition during generation, effectively reconstructing the original image from randomness.

This process begins with a random pattern and then iteratively refines this pattern into a coherent image that aligns with a given prompt. Unlike GANs, which involve a generator and discriminator working in opposition, Diffusion Models rely on a single neural network that predicts how to remove noise at each step. Over numerous iterations, this network becomes adept at creating detailed and realistic images from noisy inputs.

Diffusion Models have shown remarkable capability in generating high-quality images that are often more detailed and varied than those produced by GANs. They excel in creating diverse outputs based on textual descriptions, from realistic photographs to artistic renderings. This advancement in AI-driven image generation represents a significant shift from the GAN-based approach, offering more flexibility and potential for creative applications.

2.4 Preliminary of multimodal generative AI

Text and image-generative AI models excel in text-image-based tasks but are limited to their specific data types and cannot simultaneously process images, videos, or audio. This limitation is crucial in real-world applications where multimodal data are shared. To bridge this gap, Multimodal Large Language Models (MLLMs) [8] and Large Vision Models (LVMs) [9] have been developed. MLLMs and LVMs combine a large language model with multimodal adaptors and various diffusion decoders, allowing them to process and generate outputs across different media formats.

The concept of multimodality in these models is inspired by human communication, which often involves multiple channels. MLLMs and MLVMs are trained on extensive multimodal datasets, including image captions, video descriptions, and audio transcripts. They can recognize patterns in these data and generate coherent outputs that match the input modality. Despite these advancements, integrating new modalities into existing models remains challenging. It requires extensive data that include the new modality and often necessitate retraining from scratch, demanding significant computational resources and data quality efforts. Several recent frontier research studies have been devoted to this future direction, for example, Emu2 [10] and Google Gemini [11].

Overall, AI is moving towards more adaptable, efficient, and versatile systems capable of handling a broader range of tasks and data types, reflecting a more holistic approach to artificial intelligence.

3. Applications of generative AI in education

Generative AI (GenAI) can be used in a variety of educational contexts, from creating innovative content and generating personalized learning materials to automating assessment and feedback. For example, students can use GenAI to help with homework or explore creative writing and art, educators can use these tools to create engaging teaching materials or to provide personalized learning experiences, and institutes can use GenAI to improve their administrative and educational services. GenAI applications can be used as standalone tools or integrated into other systems or platforms. In conjunction with different forms of AI, GenAI models, especially LLMs, can enhance and support learning activities at all levels of education. A recent report published jointly by the German Research Center for Artificial Intelligence (DFKI) and mmb Institut GmbH for the Deutsche Telekom Foundation highlighted applications of AI in school education [12]. The report mentions the practical use cases of GenAI models like ChatGPT in school education. It provides a structured overview of AI systems benefiting learners, educators, and institutes and shaping everyday school life in the future. In the context of INVITE -innovation competition funded by the German Federal Ministry of Education and Research (BMBF), VDI/VDE-IT comprehensively introduces LLMs and their transformative impact on continuing vocational training [13]. Kasneci et al. presented many possibilities that can be realized using GenAI models such as ChatGPT or other LLMs [14]. They highlighted how these models can be used to create educational content, improve student engagement and interaction, and personalize learning experiences while emphasizing that the use of LLMs in education requires students and teachers to develop a set of competencies and literacies necessary to understand both the technology and its limitations, as well as the unexpected brittleness of such systems.

The applications of AI in education can be divided into three categories: student-supporting applications, teacher-supporting applications, and institute- or system-supporting applications [15]. In addition to providing support in these three categories, they also have the potential to promote learning activities for children with disabilities, support online and collaborative learning, and boost professional development and training. This section provides an overview of some of the applications of GenAI supporting learning and educational activities in these directions.

3.1 Student-supporting applications

Generative AI (GenAI) has the potential to serve as a learning companion and support educational activities at all levels. Students can use GenAI as a general search tool to get answers to their immediate questions about a specific topic, get support in completing homework or preparing for their exams, and learn new skills.

3.1.1 Support in primary to higher education

In primary school education, GenAI can help children improve their reading and writing skills by suggesting syntactic and grammatical improvements. It can also improve writing style, develop critical thinking skills, and improve reading comprehension by providing students with summaries and explanations of complex texts, making the material more accessible. In a recent study, Han and Cai present a visual storytelling prototype based on generative AI tools, such as ChatGPT, Stable Diffusion, and Midjourney, for children's creative expression, storytelling, and literacy development. It was found that generative AI could significantly enhance reading and writing skills and promote creative storytelling and literacy development in young learners [16]. Another study on the use of GenAI, specifically ChatGPT 3.5 and 4, in primary school education, showed its potential to personalize learning material and cater to students' diverse knowledge and learning abilities. The study involved 110 students and demonstrated that generative AI could support motivated learning and skill development, suggesting a promising future for its use in school education [17].

Generative AI (GenAI) can provide valuable support for middle and high school students in language acquisition and mastering different writing styles in subjects such as language and literature. GenAI can also help students prepare for their exams and assessments. GenAI tools can generate practice problems and quizzes for subjects like mathematics and physics that help improve the studied material's understanding, contextualization, and retention. They also help to improve problem-solving skills by providing students with detailed explanations, step-by-step solutions, and engaging supplementary questions. This approach clarifies the reasoning behind solutions, promotes analytical thinking, and encourages creative problem-solving strategies. There is also a potential for GenAI, specifically LLMs, to enhance science, technology, engineering, and mathematics (STEM) education with a multimodal analogical reasoning approach. A study by Cao et al. demonstrated how GenAI transforms intricate principles in mathematics, physics, and programming into comprehensible metaphors and converts them into visual form to augment the educational experience further [18]. Khanmigo.ai, for example, powered by GenAI tools, provides a whole bunch of services to support students in improving their writing and critical thinking skills, supporting them in preparation for their maths quizzes, and even helping them with real-time feedback, debate, and collaboration. Similarly, tools.fobizz.com provides a suite of GenAI-powered tools to support both educators and learners.

At the university level, GenAI models such as ChatGPT are recognized for their role in supporting research, completing assignments, and fostering the development of critical thinking and problem-solving skills. These tools can enhance students' research skills by providing access to comprehensive information and relevant resources on specific topics. Students can use these tools to create summaries, organized outlines, and initial sketches of their intended research subject. These tools can also help students efficiently comprehend important concepts and simplify their writing processes. Furthermore, LLMs can be used to find unexplored areas and current research trends, enhancing students' understanding of the subject matter and facilitating analytical analysis [14]. For example, Castillo-Segura et al. demonstrate the potential of generative AI in accelerating the research process, specifically in conducting systematic literature reviews in academia. This research compares six GenAIs (Forefront, GetGPT, ThebAI, Claude, Bard, and H2O) with their respective large language models (LLMs) when classifying 596 articles in the screening phase of a systematic review in the field of medical education. It has been observed that Generative AI tools can significantly reduce the time and effort required for systematic literature reviews (SLRs) by facilitating article identification and classification, demonstrating the practical benefits of AI in academic research [19]. Jonsson & Tholander present a study of a group of university students using generative machine learning to translate from natural language to computer code. The study explores how AI can be understood in terms of co-creation, focusing on how AI may serve as a resource for understanding and learning and, on the other hand, it affects the creative processes [20].

3.1.2 Support in professional education

The integration of AI into daily life and the workplace is increasing, making it crucial to prepare students for future technological demands. According to the World Economic Forum, AI integration will result in a mixed job outlook by 2027, with 25% of companies anticipating job losses and 50% expecting job growth. This trend highlights the significance of providing students with skills in emerging technologies [21]. Familiarity with AI and chatbots may become essential for entering the workforce. Furthermore, technology companies like Google and Microsoft have already announced plans to integrate AI into their products, indicating the increasing prevalence of AI in various products and services [22, 23]. Given these developments, the use of AI technologies in education prepares students for future job markets and offers opportunities to enhance learning beyond traditional methods.

Large language models (LLMs) have the potential to enhance professional training in a wide range of fields by providing tailored support for the development of specific competencies that are critical for different professional environments. By fine-tuning LLMs on domain-specific corpora, these models can generate industry-specific language and help learners acquire the specialized vocabulary and stylistic nuances needed to write technical reports. This extends the capabilities of LLMs beyond general language processing to specialized training tools that can simulate real-world professional writing and communication scenarios. For example, the adaptability of LLMs to specific professional needs, such as programming, report writing, project management, decision-making, etc., underlines their usefulness in enhancing skills critical to modern workplaces. Tools such as GitHub Copilot, powered by OpenAI's Codex, show how AI can assist with real-time coding tasks, suggesting code completions and providing programming insights directly within the coding environment [24]. Such tools can help generate code, understand programming languages, and even offer debugging support, making them invaluable tools for novice and experienced programmers. The role of LLMs in professional training also emphasizes the importance of domain-specific tuning and the integration of professional expertise to maximize their effectiveness [14, 25]. In the medical field, for example, LLMs can simulate patient interactions, help with the generation of medical reports, and even assist in medical research by providing up-to-date information and generating hypotheses based on current medical literature [26, 27]. Pavlik discussed the potential of generative AI platforms such as ChatGPT for journalism and media education, highlighting the importance of understanding the capabilities and limitations of such technologies [28]. However, integrating LLMs into professional education requires a collaborative approach involving educators, industry professionals, and AI developers. This collaboration ensures that LLMs are fine-tuned for technical accuracy, ethical considerations, professionalism, and the nuanced understanding required in specific fields.

3.1.3 Support in learning with disabilities

Generative AI can help people with disabilities or neurodivergent in the context of education, work, and leisure. By developing inclusive learning strategies, GenAI can help create adaptive writing tools, translate complex texts into more accessible formats, and highlight key content across different media, ensuring that educational content is accessible to all students, regardless of their disabilities. For example, GenAI-powered applications such as "goblin.tools" can help children with Attention Deficit Hyperactivity Disorder (ADHD) to simplify tasks they find overwhelming or difficult [29]. LLMs can be seamlessly integrated with speech-to-text and text-to-speech solutions to assist the students with visual impairments. Earlier research into the use of video models to teach generative spellings to a child with autism spectrum disorder also showed a potential of employing GenAI for the generation of such video models [30]. For example, GenAI models can be integrated into virtual reality (VR) and extended reality (XR) technologies and can provide a wide range of innovative use cases and interaction concepts that can help to reduce barriers for people with specific physical or mental needs, e.g., through simulations, gamification, or training scenarios [31, 32].

Because GenAI has the potential to reflect and perpetuate societal biases, including those related to disability, the use of these technologies must be undertaken with careful consideration of the ethical implications, potential biases, and the need for professional oversight to ensure that the benefits of LLMs are shared equitably among all learners, including those with disabilities [33]. There is also a need to work with therapists, educators, and other professionals to meet the specific needs of learners. The interactions of people with disabilities with LLMs also highlight the need to involve those with lived experience of disability in the development and training of LLMs to ensure that these models serve as empowering tools rather than sources of further marginalization [34].

3.2 Teacher-supporting applications

Teachers use GenAI to enhance their pedagogical activities and ensure students develop the desired learning outcomes. GenAI helps them create tailored learning materials, generate assignments and quizzes, provide feedback, and assist in developing or assessing student exams [14].

3.2.1 Support in creation of learning material

Generative AI (GenAI), specifically LLMs, can be used to create learning content, exercises, quizzes, presentation slides, etc., for a wide range of subjects and educational levels that can be adapted to meet the diverse needs of students [35]. For example, LLMs can support curriculum development, teaching methodologies, personalized study plans and learning materials, student assessment, and more in medical education [36]. Similarly, Rüdian & Pinkwart presented the use of LLMs (ChatGPT 3.5) in generating learning content for a concrete micro-learning template in German language teaching. Teachers provide a topic as input, and the approach then elicits the required information with instructional prompts and combines responses into a language learning unit. The quality of the generated learning units was assessed for correctness and appropriateness. The results showed the best performance, but the need for "teacher-in-the-loop" is suggested [37]. Based on some existing ideas, teachers can also use GenAI to create innovative discussion topics, animations, and short stories to enhance student engagement or group discussion. Considering the example of creating model texts, LLMs can be asked to "create a discussion about the use of AI in school" and the generated texts can then be adapted by the teacher according to the didactic goal of the learning content [12].

3.2.2 Support in teaching activities

Generative AI (GenAI) can help teachers provide personalized learning experiences for students by analyzing their responses to specific learning tasks. GenAI can give feedback, hints, or suggestions for learning tasks or generate materials matching the student's learning needs or skills. For example, LLMs can support teachers in creating inclusive learning activities, questions, and assessment exercises or quizzes targeted to students at different levels of knowledge, ability, and learning styles. Phung et al. present a study on programming showing how GenAI can improve STEM education. Their research highlights the broader capabilities of GenAI to support learning and human tutors in different programming education scenarios. It does so by providing personalized digital tutoring, generating hints, and creating tasks and explanations [38]. Such support can save teachers time and effort in creating personalized materials and allow them to concentrate on other aspects of teaching.

3.2.3 Support in grading of assignments

Generative AI (GenAI) can support teachers by automating the grading of assignments or exams and providing immediate feedback to students. For instance, LLMs can be used to identify potential strengths or weaknesses in written essays or other writing assignments and provide individualized feedback to students [39, 40]. It speeds up the individual evaluation process and allows for more consistent and objective grading [41]. Furthermore, LLMs can also be used to detect plagiarism, which can help to prevent cheating on submitted writing assignments. Additional AI tools can also be used to analyze assessment data to identify trends, such as common areas where students struggle, enabling targeted interventions to support learning.

3.3 System-supporting applications

3.3.1 Support in administration activities

Large language models (LLMs) can support educational institutions through chatbots to provide instant answers to questions on various administrative topics. For example, LLMs can be used to respond to queries from potential applicants and provide them with up-to-date information. These models can help existing students register for courses and provide administrative information, such as courses, exams, schedules, etc. Students can also use LLMs-based chatbots to find news or other information. LLMs-based chatbots can also be set up for international students and staff to provide multilingual information to students. In addition, LLMs can generate offers or advertisements of learning opportunities based on various factors, such as target audience, age group, gender, location, etc.

Another potential application is the automated tagging of learning resources to provide metadata for effective management and efficient discovery of learning resources [42]. LLMs can be used to develop solutions for automated or semiautomated generation of metadata fields from learning resources using explicitly defined metadata standards. It will significantly facilitate the implementation of personalized learning or intelligent tutoring systems by making it easier to find appropriate learning content without human intervention.

3.3.2 Support in tutoring and mentoring

Generative AI (GenAI) can power intelligent tutoring systems that provide personalized guidance and support to students. These systems can analyze student responses, identify misconceptions, and generate customized explanations or additional practice materials to address individual learning needs. AI tutors can adapt their teaching strategies based on student progress and learning styles, creating a more effective and tailored learning environment. The integration of LLMs into existing learning management systems (LMS) can provide tutoring or mentoring support to students as an educational chatbot. One such example is the "tech4compKI," a Federal Ministry of Education and Research (BMBF)-funded project of the Educational Technology Lab, DFKI Berlin, where students ask questions and LLM-supported chatbot "BiWi AI Tutor" retrieves information from material (structured knowledge of the module, learning material from lectures and seminars as well as organized information) and analyzes it to answer questions.

3.3.3 Support in collaborative and remote learning

Generative AI (GenAI) can simulate a collaborative learning environment by acting as a peer to provide support in various collaborative learning tasks, such as peer discussion, to explore a research question or a discussion topic. These models can give immediate feedback on writing artifacts or generate new ideas through peer discussion. One example is PEER (Plan, Edit, Explain, Repeat), a collaborative language model trained to imitate the entire writing process. The model can draft, make suggestions, propose edits, and explain its actions. The model also uses self-learning techniques to adapt to new areas of learning while demonstrating strong performance in different domains and editing tasks [43]. The use of GenAI models in online education systems has the potential to transform remote and group learning by providing interactive, responsive, and tailored responses to each student. In group and remote learning environments, GenAI models can provide a structure for discussion, offer real-time feedback, and give personalized guidance to students, thereby increasing student engagement and participation while adapting to the dynamic nature of group discussions and debates. For example, LLMs can be used to manage conversations and balance the preferences of group members in collective decision-making tasks (such as scheduling meetings) with fair consideration for all participants by extracting individual preferences and suggesting options that satisfy most group members [44].

4. Integration of generative AI in educational systems

4.1 Teacher's journey from awareness to creation

Integrating GenAI tools into educational settings is not just an emerging trend; it's rapidly becoming an essential part of modern teaching methodologies across multiple disciplines. However, teachers' perspectives towards the adoption of these tools vary significantly, predominantly based on their awareness, understanding, and experience with GenAI technologies [45, 46]. This discussion explores educators' perceptions regarding integrating GenAI in education, focusing on transitioning from awareness to classroom creation. Possible stages of adoption and integration are as follows.

Awareness: The initial phase involves teachers becoming aware of GenAI's potential educational applications. At this stage, the primary emotion is curiosity mixed with a hint of skepticism, especially among those with limited background in technology. The challenge lies in transforming this curiosity into a genuine interest capable of driving further exploration. As GenAI continues to evolve, the argument for its inevitability in education strengthens; this phase is crucial for setting the groundwork for future engagement.

Learning: The learning phase is marked by teachers actively seeking knowledge about GenAI, its capabilities, and how it might be leveraged to enhance teaching and learning experiences. This stage is often accompanied by a range of emotions, from excitement at the possibilities to frustration over the learning curve of new technology. Professional development opportunities, workshops, and online resources support educators through this phase. Here, the divide between tech-savvy educators and those from non-IT backgrounds becomes most apparent, necessitating tailored learning paths to ensure inclusivity.

Familiarity: As educators gain hands-on experience with GenAI tools, their comfort level increases. This familiarity phase is characterized by growing confidence in using these technologies for specific, often limited, tasks within the educational context. For many, this is the stage where the potential of GenAI to transform aspects of teaching and research becomes tangible. Experiences shared among teachers through forums, collaborations, and professional networks further nurture this growing familiarity, turning apprehension into acceptance.

Creation: The final phase sees teachers actively integrating GenAI tools into their curriculum as an adjunct technology and a central component of their instructional strategies. This creation phase is where the full potential of GenAI integration is realized, with educators innovating new ways to engage students, personalize learning experiences, and streamline administrative tasks. Here, the emphasis shifts from understanding GenAI to leveraging its capabilities to foster more effective, efficient, and exciting educational environments.

4.2 Learner's perspectives towards GenAI

Understanding the evolving landscape of educational technologies, especially with the integration of GenAI, mandates considering learners' perspectives [47, 48]. As they navigate these technological advancements in their educational journey, their attitudes and skills development play a critical role. Here's a focused exploration of the potential learner's perspectives towards GenAI integration in education:

Enhancing learning efficiency: At the outset, learners perceive GenAI as a revolutionary tool, much like a "search engine 2.0," which propels the efficiency of study practices to new heights. Unlike traditional search engines that return a vast array of links requiring further analysis, GenAI, such as ChatGPT, provides concise, tailored answers, significantly reducing the effort and time needed for research. Students may foster a positively inclined attitude towards these technologies, recognizing their potential as a study aid and an essential competency for their future careers. They understand that integrating GenAI technologies into their learning strategies can vastly enhance their understanding and retention of information.

Navigating the accuracy of GenAI: An essential perspective that learners must adopt is the understanding that GenAI tools are not infallible for all their advancements. The ability to discern and validate the information provided by GenAI becomes crucial. Hence, developing fact-checking skills is imperative for students, ensuring they can distinguish between accurate information and potential inaccuracies generated by these tools. This skill is particularly significant because GenAI sometimes creates plausible but erroneous content. Encouraging a critical approach towards accepting information will enable learners to utilize these technologies effectively, ensuring their reliance on GenAI complements rather than compromises their educational integrity.

Personalized learning support: GenAI technologies have the potential to revolutionize the concept of personalized learning through virtual 1:1 coaching. In this vision, GenAI is an ever-present tutor, available to address learners' queries instantaneously, guide them through complex problems, and provide tailored learning experiences. This perspective views GenAI as a tool and a learning partner capable of fostering the self-paced practice of foundational skills in learners. Such 1:1 coaching can fill gaps in understanding, offering explanations and resources customized to each learner's needs and pace. For this potential to be fully realized, the application of GenAI in education must be guided by ethical standards and pedagogical principles, ensuring that these technologies genuinely support and enhance the learning experience.

4.3 Administrative roadmap to integrate GenAI

Strategic planning for GenAI integration: Strategic planning forms the bedrock of successful GenAI integration within educational settings. Administrators must embark on a detailed analysis of the current technological infrastructure, identifying gaps and envisioning a roadmap for incorporating GenAI tools. This includes addressing hardware and software needs, internet bandwidth, and cybersecurity measures. The strategic plan should align with the institution's broader educational goals, ensuring that GenAI adoption enhances, rather than disrupts, the learning experience. It may involve setting up pilot programs to test the effectiveness of GenAI tools in specific subjects or activities and gathering data to inform broader implementation strategies.

Comprehensive resource allocation: Beyond the initial enthusiasm for GenAI's potential, administrators face the practical challenge of resource allocation. It goes beyond budget considerations to encompass a holistic approach that addresses equity, access, and sustainability. For instance, ensuring that all students, including those from underprivileged backgrounds or with special needs, have access to these technologies is crucial. Administrators may discuss partnerships with tech companies for donations or discounts, grants, and other financing models to support this equitable access. Moreover, resource allocation is not solely about technology procurement; it also encompasses investing in human capital—recruiting staff adept at blending GenAI tools with teaching methodologies or providing ongoing training for current educators.

Ethical considerations and digital citizenship: The ethical deployment of GenAI in education requires administrators to grapple with questions of data privacy, intellectual honesty, and the potential for technology misuse. Establishing a framework that promotes ethical use while encouraging innovation is a balancing act. Discussions may revolve around developing comprehensive policies that govern data use, consent protocols, and transparency measures. Furthermore, fostering a culture of digital citizenship, where students learn to use GenAI responsibly, becomes a shared mission. This could include curriculum updates to cover topics like digital ethics, privacy rights, and the implications of AI on society.

Professional development tailored to GenAI: Professional development is a cornerstone of effective GenAI integration, necessitating discussions on equipping teachers with the necessary skills and confidence to use these tools in the classroom. Training programs might need to be overhauled to include GenAI competencies, ranging from technical know-how to pedagogical strategies for integrating AI tools into lesson plans. Moreover, creating a supportive community where teachers can share experiences, challenges, and best practices is vital. Such initiatives could include internal workshops, online forums, or collaboration with external experts and institutions. The goal is to foster an environment where teachers feel empowered to innovate while ensuring that GenAI enhances the educational experience.

Policy formulation and student preparedness: Crafting policies that reflect the complexities of GenAI use in education is another pivotal discussion point. These policies must address academic integrity, ensuring students leverage GenAI as a learning aid without compromising their intellectual development. For instance, guidelines must clearly define how and when GenAI can be used for assignments, projects, and research. Additionally, preparing students for a future where AI is ubiquitous involves adjusting curricula to include critical thinking, problem-solving, and digital literacy skills. Administrators must consider how education can adapt to prepare students to use AI and understand its impact on society and the workforce.

Engaging the broader educational community: Ultimately, successful GenAI integration depends on the engagement of the entire educational community—students, parents, teachers, and staff. Administrators should lead efforts to educate these stakeholders on the benefits and challenges of GenAI, setting realistic expectations while addressing concerns and soliciting feedback. This may involve community forums, informational newsletters, and transparent reporting on pilot programs and initiatives. By fostering an inclusive dialog, administrators can build trust and enthusiasm for the transformative potential of GenAI in education.

5. Potential risks and challenges

5.1 Imperfection and polluted content

The proliferation of generative AI has risks and challenges, particularly in the educational context. A primary issue is the quality control of the content generated. While these AI systems can create high-quality educational materials, they can also produce inaccurate or misleading information, especially in scenarios not covered in their training datasets. This problem is related to AI "hallucination," where the AI produces content that significantly deviates from factual accuracy, leaning more towards the AI's interpretations rather than an accurate representation of the information. These hallucinations occur because AI systems, in their current form, do not possess proper knowledge or understanding [49, 50]. In practical terms, if an AI model is uncertain about the factual accuracy of a piece of information, e.g., a historical event, it may provide an educated guess rather than an accurate answer. Compared to carefully curated textbooks, AI-generated content, commonly sourced from the vast and varied Internet, often lacks the rigorous vetting necessary for educational purposes. This discrepancy is significant because the Internet, while a treasure trove of information, can also contain discriminatory and unethical language.

Generative AI's (GenAI's) ability to alter or fabricate images and videos, creating highly realistic 'deepfakes,' poses a unique challenge to educators and students alike. These deepfakes are increasingly indistinguishable from authentic materials, making it easier to produce and disseminate 'fake news' and other forms of misleading information. The absence of stringent regulations and robust monitoring systems means potentially biased or erroneous AI-generated materials are becoming more prevalent online, influencing one of the primary knowledge sources for learners globally. This situation mainly concerns young learners, who may need more background knowledge to discern inaccuracies or biases in AI-generated content. Moreover, there's a legal aspect to consider: using content without proper consent or attribution can lead to copyright issues and undermine the integrity of educational resources. This issue also creates a recursive challenge for future AI models, as they might train on their AI-generated content from the Internet, perpetuating and amplifying existing biases and errors. This cycle underscores the need for more effective oversight and ethical standards in AI-generated content, particularly in educational contexts. This challenge is exacerbated in an educational setting where critical thinking and accurate information are foundational.

Addressing those mentioned challenges requires a multifaceted approach. A potential solution lies in applying reinforcement learning with human feedback (RLHF), a technique where AI models are fine-tuned based on evaluations of their outputs by human experts. This method allows the models to iteratively improve their content generation to align more closely with factual accuracy and ethical standards. Furthermore, developing and utilizing foundation educational datasets for fine-tuning LLMs can ensure that the AI systems have a solid base of verified and unbiased information, specifically tailored for educational purposes. These datasets would consist of meticulously curated and peer-reviewed educational content, encompassing various subjects and perspectives to minimize biases and inaccuracies. Additionally, implementing robust monitoring systems and ethical guidelines for AI-generated content can help safeguard against disseminating misleading information. Educators and developers must collaborate to integrate these solutions, ensuring AI tools enhance learning experiences while maintaining integrity and factual correctness. This approach addresses the immediate concerns of content quality and lays the groundwork for responsible and ethical AI use in education.

5.2 Ethical and societal concerns

Integrating AI into education requires careful consideration of ethical implications, especially in contexts where comprehensive governmental regulations may not fully address the rapid advancements in AI technologies [51, 52]. A pivotal concern is finding a harmonious balance between leveraging the benefits of AI-driven educational tools and safeguarding student privacy. The effectiveness of these tools largely depends on their ability to analyze extensive datasets, which poses significant questions regarding how student information is utilized. Educators, therefore, bear the critical responsibility of rigorously evaluating AI technologies before their implementation. This entails thoroughly examining the ethical frameworks governing AI tools to ensure they are in harmony with the core values of educational institutions. It also involves fostering collaborative efforts with administrators, parents, and other stakeholders to create a unified set of ethical standards. Additionally, educators must implement robust mechanisms for continuous assessment to ensure that AI applications consistently enhance the educational experience without detracting from it. This commitment includes ongoing professional development to stay informed on the evolving challenges and technological progress in AI's role in education.

An ethical approach to integrating AI in education further includes considerations of accessibility and equity. It is imperative to guarantee that AI's incorporation into the classroom does not deepen existing educational inequalities. Effective strategies must be established to ensure equitable access to AI technologies, aiming to reduce the achievement gap and prevent the expansion of the digital divide. This inclusive approach guarantees that students from diverse backgrounds can equally benefit from the innovations in educational technology [53]. Moreover, integrating a human-in-the-loop system is crucial to complement AI's capabilities [54, 55]. This provides a mechanism for human oversight that ensures AI-driven decisions are appropriate and ethical. This human presence facilitates more accurate and context-aware responses and introduces an essential layer of accountability and transparency in AI applications in education.

Artificial intelligence (AI) systems can inherit and amplify biases present in their training data or design. In an educational context, this could lead to unfair treatment of students based on race, gender, socioeconomic status, or other factors.

Incorporating mechanisms for human feedback is vital for identifying and eliminating biases and inaccuracies within AI systems. Educators and users should be empowered to report and correct biased information and errors, thereby improving the reliability and fairness of the system. This feedback loop is essential for cultivating an AI ecosystem that is both reflective and responsive to the diverse needs and values of its user base, promoting an educational environment that is equitable, inclusive, and grounded in ethical principles.

5.3 Environmental costs

Training large-scale generative AI models' computational and environmental costs are considerable [56]. These models require extensive computational resources, leading to significant energy consumption and carbon dioxide (CO_2) emissions. These models have demonstrated remarkable performance in various natural language processing tasks. Still, their increasing size has posed challenges for deployment and raised concerns about their environmental and economic impact due to high energy consumption. As the field progresses towards more complex and capable models, managing these costs and their environmental impacts is crucial. In the educational sector, this presents an opportunity to integrate discussions about sustainable AI practices into the curriculum, fostering a generation of students who are not only technologically proficient but also environmentally conscious.

Fortunately, scientists have achieved a breakthrough by proposing small language models (SLMs) and recent 1-bit LLMs [57–60]. An SLM refers to a version of a language-processing AI with fewer parameters than larger models. In machine learning and AI, parameters are the parts of the model learned from the training data and determine the model's behavior. Smaller language models are designed to be lighter and more efficient, making them easier to deploy on devices with limited computational resources, such as mobile phones or embedded systems. While they may not achieve the same level of performance or accuracy as larger models on complex language tasks, small language models can still understand and generate human-like text, perform classification tasks, and more.

6. Conclusions

This chapter delves into the technological underpinnings, diverse applications, challenges, and ethical considerations of integrating generative artificial intelligence into educational systems. It comprehensively explores how GenAI can enhance personalized learning, streamline administrative tasks, and offer innovative teaching and assessment tools while highlighting the need for a critical approach to ensure accuracy and equitable access. The chapter emphasizes the transformative potential of GenAI in education, from improving student engagement and learning outcomes to fostering professional development and addressing the digital divide.

In particular, GenAI holds great promise for education through personalized learning, automated content generation, virtual tutoring, language learning, creativity enhancement, and automated assessment. However, it also underscores the importance of addressing the imperfection and environmental impact of GenAI technologies and the necessity of continuous evaluation and adaptation to leverage GenAI responsibly and effectively in educational contexts. Effective implementation requires seamless integration of GenAI into existing curricula, training of teachers, addressing ethical concerns, design with a focus on student needs, and continuous evaluation of the impact on educational outcomes.

Additionally, it addresses ethical concerns and the necessity for robust policy frameworks to ensure equitable access and prevent misuse of AI technologies in education. The authors emphasize the aim of guiding educators and policymakers in making informed decisions to leverage GenAI to improve educational outcomes and foster a more inclusive, adaptive, and future-ready learning environment.

Author details

Sheikh Faisal Rashid[1*], Nghia Duong-Trung[1] and Niels Pinkwart[1,2]

1 Educational Technology Lab, German Research Center for Artificial Intelligence (DFKI), Alt-Moabit, Berlin, Germany

2 Humboldt Universität zu Berlin, Unter den Linden, Berlin, Germany

*Address all correspondence to: faisal.rashid@dfki.de

References

[1] Farrelly T, Baker N. Generative artificial intelligence: Implications and considerations for higher education practice. Education Sciences. 2023;**13**(11):1-14. [Online]. Available from: https://www.mdpi.com/2227-7102/13/11/1109

[2] Lo CK. What is the impact of ChatGPT on education? A rapid review of the literature. Education Sciences. 2023;**13**(4):1-15. [Online]. Available from: https://www.mdpi.com/2227-7102/13/4/410

[3] Pinkwart N, Paaßen B, Burchardt A. Chancen, Potenziale und Grenzen von ChatGPT in der Bildung – Stellungnahme des DFKI Labor Berlin. 2023. Available from: https://www.dfki.de/web/news/chancen-potenziale-und-grenzen-von-chatgpt-in-der-bildung-stellungnahme-des-dfki-labor-berlin [Accessed: Feb 01, 2024]

[4] Department of Education UK. Policy paper - Generative artificial intelligence (AI) in education. 2023. Available from: https://www.gov.uk/government/publications/generative-artificial-intelligence-in-education/generative-artificial-intelligence-ai-in-education#understanding-generative-ai [Accessed: Feb 10, 2024]

[5] Cornell University. CU committee report: Generative artificial intelligence for education and pedagogy. 2023. Available from: https://www.teaching.cornell.edu/generative-artificial-intelligence/cu-committee-report-generative-artificial-intelligence-education [Accessed: Feb 10, 2024]

[6] UNESCO. Guidance for Generative AI in Education and Research. France: United Nations Educational, Scientific and Cultural Organization; 2023

[7] Cao Y, Li S, Liu Y, Yan Z, Dai Y, Yu PS, et al. A comprehensive survey of AI-generated content (AIGC): A history of generative AI from GAN to ChatGPT. arXiv preprint arXiv:2303.04226. 2023

[8] Wu J, Gan W, Chen Z, Wan S, Yu PS. Multimodal large language models: A survey. arXiv preprint arXiv:2311.13165. 2023

[9] Zhou K, Yang J, Loy CC, Liu Z. Learning to prompt for vision-language models. International Journal of Computer Vision. 2022;**130**(9):2337-2348

[10] Sun Q, Cui Y, Zhang X, Zhang F, Yu Q, Luo Z, et al. Generative Multimodal Models Are in-Context Learners. arXiv preprint arXiv:2312.13286. [Online] 2023

[11] Team G, Anil R, Borgeaud S, Wu Y, Alayrac J-B, Yu J, et al. Gemini: A family of highly capable multimodal models. arXiv preprint arXiv:2312.11805. 2023

[12] DFKI, MMB. Schule und ki - lehren und lernen mit künstlicher intelligenz. 2023. Available from: https://www.dfki.de/web/news/telekom-stiftung-veroeffentlicht-leitfaden-schule-und-ki [Accessed: Feb 15, 2024]

[13] Hübsch T, Vogel-Adham E, Vogt A, Wilhelm-Weidner A. Sprachgewandt in die zukunft: Large language models im dienst der beruflichen weiterbildung. In: ein beitrag der digitalbegleitung im rahmen des innovationswettbewerbs invite. Berlin: VDI/VDE Innovation + Technik GmbH; 2024. p. 46 S

[14] Kasneci E, Sessler K, Küchemann S, Bannert M, Dementieva D, Fischer F, et al. ChatGPT for good? On

opportunities and challenges of large language models for education. Learning and Individual Differences. 2023;**103**:1-12

[15] Holmes W, Bialik M, Fadel C. Artificial Intelligence in Education: Promises and Implications for Teaching Learning. Boston, MA: The Center for Curriculum Redesign; 2019

[16] Han A, Cai Z. Design implications of generative AI systems for visual storytelling for young learners. In: IDC '23: Proceedings of the 22nd Annual ACM Interaction Design and Children Conference. New York, NY, United States: Association for Computing Machinery; Jun 2023. pp. 470-474. DOI: 10.1145/3585088.3593867

[17] Jauhiainen JS, Guerra AG. Generative AI and ChatGPT in school Children's education: Evidence from a school lesson. Sustainability. 2023;**15**(18):1-22. [Online]. Available from: https://www.mdpi.com/2071-1050/15/18/14025

[18] Cao C, Ding Z, Lee G-G, Jiao J, Lin J, Zhai X. Elucidating stem concepts through generative AI: A multi-modal exploration of analogical reasoning. arxiv, vol. abs/2308.10454. 2023

[19] Castillo-Segura P, Alario-Hoyos C, Kloos CD, Fernández Panadero C. Leveraging the Potential of Generative AI to Accelerate Systematic Litera-ture Reviews: An Example in the Area of Educational Technology, 2023 World Engineering Education Forum - Global Engineering Deans Council (WEEF-GEDC). Monterrey, Mexico: IEEE; 2023. pp. 1-8. DOI: 10.1109/WEEF-GEDC59520.2023.10344098

[20] Jonsson M, Tholander J. Cracking the code: Co-coding with AI in creative programming education. In: Proceedings of the 14th Conference on Creativity and Cognition. New York, NY, United States: Association for Computing Machinery; Jun 2022. pp. 5-14. DOI: 10.1145/3527927.3532801

[21] Future of Jobs Report 2023. Up to a quarter of jobs expected to change in next five years. 2023. Available from: https://www.weforum.org/press/2023/04/future-of-jobs-report-2023-up-to-a-quarter-of-jobs-expected-to-change-in-next-five-years [Accessed: Feb 10, 2024]

[22] Wright JV. A new era for AI and google workspace. 2023. [Online]. Available from: https://www.workspace.google.com/blog/product-announcements/generative-ai [Accessed: Feb 10, 2024]

[23] Spataro J. Introducing microsoft 365 copilot – your copilot for work. In: Proceedings of the XXXVII Brazilian Symposium on Software Engineering. New York, NY, United States: Association for Computing Machinery; Sep 2023. DOI: 10.1145/3613372.3614197. Available from: https://www.blogs.microsoft.com/blog/2023/03/16/introducing-microsoft-365-copilot-your-copilot-for-work/ [Accessed: Feb 10, 2024]

[24] Github Copilot. Available from: https://www.github.com/features/copilot [Accessed: Feb 15, 2024]

[25] Zhao WX, Zhou K, Li J, Tang T, Wang X, Hou Y, et al. A survey of large language models. arxiv. vol. abs/2303.18223. 2023

[26] Zhang P, Boulos MNK. Generative AI in Medicine and Healthcare: Promises, Opportunities and Challenges Future Internet. 2023;**15**(9):286. pp. 1-15. DOI: 10.3390/fi15090286

[27] Eysenbach G. The role of ChatGPT, generative language models, and artificial intelligence in medical

education: A conversation with ChatGPT and a call for papers. JMIR Medical Education. 2023;**9**:1-13

[28] Pavlik J. Collaborating with ChatGPT: Considering the implications of generative artificial intelligence for journalism and media education. Journalism Mass Communication Educator. 2023;**78**:84-93

[29] Buyser BD. Goblin Tools. 2024. [Retrieved 12.02.2024]. [Online]. Available from: https://goblin.tools

[30] Kinney EM, Vedora J, Stromer R. Computer-presented video models to teach generative spelling to a child with an autism spectrum disorder. Journal of Positive Behavior Interventions. 2003;**5**:22-29

[31] Engel C, Schmalfuß-Schwarz J, Gollasch D, Branig M, Dirks S, Weber G. Workshop on designing accessible extended reality: An opportunity for people with disabilities and disorders. In: Mensch Und Computer 2023 – Workshopband. New York, NY, United States: Association for Computing Machinery; 2023

[32] KiwiTech. Applications of generative AI in augmented and virtual reality. 2023. Available from: https://www.medium.com/@KiwiTech/applications-of-generative-ai-in-augmented-and-virtual-reality-20cecec50886 [Accessed: Feb 10, 2024]

[33] Gollasch D, Branig M, Gerling K, Gulliksen J, Metatla O, Spiel K, et al. Designing technology for neurodivergent self-determination: Challenges and opportunities. In: Nocera JA, Lárusdóttir MK, Petrie H, Piccinno A, Winckler M, editors. Human-Computer Interaction – Interact 2023. Cham: Springer Nature Switzerland; 2023. pp. 621-626

[34] Gadiraju V, Kane SK, Dev S, Taylor AS, Wang D, Denton E, et al. "I wouldn't say offensive but...": Disability-centered perspectives on large language models. In: Proceedings of the 2023 ACM Conference on Fairness, Accountability, and Transparency. New York, NY, United States: Association for Computing Machinery; Jun 2023. pp. 205-216

[35] Atlas S. ChatGPT for Higher Education and Professional Development: A Guide to Conversational AI. 2023. Available from: https://www.digitalcommons.uri.edu/cba_facpubs/548 [Accessed: Feb 15, 2024]

[36] Abd-alrazaq AA, AlSaad R, Alhuwail D, Ahmed A, Healy P, Latifi S, et al. Large language models in medical education: Opportunities, challenges, and future directions. JMIR Medical Education. 2023;**9**

[37] Rüdian S, Pinkwart N. Auto-generated language learning online courses using generative AI models like ChatGPT. In: 21. Fachtagung Bildungstechnologien (DELFI). Bonn: Gesellschaft für Informatik e.V; 2023. pp. 65-76

[38] Phung T, Pădurean V-A, Cambronero J, Gulwani S, Kohn T, Majumdar R, et al. Generative AI for programming education: Benchmarking ChatGPT, GPT-4, and human tutors. In: Proceedings of the 2023 ACM Conference on International Computing Education Research. Vol. 2. New York, NY, USA: Association for Computing Machinery; 2023

[39] Pinto G, Cardoso-Pereira I, Monteiro D, Lucena D, Souza A, Gama K. Large language models for education: Grading open-ended questions using ChatGPT. In: SBES '23: Proceedings of the XXXVII Brazilian Symposium on Software Engineering. New York, NY,

United States: Association for Computing Machinery; 2023. pp. 293-302. DOI: 10.1145/3613372.3614197

[40] Lagakis P, Demetriadis S, Psathas G. Automated Grading in Coding Exercises Using Large Language Models. In: Auer ME, Tsiatsos T, editors. Smart Mobile Communication & Artificial Intelligence. IMCL 2023. Lecture Notes in Networks and Systems. Vol. 936. Cham: Springer; 2023. pp. 363-373. DOI: 10.1007/978-3-031-54327-2_37

[41] Agostini D, Picasso F. Large language models for sustainable assessment and feedback in higher education: Towards a pedagogical and technological framework. In: Proceedings of the First International Workshop on High-Performance Artificial Intelligence Systems in Education Co-Located with 22nd International Conference of the Italian Association for Artificial Intelligence (AIxIA 2023). CEUR Workshop Proceedings (CEUR-WS.org). 2023

[42] Rashid SF, Goertz L, Reichow I. Metadata for learning processes – Results of an international interview study. Preprints. 2024. [Online]. DOI: 10.20944/preprints202402.0889.v1

[43] Schick T, Dwivedi-Yu J, Jiang Z, Petroni F, Lewis P, Izacard G, et al. Peer: A collaborative language model. arxiv. vol. abs/2208.11663. 2022

[44] Papachristou M, Yang L, Hsu C-C. Leveraging large language models for collective decision-making. arxiv. vol. abs/2311.04928. 2023

[45] Kaplan-Rakowski R, Grotewold K, Hartwick P, Papin K. Generative AI and teachers' perspectives on its implementation in education. Journal of Interactive Learning Research. 2023;**34**(2):313-338

[46] Barrett A, Pack A. Not quite eye to AI: Student and teacher perspectives on the use of generative artificial intelligence in the writing process. International Journal of Educational Technology in Higher Education. 2023;**20**(1):59

[47] Chan CKY, Hu W. Students' voices on generative AI: Perceptions, benefits, and challenges in higher education. arXiv preprint arXiv:2305.00290. 2023

[48] Lee AVY, Tan SC, Teo CL. Designs and practices using generative AI for sustainable student discourse and knowledge creation. Smart Learning Environments. 2023;**10**(1):59

[49] Rawte V, Sheth A, Das A. A survey of hallucination in large foundation models. arXiv preprint arXiv:2309.05922. 2023

[50] Ji Z, Lee N, Frieske R, Yu T, Su D, Xu Y, et al. Survey of hallucination in natural language generation. ACM Computing Surveys. 2023;**55**(12):1-38

[51] Sharples M. Towards social generative AI for education: Theory, practices and ethics. Learning: Research and Practice. 2023;**9**(2):159-167

[52] Head CB, Jasper P, McConnachie M, Raftree L, Higdon G. Large language model applications for evaluation: Opportunities and ethical implications. New Directions for Evaluation. 2023;**2023**(178-179):33-46

[53] Holmes W, Porayska-Pomsta K. The Ethics of Artificial Intelligence in Education: Practices, Challenges, and Debates. New York: Taylor & Francis; 2022. Available from: https://www.taylorfrancis.com/books/edit/10.4324/9780429329067/ethics-artificial-intelligence-education-wayne-holmes-ka%C5%9Bka-porayska-pomsta

[54] Mosqueira-Rey E, Hernández-Pereira E, Alonso-Ríos D, Bobes-Bascarán J, Fernández-Leal Á. Human-in-the-loop machine learning: A state of the art. Artificial Intelligence Review. 2023;**56**(4):3005-3054

[55] Amirizaniani M, Yao J, Lavergne A, Okada ES, Chadha A, Roosta T, et al. Developing a framework for auditing large language models using human-in-the-loop. arXiv preprint arXiv:2402.09346. 2024

[56] Faiz A, Kaneda S, Wang R, Osi R, Sharma P, Chen F, et al. LLMcarbon: Modeling the end-to-end carbon footprint of large language models. arXiv preprint arXiv:2309.14393. 2023:1-30

[57] Schick T, Schütze H. It's not just size that matters: Small language models are also few-shot learners. arXiv preprint arXiv:2009.07118. 2020

[58] Kang M, Lee S, Baek J, Kawaguchi K, Hwang SJ. Knowledge-augmented reasoning distillation for small language models in knowledge-intensive tasks. Advances in Neural Information Processing Systems. 2024;**36**

[59] Mitra A, Del Corro L, Mahajan S, Codas A, Simoes C, Agarwal S, et al. Orca 2: Teaching small language models how to reason. arXiv preprint arXiv:2311.11045. 2023

[60] Ma S, Wang H, Ma L, Wang L, Wang W, Huang S, et al. The era of 1-bit LLMS: All large language models are in 1.58 bits. arXiv preprint arXiv:2402.17764. 2024

Chapter 3

Artificial Intelligence for People with Special Educational Needs

Esmaeil Zaraii Zavaraki

Abstract

Artificial intelligence have very high capacities and capabilities in special education and have been strengthened and upgraded compared to the past. Especially in the field of education, teaching and learning, remarkable progress has happened and is happening. There are examples of artificial intelligence based on assistive technologies that can help people with special educational needs access and interact with educational content. In this chapter of the book, the opportunities and capabilities of artificial intelligence for people with special educational needs, particularly in teaching and learning process, have been introduced and analyzed. The types of assistive technologies based on artificial intelligence for disabled people have been explained. Also, in this chapter of the book, considering the importance of approaches and educational design models in the teaching and learning process of people with special educational and learning needs, the blended learning approach and model in special education have been introduced and its basic components and sub-components have been explained. At the end of the chapter of the book, case studies that have been done by author or by author's supervision in this field have been mentioned.

Keywords: special education, educational technology, assistive technologies, artificial intelligence, special teaching and learning

1. Introduction

Perhaps the following classification of types of education and training can be provided:

- General or regular Education
- Special Education
- Organizational Education/Training
- Adult Education
- Lifelong learning and continuous professional development

What is meant from the types of education and training mentioned in this chapter of the book is special education. Special education is the study and practice of educating students in a way that adapts their individual differences, disabilities, and special needs. In other word, Special education is the study and ethical application of best practices to improve learning and performance of people with special educational and learning needs through the special strategic of analysis, designing, developing, production, implementation, management, support, assessment and evaluation of learning and instruction processes and resources [1]. Special education aim is to provide adapted education for people with special educational needs such as mental retardation, visual impairment, hearing impairment, physical-motor impairment, learning difficulties, emotional-behavioral impairment, speech-communication impairment, special diseases, multi handicapped and talented and gifted [2]. Some scholars of education may categorize gifted education under the umbrella of "special education". Although there are still differences of opinion in this field. People with special educational needs include a wide variety of people with different cognitive, physical, emotional and behavioral learning needs.

New learning approaches such as blended learning approach and new technologies such as artificial intelligence have changed the fundamental nature of teaching - learning process particularly in special education areas. They have made possible learning activities that were not previously practical or feasible. A blended learning environment provides opportunities for the students with special educational needs to interact with their teachers and content. Various researchers have noted that there are limited studies investigating learning through blended learning approach among students with special educational needs [1, 3–12]. According to Allahi [13], Jangizehi [14], Mathews [15], Stamer [16], Toofaninejad [17], Moradi [18], Zaraii Zavaraki & Schneider [11], Zaraii Zavaraki [1], and Linda [12], blended learning approach and new technologies have some potential for special education teachers and their students.

2. Artificial intelligence in education

Artificial intelligence is the theory and development of computer systems capable of performing tasks that historically required human intelligence, such as recognizing speech, making decisions, and identifying patterns.

Some of Artificial intelligence applications are:

- Advanced web search engines
- Recommendation systems
- Understanding human speech
- Generative or creative tools
- Competing at the highest level in strategic games

The traits described below have received the most attention and cover the scope of AI research:

Reasoning, problem-solving, Knowledge representation, Planning and decision making, Learning, Natural language processing, Perception, Robotics, Social intelligence, General intelligence.

Artificial Intelligence technology is used in most of the essential applications, including:

- Search engines
- Targeting online advertisements, recommendation systems
- Driving internet traffic, targeted advertising
- Virtual assistants
- Autonomous vehicles
- Automatic language translation
- Facial recognition
- Image labeling

The rapid development of Artificial Intelligence is having a major impact on all of areas of educational systems. Advances in AI-powered solutions carry enormous potential for the achievement of the sustainable development goals. Artificial Intelligence has the potential to address some of the biggest challenges in education today, innovate teaching and learning practices, and ultimately accelerate the progress towards sustainable development goal 4, quality education. Sustainable Development Goal 4 aims at ensuring inclusive and equitable quality education and promote lifelong learning opportunities for all [19]. Over the past decade, the use of AI tools to support or enhance learning has grown exponentially [20]. This has only increased following the COVID-19 school closures. However, evidence remains scarce on how AI can improve learning outcomes and whether it can help learning scientists and practitioners to better understand how effective learning happens [21]. AI applications designed for education have been divided into three main categories: system-facing, student-facing and teacher-facing [19]. However, for policy-makers, UNESCO propose a set of four needs-based categories of emerging and potential applications:

- Education management and delivery
- Learning and assessment
- Empowering teachers and enhancing teaching
- Lifelong learning [19].

Educational chatbots, OU Analyze, Swift, The ALP, The UniTime, Intelligent tutoring system, Dialog-based tutoring systems, Exploratory learning environments, Automated writing evaluation, AI-supported reading and language learning, Smart robots, Teachable agents, Educational virtual and augmented reality, Learning network orchestrators, AI-enabled collaborative learning, AI-driven discussion forum monitoring, AI-human 'dual teacher' model, AI-powered teaching assistants, AI-driven lifelong learning companion, AI-enabled continuous assessment, AI-enabled record of lifelong learning achievements, and opportunities for AI to advance inclusion and equity in education are some of applications of AI in education [19].

2.1 Artificial intelligence in special education

Artificial intelligence and its applications in the process of education, teaching and learning of people with special educational needs have been proposed at the global level under the two keywords of inclusive education and assistive technologies. In the continuation of this chapter, the inclusive education and then the assistive technologies have been introduced and explained.

2.1.1 Opportunities for artificial intelligence to advance inclusion and equity in education

In addition to focusing on equitable access to Artificial Intelligence technologies for all, we also need to consider the potential of Artificial Intelligence to help achieve SDG 4, to help 'ensure inclusive and equitable quality education and promote lifelong learning opportunities for all' [22].

In this challenging context, many Artificial Intelligence technologies might be used, or further developed, to help improve education – especially for older people, refugees, marginalized or isolated communities, and people with special educational and learning needs. To begin with, the UNESCO's ROAM framework ('Rights, Openness, Access and Multi-stakeholder Governance') should be applied, to ensure that the application of Artificial Intelligence in education addresses broader human rights and emerging ethical issues in a holistic manner [23].

For example, and in particular, Artificial Intelligence in education should be made accessible to all citizens, especially for vulnerable groups such as students with learning disabilities), without exacerbating existing inequalities.

Some of examples of Artificial intelligence applications for inclusion and equity in education are:

- The Global Digital Library.
- Dytective, an AI-powered screening tools.
- AI-powered artificial voices.
- AI and augmented reality applications.
- AI-enabled 'smart' robots,
- Telepresence robots.
- AI-powered intelligent tutoring systems.

2.1.2 Assistive technologies based on artificial intelligence

Assistive technologies are tools, devices and services that help people with special educational needs to achieve better functioning and independence at home, school and community. Assistive technologies have high capacities due to the potentials of artificial intelligence and have been strengthened and upgraded compared to the past. Especially in the field of special education, teaching and learning, remarkable progress has happened and is happening.

There are examples of AI-based assistive technology that can help people with special educational needs access and interact with educational content. Some of these examples are:

- Speech recognition software: This technology uses artificial intelligence to convert spoken words into written text. This can be especially helpful for students with physical–motor impairment who may have difficulty typing or writing by hand.

- Text-to-speech software: This technology uses artificial intelligence to convert written text into spoken words, making digital content easier to access and understand for students with visual impairments or learning disabilities.

- Augmentative and alternative communication devices: These devices use artificial intelligence to help students with communication disorders to express themselves. They may include features such as speech synthesis, predictive text, and eye tracking to help students communicate more effectively.

- Virtual assistants: Artificial intelligence-based virtual assistants such as Google Assistant can be used to help students with special educational needs access information, complete tasks, and interact with digital devices using voice commands. Chat GPT as an assistant is another example that has been in use in recent months.

- Adaptive learning software: This technology uses artificial intelligence to personalize learning experiences for students, adapting content and activities to their unique needs and abilities. This can be particularly useful for students with learning disabilities or other special needs who may need different types of support to learn effectively.

- Computer vision technology: This technology uses artificial intelligence to interpret the visual input from the camera. It can be used to help visually impaired students navigate their surroundings, recognize objects, and read printed materials. For example, a computer vision system can be used to read text aloud to a visually impaired student or to identify and describe objects in the student's environment. The Horus system is an example of this technology. The Horus system has integrated capabilities such as routing and independent movement during game activities, being aware of the location of friends in class and school, text recognition and reading, face recognition and object recognition. This technology has the ability to navigate and identify obstacles in 3D, which helps the safe movement of a visually impaired person, and it also has the ability to read written text and Braille, which can help a person independently in their education and learning activities.

- Predictive analytics: This technology uses artificial intelligence to analyze data and predict future outcomes. In education, it can be used to identify students who are at risk of falling behind or dropping out and provide targeted interventions to help them succeed. For students with special needs, predictive analytics can help educators and caregivers identify patterns and trends in their behavior and performance and develop more effective strategies to support them.

- Brain-computer interfaces: These devices use artificial intelligence to convert brain signals into computer commands, allowing users to interact with digital devices using their thoughts. They can be especially useful for students with physical –motor impairment who may not be able to use traditional input devices such as keyboards. For example, a brain-computer interface might allow a student with cerebral palsy to control a computer using their brain signals.

- Social robots: These robots use artificial intelligence to interact with humans in social and emotional ways, providing companionship and support. They can be especially helpful for students with autism spectrum disorder, who may have difficulty with social interactions and communication. Social robots can be programmed to respond to facial expressions, body language and other social cues, helping students with autism spectrum disorder develop social skills and build relationships.

- Virtual Reality (VR), Augmented Reality (AR), Mixed Reality (MR) and Extended Reality (ER): These technologies use artificial intelligence to create immersive and interactive digital environments. They can be used to provide students with special educational needs hands-on learning experiences that may not be possible in the physical world. For example, a virtual reality simulation can allow a student with physical –motor impairment to explore a historical site or laboratory, while an augmented reality application can provide visual and auditory feedback to a student with autism spectrum disorder educate during a session social skills.

Overall, AI-based assistive technology has the capacity to change the way students with special needs are supported. Using the power of machine learning and other artificial intelligence techniques, we can create more effective and personalized tools and interventions to help these students learn, communicate, and thrive.

Elon Musk's company has obtained the license to implant a chip in the human brain. Neuralink announced that it has received approval from the US Food and Drug Administration to conduct the first human trials of its chip. The company says it wants to help restore sight and mobility to the disabled by connecting the brain to a computer.

One of the important missions of educational technologists and learning specialists is to identify emerging technologies and introduce their strengths, weaknesses, opportunities and threats in the process of education, teaching and learning. In the meantime, it is very important to discover the platforms and how to use these technologies intelligently and effectively to improve education and learning. One of the emerging technologies that has emerged in recent years is Metaverse technology, which may be called Farajehan technology in Persian. It seems that this technology has considerable capacities and platforms in improving the teaching and learning process. Whether it is normal education, special education, organizational internship, adult education, continuous learning, lifelong learning, and professional development, which requires extensive research. Metaverse is a multi-user environment that integrates physical and virtual reality. Virtual reality, augmented reality, mixed reality, extended reality, recording of daily activities of a person using a digital device1 and mirror world can be considered among the Metaverse technologies that have wide applications in the process of education, teaching and learning, especially for people with special educational needs. Computer, network, game console, headset and virtual reality glasses, mixed and augmented, virtual reality gloves are some of the necessary tools to use this technology. Some of the advantages that Metaverse technology has in the teaching and learning process are: establishing educational justice, experiencing comprehensive interaction, visualization, increasing participation, reducing costs, continuous improvement, unlimited time and space, fast sharing, personalization and strengthen communication. It seems that we need more researches at the national and international level to discover more of this emerging technology [24].

2.2 Blended learning approach and models

In the process of education, teaching and learning, we need pedagogy, andragogy and heutagogy to take advantage of the capacities of any type of technology, including artificial intelligence technology. Therefore, choosing the right approach and models in this field is very necessary and vital. Without having a suitable approach and model, we will not be able to make the most of the capacities of technology, especially new and emerging technologies in education, especially in the education of people with special educational needs. Therefore, in the rest of this chapter, we will discuss the blended learning approach and its conceptual and procedural models.

Blended learning refers to the systematic integration of online and face-to-face engagement to support and enhance meaningful interaction between students, teachers and resources [25]. Blended learning gives students with special educational needs the benefits of both online learning and in-person instruction. According to Christensen, Horn, and Staker ([26], p. 9), blended learning is defined as: "a formal education program in which a student learns at least in part through online learning with some element of student control over time, place, path, and/or pace and at least in part at a supervised brick-and-mortar location away from home. The modalities along each student's learning path within a course or subject are connected to provide an integrated learning experience." Students in formal blended learning educational programs learn online part of the time, yet have the benefit of face-to-face instruction and supervision to maximize their learning and to best fit their own needs ([27], p. 5).

Blended learning may also allow teachers to spend less time giving whole-class lessons, and more time meeting and interacting with students with special educational needs individually or in small groups to help them with specific concepts, skills, questions, or learning problems (**Figure 1**, [1]). In blended-learning situations, students with special educational needs are required to use digital and online technologies and they naturally acquire more technological literacy and greater confidence using new technologies, which is very essential in twenty-first century life and in future [11].

The author of the chapter, made extensive efforts to design and develop the educational model based on blended learning approach for students with special

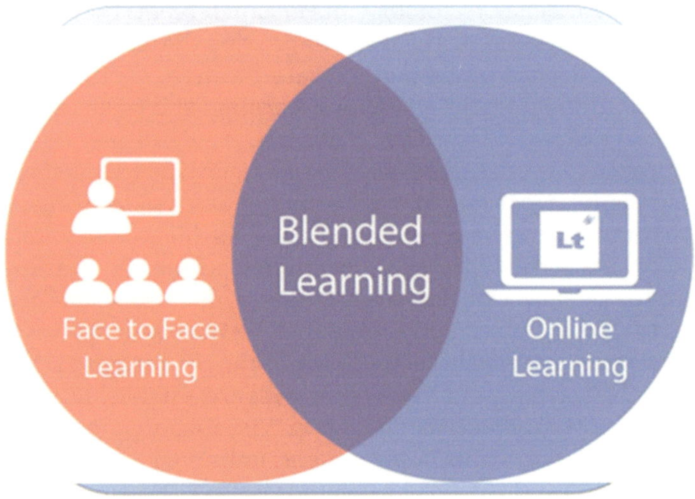

Figure 1.
The conceptual learning model based on blended learning approach for students with special educational needs [11, 28].

educational needs. According to **Table 1**, 6 main components and 49 sub-components were found. These components underneath the umbrella of two other infrastructure components were considered as face-to-face learning and computer and network-based learning. First, a conceptual model that included all of the main components was designed and then the procedural model was designed and developed by the investigator. The conceptual model and the procedural model showed in the **Figures 2** and **3**.

According to the **Figure 2**, 6 main components of the educational model based on blended learning approach for students with special educational needs are: analysis, designing, production, implementation, management and support, assessment and evaluation and revision and modification were obtained. These six main components underneath the umbrella of two other infrastructure components were considered as face-to-face learning and computer and network-based learning.

According to the **Figure 3**, 49 sub-components of the educational model based on blended learning approach for students with special educational needs are:

goal, learner, teacher, the content, media, message, environment and learning space, educational resources and learning resources, technology, the context, the combination of learning theories, purpose, environment and learning space, methods, strategies, media, message, technology, principles of universal design, presence, learning activities, measurement techniques, technology, media, training, material, guides, enrichment, interaction, participation, access, engagement, facilitate, coordination, presence, cooperation, learner support, teacher support, course support, portfolios, self-assessment, peer-assessment, teacher-assessment, intelligent assessment, diagnostic evaluation, formative evaluation, summative evaluation, fallow up evaluation, evaluation of learners' interaction.

As showed in **Table 1** and **Figure 3**, each of the sub-component including several minor sub-components. For example, the sub-component of learner including minor sub-components of mental retardation, visual impairment, hearing impairment, physical-motor impairment, learning difficulties, emotional-behavioral impairment,

Row	Components	Sub-components
1	Analysis	Goal, Learner (mental retardation, visual impairment, hearing impairment, physical-motor impairment, learning difficulties, emotional-behavioral impairment, speech-communication impairment, special diseases, talent and gift), Teacher, The content, Media (interactive, non-interactive), Message, Environment and learning space, Educational resources and learning resources, Technology (print, visual, audio, audio-visual, computer-based, network-based, technology integration), The Context (social, cultural, economic, political).
2	Designing	The combination of learning theories (Behaviorism, Cognitivism, Constructivism, and Connectivism), Purpose (general, partial and behavioral), Environment and learning space, Methods (based on class, computer-based, multimedia-based, social media -based, web-based), Strategies (educational, learning), Media (interactive, non- interactive), Message, Technology (print, visual, audio, audio-visual, computer-based, network-based, technology integration), Principles of universal design (fair use, flexibility of learning activities in use, simple and direct, comprehensible information, Tolerance against errors, Low physical effort, Space and size for use, community of learners, educational atmosphere), Presence (cognitive, teaching, social, emotional), Learning activities (based on information and communications technology, without information and communications technology, based on artistic, individual and group activities), Measurement techniques (self-assessment, peer evaluation, teacher assessment, portfolio, intelligent evaluation).
3	Production	Technology (print, visual, audio, audio-visual, computer-based, network-based, gaming, simulation, augmented reality, virtual reality, technology integration), Media (interactive, non- interactive), Content (textbook, tuition, announcement, newsletter), Training, material, Guides (Teacher's guide, Learner's guide, Parent's guide, Tutorial assistant guide).
4	Implementation, Management and Support	Enrichment, Interaction (learner - teacher, learner-learner, learner-content) Participation (learner, teacher, parent, assistant, manager, therapist, counselor, social worker), Access (limited to the classroom, limited to school, unlimited, throughout the day), Engagement (question and answer, discussion, activity, observe, report presentation), Facilitate, Coordination, Presence (cognitive, teaching, social, emotional), Cooperation, Learner support (administrative, service, advisory, medical, therapeutic, educational, technical), Teacher support (educational, administrative, service, professional, specialized), Course support (educational, technical).
5	Assessment and Evaluation	Portfolios (traditional, electronic), Self-assessment, Peer-assessment, Teacher-assessment, Intelligent assessment, Diagnostic evaluation, Formative evaluation, Summative evaluation, Fallow up evaluation, Evaluation of learners' interaction (in the learning environment, in the learning Space. (
6	Revision and Modification	Revision and Modification of the teaching – learning process.

Table 1.
Component and sub-components of educational model based on blended learning approach for students with special educational needs [1].

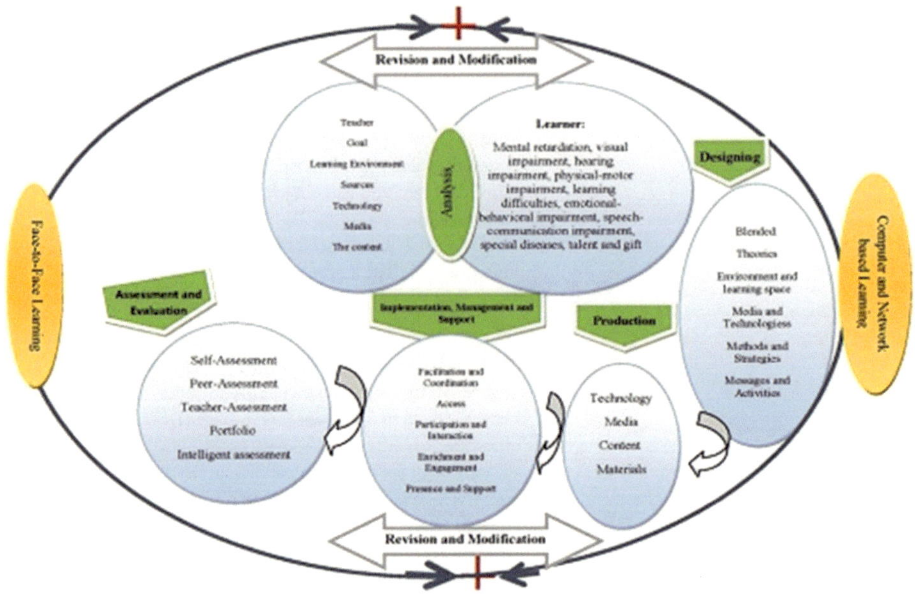

Figure 2.
The conceptual learning model based on blended learning approach for students with special educational needs [1, 28].

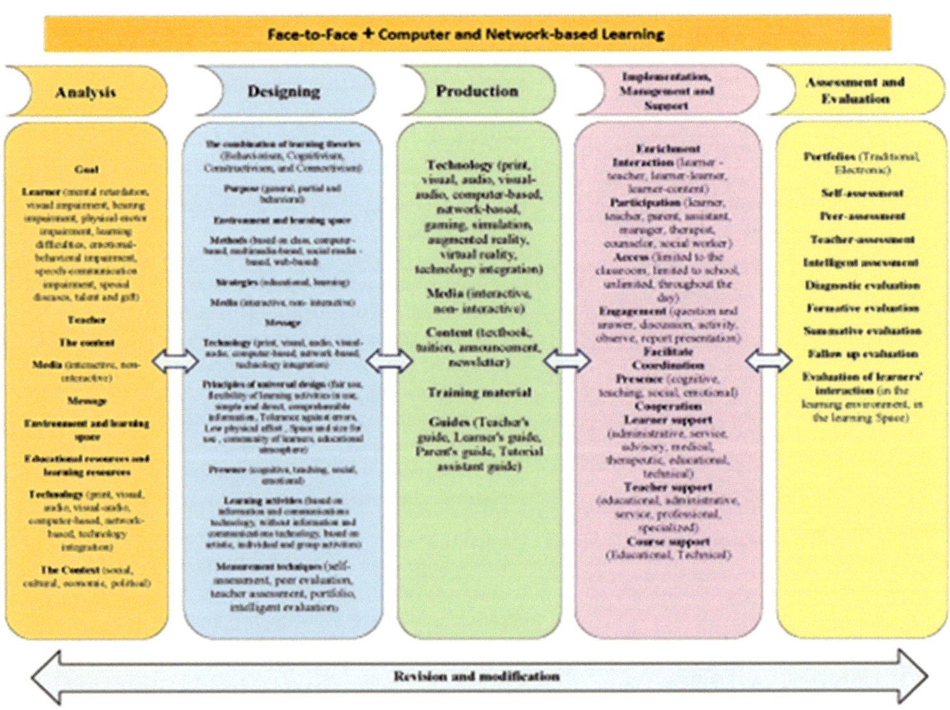

Figure 3.
The procedural learning model based on blended learning approach for students with special educational needs [1].

speech-communication impairment, special diseases, talent and gift, and the sub-component of learner support including minor sub-components of administrative, service, advisory, medical, therapeutic, educational and technical support [1].

2.3 Case studies

To check the validity of the blended learning approach and its models in special education, several researches were designed and implemented by the author of the current chapter of the book or with his supervisor in the last years. In the rest of this chapter of the book, some of these researches are mentioned.

Allahi [13] conducted a research entitled "design and validation of a mobile phone learning pattern for visually impaired students". The aim of his study was to design an educational model of using a mobile phone and its effects on academic achievement motivation, learning and participation ratio of blind students in English courses. He developed an educational conceptual model of using the mobile phone for blind people with 4 components of human, training, technology and support factors as well as a procedural model with 4 stages of preparation, design, implementation and evaluation.

Toofaninejad [17] conducted a research entitled "designing instructional pattern of the learning environment enriched by virtual social network and its impact on learning rate and the social skills of the students with hearing impairment in science course". He discovered 7 final categories including engagement, interaction, feedback, content, sources, evaluation, and support. He also were extracted 24 sub categories. Based upon discovered categories, a conceptual model was designed that is encompassed all mentioned categories. Next, procedural model, which it is a practical model, was designed and developed. The investigator also showed that the instructional pattern of the learning environment enriched by the virtual social network by 99% of confidence interval has a positive effect on learning and social skills of DHH students.

Zareei [29] conducted a research entitled "design and validation of virtual social networks of Iran's schools". He discovered four main components of the model including network learning, network facilitation, network management, network technology, and the following sixteen sub-components: information valuation, content co-creation and reproduction, interaction, self-assessment and peer-assessment, excitement, learning support, process evaluation, empowerment, learning analytic, network information, network content, network interaction, network evaluation.

Moradi [30] conducted a research entitled "design and validate an educational model based on assistive technology and its effect on the amount of motivation for academic achievement, learning and academic satisfaction of students with physical-motor impairment in English language lesson". He discovered five important components of analysis, design, production, implementation, support and evaluation. Also, 34 sub-components were extracted for the main components. The researcher designed and developed a conceptual model and the procedural pattern. Also, the results showed that the educational model based on assistive technology for teaching English language lesson for students with physical-motor impairment is effective.

Zaraii Zavaraki and Schneider [11] conducted a research entitled "blended learning approach for students with special educational needs: a systematic review". The analysis revealed that approach, environment, learner, tools, support and evaluation are categories of extracted in blended learning approach for students with special educational needs. The elements of each category are introduced and explained in the article. However besides presenting the evidence found in literature, our analysis highlights that researching the effect of blended learning approach on special educational needs students remains an under-explored area of study. Further well-designed research into the use of blended learning approach in special education is therefore needed.

Zaraii Zavaraki [1] conducted a research entitled "designing and validating of blended learning model with emphasis on digital technologies for students with special educational needs". The purpose of this study was to design and validate an educational model based on blended learning approach for students with Special Educational Needs. A mixed method in a type of the exploratory was used. To obtain the components of the model, an inductive content analysis was performed and for internal validity of the model, descriptive survey method was used. First, a conceptual model that included all of the main components was designed and then the procedural model was designed and developed. Findings of internal validation from the viewpoint of experts showed that the conceptual and procedural models were considered to be comprehensive, suitable, applicable, enriching, appropriate, promote new educational approaches, can help in the development of educational strategies, suitable and can improve the quality of teaching and learning processes of students with special educational needs at the international level. Therefore, it is suggested to use these models as innovative strategies for students with special educational needs.

Delavaryan [31] conducted a research entitled "designing a technology-based instructional program for teaching English to mentally retarded students and its impact on academic achievement motivation, learning and retention". The collected data were analyzed in multi-variable MANCOVA method. The results of statistics analysis showed that the technology-based instruction has caused a meaningful increase in academic achievement motivation of experiment group students and also the development and improvement of their learning and retention in English Language Lesson.

Khateri [32] conducted a research entitled "the impact of augmented reality technology on academic achievement motivation and learning of second-grade dyslexic students". In this study, the sample size was 20 students divided into experimental and control groups. At first, a pre-test learning and academic achievement motivation is taken from both groups and then an Augmented Reality Program was used for the experimental group. The results show that augmented reality utilization improve learning and academic achievement motivation the students with dyslexia disorder.

Ghasemi Sameni [33] conducted a research entitled "designing and validating of instructional model in computer game environment and its effect on cooperative learning of high-functioning autism students". Once content analysis and code extraction had been completed, the components and sub-components were provided in the form of a model. A conceptual model was constructed accordingly. Then, a procedural model was developed. It was found that the difference between the experimental group and control group in cooperative learning and its components (positive interdependence, individual accountability, group processing, social skills, and interaction) was significant after weighting the pre-test scores. Therefore, it can be said that the instructional model in computer game environment had a significant effect on the cooperative learning of high-functioning autistic students.

Masnavi [34] conducted a research entitled "Design and validation of a model of mobile multimedia learning environment and its effect on learning and learning transfer of social skills of students with high-performance autism spectrum disorder".

Quantitative and qualitative research methods were used to conduct the research. The results showed that the program designed based on the model of mobile multimedia learning environment was effective on learning and learning transfer of 3 subjects and had an effect on one subject learning but on his learning transfer did not have. Using these results, it can be said that the use of the proposed mobile multimedia learning environment model was effective on learning and transfer of learning, learners with autism spectrum disorder with high performance.

Bakhtiarvand [35] conducted a research entitled "designing and validating an instructional model of picture exchange communication system based on technology and investigating its effectiveness on social and communication skills and behavioral problems in children with high-functioning autism disorder". Mixed research method was used to conduct the research. Using these results, it can be said that the use of the proposed educational model of technology-based PECS is effective on social and communication skills and reducing the behavioral problems of high-functioning autistic children.

Kabiri [36] conducted a research entitled "designing and validating of educational model for applying technology in philosophy program for children and its impact on the critical thinking of gifted students in the second period of elementary school". In order to achieve the goal of the research, a mixed method of sequential exploratory type was used. The results obtained from the data analysis related to the research hypothesis showed that after removing the pre-test effect, there was a significant difference between the average critical thinking scores of the two groups in the post-test stage. The results showed that this educational intervention had a positive impact on all the components of critical thinking. According to the results of the research, it can be said that the educational model for applying technology in the philosophy program for children is effective on the development of critical thinking of gifted students in the second period of elementary school.

Ghanat [37] conducted a research entitled "the effect of computer game on the recognition of facial emotions in students with hearing impairment in the elementary school of Arak city". In this study, the experimental group received 12 sessions of the Emotion Sorting computer game, which was designed for this group of children with the aim of improving facial emotion recognition skills and in accordance with the pattern and principles of game design. On the other hand, in the control group, it was done in the usual way. The results of this study showed that the emotion matching computer game with the aim of recognizing facial emotions in the four levels of discrimination, naming, selection and matching was effective in children with hearing impairment and improved this skill, the most impact being on the emotion matching component with 71% was. The results of this study can provide important information about the effectiveness of computer games as a tool to improve cognition and emotional understanding in children with hearing loss. This information can be used to guide the development of interventions and programs to improve the social and emotional well-being of children with hearing loss.

Abbas Zadeh Rogoshui [38] conducted a research entitled "the effect of digital educational game on the visual perception of dyslexic first grade students". The findings showed that the digital educational game has an effect on improving the subjects' visual perception and all five students were able to identify and write Farsi signs independently after participating in the project. Therefore, it can be concluded that the digital educational game is a new tool with the combination of images, sound and interactive activities in the game environment of the subject, which creates and maintains the focus of learning and the continuity of the signs compared to the traditional model, and it can be used to improve and learn better Farsi signs benefit students with dyslexia.

Zaraii Zavaraki and Alimardani [28] conducted a research entitled "the role of blended learning approach on interaction process of students with special educational needs". The purpose of this study was to determine the role of blended learning approach on interaction process of students with special educational needs. The data gathering tool was a questionnaire of 52 items that 51 questions was created according to7 point Likert scale and 1 question was created according to open-ended question. Three items of this questionnaire were related to the interaction variable and the other items included other variables such as learning enrichment, engagement, participation, facilitation, coordination, sense of presence, cooperation, support, motivation, academic achievement and learning rate. The validity of the questionnaire was evaluated by teachers and its reliability was 0.98 with Cronbach's alpha. Findings from the viewpoint of teachers showed that the blended learning approach has a great role on interaction process of students with special educational needs.

Faizi [39] study is in progress) Conducting a research titled "designing and validating of learning program using artificial intelligence and its impact on problem solving skills of students with math learning disorder".

Ahmadi [40] study is in progress) Conducting a research titled "designing an educational program based on social robot technology and its impact on social skills and academic achievement motivation of high-functioning autism spectrum disorder students".

3. Conclusions

In this chapter of the book, the author tried to introduce the capacities and capabilities of artificial intelligence in education, especially special education, and share its common uses in the process of education, teaching and learning of people with special educational needs with the esteemed readers. Also, to introduce various assistive technologies based on artificial intelligence in special education. Considering the importance of the approach and model in the education process, the blended learning approach and related models were introduced. The case studies that have been done and are being done in the field of artificial intelligence and its applications in special education were also discussed by the author and his research team. It seems that we need more researches at the national and international level to discover more of this emerging technology.

Acknowledgements

I sincerely thanks from all of the colleagues and students who have cooperated in the research team of the author of this chapter of the book especially the students, teachers, parents and experts who in any way that have provided basic knowledge in the field of artificial intelligence technology.

DOI: http://dx.doi.org/10.5772/intechopen.1004158

Author details

Esmaeil Zaraii Zavaraki
Faculty of Psychology and Educational Sciences, Educational Technology Department, Allameh Tabataba'i University, Tehran, Iran

*Address all correspondence to: zavaraki@atu.ac.ir

References

[1] Zaraii ZE. Designing and validating the blended learning model with emphasis on digital technologies for students with special educational needs. Psychology of Exceptional Individuals. 2019;**9**(34):51-78

[2] Zaraii, Zavaraki E, Velayati E, Moradi R. In: Farhang RE, editor. Special Education with Technology Approach. Iran; 2017

[3] Gerich J, Fellinger J. Effects of social networks on the quality of life in an elder and middle-aged deaf community sample. Journal of Deaf Studies and Deaf Education. 2012;**17**(1):102-115

[4] Cuculick JA. Facebook among Deaf College Students: Deaf-Gain and Funds of Knowledge. USA: University of Rochester; 2014

[5] Gregor SE. Social networking: closing the achievement gap between regular and special education students [thesis]. USA: Nova Southeastern University; 2014

[6] Hill CL. An investigation of the connections between use of Facebook and the self-esteem/well-being of students with disabilities in the University of Iowa Reach Program [thesis]. USA: The University of Iowa; 2015

[7] Oyewumi A, Isaiah O, Adigun O. Influence of social networking on the psychological adjustment of adolescents with hearing impairment in Ibadan Oyo State, Nigeria [thesis]. Nigeria: University of Ibadan; 2015

[8] Kožuh I, Hintermair M, Holzinger A, Volčič Z, Debevc M. Enhancing universal access: Deaf and hard of hearing people on social networking sites. Universal access in the information society. 2015;**14**:537-545

[9] Saunders KC. A double-edged sword: Social media as a tool of online disinhibition regarding American sign language and deaf cultural experience marginalization, and as a tool of cultural and linguistic exposure. Social Media+ Society. 2016;**2**(1):2056305115624529

[10] Toofaninejad E, Zaraii Zavaraki E, Dawson S, Poquet O, Sharifi DP. Social media use for deaf and hard of hearing students in educational settings: A systematic review of literature. Deafness & Education International. 2017;**19**(3-4):144-161

[11] Zavaraki E, Schneider D. Blended learning approach for students with special educational needs: A systematic review. Journal of Education & Social Policy. 2019;**6**(3):1-2

[12] Daniela L, editor. Inclusive Digital Education. USA: Springer Nature; 2022

[13] Allahi Z. Design and validation of a mobile phone learning pattern for visually impaired students [thesis]. Tehran, Iran: Allameh Tabataba'i University; 2016

[14] Jangi ZH. Design and validate the model of educational multimedia elements and its effect on learning, motivation for academic achievement and participation of mentally retarded students in math lessons [thesis]. Tehran, Iran: Allameh Tabataba'i University; 2016

[15] Mathews KM. Transformative models in K-12 education: The impact of a blended universal design for learning intervention. An experimental mixed methods study [thesis]. USA: University of San Diego; 2016

[16] Stamer T. Promoting learning strategies in students with learning disabilities through blended learning [thesis]. USA: McKendree University; 2017

[17] Toofaninejad E. Designing instructional pattern of the learning environment enriched by virtual social network and its impact on learning rate and the social skills of the students with hearing impairment in science course [thesis]. Tehran, Iran: Allameh Tabataba'i University; 2017

[18] Moradi R, Zaraii Zavaraki E, Sharifi-Daramadi P, Nili-Ahmadabadi MR, Delavar A. The impact of an instructional model with assistive technology on achievement satisfaction of people with physical-motor impairments. Research and Development in Medical Education. 2018;7(2):95-101

[19] Miao F, Holmes W, Huang R, Zhang H. AI and Education: A Guidance for Policymakers. France: UNESCO Publishing; 2021

[20] Holmes W, Bialik M, Fadel C. Artificial Intelligence in Education. USA: Globethics Publications; 2023

[21] Zawacki-Richter O, Marín VI, Bond M, Governor F. Systematic review of research on artificial intelligence applications in higher education–where are the educators? International Journal of Educational Technology in Higher Education. 2019;**16**(1):1-27

[22] UNESCO Institute for Statistics. The World Needs Almost 69 Million New Teachers to Reach the 2030 Education Goals. (Fact Sheet No. 39). France: UNESCO Institute for Statistics; 2016

[23] Hu X, Neupane B, Echaiz LF, Sibal P, Rivera LM. Steering AI and Advanced ICTs for Knowledge Societies: A Rights, Openness, Access, and Multi-Stakeholder Perspective. France: UNESCO Publishing; 2019

[24] Zaraii ZE. Introducing Metaverse as a new technology. Educational Technologies in Learning. 2022;**5**(18):7-9

[25] Garrison DR, Kanuka H. Blended learning: Uncovering its transformative potential in higher education. The Internet and Higher Education. 2004;**7**(2):95-105

[26] Christensen CM, Horn MB, Staker H. Is K-12 Blended Learning Disruptive? An Introduction to the Theory of Hybrids. Clayton Christensen Institute for Disruptive Innovation. USA; 2013

[27] Powell A, Watson J, Staley P, Patrick S, Horn M, Fetzer L, et al. Blending Learning: The Evolution of Online and Face-to-Face Education from 2008-2015. Promising Practices in Blended and Online Learning Series. International association for K-12 online learning; 2015

[28] Zavaraki EZ, Alimardani F. The role of blended learning approach on interaction process of students with special educational needs. In: EdMedia+ Innovate Learning. Association for the Advancement of Computing in Education (AACE). USA; 2023. pp. 1243-1247

[29] Zareei M. Design and validation of virtual social networks of Iran's schools [thesis]. Tehran, Iran: Allameh Tabataba'i University; 2018

[30] Moradi R. Design and validate an educational model based on assistive technology and its effect on the amount of motivation for academic achievement, learning and academic satisfaction of students with physical-motor impairment in English language lesson

[thesis]. Tehran, Iran: Allameh Tabataba'i University; 2018

[31] Delavaryan F. Designing a technology-based instructional program for teaching English to mentally retarded students and its impact on academic achievement motivation, learning and retention [thesis]. Tehran, Iran: Allameh Tabataba'i University; 2019

[32] Khateri E. The impact of augmented reality technology on academic achievement motivation and learning of second-grade dyslexic students [thesis]. Tehran, Iran: Allameh Tabataba'i University; 2019

[33] Ghasemi SM. Designing and validating of instructional model in computer game environment and its effect on cooperative learning of high-functioning autism students [thesis]. Tehran, Iran: Allameh Tabataba'i University; 2021

[34] Masnavi A. Design and validation of a model of mobile multimedia learning environment and its effect on learning and learning transfer of social skills of students with high-performance autism spectrum disorder [thesis]. Tehran, Iran: Allameh Tabataba'i University; 2021

[35] Bakhtiarvand M. Designing and validating an instructional model of picture exchange communication system based on technology and investigating its effectiveness on social and communication skills and behavioral problems in children with high-functioning autism disorder [thesis]. Tehran, Iran: Allameh Tabataba'i University; 2022

[36] Kabiri A. Designing and validating of educational model for applying technology in philosophy program for children and its impact on the critical thinking of gifted students in the second period of elementary school [thesis]. Tehran, Iran: Allameh Tabataba'i University; 2022

[37] Ghanat M. The effect of computer game on the recognition of facial emotions in students with hearing impairment in the elementary school of Arak city [thesis]. Tehran, Iran: Allameh Tabataba'i University; 2023

[38] Abbas ZR. The effect of digital educational game on the visual perception of dyslexic first grade students [thesis]. Tehran, Iran: Allameh Tabataba'i University; 2023

[39] Faizi F. Designing and validating of learning program using artificial intelligence and its impact on problem solving skills of students with math learning disorder [thesis]. Tehran, Iran: Allameh Tabataba'i University; 2023

[40] Ahmadi A. Designing an educational program based on social robot technology and its impact on social skills and academic achievement motivation of high-functioning autism spectrum disorder students [thesis]. Tehran, Iran: Allameh Tabataba'i University; 2023

Chapter 4

Impact of Artificial Intelligence in Achieving Quality Education

Agatha Aballa Nkechi, Akintayo O. Ojo and Obinna A. Eneh

Abstract

This chapter highlighted the impact of artificial intelligence (AI) in achieving quality education. The purpose of this chapter was clearly articulated by investigating the intersection of AI and education by considering its role in educational technology through AI-powered learning management system, personalized learning, virtual and augmented reality among others. The various ways of enhancing teaching methods using AI-driven tutoring systems, adaptive learning platforms, intelligent classroom assistants were constructively considered. Literature review showed the various ways that artificial intelligence have been used in addressing inequality in education by helping to mitigate learning disparities in diverse student populations, overcoming language barriers through AI translation and bridging the digital divide. The challenges and ethical considerations in educational AI system were identified as Bias in AI algorithms and privacy concerns. The various strategies mitigating these challenges relating to AI use in education to achieve quality education were also captured. However, the policy implications and governance regarding AI use in education including international collaboration for standardization to assist in enhancing responsible AI implementation to achieve quality education were also properly presented. The outcome of the investigations showed that with the active collaboration of the major stakeholders in education in its implementation, artificial intelligence has improved the quality of education globally.

Keywords: quality education, artificial intelligence, AI-translation, learners, educators

1. Introduction

Artificial intelligence has been a topic of interest and has raised curiosity in various disciplines of which education is one of the prime spaces of its applications. Artificial intelligence (AI) is described as the designing and building of intelligent agents which depend on recognition by the senses from the environment and acts in a way that affects the environment. It is a branch of computer science that use algorithms and machine learning techniques to replicate or simulate human intelligence [1]. The Oxford dictionary of phrase and fable further defined artificial intelligence as the theory and development of computer systems which has the ability to carry out tasks that normally require human intelligence, decision making as well as language translations. There are three types of artificial intelligence namely: Narrow AI, General AI and Artificial superintelligence. Narrow AI is the most common and realized form of artificial intelligence. It is goal-oriented and employs machine learning techniques to actualize a task. General AI is also known as deep AI, it is one that is deemed on par with human capabilities which can discern the needs or emotions of other intelligent beings. While the Artificial Super intelligent is AI which is more capable than humans [2]. AI is considered the most admirable among all other technological revolutions in the world. Quality education describes a holistic educational approach which provides learners with academic knowledge, critical thinking, skills for solving daily problems, creativity and provide them with a sense of social responsibility. Any form of education which is considered quality must be all inclusive, equitable and accessible to all learners, must promote lifelong learning and should be able to prepare its recipients for active participation in the society. Quality education is the basis for individual growth and societal development globally. It provides knowledge for individuals empowering them with critical thinking skills and creativity, encouraging personal developments and learning throughout the individual's life. It helps to create economic opportunities and employment for individuals thereby promoting their income potentials. Quality education speeds up societal progress by promoting social integration, bringing down societal inequality while enhancing the creation of an informed citizenry who are properly engaged. Nations whose citizens have received quality education are better prepared and equipped to address the challenges of poverty, health and environment. Such nations have been known to develop better innovations and adapted favorably to the demands of a rapidly evolving global economy. Quality education has served as the driving force for sustainable development and progress in the society while it shapes and empowers individuals as informed citizens. UNESCO [3] emphasized the global significance of quality education for individual and societal development using the SDG4—Education 2030, Incheon Declaration (ID) and framework for action which states that "For the implementation of sustainable Development Goal 4, ensure inclusive and Equitable Quality Education and Promote Lifelong Learning Opportunities for All". Hence, the purpose of this chapter, is to explore the impact of artificial intelligence (AI) in achieving quality education.

2. Main body

2.1 Impact of artificial intelligence in achieving quality education

Artificial intelligence has transformed the entire globe, it is important that this chapter provides a brief discussion of the many applications of artificial intelligence across various industries and domains that showcase its versatility in improving efficiency and decisions before giving a detailed description of its impact in achieving quality education. The domains are as briefly discussed below:

a. *Education*: artificial intelligence has been employed in the education sector to support personalized learning, instructional system, educational analytics, among others.

b. *Health care*: AI has been found very assistive in disease diagnosis, personalized medicine and drug delivery.

c. *Finance*: AI is used by financial institutions for detection of fraud, algorithmic trading as well as enhancing services rendered to customers in banks.

d. *Retail services*: AI has improved retail services by its ability to enable recommendation systems, demand forecasting and supply chain optimization.

e. *Autonomous vehicles*: the emergence of self-driving cars and drones for navigation and decision making has been traced back to Artificial Intelligence which powers them.

f. *Manufacturing*: AI is used in manufacturing for predictive maintenance procedures, quality control as well as process optimization.

g. *Customer services*: customer support interactions have been greatly improved by the use of Chatbots and virtual assistants.

h. *Marketing*: AI has improved marketing strategies by assisting in targeted advertising, social media analysis and in customer segmentation.

i. *Cyber security*: AI is used to detect threats, anomaly detection and to secure digital systems.

j. *Human resources*: AI helps employers in engaging employees, talent acquisition and workforce analytics.

k. National language processing: AI has been used in chatbots, language translations and in sentiment analysis.

l. *Entertainment*: AI has been employed in games, virtual reality experiences and content recommendation.

m. *Agriculture*: precision farming, monitoring of crops and prediction of yields has been improved by Artificial intelligence.

n. *Energy*: AI is employed for smart grid management, preventive maintenance and optimization for energy consumption.

o. *Legal*: AI assist legal practitioners to review documents, analyze contracts and carry out legal research.

However, in exploring the impact of AI in achieving quality education. It is necessary to look at various areas where AI and education intersect by considering the following subsections:

2.2 The role of AI in educational technology

2.2.1 AI powered learning management system

According to [4], learning analytics can substantially increase the way the field of learning science understands learning both theoretically and practically. Hence artificial intelligence in learning management system through data analysis can optimize content delivery by personalizing learning paths, access learner's performance and improve engagement with adaptive learning techniques.

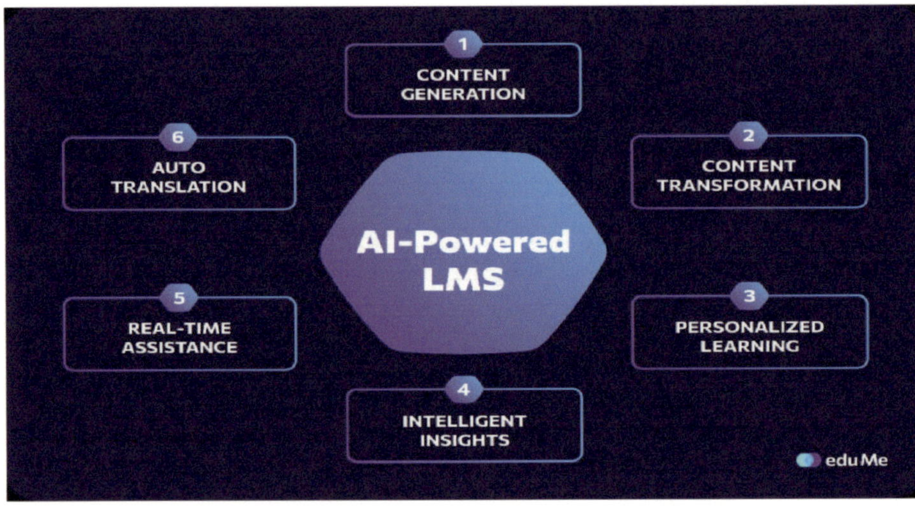

Figure 1.
Roles of AI-powered LMS.

The significant roles played by AI-powered LMS centers around the improvement of content generation and transformation, personalized learning, intelligent insights, real-time assistance, and auto translation as shown in **Figure 1**. Artificial intelligence easily adapts to learners needs and provides contents and feedbacks which are tailored to actualize learning outcomes in the best possible ways. With AI, the teachers are assisted to create more effective lesson plans and assessments.

Artificial intelligence can become a student companion and assistant. This is so because AI understands the strength and weakness of an individual learner. AI has the capacity of analyzing learner's pace and can identify the latent skills in learners which can be used to guide them during the teaching and learning process.

On the part of the teacher, AI can assist the teacher to properly guide the learners, to know the right time for evaluation of their teaching, the type of questions to be chosen for evaluation on a particular topic, the rate of lecture delivery and the choice of topic that will benefit redundant learners. With AI, teachers can quickly outsource their targets and actualize the outcomes set for each learner's performance. There are many examples of AI-powered LMS but a few of them has been described in this chapter.

- *Zavvy*: this is an AI powered LMS which provides features that can be used to create engaging training courses in a very short duration. It assists in course creation, launches workshops and on-demand courses faster than traditional learning platforms. It can handle repetitive basics while allowing more time for the improvement of course content. Zavvy platform is used in translating training contents into multiple languages and can help its users manage their courses into one centralized location there by enhancing a course library management, thus bridging knowledge gaps.

- *Paradiso*: this is an AI-powered authoring tool which creates e-learning courses. Its templates help learners generate content, videos, images and presentations to be transformed into training materials. This technology help learners manage instruction design, graphic design, voice-overs, videos and assessments which can be customized for individual learners to enhance their specific development goals.

- *Docebo*: this provides a variety of efficient and customized learning programs with special features which permits learners to create audience specific pages using drag and drop functionality. Docebo encourages customized learning experiences with virtual coaching, content suggestion as well as auto-tagging thereby encouraging personalized learning. This LMS offers gamification, auto assignment of training depending on the required skills of the learners. It also helps learners to streamline non-formal learning and automates learning management. With this platform, various business systems are brought within one learning environment using its over four hundred integrations with the inclusion of web conferencing, payment gateways among others.

- *Cipher learning*: this AI-LMS helps learners create courses and assessments. It encourages personalized learning. Learners can build course outlines, descriptions and content in very many languages. Learner's progress can easily be tracked.

- *Moodle*: this is an online learning management system built on an open-source platform. This is best suited for organizing distance learning in which the learner is privileged to independently choose the sequence and schedule of learning.

- *EdAPP*: this is a mobile learning platform with efficient AI features that can be used for training. With this platform users have their personalized course content, course creation and generation of whole lesson highly simplified. Learners can easily analyze online resources and obtain relevant information within a possible short time frame. Schools and organization employers can use EdAPP to manage training of their employees anytime and from various locations.

- *Absorb LMS*: this is designed for universal learning. With this system, organizations are empowered to train their staff with modern skills to keep them updated and relevant in modern societies. It enables organizations maximize staff efficiency by inspiring learning and increasing business productivity. It is a cloud based LMS which engages learners, enhances content retention and increases training programs.

2.2.2 Examples of AI-powered intelligent classroom assistants

- *Squirrel AI*: this is an adaptive learning platform that uses AI algorithms to personalize learning for each learner. It provides tailored recommendations and feedback to improve learners understanding.

- *Edmodo*: this uses AI to provide learners with personalized learning experiences. It offers features such as quizzes, assignments and collaborative tools to enhance student engagement and performance.

- *Brainy*: this is a community-driven assistant that can help learners ask and answer academic question. It uses AI to match questions with relevant answers helping them find solutions quickly and efficiently.

- *Cognii*: this is a virtual learning assistant that relies on conversational technology to assist learners from open resource format. They are designed to provide instant feedback with written responses. It utilizes Natural Language Processing algorithm to evaluate essays, giving personalized feedback thereby improving learners writing skills.

- *IBM Watson Education*: this assistant provides AI powered solutions for classrooms. Its tools leverage AI to enhance individualized learning experiences, enhancing learning outcomes and optimizing teaching techniques. It provides virtual tutors, adaptive learning systems as well as analytic tools.

2.2.3 AI enhances individualized or personalized learning

This is so because AI can adapt to individual learner's related requirements there by reducing the unproductive works of the teachers and help to increase the quality of teaching and learning.

AI has also improved the grading system used in the educational sector. This has been possible by automating the grading system in multiple choice questions. This has reduced to a great extent time wastage and stress on the part of the teachers.

AI promotes global access to quality education by providing global classroom for learners in every part of the world. The gap of communication between learners and instructors can be comfortably filled by the use of artificial intelligence. Real time subtitles in AI can help blind learners with sight, hearing difficulties and those with different languages to understand and assimilate the learning process effectively. Moreover, AI assists learners from all over the world to connect themselves for brainstorming and further interaction thereby increasing knowledge spread among the students. Vanlehn [5], in an experiment to compare the effectiveness of human tutoring and intelligent tutoring system with other tutoring systems, observed that intelligent tutoring system possess step-based form of interaction while other tutoring systems have answer-based user interfaces. The experiment concluded that the effective size of intelligent tutoring system was as nearly as effective as that of the human tutoring. Hence AI algorithms can be used to customized educational content and learning paths thereby enhancing personalized learning which in turn improves the quality of learning.

AI algorithms analyze large amounts of data to ensure that learning experiences are learner centered and are tailored to the learners needs and preferences.

However, personalization in education using artificial intelligent are reflected in the following aspects which are described below:

- *Adaptive learning platforms*: this provides opportunity for the learner's strengths and weaknesses to be assessed by AI algorithms and the content delivery are adapted to focus on the areas that need to be improved. This ensures that learning pace and style proceeds based on individual learner. Dam [6] showed from his work that the integration of AI methods into adaptive learning system enables it to dynamically adapt feedback, pacing, content and teaching methods to the learners need. According to him, advances in multimodal learning analytics, Natural Language Processing (NLP) and affective computing can improve personalization and adaptability of AI powered learning system. The adaptation is achieved in AI system by analyzing learner's performance, preferences and behavior to dynamically adjust the experience in order to promote an effective learning path. AI performs an important role in adaptive navigation support within educational hypermedia systems. These systems utilize AI algorithms to individualize the learning experience in the following specific ways:

1. *User modeling*: by analyzing learners' preferences, behavior and style of learning, AI creates accurate user models. The knowledge obtained from such models will then be learning used to tailor the navigation suggestions to the learners need.

2. *Adaptive content delivery*: AI helps to evaluate learner's proficiency and adaptive learners' content at a level where the learner finds difficult and makes sure that the instructional materials suit the current understanding of the learner; this ensures the optimization of learning outcome.

3. *Use of AI-powered recommendation system*: this ensures the provision of personalized suggestions for learning paths. This in turn assist learners to navigate through the entire learning content more efficiently thereby improving learning comprehension and engagement of the learning process.

4. *Natural Language Processing (NLP)*: using its NPL ability, AI is able to understand and answer learner's questions correctly. This provides learners with an effective interactive and conversational learning experience.

5. *Provision of instant feedback*: AI provides real-time feedback on the progress of the learner pointing out the learner's area of strength and weakness. Such feedbacks affect the adaptive navigation by moving the learner to the needed instructional content which will address the identified gaps in learning.

6. *Dynamic pathways*: AI has the ability to dynamically adjust the learning path according to the learner's performances. The system can suggest a more challenging material if the learner's performance becomes excellent in a particular area while providing additional support in areas where help is needed.

7. *Continuous learning improvement*: the continuous interaction of the learner with AI system makes it possible for the system to identify the learner and to continuously refine his model, making specific recommendations for him. This creates opportunities for interactive learning which ensures accuracy of adaptive navigation support that is tailored specifically to individual learner.

8. *Intelligent classroom assistants*: AI assistants are used by teachers for routine administrative tasks and classroom management. AI intelligent assistants help by automating administrative tasks, streamlining workflows and enhancing efficiency. These assistants can manage data entry, schedule appointments and routine communications thus allowing human professionals to focus their attention on more complex roles. Thus, saving meaningful time and minimizing human errors associated with manual tasks. In classroom management, AI assistants provide educators with teaching tools to improve their teaching strategies. They can assist with grading and analysis of the learners thereby providing the instructor with the proper insight into the learner's performance. However, it is important to strike a balance in order to prevent over reliance on AI assistance as this may diminish the human touch in education and administration which is extremely needed in order to produce learners with exceptional outcomes.

- *Learning analytics*: artificial intelligence has been used to track the learner's progress. This helps to provide insights as to the learning pattern of each student. This knowledge can be used by instructors to pinpoint learner's challenges, refine the instructional materials/contents and hence can provide adequate and timely help where it is needed to improve the entire learning experience.

- *Automated feedback*: AI algorithms helps to generate immediate and automated feedbacks on assignments, quizzes and examinations to enable the learners see and understand where they have made mistakes and correct them. This in turn reinforces the learning objectives.

- *Accessibility and inclusivity*: the various learning needs of the students can be taken care of easily by artificial intelligence which can customize the learning content. The diverse group of learners are considered including disabled learners and those with learning challenges. This enhances inclusivity of all learners in the learning process.

- *Adjustments of learning paths*: artificial intelligence helps to adjust the learning path of the learner depending on his/her progress. This helps to prevent boredom and improves learner's ability to learn.

- *Prediction of learners needs*: based on the historical data of the learner, AI has the ability to predict the future learning needs of the student. This helps the instructor to be very proactive in addressing the learning challenges. This approach helps to fill the gaps in learning and enhances quality education.

- *Cognitive tutoring system*: AI tutoring system encourages individualized interaction with learners. This enables tutors to adapt their teaching strategies to fit individuals learning styles. The cognitive tutoring system powered by AI provides real-time assistance and better comprehension and retention by the learners.

2.2.4 Virtual and augmented reality in education

In describing the role of virtual and augmented reality in creating immersive learning experiences, Dede [7] pointed out that interactive media enable various degrees of digital immersion and that the more a virtual immersive experience is based on design strategies that combine actional symbolic and sensory factors, the greater the participants suspension of disbelief of being inside an enhanced digital setting. He emphasized that emersion in a digital environment enhances quality education by allowing multiple perspectives situated in learning and transfer. Virtual and augmented reality contribute in developing quality education by making available immersive individualized and accessible learning experiences. This is made possible by enabling remote learning and removal of geographical barriers. The availability of virtual classrooms makes possible for experiences to be shared among learners as well as encourage collaborative projects among them. This in effect fosters a sense of global connection among students. Augmented reality offers contextual insights in education by overlaying digital information onto the real world. It has the immense ability of fostering deeper understanding of learners by enriching textbooks with 3D models, audiovisual aids and supplementary information. Virtual reality uses simulated environments to immerse learners into a world of experimental learning. This is a powerful tool used in producing quality and realistic learning in science subjects or in exploring historical events.

2.2.5 Enhancement of teaching methods by artificial intelligence

Teaching methods have been greatly enhanced by using AI-driven tutoring systems. The effectiveness of AI tutors in providing personalized assistance to learners cannot be over emphasized. Anderson et al. [8] observed from their experiments using a computer based instructional technology called Cognitive Tutor that the best tutorial interaction style for learners was the one in which the tutor provided immediate feedback with short and direct error messages. They pointed out that tutors worked better when presented as non-human tools. Hence, AI-driven tutor system through the provision of personalized assistance to learners is playing a pivotal role in achieving quality education.

2.2.6 Addressing educational inequality with artificial intelligence

AI has been employed to mitigate learning disparities in diverse student populations. Warschaver and Matuchniak [9] showed that new technologies and digital world can be powerful tools for transforming learning and can be used in several ways to promote equality and justice in the society by providing access to knowledge and improved communication to every one irrespective of their geographical location and socio-economic status. AI has been classified as one of the latest technologies and the most admired among them which can be used to address educational inequality globally. By providing personalized and adaptive learning experiences which are tailored to individual learners, by analyzing diverse learner's data, discovering learning gaps and offering targeted exercises, assisting instructors in creating inclusive learning contents that takes care of varied learning styles, diverse cultures and languages, AI provides an equitable educational environment for learners which in turn helps to achieve quality education.

2.2.7 Overcoming language barriers through AI translation

AI has played vital roles in breaking language barriers and promoting inclusivity in an attempt to improve the quality of education. For instance, AI-driven translation tools facilitate multilingual communication, empowering both teachers and students to easily access teaching and learning resources in their native languages. AI powered language learning platform personalize instructions adapting to the learner's proficiency levels and styles thereby contributing to a more inclusive and effective educational experiences. AI-driven language translation tools enable learners and instructors from diverse nations of the world and languages to collaborate on different projects and learn from each other. Agbonika [10], in his findings, reaffirmed that by harnessing the power of AI, the status quo of Northern Nigeria can be disrupted and her educational system transformed, empowering her students to reach their full potential thereby contributing to a more equitable and prosperous future for the region.

2.2.8 Bridging the digital divide with artificial intelligence

AI initiatives are used to improve digital accessibility in underserved communities. The AI-driven mobile apps are used to provide educational resources, bridging the digital divide by offering learning materials to learners who have limited access to traditional schooling. For areas with limited medical resources, AI-powered Chatbots have been used to make available healthcare information and support. All these initiatives aim at leveraging AI technology to address inequality and improve access to essential services.

2.3 Real-world applications and case study success stories of AI in achieving quality of education

There have been notable instances showing how AI has impacted positively and improved the quality of educational outcomes. Seo et al. [11] conducted a study to determine the impact of artificial intelligence on learner-instructor interaction in online learning by using speed Dating with storyboards to analyze the authentic voices of 12 students and 11 instructors on diverse use of possible AI systems in online learning. Their findings showed that participants envisioned that adopting AI system in online learning can enable personalized learner-instructor interaction at scale though at the risk of violating social boundaries. Their findings have implications for the design of AI systems to ensure explainability, human-in-the-loop, careful data collection and presentation.

Chen et al. [12] conducted a study to access the impact of AI on education with the scope limited to AI application and effects on administration, instruction and learning, leveraging the use of literature review as a research design. Their study discovered that using AI, instructors were able to review and grade students more effectively and efficiently and achieved higher quality in their teaching activities. The curriculum and learning contents became customized and adapted to the learners needs which improved their learning experiences and overall quality of learning.

In an attempt to explore some real-life examples of how AI was used to improve the quality of education and how students' successes were achieved, [13], described how an AI program developed by a Stanford researcher was used to provide students with assistance when they get stuck in self-paced digital learning. The study tested a machine-learning program that would predict when a student was likely to get stuck and start wheel-spinning as well as the point it would recommend a relevant solution. The training was done by analyzing the performance data from over a hundred school children who had used tablets to learn English reading skills using videos and mini-games. The result showed that the program clearly predicted when a learner would fell into wheel-spinning before the start of a new lesson. His conclusion showed that AI can be used to identify learner's problems and make it easier for a limited number of human instructors to assist a large number of learners.

Hwang et al. [14] proposed an intelligent tutoring system to assist fifth-grade students learn multiplication and division of mathematical units. They found out that students mathematical learning performance and learning motivation was greatly improved by the intelligent tutoring system (ITS) and that AI improved students grades by 30% while reducing their anxiety by 20%.

According to Verma [13], an AI-powered Chatbot called Jill Watson developed by IBM's Watson at the Georgia institute of technology which was employed as a teaching assistant for a course with 300 students produced a 97% success rate in answering ten thousand students' inquiries each semester with a very remarkable human-like efficiency.

The result of the study conducted by [15], with the objective of analyzing the impact of AI components and computational sciences on student performance using web of science (WOS) and Scopus databases showed the positive impact that AI and computational sciences have on student performance. They found a rise in the students' attitude towards learning and their motivation especially in the Science, Technology, Engineering and Mathematics.

Camesaria et al. [16] conducted a systematic review of literature which focused on the analysis of the application of AI in the assessment of primary and secondary school students through collections of published articles in the most popular databases from 2010 onwards. They concluded that despite the complexity of AI, the potential of AI-related tools to improve the quality of teaching particularly student assessment at lower levels is very high.

2.4 Challenges and ethical considerations facing use of AI in achieving quality education

2.4.1 Bias in AI algorithms

Bias in AI algorithm can stem from biased training data or the algorithm design itself. It is important to identify and address these to ensure that AI applications produce fair and equitable outcomes.

However, the essential factors for mitigating this challenge include the following factors

- Ethical considerations
- Regular audits
- Diverse dataset representation

Bias in educational artificial intelligence algorithms can lead to several challenges. If there is any form of bias in the training data employed in developing the AI algorithm, then it is certain that it will perpetuate and increase the existing inequality in education. This may affect certain groups and lead to unfair advantages and disadvantages. Biased AI algorithms can limit opportunities for certain groups preventing their access to quality education. This can lead to a lack of diversity in educational resources and content. Biased AI algorithms is related to transparency and accountability. However, to mitigate this challenge, a thorough understanding of how AI algorithms make decisions and efforts to ensure that they are fair and unbiased will help to build trust in this educational technology. Moreover, collaborative efforts from educators, policymakers, and developers to mitigate this challenge remains very essential to creating inclusive and equitable educational technologies.

2.4.2 Privacy concerns in educational AI

As AI technology expands across the diverse industries, it opens up a great many privacy concerns thereby putting the traditional norms of personal data protection to many challenges. AI privacy dilemma involves the following main issues:

- The insatiable appetite of AI for extensive personal data to supply its machine learning algorithm has brought about serious concerns regarding data storage, usage and access. These have raised questions regarding data sources, storage and accessibility which the traditional data protection cannot provide answers to. The remarkable ability of AI to analyze and make complex analysis within the received data poses privacy concerns. AI's potential to infer sensitive information relating to an individual's location, habits and personal preferences, poses risks of unauthorized data dissemination. The fact that AI also has the potential for identity theft and unwarranted surveillance poses a set of worries and challenges that requires proactive attentions. AI developments have given rise to ethical guidelines and best practices to reduce privacy risks and several leaders in industries have risen to address these concerns. Several esteemed bodies have proposed ethical benchmark to which the partnerships on AI (PAI) which is a coalition of leading companies, organizations and persons impacted by AI stand out a cornerstone. By bringing different stakeholders from tech-giants to AI users together, PAI creates a platform that share collaboration between entities that might not typically interact. The aim is to establish a common ground with PAI as a unifying agent for positive change among AI users. Collecting student's data with artificial intelligence has raised several ethical concerns related to privacy, consent and potential misuse of information. Therefore, it is necessary to ensure transparent communication regarding data collection purposes, obtain users consent and implement robust sensitive measures to protect sensitive information. Addressing the challenges of biases in AI algorithms is important to mitigate unfair treatment or discrimination based on student data. Therefore, a balance should be maintained between leveraging on AI for educational advantages while respecting the user's privacy.

2.5 Ensuring fair access to AI-enhanced education

To ensure fair access to AI-enhanced education, the following strategies must be considered in combination:

Digital inclusion: access to the necessary hardware and internet connectivity must be provided for all learners.

Integration into educational content: AI educational tools must be integrated into the curriculum of the learners in order to expose them to the modern educational technology.

Training educators: the educators must be trained on AI tools and methods of use to ensure efficient and effective implementation.

Equitable distribution of AI tools: AI tools and resources must be distributed evenly among schools and communities.

Accessibility features: AI applications and tools must be developed with accessibility features that can accommodate and sustain the different learning needs of the students.

Community engagement: communities must be carried along and included in decision-making to help in addressing local needs and ensure inclusivity.

Regular assessment: the need for regular monitoring and evaluation of the impact of AI enhanced education on both the teachers and learners within different geographical locations will help to identify and address disparities.

Addressing privacy and security concerns: there must be measures put in place to respect the learner's privacy and security when using AI tools.

2.6 Future trends and innovation in educational AI

2.6.1 Emerging technologies in educational AI

There are many cutting-edge technologies likely to shape the future of AI in education. These cutting-edge technologies possess the potentials of reshaping education by making a more personalized, efficient and adaptive learning experiences. This in turn enhances a highly advanced era in AI education. The following technologies described below hold the potential to significantly shape the educational AI.

Brain-computer interface: commonly abbreviated as (BCIs), can enable direct thought-based interaction with the educational curriculum which in turn could revolutionize learning when there is constant direct communication between the brain and computers.

Edge AI: this involves bringing AI processing nearer to the source of data enabling real-time and localized AI applications which is advantageous in constrained environments.

Neuro-informed learning systems: this educational approach draws insight from neuroscience to inform the design and implementation of learning experiences. It leverages an understanding of the brains ability to learn and process data to optimize educational strategies and techniques to enhance individualize instruction, cognitive load management, feedback mechanism, metacognition development among other key elements. AI system could incorporate neuroscientific principles to understand and care for learner's cognitive processes in a much better way.

Explainable AI (XAI): this is the capability of an AI system to provide explanations that are understandable and clear regarding its decisions and actions. The aim of XAI is to make AI systems more accessible and interpretable enabling users to comprehend reasons behind its specific outputs. This applies more in the domains of healthcare and finance where trust is highly needed.

Block chain for credential verification: this is an emerging technology in educational AI which can secure and streamline the verification of academic credentials to enhance the authenticity of educational records and to prevent frauds.

Emotion recognition: this can be incorporated in the AI systems to help recognize and respond to learners' emotions. This will enhance the learning experience by adapting the curriculum based on the learner's emotional state and provide support when needed.

Computer vision: when computer vision is integrated into education, remote learning experiences are encouraged and enhanced. This can be achieved by enabling the AI systems to analyze learners' facial expressions and engagement levels, while providing necessary feedback to the teachers.

Educational robots: robots which are designed with AI has the potentials of providing interactive learning experiences and they can help with important learning tasks such as language learning, coding and problem solving.

Federated learning: this is a decentralized educational approach which allows AI models to be trained across multiple servers or devices without the exchange of raw data. This helps to improve security and privacy by localizing student's sensitive data.

Quantum Computing: this computing system has the potential of solving complex problems in machine learning and optimization. It is believed that the quantum computing technology can revolutionize the AI application in education.

AI wearable technologies: incorporating AI with human wearable devices or other technologies can improve human capacity. This can produce tools or devices that can provide assistants to handicapped or disabled learners and provide additional support in education.

2.6.2 Potential disruptions and transformations by artificial intelligence

AI may disrupt traditional educational models and foster innovative approaches. AI has the potential of revolutionizing education by personalizing learning, automating educational administrative tasks and enhancing innovative pedagogical techniques. AI powered adaptive learning platforms can tailor educational curriculum to individual learners needs promoting better learning. It can automate grading, creating more free time for instructors to concentrate on personal mentorship. AI-driven simulation environments and virtual reality provides immersive learning environments for improving practical skills. However, the challenges associated with AI use relating to ethical considerations, privacy of data and digital divide must be adequately managed to ensure equal accessibility and responsible implementation. Christenson et al. [17] explored the concept of destructive innovation in education. They emphasized the potential of technology instruction to transform traditional educational models. They argued that customizable student-centric approaches can meet diverse learning needs better and challenge established norms. They also described the impact of disruptive technologies on K-12 education and suggested that accepting innovations is the key to addressing the limitations of conventional education.

2.7 Anticipated challenges of educational AI and preparation for the future

Future AI developments may pose several challenges such as described in this chapter. However, addressing these challenges will entail collaborative efforts from researchers, government, leaders in industries and the entire society to ensure that AI technologies are deployed beneficially and responsibly. Some of the anticipated challenges of AI developments for educational use include the following:

1. *Ethical concerns*: AI usage ethical concerns are usually related to bias, transparency and accountability. Experience has shown that ensuring fairness and avoidance of discrimination in AI systems has remained critical challenges.

2. *Redundancy*: AI has the potential of automating the grading system and other administrative tasks in education. However, AI-driven automation could replace and displace jobs leading to unemployment in certain sectors thereby preparing the workforce for new skill requirements which is not very convenient.

3. *Lack of standards and regulations*: the rapid pace of AI development can outstrip regulatory frameworks leading to gap creation in oversight. Therefore, it is important to establish global standards for AI deployment to ensure responsible practices.

4. *Privacy and security risks*: as the vast amount of personal data handled by AI system increases, an equal increase in risk of privacy breaches arises. With this, the possibility of safeguarding sensitive information and preventing unauthorized access becomes an increasing risk.

5. Dependency and bias: prolonged and continuous usage of AI system can lead to over reliance and dependency on the system. This factor alongside the bias embedded in AI system can bring about errors in decision making as well as reinforce already existing social inequalities. Nevertheless, there have been ongoing efforts to manage these challenges.

6. *Unintended consequences*: AI systems produces unintended behaviors resulting in consequences that are unintended especially in complex and dynamic

environments. Anticipating and preventing these unintended outcomes has posed significant challenges.

7. *Human-AI collaboration*: it has been a huge challenge to create an effective collaboration between humans and the AI system. This is because all the efforts to do so have resulted in diminishing human autonomy and creating over-dependence on the AI-system. Hence, careful design and implementation is required in order to mitigate this challenge.

8. *Energy consumption*: developing energy efficient AI solutions has been a very big challenge and this has impacted negatively on the environment. This is so because substantial computing power require large amount of energy in training sophisticated AI models. This can be prevented by developing energy efficient AI solutions which are yet to be achieved.

9. *Public understanding and perception of AI*: there has been widespread fear and resistance to AI implementation by the public due to their perceptions and misunderstanding regarding AI and its usage. This has affected the responsible adoption of AI both in education and other sectors. Hence enhancing public perception through proper public awareness campaigns and education regarding AI capabilities and limitations can increase its adoption by the public.

10. *International cooperation*: the global nature of AI development requires that international standards and norm be established to manage all its challenges which are related to ethical standards, data sharing and security issues.

2.8 Strategies for addressing potential challenges arising from future AI developments

In view of the anticipated potential challenges arising from future AI developments; the following strategies have been suggested for educators in order to prepare themselves ahead of these challenges

Ethical frameworks and guidelines: ethical guidelines should be developed and enforced in order to address challenges of bias, transparency and accountability. The industrial sector should be encouraged to embrace self-regulation and adhere to ethical standards.

Training and education of the workforce: public education awareness campaigns should be launched to enlighten the public on AI technologies and thereby impact on the society. Misconceptions regarding AI should also be cleared by fostering open dialogs with the public.

Educational and training programs should be organized regularly in order to equip the workforce with the required skills relevant to face the AI driven future labor market.

- *Privacy and security measures*: the use of privacy-preserving technologies and encryption in AI applications should be strengthened with strict privacy measures properly enforced.

- *Regulatory policies*: it is important to form clear regulatory policies that can adapt to the rapid pace of AI advancements. Harmonized standards for

AI developments and implementations should be created by international cooperation.

- *Transparency*: transparency in AI algorithms should be advocated and standards for providing explanations for AI driven decisions should also be put in place especially in the critical domains such as in the health care and finance sectors where trust is highly needed

- *Diversity and inclusion*: this can be used to reduce bias in AI algorithms when diversity is promoted among the AI development teams. When proper representations from different and varied backgrounds are included in decision making, then the challenge related to decision-making processes in AI will be addressed.

- *Energy efficiency*: researches geared towards the development of energy efficient AI models and algorithms should be encouraged and developed to address the challenge of energy consumption and its consequent environmental impact. The use of renewable sources of energy in AI data centers should also be promoted.

- *International collaborations*: by fostering international collaboration and global cooperation in the policies and standards of AI, the challenges can be properly managed. Establishing forums or platforms that are internationally coordinated will help to share ideas on the best practices globally.

2.9 Need for regulatory frameworks for artificial intelligence in education

These is a critical need for establishing regulatory frameworks for AI in education to ensure its responsible and ethical use in education in order to achieve quality education. These frameworks provide protection against misuse. This is because regulatory frameworks establish guidelines for ethical development of AI algorithms, usage and data handling. Thus, they foster responsible learning environments and help to prevent bias in education AI.

According to European commission [18], "a framework for trustworthy AI provide guidelines which articulates a framework for achieving trustworthy AI based on fundamental right as enshrined in the charter of fundamental Rights of the European union (EU charter) and in relevant international human rights law". The commission emphasized that any trustworthy AI is composed of three components which include the following:

1. It should be lawful, complying with all laws and regulations which are applicable.

2. It should be ethical, ensuring adherence to all ethical principles and values.

3. It should be robust both from a technical and social perspective. Indicating that such AI systems should perform in a safe, secure and reliable manner and safeguards should be foreseen to avoid unintended adverse effects.

The following guidelines and best practices among others have been Suggested for maintaining standards in AI applications:

1. *Ethical committee formation*: establishment of ethical review boards or committees is a good practice which will assess the impact of AI and its ethical implications before applying in teaching and learning.

2. *Regular monitoring and evaluation*: regularly monitoring and evaluating of the effects of AI applications on the learners, instructors and the entire educational system will help to predict and identify possible unintended consequences and enable stakeholders to effect corrective measures.

3. *Involvement of stakeholders*: all the major stakeholders in education such as students, parents, educators and the community must be actively involved in the decision-making process in AI education application as this will foster collaborative cooperation and decision making.

4. *Continuous training of stakeholders*: educators and administrators should be provided with continuous training and upgrading to remind them of the ethical considerations, responsible use of AI and potential biases associated with its applications.

5. *Regular audit*: auditing of the AI system should be performed at regular intervals to identify and correct any challenges and concerns. Such regular auditing process will promote any ongoing improvement in AI technology.

6. *Improving accessibility*: there is need to design AI system with more accessibility to learners. This will provide an AI system that is adapted to the personalized needs and ability of the learner.

7. Incorporation of human oversight: this will ensure that decisions are made in line with the required ethical standards thus enhancing interventions when needed.

8. Transparency and explainability: the AI system when designed and made explainable, enable stakeholders to understand how decisions are made. This in turn promotes accountability and trust.

9. *Informed consent*: it is very important that user's consent be sorted especially regarding sensitive data. It is also very necessary to let the users know and understand how data obtained from them will be handled. This is important for building trust in the AI system.

2.10 Stakeholders perspectives on the impact of AI in achieving quality education

Byers [19] reported on some interviews conducted among educators on their embrace of the usefulness of AI in achieving quality education and its potential problems when used in the classroom. He quoted a high school principal saying that the focus is driven towards how to use AI to elicit better educational experience for the students. The report further described the use of AI in education as an evolving phenomenon at every level of the academia which has made educators and administrators to seek ways of determining program's potential use in the classroom and whether its use could lead to possible forms of cheating, plagiarism or other forms of academic misconducts which are detrimental to students.

The U.S department of Educator's AI main focus has been on how AI permits the educators and their students to experience new forms of interaction, loops of feedback enhancement and how the technology has been making easier the work of the teachers [19]. Commenting further on the positive impacts of AI use in the classroom, the high school principal acknowledged that learning software embedded in AI such as Grammarly, provides help for writing and that other programs such as Magic eraser for image alteration and Tetra can assist in note taking during virtual meetings. He stated that the technology is making people more efficient especially as it applies to teachers creating lesson plans and scoring tests leading to more time to focus on students' school experience.

An interview with a professor of humanities reported by the same author, exposed the fear and notion nurtured by some educators regarding AI technology. The professor questioned information gathered by AI system and who has access to them. He used the phrase "garbage in, garbage out" implying that if the source material for AI is not quality, then the content will not yield quality, and that the education system is running the risk of over trusting in the AI system. Another educator expressed his concern on students use of AI programs such as Chat GPT to complete an assignment, emphasizing that such works often lack the personal touch and author's voice which makes the work original, and that disciplinary measures are not yet in place on how to manage AI use in assignments. He concluded that the inclusion of AI in education may cause him to rethink some of his assignments to ensure the originality of the work.

Ryu and Han [20] studied Korean school teachers' perception of AI education and reported that teachers with experience in leadership recognized that AI would help to improve creativity.

The study carried out by Marrone et al. [21], highlighted an analysis on the students view on the relationship between AI and creativity as four key concepts captured as social, affective, technological and learning factors. According to them, the students with a higher understanding of AI gave more positive thoughts about integrating AI into their classrooms, while those with low understanding of the system expressed fear of AI technology inclusion in education and hence explained that AI could never match human creativity. Nevertheless, the majority of the learners expressed satisfaction for the benefits they enjoyed from AI enhanced and quality education through personalized learning, adaptive assessments and real-time feedback. AI system also facilitated collaborative learning among learners from varied geographical locations, promoted critical thinking and offered diverse educational resources which contributed to a better engagement and holistic environment.

Parental and community involvement in AI educational initiative ensures that AI education is made relevant and accessible to a wide range of learners. Parental involvement plays a crucial role in a child's academic achievement and overall well-being. Parental involvement is a powerful force which shapes the student's academic success, social development and the entire well-being. It creates a supportive and enriching educational environment which empowers the students to reach their full potentials. Recent reports from Tech novation on family interest on artificial intelligence revealed that 86% of parents want new ways to learn critical computing skills outside the traditional classroom in order to provide their children with more guidance on at-home education. With immersive AI curriculum for children and their families, parents learn with their children and help to create AI-based products that solve problems in their communities. Hence, involving parents in the AI education initiatives helps to create a holistic learning environment for the learners. It also helps in bridging the generational gap in understanding technology as well as fosters parental support for their children. Community involvement in AI education initiatives encourages collaboration, knowledge sharing, diversity and supportive network for the learners. Diverse perceptions from members of the community provides grounds for more initiatives and helps to identify and address local challenges associated with AI initiatives.

2.11 Use of AI in addressing language barriers in education

AI has been a vital tool for addressing language barriers in education by making available the following relevant language resources and services available to both learners and educators.

- *Provision of language translation services*: AI translation tools can assist language translators to point out and correct mistakes, make suggestions, recommend alternatives and provide translation in real-time. AI provides enabling environment for communication among people with diverse languages to understand each other. AI translation tools can augment human translators by providing instant translations of spoken or written content. AI translators work by ensuring that contents are broken down into smaller segments and algorithms are then applied to each segment. This analyzes the content to generate a translation depending on patterns and set rules. AI translators can also improve future translations accuracy and consistency by learning from previous translations. Example of AI translators include: Google Translate, Microsoft Translator and DeepL.

- *Provision of Natural Language Processing resources*: AI provides natural language tools such as google Cloud Natural Language API, Gensim, SpaCy, IBM Watson, Natural language Toolkit, chatbots, search engines among others to help resolve ambiguities in language. The language APP provided by AI can help learners with diverse languages with list of vocabularies, listening exercises, gramma lessons, explanations and interactive language exercises.

- *Language teaching*: AI language Apps are used by teachers to translate texts, generate summaries at various levels. Multilingual learners also use AI tools and platforms to interpret words, construct sentences, write essays, generate communications and learn different language skills. AI enhance language evaluation by providing automated grading and feedback. AI language processing tools and Chatbots provide enhanced immersive language practice. Learners' speech can be evaluated and real-time feedbacks on punctuation, intonation and rhythm can be provided for him/her and the feedback in turn gives the learner the opportunity to improve the speaking skills accurately within the shortest time.

2.12 Current policies and regulations regarding the use of AI in education

Policies and regulations of AI in education differ from country to country and regions, however, irrespective of how these policies differ, some common considerations have been identified which play vital roles in teaching and learning. They are as follows:

Policies on ethical considerations: all stakeholders in education must be educated on the ethical use of AI in education. This is important because maintaining student privacy must be considered a primary ethical issue. This is supported by the fact that AI requires data to function and educational AI collects and analyzes sensitive data related to the learner's personal information which must be kept confidential and must not be shared without their consent. Even with consent, it is not right to prompt public models with data which can be traced to the learner.

Policies on bias and fairness: all stakeholders in education must be trained to identify and respond to bias associated with AI algorithms appropriately. This is because AI is usually trained using existing database which can lead to bias. Hence educators and learners must be prepared to recognize inaccurate information and racism as well as how to manage them.

Accessibility and equity: AI technology must be universal and accessible to all categories of educators and learners irrespective of their location, culture, ability, disability and social-economic status.

Instructor-learner relationship: AI technology must not be allowed to replace the human emotional, social and moral components which cannot be exhibited by AI. This relationship is very essential to enhance improve quality of learning in education.

Policy on balance use of AI: all stakeholders in education must be trained on the responsible use of AI in teaching and learning. This is because AI technologies use over time can pose the risk of dependency which can lead to hindered critical thinking skills which will affect the cognitive ability of the learners. This will also affect their problem-solving ability which can hinder learning. Hence the policy on maintenance of balance in use of educational AI.

2.13 Impact of generative AI on the quality of education

Generative AI includes all AI systems which has the potential of generating new contents such as written text, images and videos. Examples of Generative AIs include GPT, Pix 2 Pix, Cycle GAN, Style GAN, Text to image synthesis among others. Generative AI has significantly impacted the quality of education in the following ways.

Analyzing data: AI algorithms has the potential of analyzing large amount of educational data to point out areas that require improvement, identify patterns and various trends thereby educating stakeholders in education to reach useful decisions and help educators refine their pedagogical strategies.

Easy accessibility: generative AI makes it easy for learners to access learning resources irrespective of the category they belong. They help learners with various disabilities with alternative formats in form of text-to-speech features-, audio- and audio-visual descriptors for enhanced quality of learning, this on the other hand makes teaching and learning more inclusive.

Bridging language barriers: generative AI provides immersive experiences, language translation and language acquisition for all learners globally. These has helped in bridging the language gap in teaching and learning thereby making teaching and learning experience more qualitative.

Provision of automated feedbacks: generative AI provides a tutoring system that enable instant or automated feedback for learners. This in effect provide corrections for errors and provide guidance and support. These tutoring has the potential of augmenting the conventional pedagogy technologies and providing a better understanding of the learning process.

Individualized learning/instruction: generative AI enhances education quality by helping the learners learn at their own pace, according to their specific needs and circumstance, preferences and learning style. This helps to encourage engagement and better comprehension.

Task management: generative AI platforms can help instructors to manage their administrative tasks by automating assignment and quiz gradings, managing learners records thereby permitting the teachers to concentrate more on teaching and mentoring of the learners.

Creation of educational content: generative AI helps educators and instructors in generating assignments, quizzes, exercises and lesson content preparations and textbooks. This in turn helps him/her to present a very high-quality content to the learners.

3. Conclusion

The emergence and inclusion of artificial intelligence in education has contributed immensely in the achievement of quality education globally and this has benefited all the stakeholders. With AI system integration in teaching and learning content, the learners are better equipped mentally and psychologically with material resources to face learning challenges. The educators are assisted with administrative tasks which affords them the opportunity to attend to other important pedagogical activities. AI in education has given rise to automated grading system, real-time feedback, personalized learning, intelligent classroom assistants, intelligent Tutoring system among other advantages and these have enhanced the quality of education. However, AI system, has posed several challenges related to biases in AI algorithms, privacy concerns, redundancy, over dependency among others. These challenges have been addressed and strategies are being put in place to prepare for future challenges. Hence, AI in education has impacted positively in achieving quality education on a global scale.

Acknowledgements

The authors sincerely appreciate the contributions of the ICT staff of the Federal College of Dental Technology and Therapy Enugu Nigeria for their support during the full chapter development. We appreciate the useful contributions of some of our colleagues in the Biomedical Engineering department of the college in providing basic resources for the chapter. Special thanks to all the staff of IntechOpen publishers especially Dr. Ivana Barac for continuous effort to contact and encourage the authors. Thank you all.

Author details

Agatha Aballa Nkechi*, Akintayo O. Ojo and Obinna A. Eneh
Department of biomedical Engineering, Federal College of Dental Technology and Therapy, Trans-Ekulu, Enugu, Nigeria

*Address all correspondence to: aballaagathankechi@fedcodtten.edu.ng

References

[1] Helm JM, Swiergosz AM, Heather HS, Karnuta JM, Jonatha LS, Viktor EK, et al. Machine learning and artificial intelligence: Definition, application and future direction. Current Reviews in Musculoskeletal Medicine. 2020;**13**:69-76. DOI: 10.1007/5/2178-020-09600-8

[2] Hassain H, Silva ES, Unger S, Tajmazinani M, MacFeely S. Artificial intelligence (AI) or Intelligence Augmentation (IA): What is the future? AI. 2020;**1**:143-155. DOI: 10.3390/ai/020008

[3] UNESCO 2015 Incheon Declaration and framework for Action for the Implementation of Sustainable Development Goal 4. Paris, France: Science Research Publishing. Available from: https://www.scrip.org>reference

[4] Baker R, Siemens G. Educational data mining and learning analytics. In: Sawyer I, editor. Cambridge Handbook of the Learning Sciences. 2nd ed. New York, NY: Cambridge University Press; 2014. pp. 253-274

[5] Vanlehn K. The relative effectiveness of human tutoring, intelligent tutoring systems and other tutoring systems. Educational Psychologist. 2011;**46**(4):197-221. DOI: 10.1080/00461520

[6] Dam S. How Does AI Help in Adaptive learning AZoAI. 2023. Available from: https://www.Azoai.com/article/How-Does-AI-Help-In-Adaptive-learning.aspx [Accessed: January 12, 2024]

[7] Dede C. Immersive interfaces for engagement and learning. Science. 2009;**323**(5910):66-69. DOI: 10.1126/Science.1167311

[8] Anderson JR, Corbett AT, Koedinger KR, Pelletier R. Cognitive Tutors: Lessons learned. The Journal of the Learning Sciences. 1995;**4**(2):167-207

[9] Warschaver M, Matuchniak T. New technology and digital worlds: Analyzing evidence of equality in access, use and outcomes. Review of Research in Education. 2010;**34**(1):179-225. ISSN 0091-732. Available from: https://www.Learntechlib.org/p/108396

[10] Agbonika H. Disrupting the status que: Can artificial intelligence transform education in northern Nigeria? Public policy/governance/AI governance enthusiast. Policy delivery international limited. Powered by AI the LinkedIn Community

[11] Seo K, Tang J, Yoon D: The impact of AI on learner-instructor interaction in online learning, International Journal of Educational Technology in Higher Education. 2021;**18**:article number: 54. Available from: https://www.educationaltechnologyjournal.springeropen.com/articles>1.10.1186/s41239-021-00292-9

[12] Chen L, Chen P, Lin Z. Artificial intelligence in education. A review. Journal and Magazine. **8**:1-25. Available from: https://www.ieeeexplore.ieee.org

[13] Verma N. How effective is AI in education? Ten case studies and examples, artificial intelligence research. Axonpark.com. 2023. Available from: https://www.axonpark.com/how-effective-is-ai-in-education

[14] Hwang GJ, Xie H, Wah BW, Gasevic D. Vision, challenges, roles and research issues of AI in education,

computer and education. Artificial Intelligence. 2014;**1**:100001

[15] Garcia-Martinez I, Fernandez-Batanero J, Fernandez-Cerero J, Leon PS. Analyzing the impact of artificial intelligence and computational sciences on student performance. Systemic review and meta-analysis. Journal of New Approaches in Educational Research. 2023;**12**(1). Available from: https://www.naerjournal.ua.es

[16] Camesaria M, Rigueira-Diaz X, Janeiro A, Torres J, Prado I, Kreibel D. Impact of artificial intelligence on assessment methods in primary and secondary education: Systemic literature review. ScienceDirect. 2023;**28**(2):93-103

[17] Christenson CM, Horn MB, Johnson CW. Disrupting Class: How Disruptive Innovation Will Change the Way the World Learns. Expanded Edition. Washington, USA: McGraw-Hill e-book. ISBN: 978-0-07-175910-6

[18] European commission 2019, independent high-level expert on artificial intelligence: Ethics Guidelines for Trustworthy AI

[19] Byers J. Artificial intelligence: Educators embrace usefulness of artificial intelligence in classroom, note potential problems: The rise of AI. In: CNHI Special Reports. Pittsburgh, USA: CNHINEWS; 2023

[20] Ryu M, Han S. The educational perception on artificial intelligence by elementary school teachers. Journal of the Korean Association of Information Education. 2018;**22**(3):317-324. DOI: 10.14352/jkaie.2018.22.3.317

[21] Marrone R, Taddeo V, Hill G. Creativity and Artificial intelligence-A student perspective. National Library of Medicine (NIH), National center for biotechnology information. PubMed Central Journal of Intelligence. 2022;**10**(3):65. DOI: 10.3390/intelligence/0030065

Chapter 5

An Approach to Assist Learners to Build Their Own Curriculum in Personal Learning Environment Context, Based on the AI Concepts

Belhassen Guettat, Ramzi Farhat and Syrine Karoui

Abstract

With the emergence of information technologies and the proliferation and diversification of learning tools, educational institutions have diversified their approaches and resources in order to improve learners' performance and education quality. Thus, they have invested in the learning environments which offer learners' personalization learning possibilities. But this option depends generally on institutional constraints. Considering this limit, personal learning environments (PLEs) have come to allow learners to individually develop their learning environment by selecting the right curriculum, resources, and appropriate activities. This concept is in vogue, especially in the lifelong learning context. Mostly, the setting up of such an environment is not based on educational concepts (choice of objectives, selection of appropriate programs and activities). As a result, we are faced with undesirable situations: learning is not aligned with learner prerequisites; training layout does not align with the content and learner expectations. The question arises: how can learners define their learning objectives, set up their own activities, and follow their training? In this context, we propose an approach to assist learners to build their own curriculum, which is supported by an assisted PLE developed on the basis of artificial intelligence (AI) concepts and using a dynamic questionnaire.

Keywords: learning curriculum, personalization, personal learning environment (PLE), lifelong learning, machine learning algorithms

1. Introduction

Our contribution consists of defining an approach to assist engineering learners to build their own curriculum without pedagogy prior knowledge and to take autonomous control of their learning. It promotes lifelong learning, which aligns perfectly with the goals of sustainable development (SDGs). Three phases characterize our approach: the first concerns assistance in the identification of learning objectives and the recommend appropriate curriculum, the second will be assistance in the search for appropriate activities to the objectives already set, and the last will be responsible for maintaining and managing the learner's profile. In this paper, we will only focus on the first phase; the purpose is to build a learner's own curriculum. To do this, a consistent environment must be able to provide assistance to engineering learners to identify and choose their learning objectives. It is based on a dynamic questionnaire that takes into consideration the profiles and feedback learners' information. An educational learning objective according to IMS-LD standard [1] is represented by a couple formed by a concept (C) and a learning level (N): C being a concept belonging to a domain ontology θ and N is one among the taxonomic levels in pedagogy [2]. Once the couple is identified, our environment will be able to offer one or more curricula from a curricula corpus issue from different sources: educational and training institutions and other resources (Cloud, OER, Moocs). A classification strategy using unsupervised machine learning algorithms (ascending hierarchical classification also called clustering) will then be applied to recommend appropriate curricula. Before going into the details, it would be appropriate to review the basic PLE theoretical foundations and the related work carried out in this area. We will first start by explaining the PLE concept, the PLE-related work, and then we will present our assistance approach, and subsequently our recommender system. We will end by presenting our experiment, and the results are obtained.

2. Theoretical foundations of a PLE

Personal learning environment can be considered as a concept related to the use of learning technologies emphasizing learner ownership of tools and resources. The questions discussed are how does the learner use technology to manage his/her learning? How is individual activity captured? What are the distinctive characteristics of personal learning environment? This study supports the idea that PLE can be considered as a complex activity system using the activity theory (AT) framework [3].

2.1 Activity theory (AT) as an integrated framework

The PLE concept emphasizes the appropriation of tools and resources by learners. The view of learning as a mediation tool or collective activity is the basic principle of activity [4, 5]. Activity theory (AT) has been used as a framework for exploring pedagogical innovations and as a conceptual framework for analyzing and designing support systems for collaborative learning [6–10], mobile learning [4], and learning technologies evaluation [11, 12].

The study conducted by Buchem on a wide range of PLE publications supports the idea that a PLE can be considered as a complex system of activities and can be analyzed using the framework of activity theory in order to describe its main components [3, 9] (**Figure 1**).

2.2 PLE definitions

Several works have been carried out to bring a certain maturity to the PLE concept and above all a definition; we mainly cite the works of Buchem [3, 13–15]. From this review of the literature, we have identified several definitions of the term PLE: some have an educational vocation and others a technological vocation (**Tables 1** and **2**).

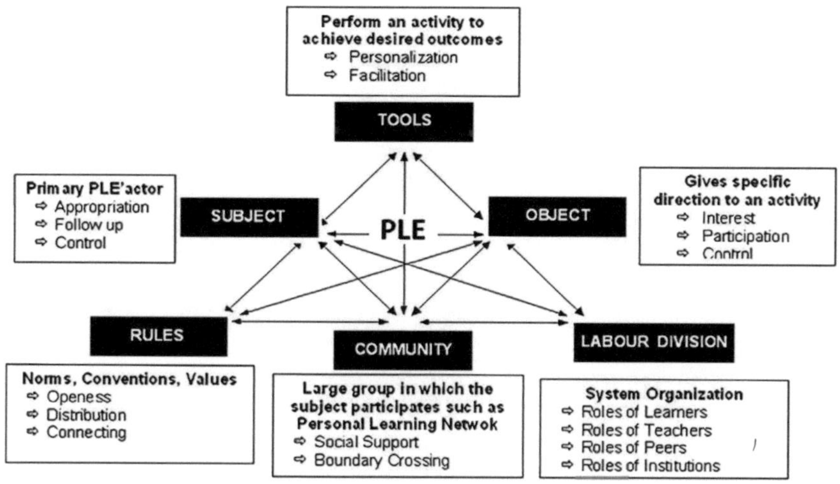

Figure 1.
Summary of the PLE elements and its main dimensions.

This diversification in the definitions led us to analyze them more closely and to propose our own definition which has a techno-pedagogical vocation: "A PLE is a learner-centered approach, based on web technologies and allowing support, control and appropriation of learning independently of technical and institutional constraints" [15].

2.3 PLE characteristics

Based on the views of Chattiand et al. [14], Martindale et al. [31], Drachsler et al. [32], Jafari et al. [33], Johnson et al. [21], Lubensky [17], and Guettat et al. [15], PLEs should have the following characteristics:

- PLEs are open systems controlled by learners independent of the educational establishment.
- PLEs are customizable by learners.
- PLEs concentrate all the tools useful for the learner in a single environment.
- PLEs promote informal learning and lifelong learning.

As a result, it becomes clear that PLEs represent a turning point, from a model where learners simply consume information to one where learners become autonomous and create connections with a variety of resources that they select and curate themselves.

2.4 PLE objectives

Although some of the fundamental needs of users of PLEs have not yet been clearly defined, two major objectives have nonetheless emerged in the literature: a PLE must be centered on learners and should enable lifelong learning [3]. These two goals align with the Sustainable Development Goals (SDGs).

Year	Author(s)	Definition
2006	Van Harmelen [16]	"A PLE is a system that helps learners to take charge of their own learning: set their own learning goals, manage their learning (content and process), communicate with others and thus achieve their learning goals."
2006	Lubensky [17]	"A PLE is the ease for an individual to access, aggregate, configure and manipulate digital objects during his/her learning."
2006	Milligan, Johnson, Sharples, Wilson, Liber [18]	"A PLE facilitates choice and control for a learner and enables the selection and combination of formal and informal learning opportunities issue from various sources."
2007	Attwell [19]	"A PLE should not be considered a software application, but a new approach in the use of new technologies in learning."
2007	Chen, Huang, Li [20]	"A PLE is a learning-oriented strategy that promotes each learner's learning ability in a web-based learning environment."
2007	Downes [13]	"A PLE is recognition that the "one size fits all" approach which characterizes LMSs will no longer be sufficient to meet the varied needs of the learners. It is not a software application in itself, but rather a characterization of an e-Learning approach."
2008	Johnson, Liber [21]	"A PLE is the desire to create learning centered on the learner; who is seen as the provider of his/her learning and that personal learning is fundamentally a learner-driven model of education where the traditional role centered on institutions is contested."
2008	Wilson, Liber, Johnson, Beauvoir, Sharples, Milligan [22]	"A PLE can be considered as a concept related to the use of technology for learning with an emphasis on learner appropriation of tools and resources."
2008	Schaffert, Hilzensauer [23]	"A PLE is the idea of a user-centered approach to learning, using social tools and software."
2008	Aviram, Ronen, Somekh, Winer, Sarid [24]	"A PLE is an educational model used for the self-regulated development of learners who are able to make informed, considered and strategic choices and plan their own learning as well as adapt the learning process according to their own needs, interests and their preferences."
2010	Buchem [3]	"A PLE can be considered as a concept related to the use of technology for learning with an emphasis on learner appropriation of tools and resources."
2010	Mcloughlin, Lee [25]	"A PLE is a concept adopting web 2.0 to promote continuous learning, informal learning and self-directed learning. It is an approach and not an application that allows the learner to take control of own learning environment."
2011	Chatti, Jarke, Specht, Schroeder [14]	"A PLE is a model for learner-centered learning that takes small, loosely-glued instructional chunks, characterized by their free-form use by a set of learner-controlled learning tools."
2022	Sarah, Serhan, Noraffandy [26]	"A PLE can be identified as a lifelong learning environment."

Table 1.
PLE definitions from an educational point of view.

2.5 PLE needs

With the trend in PLE being clearly visible today, questions continue to arise [14]: Who would need a PLE? Is there any feedback on PLEs? We need PLEs for lifelong

Year	Author(s)	Definition
2001	Olivier, Liber [27]	"A PLE is a single-user e-Learning system that provides access to a variety of learning resources, and which may provide access to learners and teachers who use other PLEs and/or other VLE (Virtual Learning Environments)."
2007	Wilson, Liber, Johnson, Beauvoir, Sharples, Milligan [22]	"A PLE is not a piece of software. It is an environment where people, tools, communities and resources interact flexibly."
2007	Anderson [28]	"A PLE is a web interface in the owners' digital environment."
2007	Siemens [29]	"A PLE is a collection of tools brought together under the conceptual notion of openness, interoperability and learner control."
2008	Attwell, Costa [19]	"A PLE provides both the framework and technologies to integrate personal learning and the work."
2009	Educause [30]	"A PLE describes the tools, communities and services that constitute the different educational platforms learners use to direct their own learning and to achieve educational goals."
2010	Martindale, Dowdy [31]	"A PLE is a specific tool or collection of defined tools used by the learner to organize and control their own learning."
2011	Chatti, Jarke, Specht, Schroeder [14]	"A PLE is a way of creating new web applications by combining (by aggregation and/or integration) existing data and services from different sources."
2022	Sarah, Serhan, Noraffandy [26]	"A PLE is open access online learning with learner-based guidance, self-direction and self-regulation."

Table 2.
PLE definitions from a technological point of view.

learning. The need is also the response to approaches requiring that the learning environment be under the control of the learner and that activities can be carried out offline. These needs have encouraged certain educational establishments to raise awareness among learners and teachers of the interest and contribution of PLEs, by instilling in them this PLENK 2010 culture [13].

2.6 PLE statistics

A study carried out by Sarah et al. [26] and published in 2022 in the "International Journal of Information and Education Technology," presented statistics and graphs showing the keen interest shown in this concept by researchers (**Figure 2**).

2.7 PLE limits

Following this brief overview of PLEs, we noted the lack of consensus on the terminology and even on the definition, hence the need for a more stable theoretical framework. In addition, the tools making up a PLE are heterogeneous; the need to orchestrate them is therefore desired. We also noted the limited number of methods and approaches that respond to the various issues raised [15]. We believe that, with the emergence of information technologies, PLEs work will progress further to compensate for the scientific deficiencies observed.

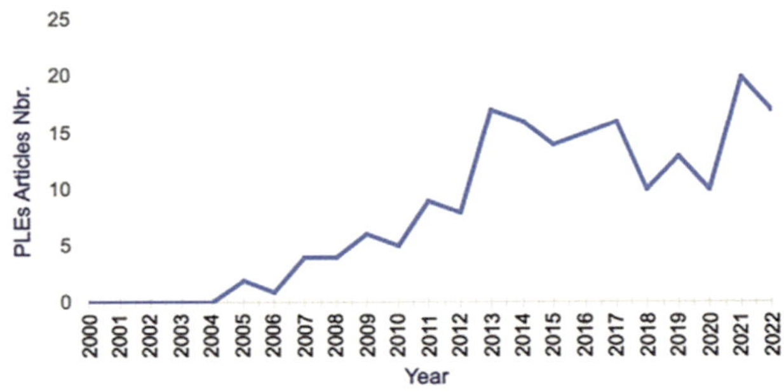

Figure 2.
Number of PLE article publications per year.

2.8 Synthesis

After analyzing publications made on PLEs, and reviewing the studies and work carried out by many researchers [3, 15, 26], we deduce the learner could obtain his/her educational and technological independence toward institution, by having the possibility of building his/her own learning environment and choosing his/her own resources and activities in order to achieve learning and personal objectives. However, several questions arise: how will the learner identify his/her objectives and therefore his/her learning curriculum?

Two answers can be considered: either the teacher guides the learner in his/her choices, in that case, we return to institutional learning environments with tutors and curricula dependency, or the learner, through the use of an intelligent system, could compose his/her personal curriculum. This second alternative, which has not been the subject of previous work, will be the subject of our research [15].

3. Lifelong learning concepts

Faced with the new landscape of educational technologies, learners continually face challenges in their learning. The speed of change as well as the growth of needs motivate learners to maintain the direction and extent of their lifelong learning. PLEs can be the appropriate solution to these situations. These environments give learners the freedom to learn beyond course boundaries and institutional constraints and customize their own learning environments before and during training. Additionally, e-portfolios used by learners as a tool to trace their learning provide future employers an overview of the individual's learning history and results, skills, and achievements. With PLEs, they allow learners to demonstrate their professional abilities in a continuous learning framework [34].

3.1 Lifelong learning vision

Lifelong learning is the "Ongoing, voluntary, and self-motivated pursuit of knowledge for either personal or professional reasons. Therefore, it not only enhances social inclusion, active citizenship, and personal development, but also competitiveness and employability" [13, 35, 36]. The diffusing of the lifelong learning vision signals the need for more personal, social, and participatory approaches that support learners in becoming active users and coproducers of his/her learning resources [35, 37]. The emphasis on the shift from formal to informal e-learning through knowledge management and sharing has been placed, with particular attention on the PLE as a learner-centered space. Nevertheless, the investigations are motivated by the many educational theories, implications, and challenges that PLE concept has posed [25].

3.2 Learner-centered learning

In a landscape marked by the evolution and emergence of educational technologies, and innovation in learning modes, models, and methods, the learner is obliged to assume his/her tool choices to use and contributions intended to make in learning. Therefore, we need a learning model centered on learner, adaptable, flexible, and specific, depending on the context, such that the learner will be able to control his/her individual choices in terms of the technologies to use by aligning them with his/her personal needs, interests, learning style, preferences, and context. In this way, learners will know how to build and manage a personal and self-reflective learning environment rather than operating an environment constructed, managed, and imposed by the teacher and/or institution [19].

3.3 PLEs roles in the lifelong learning

The PLEs give students the freedom to learn beyond course boundaries and to personalize their own learning environment. They allow learners to learn anytime and anywhere. E-portfolios are currently used by learners in many education institutions as a tool to document and reflect on their learning. They provide future employers with a snapshot of the learner's learning history, learning achievements, and reflective practice [38].

3.4 Our critical analysis

Today's learning systems should break away from traditional learning methods because they can no longer satisfy everyone, especially with the perpetual evolution of technology. Other measures should be found to motivate learners to learn not only when they are in academic training but also when they are independent. In our context, we are interested in lifelong engineering learners whose appropriation of learning can constitute a challenge for them. The solution that seems to be most appropriate is PLE. However, putting up personal learning environments requires solving a number of problems: how can the learner build his own personal curriculum? How does the learner profile will be maintained?

4. Our assistance approach

4.1 Overview

As part of our research in the field of PLE started in 2008, we have developed an innovative approach [15, 36], allowing learners to build their personal learning environment, by building their own curriculums and choosing their appropriate learning activities. Such an approach will promote lifelong learning. To do this, we defined architecture with three components: the "Curriculum builder," the "Learning activities recommender," and the "Profile manager" (**Figure 3**).

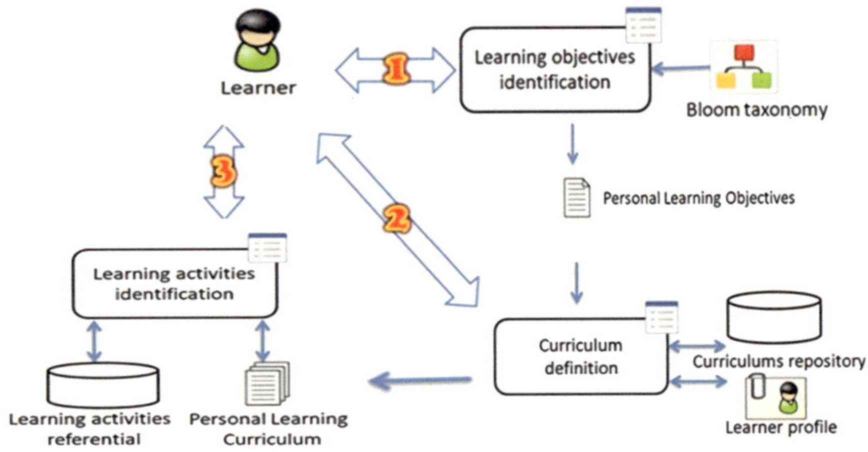

Figure 3.
Overview of our assistance approach.

4.2 Modeling learner

Nowadays, there are several specifications that aim to describe the learner in learning environments: IEEE PAPI [39], IMS-GLC-LIP [40], IMS-LTI [41], IMS-GLC-RDCEO [42], and IMS-GLC-LIS [43]. However, no study has been conducted to assess whether any of those specifications are appropriate to the PLEs.

So, we are concerned with finding a specification useful in the case of PLE in general and for our approach in particular. We have identified a requirement set of the learner model:

- Personal information (used to identify the learner and to interact with tools and learning environments),
- Previous knowledge (used to build his personal curriculum),
- Learning traces (used to manage the learning process),
- Learning objectives (used to store personal curriculum), and
- Preferences of the learner (used to select appropriate learning activities).

Based on our study, we demonstrate how the IEEE PAPI is suitable for the case of our approach and in general for the PLEs (**Table 3**) [44].

4.3 Assistance for identifying learning objectives

This component helps learners to choose their learning objectives. We start by offering them a list of concepts so they can choose one, for example, mechanics, computer science, management, mathematics, or medicine. Each concept has a sub-concepts list. For the "Computer Science" concept we propose "Algorithmics," "Office Automation," "Programming," "Databases," "Computer Architecture," "Operating Systems," and "Computer Networks." The choice of objectives will be based on an interactive dialogue with the learner using a dynamic and user-friendly questionnaire (**Figure 4**).

PAPI Category PLE requirements	PAPI contact	PAPI relations	PAPI security	PAPI preferences	PAPI performance	PAPI portfolio
Learner informations	☑		☑			
Previous knowledge						☑
Learning traces		☑			☑	
Learning objectives					☑	
Learner preferences				☑		

Table 3.
Adequacy of IEEE-PAPI for PLE.

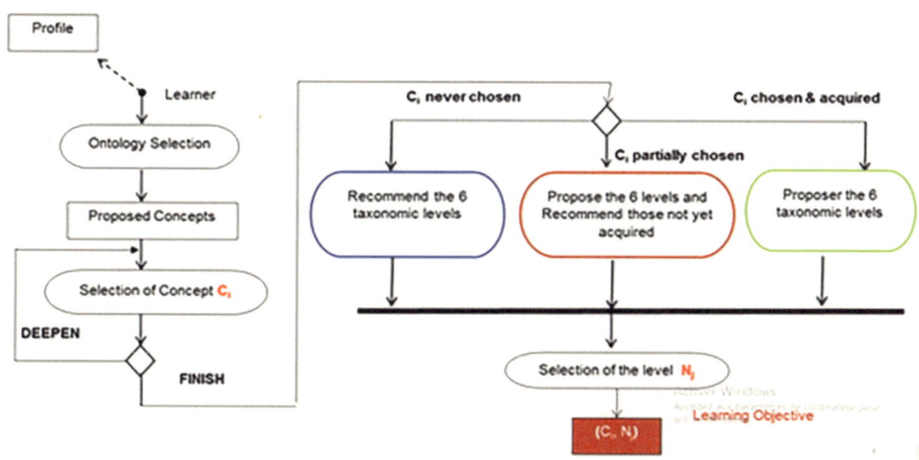

Figure 4.
Objective identification process.

4.4 Assistance for curriculum selection

Once the choice is made (concept Ci, taxonomic level Nj), the next step will consist of finding adapted curriculums. Two possible situations: In the first, an exact match is found between the curriculum's general objective and the learner learning objective, and in the second situation, we cannot find the right curriculum associated with the concept Ci.

4.4.1 First situation: exact match found

The selected curriculum will be used to identify the learning activities that must be accomplished by the learner. For example, we are looking for a course in "Computer Science" with a taxonomic level equals 2 ("Comprehension"); we found a bachelor's degree curriculum in computer science that matches. But in such a situation, several equivalent curricula may be found. Faced with such a situation, we will use concepts from Artificial Intelligence (AI) to apply one of the classification algorithms either to aggregate pieces of curriculums found or to make a classification to recommend curricula to the learner [45]. Based on our contribution which improved the IEEE-PAPI learner model in a PLE context, we are detecting significant and useful variables (features) for unsupervised machine learning algorithms—ascending hierarchical classification also called clustering (**Table 4**).

Variable	Codification	Description
V1	LANG	Learner's preferred Language: Fr, Ang, Ar, All, Esp.
V2	TYPF	Desired type of training: quick, medium, long.
V3	NBUC	Uses number of a given curriculum.
V4	NBAC	Number of completions on a given curriculum.
V5	RACC	Completion ratio on a given curriculum (RACC = NBAC/NBUC)
V6	NBOB	Number of objectives in a given curriculum.
V7	NATC	Average of marks awarded by learners on a given curriculum.

Table 4.
Sample of variables (features).

In **Figure 5**, we present a diagram describing the process of obtaining a personal curriculum.

In **Figure 5**, we can see how the learner profile will be updated once the curriculum is selected. We know the curriculum is a couple formed by an objective (Ontology Concept C) and a taxonomic level of learning N. According to our learner model based on IEEE-Papi learner, we will access the "Learning Objectives" class to update it, either by adding this objective if it is not part of the learner's profile or by updating this objective if the learner has evolved in learning by moving from one taxonomic level to another.

Previously, the classes representing the learner profile (**Table 4**) will also be updated according to the data provided by the learner, which is associated with the features involved in the ascending hierarchical classification algorithm (machine learning algorithm).

4.4.2 Second situation: right curriculum not found

For example, we are looking for training in BCNF (Boyce Codd Normal Form), but our system found nothing in the corpus. In this case, we need to go down the ontology and go to the "Normalization" node. It would then be necessary to work on the content of each curriculum concerning this node using its XML file and see if the associated block with the BCNF concept exists. The same thing here, we can find several equivalent blocks corresponding to our concept Ci and we must choose the most appropriate according to a classification strategy with always the same sample of variables.

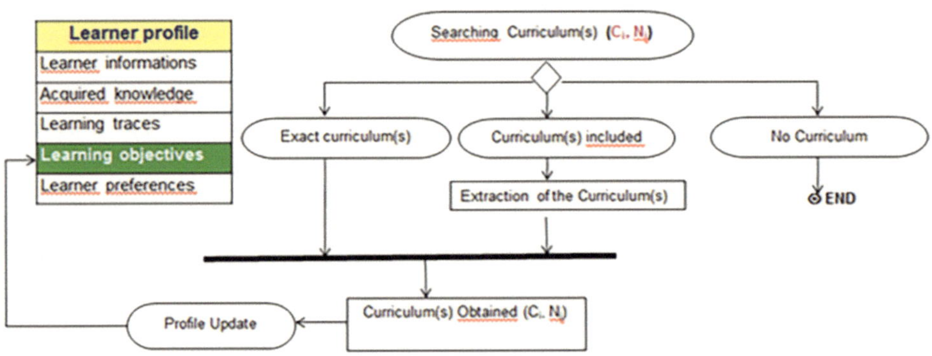

Figure 5.
Curriculum selection process.

To update the learner profile, we will use the same algorithm presented in the first situation.

5. Experimentation and results

In this section, we will experiment with a part of our approach (identification of personal learning objectives and curriculum recommendation). We have developed an assistant system, which allows any learner to use services offered without any technical or institutional constraints.

5.1 Web architecture

This is web architecture with a client using a browser (e.g., Chrome) containing our system, which will allow the learner to compose a personal curriculum and obtain the list of appropriate activities (**Figure 6**).

5.2 Web interface

In our system, which is based on our approach, we have a thin client which, through a Windows browser, offers the services planned and mentioned by our assistance system (**Figure 7**).

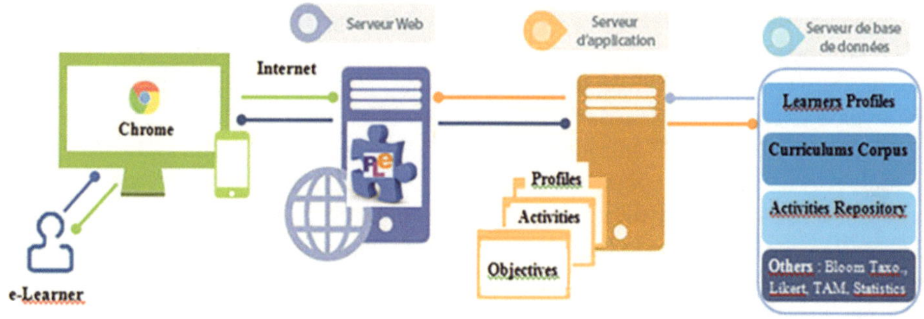

Figure 6.
Web architecture of our assistance system.

Figure 7.
Web interface of our assistance system.

By clicking on the "PLE" assistant, the learner could benefit from the offered services system: identifying learning objectives assistant, curriculums recommender, and activities recommender (**Figure 8**).

5.3 Assistance in identifying learning objectives

5.3.1 Input data set

To experiment with this component, we had the following data sets:

- Set of learner profiles with different scenarios: learner has never completed learning in a given concept, or has partially completed learning at given taxonomic levels, or has completely completed learning in a concept.

- Sample of learners requesting new learning curricula.

- A corpus of curriculums: each curriculum concerns a well-defined concept.

5.3.2 First situation: learner wants to learn "databases" level 1

Our system will offer him all the curriculums ("DB," 1) from our corpus. Which one will we recommend to him? Firstly, our system will make a filter by taking into consideration the learner's requirements and his/her profile. After that, our system will execute the unsupervised machine learning algorithm (ascending hierarchical classification also called clustering). After notification of the variables, we obtained a curriculum list from our curriculum corpus, including those dealing with "Databases" concept and the taxonomic levels. Given that the learner is interested in learning ("DB," 1), the system extracts from our corpus all the "DB" curriculums with taxonomic level equals 1 (**Table 5**).

Which curriculum our system will recommend to learner? The ML algorithm will calculate the similarity distances; before, it converts all the values in the interval [0...1], and sorts the curricula in ascending order according to d2 rubric (**Table 6**).

As we noted, the curriculum with the lowest distance will be recommended, and in our case, it is the curriculum ("DB," 1) with distance $d^2 = 0.979$.

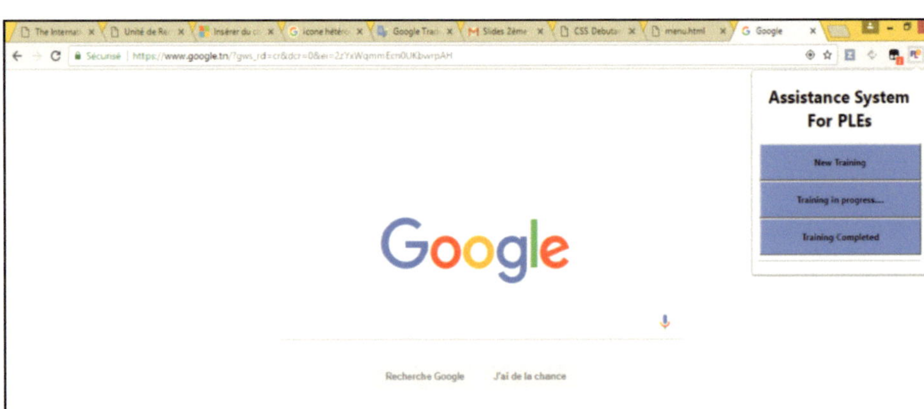

Figure 8.
Services offered by our assistance system.

Concept	Level	LANG	TYPF	NBOB	NBUC	NBAC	RACC	NATC
DB	1	1	1	2	1855	1000	0.539	06.20
DB	1	1	1	2	2500	1500	0.600	08.00
DB	1	0.66	0.66	3	1985	600	0.302	05.40
DB	1	0.5	1	2	750	300	0.400	04.80
DB	1	1	0.66	3	2265	1265	0.558	06.10
DB	1	1	0.33	5	4578	4000	0.874	08.70

Table 5.
Extraction of curriculums related to ("DB," 1).

Concept	Level	LANG	TYPF	NBOB	NBUC	NBAC	RACC	NATC	Distance (d^2)
DB	N1	1	0.33	5	4578	4000	0.874	0.98	00.979
DB	N1	1	0.66	3	2265	1265	0.558	0.69	02.060
DB	N1	1	1	2	2500	1500	0.600	0.90	02.296
DB	N1	0.66	0.66	3	1985	600	0.302	0.61	02.720
DB	N1	0.5	1	2	750	300	0.400	0.54	03.507
DB	N1	1	1	2	1855	1000	0.539	0.70	02.632

Table 6.
(d^2) sorted in ascending order by ML algorithm. Green color indicates the recommended curriculum(s).

5.3.3 Second situation: learner having "DB" levels 1 & 2 wants a "BCNF" level 1 curriculum

Our system searched in the corpus but found nothing. He turned back to his domain ontology to go back to one level. There, we found the concept "Normalization." We know well that the BCNF concept is one of the normal forms encountered in database courses, containing the "Normalization" chapter. We will therefore search all the normalization curricula and detect the presence of the specific objective relating to the Boyce and Codd normal form. This means the system will work on the curriculum content (XML file) and its metadata is made up of the following sections: Concept, level, language, training type, objective number, description, and list of specific objectives.

To find the concept, the system will process the list of specific objectives contained in the XML files. As soon as we find concept_objs = "BCNF," it will select the corresponding curriculum. After processing the already selected curriculums, we marked those which contain the concept "BCNF."

An example of an XML block containing the concept "BCNF" is shown as follows:

```
<Curriculum>
<id>BD25879          </id>
<Concept>NORM</Concept>
<Title>Normalization</Title>
<Level>1</ Level>
<Description>        Basic Concepts of Normalization</Description>
<Duration>2h00</Duration>
<Nb_Objectives>3</Nb_Objectives>
```

```
<li>
<id_objs>1</id_objs>
<concept_objs>DF</concept_objs>
<Title>Functional Dependence</Title >
< Level_objs>1</ Level_objs>
<id_objs>2</id_objs>
<concept_objs>3NF</concept_objs>
< Title >Normal Forms</ Title >
< Level_objs>1</ Level_objs>
<id_objs>3</id_objs>
<concept_objs>BCNF</concept_objs>
< Title > Boyce Codd Normal Form</Title >
< Level_objs>1</ Level_objs>
</li>
```

</Curriculum>

To find the concept, the system will process the list of specific objectives contained in the XML files. As soon as we find concept_objs = "BCNF," it will select the corresponding curriculum. After processing the already selected curriculums, we marked those which contain the concept "BCNF." Following this processing, we obtain the following four curricula (**Table** 7).

The learner is looking for a BCNF curriculum with level = 1; our system recommends two, but the one with d^2 = 0.2433 will be best recommended.

5.4 Experimentation results

Nearly, a hundred learners enrolled in the first year of IT engineering took part in the experiment. At first, they passed a pretest to divide them into two similar groups according to their level. After that, the two groups were invited for a test (on the same day: 2 hours). We asked the learners to solve the same exercise (about relational databases normalization) by creating their own PLE. Each learner in the control group has to build his/her own PLE and therefore to solve the given exercise. However, learners in the experimental group have access to our assistance system installed in their web browser. To evaluate the effectiveness of our approach, we measured the time and scores obtained by the group that used an unassisted PLE and the one that used an assisted PLE. We observed firstly the time of realization for the same activity to the two groups (control and experimental) (**Table** 8).

Concept	Level	LANG	TYPF	NBOB	NBUC	NBAC	RACC	NATC	Distance (d^2)
Normalization	N3	0,66	0,66	0,33	0,47	0,19	0,396	0,98	01,978
Normalization	N1	0,5	0,66	0,33	0,15	0,13	0,821	0,67	02,433
Normalization	N1	1	1	0,00	0,13	0,13	0,977	0,46	02,818
Normalization	N4	0,66	0,66	0,33	0,00	0,00	0,000	0,00	04,676

Table 7.
List of obtained curricula. Green color indicates the recommended curriculum(s).

	Group	Number of participants	Average (mn)	Standard deviation
Time	Control	50	108,4000	10,20022
	Experimental	50	20,8333	2,00144

Table 8.
Average of times activity.

	Group	Number of participants	Average (mn)	Standard deviation
Scores	Control	50	7,85	5,71086
	Experimental	50	14,64	1,64485

Table 9.
Average of scores activity.

The results confirm what we observed on the premises: the learner in the control group wasted a lot of time to find the appropriate resources to carry out the activity. We conclude that with an assisted PLE there is a gain in required learning time. On the other hand, we obtained the scores obtained by the two groups (**Table 9**).

We find that the mean of the control group is 7.9 with a standard deviation of 5.7. On the other hand, the results obtained in the experimental group are much better. Indeed, the average score is 14.6 (almost double) with a small standard deviation compared to that observed in the control group. This clearly shows that the use of an assisted PLE improves the learners' performance.

6. Conclusion

Nowadays, with the emergence of information and communication technologies, learners have access to different learning systems: ranging from simple resources such as tutorials, e-Books, Moocs, and OER, to learning environments, which are generally led by an institutional framework whose teaching strategies are defined in advance: curricula, predefined teaching scenarios and methods, teaching tools and logistics, trainers, etc. In such a context, the learner only exploits these environments at the request of the teacher or the institution without being involved in the structuring of courses and pedagogical choices. The only possibility offered to them during their training is to personalize their environment during learning according to their profile and preferences. Such a strategy cannot surely satisfy all learners: Demotivation during learning, poor perception of certain modules included in the curriculum, and risk of repeating certain modules already validated elsewhere. All these are because of the design and implementation of training strategies that were developed based on teaching centered on the teacher. Such systems, which suffer from several failures: lack of harmonization between education systems and the professional environment and gaps in the design of training curricula, can no longer create value within training institutions.

In certain countries where learning strategies are frequently reviewed in the direction of continuous improvement and learner satisfaction, there is a trend toward a break with traditional systems centered on the teacher. We want learning centered on the learner where the latter would be involved at all levels of training: from pedagogical choices to the choice of resources in the learning environment. Such a system aims to be flexible at all times: learning without time constraints, learning anywhere, learning with any equipment, and total autonomy of learners in the choice of curriculum elements and even in the choice of their activities learning. This break is essentially justified by a better quality of learning with better motivation of the learner, who will move from the traditional passive mode to another comfort with the appropriation of his learning, from the choice of the objectives of his curriculum until the completion of the related activities. This would be ideal for working individuals who wish to acquire scientific and professional skills as part of lifelong learning. A question arises in this case: in practice, are there learning environments can offer this type of service to these learners?

This last decade has seen the emergence of a new concept, which is currently attracting the attention of several research teams. These are personal learning environments (PLE): do not confuse them with customizable learning environments. In our chapter, we have removed this ambiguity in order to clarify the PLE concept for researchers working on it. This concept, despite the efforts made by researchers, fails to reach a maturity threshold in terms of the stability of the concept, given the different educational and technological points of view presented in several works. Most mention the learning resource needs of learners but never the method of identifying objectives and selecting training programs. As if the learner is initiated into the creation of his curriculum, it is able to choose the appropriate objectives according to his profile and able to select the appropriate learning activities. In practice, only those who have teaching skills (trainers) are capable of doing all this. At this level, we asked ourselves several questions: Is the learner capable of identifying his or her own learning objectives? Do they need to introduce them to certain educational concepts to do so? Should it depend on an institutional structure to carry out this work? In other words, could the learner, without being an expert in pedagogy, identify his objectives and build his personal curriculum? All these reflections have been detailed in this chapter thanks to a new assistance for learners in a personal learning environment context and in a lifelong learning framework and which mainly based on artificial intelligence concepts for recommending appropriate curricula and activities.

In this chapter, we presented a brief overview of the state of the art on personal learning environments: we reviewed the basic concepts; the research carried out and the fit between PLEs and lifelong learning. Then, we detailed the different phases of our assistance approach in a personal learning environment: definition and identification of learning objectives, selection of a learning curriculum, and management of the learner profile. For the learner model, we adapted the IEEE-PAPI learner standard for PLEs in general and particularly for our approach (standards adopted have until now not dealt with the case of PLEs). In the different phases of our approach, we used artificial intelligence (AI) concepts and more precisely machine learning algorithms (ascending hierarchical classification) to offer learners appropriate curricula and activities. Our approach was implemented, and the system was developed and tested on a representative sample of learners. The experimentation results are analyzed and interpreted. It was carried out on two groups of learners, a first group working freely on their own PLE to solve a given exercise, with learning tools of their choice, without guidelines, method or approach, and another group, which had the privilege of using our support system while solving the same exercise.

Obviously, this contribution remains open to other actions, either in terms of potential for improvement or in terms of extension. The first action consists of activating the assistance phase which consists of offering appropriate educational activities to the learner after choosing their curriculum; knowing that on the same curriculum, activities can vary from one learner to another depending on their preferences, profile, etc.. The same machine learning process will be applied but this time on activities and no longer on curricula.

As for perspectives, the first will concern the learner profile manager, who is responsible for storing the learner's information in a consistent manner so that his profile is up to date. A learner's activities can come from several learning environments, and to update a learner's profile, we must use a generic model to convert their data not necessarily from the same model. The goal is to prevent the learner from repeating activities already carried out in other environments.

The second perspective is an extension of our approach in terms of learning situations. We recall that we worked on two situations (finding exactly curricula aligned with learning objective and taxonomic level and finding objective and level including in others curricula).

A third situation not developed could be subject to an extension of our approach: it concerns the first approach phase of identifying objectives and the search for a curriculum: it is a situation where the learner introduces a concept Ci relating to a domain ontology and a taxonomic level N to our system, the latter will check in the corpus of curricula, but does not find a curriculum associated with the concept Ci and the taxonomic level N. However, in certain curricula, we find some pieces of curriculum (m_k), which, by aggregating them, can give rise to a curriculum approaching the desired one. Proposing a solution to this situation seems easy on the surface, but in reality, it is not going to be easy. How could our system assist a learner so that the latter can affirm whether such a piece m_k is part of his concept Ci? Generally, this is the mission of a teaching expert. In its absence, it would then be necessary to add an intelligent component (AI concepts) capable of detecting these pieces and recommending them to the learner.

Author details

Belhassen Guettat[1*], Ramzi Farhat[1] and Syrine Karoui[2]

1 National Higher School of Engineers of Tunis (ENSIT), LaTice Laboratory, University of Tunis, Tunis, Tunisia

2 Esprit School Engineering, Tunis, Tunisia

*Address all correspondence to: belhassen.guettat@wanadoo.tn

References

[1] IMS Global Learning Consortium. In: Koper, Olivier, Anderson, editors. IMS Learning Design. Best Practices and Implementation Guide. Lake Mary, USA: IMS Global Learning Consortium; 2003

[2] Bloom BS, Krathwohl DR, Masia BB. Taxonomy of Educational Objectives: The Classification of Educational Goals. New York: Longman; 1984

[3] Buchem I, Attwell G, Torres R. Understanding personal learning environments: Literature review and synthesis through the Activity Theory lens. In: Buchem I, Attwell G, Tur G, editors. Proceedings of PLE Conference 2011; 11-13 July 2011; Southampton; UK. Vol. 54, No. 2. Berlin, Germany; 2011. pp. 1-33

[4] Sharples M, Taylor J, Vavoula G. Towards a theory of mobile learning. In: Isaías P, Borg C, Kommers P, Bonanno P, editors. Proceedings of mLearn 2005; 25-28 October 2005; Brimigham University; UK. USA: IADIS; 2005. pp. 1-9

[5] Scanlon E, Issroff K. Activity theory and higher education: Evaluating learning technologies. Computer Assisted Learning Journal. 2005;**21**(6): 430-439. DOI: 10.1111/j.1365-2729.2005.00153.x

[6] Ogden CK, Richards IA. The meaning of meaning. In: Javanesegraviton, editor. A Harvest Book Harcourt. New York: Brace & World, Inc.; 1923

[7] Vygotsky L. Mind in society: The development of higher psychological processes. In: Cole M, John-Steiner V, Scribner S, Souberman E, editors. Cambridge, USA: Harvard University Press; 1980

[8] Engeström Y. Learning by Expanding: an Activity Theoretical Approach to Developmental Research. United Kingdom: Cambridge University Press; 2014. ISBN: 9781139814744. DOI: 10.1017/CBO9781139814744

[9] Holton J. The coding process and its challenges. In: Bryant, Charmaze, editors. Sage Handbook of Grounded Theory. London, UK: Sage; 2007. pp. 237-262

[10] Couros A. Developing personal learning networks for open and social learning. In: Veletsianos, editor. Book Emerging Technology in Distance Education. Ch6. Canada: AU Press, Athabasca University; 2010. pp. 109-128. ISBN: 978-1-897425-77-0

[11] Nardi BA. Studying context: A comparison of activity theory, situated action models, and distributed cognition. Context and consciousness activity theory and human computer interaction. In: Veletsianos G, editor. Book Context and Consciousness: Activity Theory and Human-Computer Interaction. Ch4. USA: MIT Press; 1996. pp. 69-102. DOI: 10.7551/mitpress/2137.001.0001

[12] Albero B. Pratiques d'apprentissage dans et hors institution: une dialectique enfin possible dans les dispositifs émergents de formation. "Recherches en Communication" Review. 2001;**15**:103-119. DOI: 10.14428/rec.v15i15.47463

[13] Downes S. Personal Learning Environments, Networks and Knowledge. PLENK. Southampton, UK: Online Discussion Forum/Blogs; 2010

[14] Chatti MA, Jarke M, Specht M, Schroeder U. Model-driven mashup personal learning environments.

International Journal of Technology Enhanced Learning. 2011;**3**(1):21-39. DOI: 10.1504/IJTEL.2011.039062

[15] Guettat B, Farhat R. Assistance system for personal learning environments. In: Proceedings 6th International Conference on Information and Communication Technology and Accessibility (ICTA'17); 19-21 December 2017; Muscat; Oman. Vol. 1. USA: IEEE Xplore; 2018. pp. 1-5. CFP1712W-POD. ISBN: 9781538613528. DOI: 10.1109/ICTA.2017.8336046

[16] Van Harmelen M. Personal learning environments. In: Proceedings of the Sixth International Conference on Advanced Learning Technologies (ICALT'06); 4-7 July 2006; Kerkrade, The Netherlands. USA: IEEE Xplore. pp. 815-816. DOI: 10.1109/ICALT.2006.1652565

[17] Lubensky R. The present and the future of personal learning environment. In: Encyclopedia of Information Science and Technology. 2nd ed. Pennsylvania, USA: IGI Global; 2008. DOI: 10.4018/978-1-60566-026-4

[18] Milligan C, Johnson M, Sharples P, Wilson S, Liber O. Developing a reference model to describe the personal learning environment. In: Nejdl W, Tochtermann K, editors. Proceedings of Innovative Approaches for Learning and Knowledge Sharing; First European Conference on Technology Enhanced Learning (EC-TEL'06); 1-4 October 2006; Crete; Greece. New York: Springer; 2006. pp. 506-511. E-ISBN: 9783540462347

[19] Attwell G. Personal learning environments: The future of eLearning. E-Learning Papers. Jan 2007;**2**(1):1-8. ISSN 1887-1542

[20] Chen CM, Huang TC, Li TH. Personalized E-learning system with self-regulated learning assisted mechanisms for promoting learning performance. In: Proceedings of the Seven International Conference on Advanced Learning Technologies (ICALT'07); 18-20 July 2007; Niigata, Japan. USA: IEEE Xplore; 2007. pp. 637-638. DOI: 10.1109/ICALT.2007.205

[21] Johnson M, Hollins P, Wilson S, Liber O. Towards a reference model for the personal learning environment. In: Markauskaite, Goodyear, Reimann, editors. Proceedings of the 23rd ASCILITE'05 Conference; 5-8 December 2004; Perth; Australia. Australia: Sydney University Press; 2006. pp. 385-389. ISBN-10: 1920898573

[22] Wilson S. Patterns of personal learning environments. Interactive Learning Environments Journal. 2008;**16**(1):17-34. DOI: 10.1080/10494820701772660

[23] Schaffert S, Hilzensauer W. On the way towards personal learning environments: Seven crucial aspects. In: Proceedings of the International Conference on ICT in Teaching and Learning (ICT'2011); 11-13 July; Hong Kong. New York: Springer Link; 2011. pp. 22-36

[24] Aviram A, Ronen Y, Somekh S, Winer A, Sarid A. Self-regulated Personalised Learning (SRPL): Developing iClass's Pedagogical Model. No. 9. Spain: eLearning Papers; 2008. pp. 1-17. ISSN 1887-154

[25] Mcloughlin C, Lee MJW. Personalized and self-regulated learning in the web 2.0 era: International exemplars of innovative pedagogy using social software. Australasian Journal of Educational Technology. 2010;**26**(1):28-43. DOI: 10.14742/ajet.1100

[26] Sarah A, Serhan L, Noraffandy Y. A systematic review and trend analysis of

personal learning environments research. International Journal of Information and Education Technology (IJIET'22). 2022;**12**(1):43-53. DOI: 10.18178/ijiet.2022.12.1.1585

[27] Olivier B, Liber O. Lifelong learning: The need for portable personal learning environments and supporting interoperability standards. In: The JISC Centre for Educational Technology Interoperability Standards, Bolton Institute (JISC'2001). UK: Bolton Institute. Available from: http://wiki.cetis.ac.uk/uploads/6/67/Olivierandliber2001.doc

[28] Anderson T. Personalized learning system and you. In: PLE'2007 Conference. Canada: Manitoba Universtity; 2007

[29] Siemens G. I acronym, therefore I exist. In: Bastiaens T, Carliner S, editors. Proceedings World Conference on E-Learning in Corporate, Government, Healthcare, & Higher Education (Elearn 2007); 15-19 October; Québec City; Canada. Maryland, USA: AACE; 2007

[30] Educause Learning Initiative (ELI). 7 Things You should Know about PLE. Educause Publisher; 2009. Available from: www.educause.edu/library/resources/7-things-youshould-know-about-personal-learning-environments

[31] Martindale T, Dowdy M. Personal learning environments. In: Veletsianos ed. Handbook Using Emerging Technologies in Distance Education. 2010. Vol. 16, pp. 177-193

[32] Drachsler H, Hummel HGK, Koper EJR. Personal recommender systems for learners in lifelong learning networks: Requirements, techniques and model. International Journal of Learning Technology. 2008;**3**(4):404-423

[33] Jafari A, McGee P, Carmean C. Managing courses, defining learning: what faculty, students, and administrators want. Educause Review;**41**(4):50-70. ISSN 1527-6619

[34] Chen T. Recommendations for creating and maintaining effective networked learning communities: A review of the literature. International Journal of Instructional Media. 2003;**30**(1):35-44

[35] Leone S. Characterization of a Personal Learning Environment as a Lifelong Learning Tool. Vol. XIV. 88 p. New York: Springer Publisher; 2013. DOI: 10.1007/978-1-4614-6274-3

[36] Guettat B, Farhat R. Du Personalisable au Personnel. In: ISLAIB, editor. Proceedings InfoLangues'2013 Conference; 13-14 April 2013; Beja; Tunisia. Tunisia: CPU; 2013

[37] European Commission. ALPINE - Adult Learning Professions in Europe: A Study of the Current Situation, Trends and Issues. Final Report. Brussels, Belgium. Zoetermeer, Netherlands: Research voor Beleid; 2008. Project number: B3276

[38] Renon F. Personal Learning Environments (PLEs): Ecologies for Building Student Capability for Lifelong Learning. Ottawa, Canada: Mac Odrum Library; Carleton University Publisher; 2013

[39] IEEE P1484.2.1/D8. PAPI Learner: Core Features. ISO/IEC JTC1 SC36, Information Technology for Learning, Education, and Training. Distributors: P, O, L Members, WG Conveners. New York, USA: IEEE; 2002

[40] IMS-GLC-LIP. IMS Learner Information Package Summary of

Changes. Lake Mary, USA: IMS Global Learning Consortium; 2005

[41] IMS-GLC-RDCEO. IMS Reusable Definition of Competency or Educational Objective Specification. Lake Mary, USA: IMS Global Learning Consortium; 2002

[42] IMS-LTI. IMS Learning Tools Interoprability. Lake Mary, USA: IMS Global Learning Consortium; 2015

[43] IMS-GLC-LIS. IMS Learning Information Services, Best Practice and Implementation Guide; 2nd version; Final Release. Lake Mary, USA: IMS Global Learning Consortium; 2011

[44] Guettat B, Farhat R. Gestion du profil d'apprenant en utilisant la norme IEEE-PAPI dans un contexte d'environnement d'apprentissage personnel. In: Proceedings InfoLangues'14; 24-26 April 2014; Béja; Tunisia. Béja, Tunisia: ISLAIB; 2014

[45] Morineau A, Lebart L, Piron M. Statistique Exploratoire Multidimensionnelle : La Classification Hiérarchique. In: Dunod, editor. 2nd ed. Malakoff, France; 1998. p. 439. ISBN-13: 978-2100040018

Chapter 6

AI-Based Edutech for Adaptive Teaching and Learning

Hwang Eunkyung

Abstract

The artificial intelligence (AI)-based problem learning system quickly and accurately performs problem setting and scoring using algorithm. In this process, the learner's level of prior learning is identified, the subject and quantity to be learned are determined and problem learning is provided for each learner. The basic use of AI-based problem learning enhances ease and fairness in performing assignment and evaluation and provides data that can strengthen interactions between instructors and students. Above all, the biggest advantage is the possibility of helping individual learners with different levels of prior learning to strengthen basic learning. To this end, instructors need to understand the technical aspects of the system, check the content system as an educational goal set by the instructor, and make efforts to supplement the necessary parts. When AI-based problem learning is used in connection with classes, a technical understanding of a system that can utilize various functions of the AI system more efficiently is required. In addition, instructional design is needed to expand thinking and strengthen capabilities through the process of structuring and understanding the contextual relationship between concepts based on the learned knowledge of students using AI-based problem learning systems.

Keywords: artificial intelligence, problem learning system, adaptive learning, adaptive class, instructional design

1. Introduction

Edutech is used in various ways to enhance the quantitative and qualitative effectiveness of the teaching and learning process based on the rapid development of the hardware of information devices and the software technology necessary for the operation and management of these devices. In particular, the use of artificial intelligence as edutech, which has recently attracted attention, is being used in various forms depending on the level of technology, such as customized education tailored to the characteristics of learners, interactive systems, learning and inquiry support, student writing analysis, and intelligent agents [1, 2].

These changes also affect the university curriculum, expanding to the use of customized learning and adaptive teaching methods in consideration of students' learning situations [3–5]. This is being attempted as an alternative to teaching and learning to meet the needs of the teaching and learning process, such as bridging the learning gap and improving basic academic ability, in the lecture field, where students who take these subjects have various majors and the difference in their prior learning and learning competencies is based on the same subject content.

2. AI-based edutech for adaptive teaching and learning

2.1 Characteristics of AI-based edutech

From Alan Turing's Turning test in 1950, the Expert system used in Dendral and MYCIN in the 1970s, and deep learning, a type of machine learning in the 2000s, to Generative AI, the rapid development of artificial intelligence, which can be seen at several symbolic stages, is expanding its role to its use as edutech. However, algorithms in which artificial intelligence (AI) learned through deep learning analyzes data on the amount of individual students and learning outcomes to present individual learning paths are one of the important features of AI-based edutech, thanks to the improved performance of parallel computers capable of processing big data at high speed [2, 3]. These algorithms are related to the Rule Base Algorithm, which sets priorities based on rules, organizes, proceeds, and verifies decision trees, and the Item Response Theory, which analyzes data in response to questions to calculate the degree of learning completion of respondents and the difficulty of individual questions [6, 7]. The flowchart in **Figure 1** briefly illustrates the principle of problem generation that plays an important role in customized or adaptive teaching and learning based on these algorithms.

First, taking Concept I as an example, if a student presents the correct answer to a given problem (Concept I-1) dealing with Concept I in the learning path, the learning of Concept I-2 proceeds, followed by the learning of Concept I-3 and Concept I-4, and

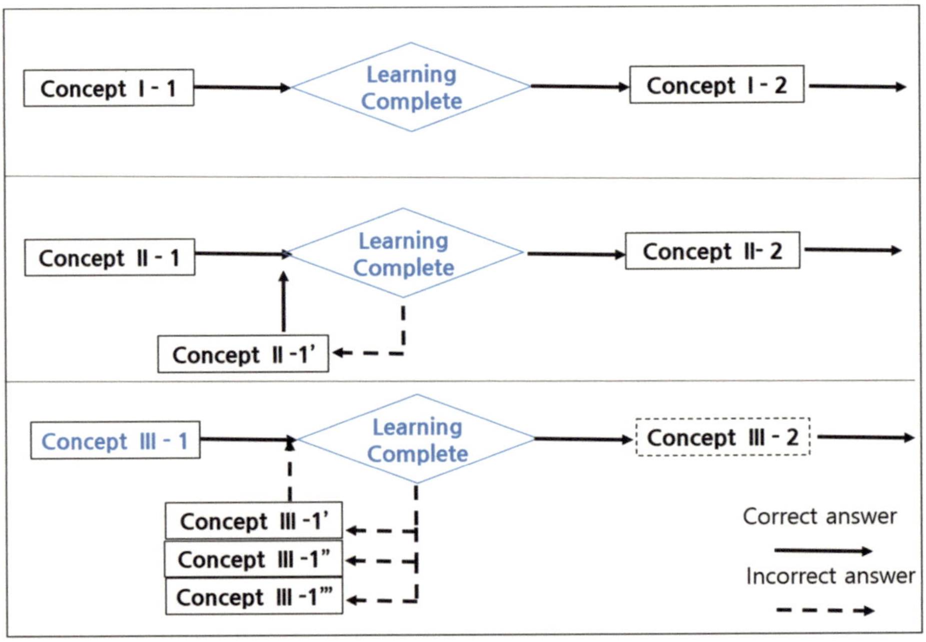

Figure 1.
Problem generation process.

learning related to Concept II can also begin. If the student gives an incorrect answer to the problem (Concept II-1), as shown in the schematic diagram related to Concept II, the algorithm will allow further learning of the content with the accompanying supplementary explanation and then make the student solve the problem of Concept II-1' similar to Concept II-1. If, as in concept III, a similar problem-solving is retried, the correct answer is not presented immediately, and the wrong answer is repeatedly submitted, the learning proceeds by continuously solving the similar problem until it is recognized as complete learning. As shown in concept III-1 in **Figure 1**, if learning completion is not achieved even through repeated attempts, the system recognizes the level of learning by the 'Learning' step, which is a lower level than 'Learning Completion', then the next learning proceeds. In the learning path shown in **Figure 1**, this problem generation method proceeds up to the main concepts I, II, and III, and learning-related information is collected by checking the number of consecutive correct answers, the number of correct answers and incorrect answers. This is used as direct data for AI algorithms to determine whether individual learners' understanding of concepts is completed or not. For example, in a system that has set three consecutive correct answers to a single concept as 'Learning Completed', if a student submits three consecutive correct answers to a question about Concept I, the system recognizes it as 'Learning Completed' and proceeds the learning path with the next concept, concept I-2. Whereas, if the incorrect answer is submitted between the submitted answers even though the correct answers have been submitted three times in this process, the learning path is conducted to solve additional problems, determine 'complete learning' using a separate scoring method, and solve the next step, concept I-2. For other topics selected in consideration of teaching and learning objectives, this principle is also basically used to generate problems that reflect priorities in the rule-based algorithm to proceed with learning. These AI problem learning systems were initially provided as independent adaptive problem learning, but recently, their scope of use has been expanded in a way that is used to implement adaptive teaching methods using problem learning in classes [8, 9].

2.2 Learning diagnostics

The evaluation method conducted by the instructor for learning diagnosis and analysis can be classified based on the factors of problem generation, scoring, and scoring result analysis as expressed in **Table 1**.

First, In the case of the traditional paper test conducted using paper prints, a professor directly creates, distributes, scores, and analyzes problems. Although the assistant supports it, the professor is involved and participates in a series of processes related to the evaluation of learning. Next, In the CBT system using screens and keyboards on computers connected to the Internet, a professor develops, creates,

	Question creation	Question distribution	Scoring	Result analysis
Paper based test system	Professor	Professor/Assistant	Professor/ Assistant	Professor/ Assistant
Computer-based test system	Professor	System	System	System
AI-based test system	System	System	System	System

Table 1.
Evaluation for learning diagnosis and analysis.

and enters a problem into the system through a professor or other inputter. Problem Distribution, scoring, and analysis are a method of receiving help from a computer connected to the Internet of systems. Recently, in conjunction with the comprehensive learning system, it also supports overall matters related to learning, such as learning materials, information, and announcements provided to students. Finally, in an AI-based test system, an AI algorithm is involved in the distribution of the generated problem and management of information about learning in an integrated manner in computer-based test systems. A subsequent learning path through analysis of the automatically scored result is determined and proposed by the AI algorithm.

The AI system identifies topics and concepts that each student needs to learn through initial diagnostic tests, generates related problems to be initially distributed based on the percentage of correct answers, and starts individual learning paths. The student's problem-solving scoring results in the learning process according to this learning path are stored as data. Basic results such as the completion rate of problem learning for each topic and concept and data such as the time required to solve problems by concept, access time, and end time are stored as diagnostic information about the learning progress of individual students and then are sometimes shared to professors and students. The information is provided to the professor as a sum and average information about all students. Additionally, diagnostic information related to students' conceptual understanding is provided by providing information such as questions with high incorrect answer rates, questions with many correct answers, and time required in order to analyze questions. These materials are converted into a relative understanding of specific topics among individual learners' entire subjects and used as basic data to determine the learner's level of prior learning, helping to determine the content and quantity of instructors' teaching and learning activities. These learning result analyses and diagnostic data are organized and shared on a dashboard for easy use by instructors. This is also possible when the CBT evaluation system itself or the CBT system is operated in conjunction with the comprehensive learning management system, but the functions are much more diverse and rich in AI-based testing systems.

In particular, it provides both instructors and individual students with comprehensive and integrated information on learning based on individual learning paths. Based on this, question-and-answer required for learning can be made more effective between students and professors, which seems to contribute to enhancing interaction. In addition, in the case of a problem learning system that provides the ability for instructors to view the student's learning screen the same, it allows one to understand the more specific request for learning.

For example, in the case of ALEKS, which is frequently used for AI-based adaptive problem learning, 'ALEKS Pie', which allows you to intuitively check the level and level of learning of each subject, and 'Progress Report', which provides information that details the learning situations are provided to professor and individual students. This information can be used to identify the level of learning through relative comparison with other students by providing average information for all students.

Also, since the AI-based problem learning system creates and distributes similar types of different problems on the same subject to students, it helps to manage and supervise large-scale or non-face-to-face test takers and to secure and strengthen fairness.

However, the questions loaded on the AI-based platform mainly deal with multiple-choice question types, and in the case of subjective questions, they mainly deal with short-answer questions that involve writing calculation results using expressions with implied concepts or writing down simple words, so professors are responsible for solving concept-related problems. To check the interpretation and understanding process or to evaluate comprehensive thinking during the problem-solving process, the professor may need to conduct an additional separate evaluation.

2.3 Self-directed learning

In addition to being used for evaluation to diagnose the learning level of students from the instructor's perspective, AI-based adaptive learning appears to play an important role as self-directed, individualized problem-solving learning from the learner's perspective. In this context, the paper-and-pencil tests, computer-based tests, and artificial intelligence-based tests as examined in **Table 1** are classified according to the scoring and analysis results that can help students learn, and the method of providing supplementary explanations can be shown in **Table 2**.

In the case of the paper test system, students usually solve test questions printed on paper, and the answers written on this questionnaire are scored by the instructor or assistant and the analysis of the scoring results. After that, the solution learning is conducted through the process of additional support such as announcing the correct answer, providing face-to-face or non-face questions and answers to necessary parts, and providing related recorded videos. In the case of CBT, if the problem generated by the instructor is loaded into the system with the correct answer, it is automatically scored, and the scoring and analysis results can be shared and announced. However, it is difficult for students to comprehensively examine information related to learning, and solutions provided in the problem-learning process are mainly presented to the extent that the solutions correspond to simple explanations. Also, Because the information that students can see is limited, in conjunction with the Learning Management System (LMS), through additional data editing steps, the problems mounted by the instructor are scored, the results are shared, and the analysis is provided through additional data editing steps. In this respect, in the case of an AI-based problem learning system, compared to a Computer Adaptive Test, which is known as an evaluation that considers learners' learning levels among CBTs, AI-based tests provide more sophisticated learning diagnostics due to deep learning of artificial intelligence and provide a customized set of problems that further reflect learners' learning levels and situations, allowing students to learn self-directed concepts using "solvable" and "to solve." Additionally, individual students can obtain and use their learning-related information in a relatively diverse and sophisticated manner through the use of dashboards in a similar manner to instructors, and they can judge

	Scoring results announcement	**Analysis results announcement**	**Review**
Paper based test system	Professor/Assistant	Professor/Assistant	Correct answer
Computer-based test system	System	System	Solution
AI-based test system	System	System	Solution/supplementary explanation adapted problem

Table 2.
Reflective learning in evaluation.

and use their learning situations and ask for help, such as question-and-answer that fits their learning needs. The supplementary explanation supported in the problem-solving process deals with a relatively detailed and varied amount of content, not just feedback. However, in the case of learning effects obtained from explanations, the degree of understanding and acquisition of knowledge is related to the student's level of understanding of prior learning, so there seems to be a limit to some extent for students to independently grasp the concept's concreteness and connection between concepts [10]. Additionally, for concept problems where incorrect answers are submitted, solving similar problems is repeated to provide additional learning.

3. AI-based adaptive learning system

AI-based adaptive problem learning can be used for personalized assignments and regular evaluations within the course operation. To this end, the basic teaching design for the operation of the semester is established based on the characteristics of the subject classification, the characteristics of the subject, the major of the student, and the grade. In addition, after reviewing the content structure of the curriculum to be covered, necessary topics must be selected from the overall content system of the AI-based system and set in the system so that they can be dealt with effectively.

3.1 Content analysis

In order to use the AI learning system as problem learning, it is necessary to grasp the hierarchical structure that deals with the essential knowledge required in the curriculum and the lower level according to the knowledge within the AI system. For example, the hierarchical structure of each piece of knowledge can be dealt with based on the major subject, the sub-subject, and the concept, which is the lower layer of each sub-topic, as shown in **Table 3**. In this regard, most programs of AI-based problem learning systems are introduced mainly by educational companies, and their programs already create problems dealing with the concepts within the system at the launch stage and provide them in input and mounted form. Therefore, the most common method of problem generation is to select and use the entire topic that is automatically presented and mounted in the system, aimed at complete learning proposed by AI. Also, based on the instructional design of the subject, the professor may select and create specific topics in consideration of the amount of problem-solving learning of students that increases as the number of topics selected increases and the completion rate of learning that may be affected accordingly. In the course of learning, the learning deadline for each topic-related problem may be set to provide a learning environment so that problems related to all concepts can be solved freely throughout the semester, or to provide an environment in which concepts related to the topics covered in the lecture can be learned according to the progress of the semester.

Major subject	Sub subject	Concept	Summary description	Difficulty	Question number
Measurement	Scientific notation	Decimal	use of an exponent to represent a large number	easy	I-1

Table 3.
An example of a hierarchical analysis of knowledge.

This content analysis strengthens the connection with the content of the knowledge system that must be dealt with to achieve the educational goals and objectives of the subject, helping to more effectively utilize the content provided by the AI-based system. The figures related to the learning progress and learning completion rates of the concepts covered in the content hierarchy of **Table 3** and the results of the formative evaluation conducted to confirm the educational effectiveness through teaching and learning performance are used to confirm whether the content provided by the system matches the educational goals.

3.2 Adapted learning system implementation

AI diagnoses the level of prior learning for the core concepts to be covered in the course. To this end, an initial diagnostic test as shown in **Figure 2** is conducted during the orientation or the first lecture for the semester. As an initial diagnostic test, the AI system distributes questions about the subject's essential concepts, analyzes them based on the results of scoring the answers submitted by the student, identifies the topics and concepts that each student needs to learn, decides related problems to be initially distributed for subsequent learning, and initiates individual learning paths. Based on this, AI assigns individual problems to individual students to learn sub-concepts related to the diagnosed subject. After that, students are assigned different concepts and a numbers of questions according to the students' individual learning situations. Students' learning time varies depending on the stage of prior learning or their understanding of learning. In addition, unlike the initial diagnostic evaluation, additional diagnosis, which is another data that determines the student's level of completion of learning for each topic, is conducted within the AI system, and the timing and evaluation of this diagnosis may be controlled by the instructor. The method of reflecting the grades for the assignment after the end of one semester can be given in various ways using the learning-related data provided, and in the case of based on the learning completion rate, the grades for the assignment can be calculated by assigning a grade such as A/B/C according to the ratio.

As a result of comparing the learning completion rate according to the deadline-setting method in relation to adaptive problem learning operation, the learning completion rate was higher when the deadline was set to induce learning than when the learning of all topics could be carried out without a limit on the learning period for one semester. In addition, in most cases, students tended to learn problems on related topics after the instructor's lecture [2].

AI-based adaptive problem-solving learning contributes to diagnosing the understanding of students with different levels of prior learning and strengthening the

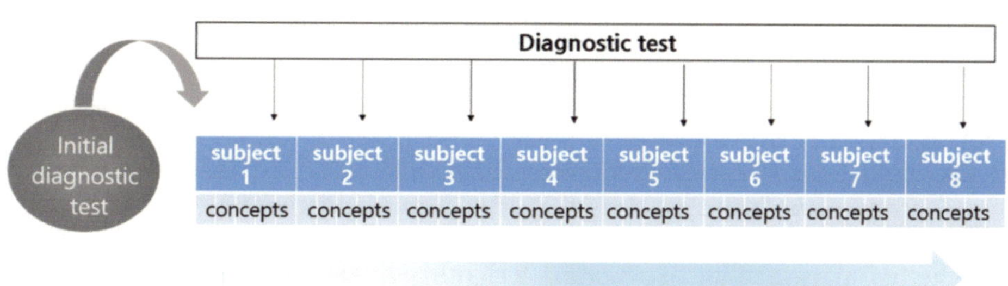

Figure 2.
A schematic diagram of system implementation.

learning of basic and core concepts in the subject, but the connection and expansion process between each learned concept seems to be lacking. In particular, additional learning can be achieved by utilizing problem-related concept explanations provided during the problem-learning process, but such learning is mainly limited to fragmentary concepts, so students without much prior knowledge may be more vulnerable to structuring and solving problems. Considering this aspect, it seems that customized problem learning should be supplemented with interactions that allow sharing of various perspectives and interpretations with instructors and other students during the learning process so that it is not biased toward answering repetitive problems or acquiring simple concepts.

4. AI-based adaptive class

4.1 Adaptive instruction

Adaptive instruction can be understood as identifying students' learning situations based on their learning data and providing appropriate teaching and learning [9, 11]. However, adaptive measures for students' learning situations in the teaching and learning field for a large number of students mainly use group learning by learning level in class, but they do not implement completely personalized class management as shown in the operation of supplementary classes to support learning for certain groups of students with low basic learning ability [12]. However, with the development of edutech, adaptive classes so far have recently been linked to AI-based adaptive systems with classes that analyzed the learning status reached in the students' conceptual learning process according to personalized learning paths, set appropriate teaching and learning goals that most students can reach, and construct classes that can deal with the use of content and concepts, and are expanding learner-centered countermeasures using teaching methods such as flipped learning and blended learning methods [3, 13, 14].

4.2 Instructional strategy

To apply AI-based problem learning in connection with adaptive instruction, it is necessary to select and set the topics to be covered in the class after further elaborating on the content systems to be used by comparing and reviewing the results of the content system analysis conducted in the previous problem learning implementation stage with the curriculum content system. In other words, rather than simply acquiring specific concepts from the content system mounted on the AI platform, it is necessary to design a content composition that creates an efficient connection relationship and structure of content systems that can effectively use each concept so that the instructor can create a mutually cooperative synergistic effect with the content system that has been previously dealt with.

It also requires a system understanding that can flexibly and selectively adjust the various features of AI-based systems needed to implement these designs. In other words, when using the problem learning system in connection with the class, first select whether to use a video or e-textbook that has already been installed in the system, whether to present and submit assignments, the duration of participation in the assignment, the timing of evaluation, and the place of evaluation. The artificial intelligence-based adaptive learning system itself also provides the use of related videos for learning basic concepts, but the use of videos produced by the participation of instructors helped in terms of class composition applying the instructor's perspective. In addition, it seems necessary to grasp the system utilization technology in order to more diverse and efficiently utilize the data on students' learning outcomes provided by the AI system. First, students autonomously learn about the subject through customized problem learning. The level of understanding through such prior learning is confirmed through a diagnostic evaluation of the learning system and the instructor checks through a report to see how much the students have learned the topics necessary for learning before the lecture and conducts the lecture. This diagnostic evaluation data was combined with the evaluation data of other individual learners and used to understand the average learning level and learning orientation of all students and reflected in the composition of the class. It is also possible to check teaching and learning outcomes by using an artificial intelligence-based evaluation system for formative evaluation.

4.3 AI-based adaptive system implementation in class

These class operations are reported to be applied in connection with classes based on the flipped learning teaching method, which is a part of blended learning that has been introduced as a teaching method that can use various edutech as a learner-centered teaching method [3, 15]. An example of the use of this adaptive problem-learning system can be represented in the same way as shown in **Figure 3**.

The course of the class is as follows. First, students learn beforehand through basic concept videos for each chapter produced by instructors and customized problem learning. In the case of a diagnostic evaluation linked to the progress of the class, a diagnostic report on the pre-learning activities on the day before class is prepared. The class began by identifying problems with high incorrect answer rates with students, and the topics related to these problems were intensively explained in the lecture. The main concepts of the content system provided by the artificial intelligence-based adaptive learning system were identified in the first class, and a summary lecture was conducted in connection with the instructor's content composition using the topics provided by the courseware. If most students have not yet progressed

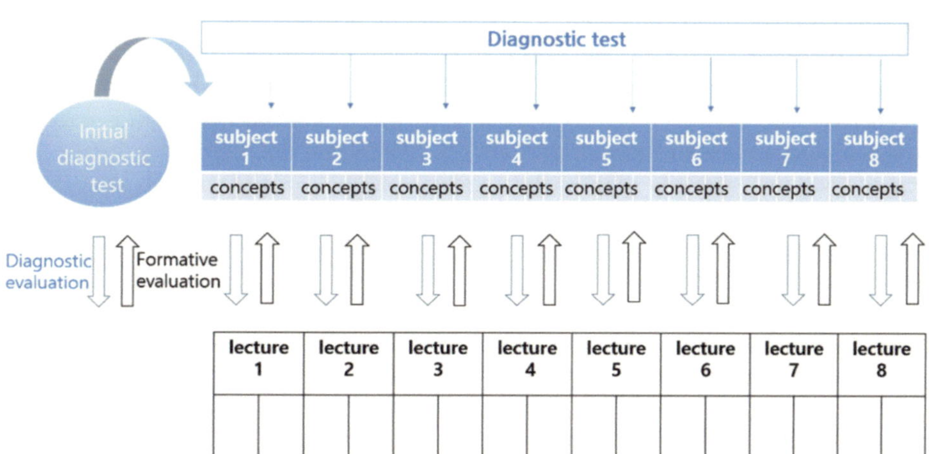

Figure 3.
Schematic class utilization of adaptive problem learning system.

on a specific topic, the lecture was used in a way that lectures were conducted from the basic contents of the related topic, not just a review. In the second class, information on the progress of students' personalized tasks is calculated in the form of a weekly report and used to encourage students to learn AI problems. In addition, since it is the second half of the chapter, each topic was viewed from a macro perspective by expanding through the system and causal relationships between the basic concepts learned by students, and presentations and group discussions on the approach in problem-solving learning was also addressed to complement the interaction between peer learners. In addition, the formative evaluation was conducted 15 min before the end of the second class, and it was completed at the end of the class.

The implementation of AI problem learning as a teaching tool linked to lectures at universities will require continuous research on teaching design that presents learning outcomes that experience opportunities for the application and expansion of concepts and foster in-depth analysis, thinking, and reasoning skills by utilizing concepts learned in terms of content. In addition, in terms of methodology, the use of learning diagnosis data will become more diverse and the ability to utilize it will increase compared to operating adaptive problem learning individually. In addition, it is common to use the hierarchical structure of concepts related to the content system to be considered and the understanding of the use of the system for individual problem learning, but its strength and importance seem to be higher.

5. Conclusions

Problem learning using AI-based systems in the teaching and learning process at university seems to be one of the good ways to understand and reinforce concepts in terms of knowledge transfer. However, the goal of basic university education is that conceptual learning should not be limited to simple knowledge understanding and problem-solving. The correlation and contextual connection of each concept should be made, and it should be utilized to play a role related to competency cultivation while expanding in various ways. Therefore, it is thought that the direction of active use of AI-based systems in university education is not to fully implement the various functions and learning completeness provided by the system, but to properly implement the knowledge learned through the system to be effectively linked to various educational activities. Finally, the points to consider for the actual operation of the AI-based system are summarized as follows. First, it can be viewed as content analysis (conceptual structure research) for system utilization. In other words, it is necessary to compare and review the content system established by the instructor, focusing on topics deemed necessary to achieve the educational goals of the subject, in connection with the content system of the system. Next, it seems that there is a need for a system adaptation process that can implement basic functions related to system operation. The costs required for purchasing and using the system should also be considered for long-term use. In the early stages of system operation, it seems necessary to cooperate with edutech experts supported at the university level to utilize the system effectively and efficiently reflecting the effort required from the instructor's side and the amount of learning required from the learner's side. In particular, when the amount of problems that must be repeatedly solved to reach the level of learning completeness required by the system is determined, there are only a few areas where the instructor can arbitrarily control the amount of learning the learner can control, so it is necessary to review the system utilization plan in this regard before the start of the semester, which should consider both learning maintenance and learning effects. In addition, since most of the languages currently used in AI-based systems are English, it is necessary to consider them at the teaching and learning design stage to reduce the additional learning burden related to language that may be felt if English is not the native language or does not use English textbooks.

Author details

Hwang Eunkyung
Ajou University, Suwon, Korea

*Address all correspondence to: seed25@ajou.ac.kr

References

[1] Holmes W, Bialik M, Fadel C. Artificial Intelligence in Education: Promises and Implications for Teaching and Learning. Globethics Publications; 2019 228 p. The Center for Curriculum Redesign. All rights reserved. Available from: https://repository.globethics.net/handle/20.500.12424/4276068

[2] Hwang GJ, Xie H, Wah BW, Gašević D. Vision, challenges, roles and research issues of artificial intelligence in education. Computers and Education: Artificial Intelligence. 2023;**1**:100001. DOI: 10.1016/j.caeai.2020.100001

[3] Eunkyung H, Jongho S. Exploratory study for introducing and applying an AI-based intelligent learning system on basic science-focusing on general chemistry class case. Korean Journal of General Education. 2021;**15**:71-86. DOI: 10.46392/kjge.2021.15.6.71

[4] Tesene MM. Adaptable selectivity: A case study in evaluating and selecting adaptive learning courseware at Georgia State University. Current Issues in Emerging eLearning. 2018;**5**(1) Art 6:62-79. Available from: https://scholarworks.umb.edu/ciee/vol5/iss1/6

[5] Vignare K, Tesene M, Lorenzo G. Case Study: Arizona State University [Case Study]. Association of Public & Land-grant Universities and Every Learner Everywhere. 2000. Available from : https://www.everylearnereverywhere.org/resources/case-study-arizona-state-university-asu/ [Accessed: January 23, 2023]

[6] Chen YW, Yang JB, Xu DL, Zhou ZJ, Tang DW. Inference analysis and adaptive training for belief rule based systems. Expert Systems with Applications. 2011;**2011**(38):12845-12860. DOI: 10.1016/j.eswa.2011.04.077

[7] Mujtaba DF, Mahapatra NR. Artificial intelligence in computerized adaptive testing. In: 2020 International Conference on Computational Science and Computational Intelligence; December 16-18, 2020. Las Vegas, NV, USA: IEEE Computer Society; 2020. pp. 649-654

[8] Martin F, Chen Y, Moore RL, Westine CD. Systematic review of adaptive learning research designs, context, strategies, and technologies from 2009 to 2018. Educational Technology Research and Development. 2020;**68**:1903-1929. DOI: 10.1007/s11423-020-09793-2

[9] Snow RE. Aptitude, learner control, and adaptive instruction. Educational Psychologist. 1980;**15**:151-158. DOI: 10.1080/00461528009529223

[10] Van den Broek P. Using texts in science education: Cognitive processes and knowledge representation. Science. 2010;**328**:453-456. DOI: 10.1126/science.1182594

[11] Duren D, Layne TE, Bray N, Irwin CC. Adapting: A study of implementing adaptive personalized instruction, innovation, and technology from an Instructor's Perspective. The Physical Educator. 2021;**78**:11-30. DOI: 10.18666/ TPE-2021-V78-I1-10064

[12] Chen X, Xie H, Zou D, Hwang GJ. Application and theory gaps during the rise of artificial intelligence in education. Computers and Education: Artificial Intelligence. 2020;**1**:1-20. DOI: 10.1016/j.caeai.2020.100002

[13] Ezzaim A, Dahbi A, Haidine A, Aqqal A. AI-based adaptive learning: A

systematic mapping of the literature. Journal of Universal Computer Science. 2023;**29**:1161-1197. DOI: 10.3897/jucs.90528

[14] Lian Y. Adaptive teaching for large classes. In: Proceedings of the International Conference on Engineering Education. Varceleona, Spain; 21-25 Jul 2003

[15] Shin JH, Choi JW, Park SY, Shon JE, Hwang EK, Ahn SH, et al. An exploratory study on the use of AI-based adaptive learning system in university class. Journal of Korean Association for Educational Information and Media. 2021;**27**(4):1545-1570. DOI: 10.15833/KAFEIAM.27.4.1545

Chapter 7

The Role of AI in Automating Grading: Enhancing Feedback and Efficiency

Johnbenetic Gnanaprakasam and Ravi Lourdusamy

Abstract

This chapter discusses the different ways in which artificial intelligence (AI) can be used to automate the grading process within the educational systems. The first part gives the background of how we got here, how grading practices have historically changed, and then how AI has progressed in integrating with these systems. The real emphasis is the potential use of AI to reduce the grading backlog (through instant feedback, learning incentives, scalability, and important notes) and more effective large and diverse student/learner management. Furthermore, it also delves into the use of AI on the subjective and creative aspects, quite a new realm of grading from the traditional ways. The chapter also provides a critical discussion about challenges associated with AI in grading (such as potential biases, fairness, and ethics), making an emphasis on the necessity to tailor such challenges in order to efficiently and responsibly deploy AI for educational purposes. Finally, it concludes with a reflection on what the next generation of AI-powered educational assessment experiences could look like and what the potential implications for educators and students may be.

Keywords: artificial intelligence (AI), automated grading systems, personalized feedback, educational assessment, natural language processing (NLP), machine learning (ML), bias and fairness in AI, data privacy in education, ethical considerations, future of educational technology

1. Introduction

1.1 Traditional grading systems

1.1.1 Historical overview of grading systems

The evolution of grading systems in education represents a complex journey, marked by efforts to balance fair assessment of student success with challenges of subjectivity and practicality. Initially, grading was rudimentary, often utilizing a simple pass/fail method. This approach, while straightforward, provided limited insights into a student's abilities and challenges. As educational systems advanced, so

IntechOpen

too did the grading methods, evolving into a more nuanced letter grade system (A, B, C, etc.). These systems offered clearer indications of performance levels but still depended heavily on educators' subjective judgments.

Dating back several centuries, the initial focus in educational settings was on oral examinations and personal evaluations by educators. Smith and Johnson offer an extensive overview of this evolution, highlighting the transition from oral to written examinations and then to a more standardized grading scale [1].

1.1.2 Challenges in traditional methods

Because of their intrinsic subjectivity, traditional grading systems present many difficulties and raise questions regarding consistency and fairness between various teachers and students. Grading takes time, especially for comprehensive assignments like essays, and can be particularly difficult in larger classes when there is a significant amount of work to be done. The assessment process is made more difficult by the consistent approach used by traditional grading techniques, which may not sufficiently take into account different learning styles or variations in student ability. Standardized examinations and scoring rubrics were created to provide more objective indicators of student achievement to solve these problems. Although these techniques address some issues with fairness, they can occasionally be restrictive and miss the distinctive contributions made by each student. Standardized testing has come under fire for prioritizing exam-taking techniques over more comprehensive learning goals [1].

Thomas, O'Brien, Sanguino, and Green draw attention to the subjectivity of these systems, emphasizing how teacher prejudices and subjective views can have a big impact on students' assessments [2]. Li draws attention to how standard grading fails to take into account various learning styles, which frequently results in a generalized approach that might not fairly represent the comprehension or advancement of any one student [3].

1.2 Need for innovation in grading

1.2.1 Limitations of conventional grading

The limitations of traditional grading systems have become increasingly clear. Cain highlights a lack of consistency and probable disparities in grading standards as severe disadvantages, with serious consequences for students' academic careers and future chances [4].

1.2.2 Emergence of digital technology in education

The incorporation of digital technology in education has brought about a significant and profound change. Kim and Lee highlight the significance of technology in facilitating the use of more impartial and thorough evaluation instruments. The emergence of AI-powered grading systems, employing algorithms and machine learning to assess student work, signifies a significant progress in rectifying the limitations of conventional grading approaches [5].

AI-powered systems offer the potential for evaluations that are more impartial and consistent by prioritizing comprehension and methodology rather than simply correct or incorrect responses. Owoc demonstrates the ability of these systems to effectively handle substantial amounts of tests, resulting in a significant decrease in the workload of instructors. This adjustment not only improves the equity of grading but also provides instructors with additional time for interactive and personalized instruction [6].

To summarize, although traditional grading systems have served as the basis for educational assessments for a long time, their inadequacies are evident. AI-powered grading systems offer a novel solution, utilizing improvements in digital technology to address the limitations of traditional approaches and establish a fairer and more efficient educational evaluation system.

2. Evolution of automated grading systems

The growth of automated grading systems is a fascinating journey, beginning with rudimentary computer-assisted evaluations in the mid-twentieth century and advancing to today's powerful AI-driven solutions. This narrative goes beyond simply documenting technological progress; it represents a change in how education is assessed and teaching methods are approached. Our thorough analysis will uncover the significant advancements that have influenced this transformation, emphasizing the interaction between technological progress and educational principles.

2.1 Introduction to computer-assisted assessments

2.1.1 Early stages and initial developments

The inception of computer-assisted examinations extends back to the early days of computer technology, first focusing on automating simple tasks like scoring multiple-choice tests. Regan provides a detailed description of these early systems, demonstrating how they utilized fundamental computational methods for tasks such as score tabulation and simple answer key management [7]. Despite their limitations due to the computer power and software capabilities of the time, these systems provided the framework for future advanced automated grading innovations.

2.2 Emergence of automated grading systems

2.2.1 Advancements in the 1980s and 1990s

The 1980s and 1990s represented a key transition in computerized grading methods. During this era, breakthroughs in computer processing and software design led to the introduction of more complicated grading systems. Garcia and Pearson highlight how these systems began implementing complicated algorithms capable of judging not just objective responses but also subjective ones like short answers and essays [8]. This shift signified a transition from solely quantitative assessment to a blend of quantitative and qualitative evaluation, enabling a broader review of student responses.

2.3 Advancements in language processing and AI

2.3.1 Turning point in AI application for grading

The adoption of powerful language processing and AI technology constituted a significant milestone in automated grading systems. Baidoo-Anu studies the combination of natural language processing (NLP) and machine learning methods, noting their significant impact on these systems' capacity to interpret textual answers [9]. This advancement brought about a more refined and sophisticated examination of written content, enabling the grading process to comprehend and evaluate complicated concepts and arguments offered by people.

2.4 Integration of pedagogical theories

2.4.1 Blending educational theory with technology

Incorporating pedagogical theories into automated grading systems has been a key advancement. Lee and Kim underline the necessity of matching automated grading with educational objectives and learning theories [10]. This fusion has resulted in grading systems that not only measure performance but also enhance the learning process by delivering pedagogically sound feedback matched with educational goals, such as increasing critical thinking and problem-solving skills.

2.5 The current landscape: AI-driven adaptive learning systems

2.5.1 State-of-the-art AI systems in education

Today's automated grading landscape is dominated by powerful AI-driven adaptive learning systems. Darvishi discusses the current innovations in this arena, emphasizing how these systems leverage advanced algorithms to offer individualized feedback and adapt to different learning patterns [11]. Capable of identifying learning gaps, personalizing content, and giving tailored interventions, these technologies dramatically increase the learning experience. They reflect the result of years of evolution in educational technology, blending AI breakthroughs, language processing, and educational theory to create a dynamic, responsive learning environment.

3. Principles of AI in grading

This chapter digs into the underlying ideas of artificial intelligence (AI) in grading systems, focusing on the core AI algorithms that have profoundly revolutionized educational assessment. It investigates the strengths and limitations of these algorithms, exploring crucial issues of accuracy and reliability in AI grading and addressing the technological and ethical components of these revolutionary systems.

3.1 AI algorithms in education

3.1.1 Core algorithms and their capabilities

AI algorithms have transformed student assessments in schools. Key algorithms such as machine learning (ML) and deep learning are pivotal in this change. Crompton, Burke, and Jones note that these algorithms excel in processing and learning from big datasets, allowing for extremely accurate evaluation of student replies [12]. They are capable of evaluating patterns in student answers, delivering personalized feedback, and modifying evaluation criteria based on individual student performance.

DOI: http://dx.doi.org/10.5772/intechopen.1005025

3.2 Evolution of AI in grading systems

3.2.1 Technological advancements in NLP and ML

Advancements in natural language processing (NLP) and machine learning (ML) have greatly pushed the advancement of AI in grading systems. Kompa, Hakim, and Palepu describe how NLP enables AI systems to interpret and analyze human language, which is critical for judging written responses [13]. Meanwhile, ML allows these systems to learn from data, improve over time, and make more accurate assessments regarding the quality and content of student work.

3.3 Accuracy and reliability of AI grading

3.3.1 Ensuring fairness and consistency

Ensuring accuracy and reliability is a fundamental priority in AI grading. Patel underlines the necessity for AI systems to be not just efficient but also fair and consistent in their assessments [14]. This involves the constant development of algorithms to decrease errors and biases and validation against broad datasets to assure equitable grading across various student populations.

3.4 Ethical considerations in AI grading

3.4.1 Addressing biases and ethical concerns

The employment of AI in grading brings forward important ethical problems. Goel underlines the significance of facing potential biases in AI systems, which may emerge from biased training data or incorrect algorithm design [15]. It is vital to guarantee AI systems are transparent and their decision-making processes are accessible to educators and students, sustaining trust and fairness in AI-assisted grading.

3.5 The future of AI in educational assessment

3.5.1 Projections and potential developments

Looking forward, AI in educational assessment is predicted to endure considerable breakthroughs. Chiu thinks that future AI systems would not only grade more efficiently but also deliver more detailed and insightful feedback to people [16]. The combination of AI with other developing technologies, such as augmented reality and virtual reality, could further increase interactive learning and tailored education.

4. Enhancing feedback with AI

4.1 Personalized feedback through AI

4.1.1 Customized responses and their impact

The integration of AI in education has ushered in a new era of individualized feedback, significantly altering student interaction with learning materials. Igbokwe studies AI systems that assess individual student replies, indicating distinct learning patterns and opportunities for improvement [17]. This individualized method enables feedback to be adapted to each student's distinct learning style and pace, resulting in more effective support for specific educational needs. The impact of this personalization is enormous, leading to higher student performance, enhanced retention of knowledge, and a deeper comprehension of the subject matter.

4.2 Immediate feedback and its advantages

4.2.1 Benefits of real-time feedback

One of the most transformational elements of AI in education is the provision of immediate feedback. Al Ka'bi addresses the advantages of real-time replies in educational environments [18]. Immediate feedback allows students to promptly rectify mistakes and reinforce accurate concepts, encouraging a more efficient and continuous learning process. This immediacy is especially advantageous in areas requiring a strong basic grasp, as it minimizes the building of misconceptions and bolsters confidence in learning.

4.3 AI-driven feedback and student engagement

4.3.1 Impact on student motivation and participation

AI-driven feedback dramatically influences student engagement and motivation. George studies how AI systems, by offering timely and meaningful feedback, might dramatically enhance student participation in the learning process [19]. This greater engagement is linked to the interactive aspect of AI feedback, which tends to be more immediate and tailored compared to older techniques. Such participation not only enhances motivation but also produces a more dynamic and participative learning environment, enabling students to actively engage in their education.

4.4 Impact of AI feedback on learning outcomes

4.4.1 Educational benefits of AI feedback

The effect of AI-generated feedback on learning outcomes is significant. Chen presents many studies indicating how AI feedback can increase academic achievement and expand subject matter understanding [20]. This efficacy is largely owing to AI systems' capacity to deliver precise, specific feedback addressing the complexities of a student's responses. Furthermore, AI feedback generally includes ideas for development and extra learning resources, making it a full educational tool.

4.5 Challenges and ethical considerations

4.5.1 Addressing potential hurdles in AI feedback systems

Despite its many advantages, the deployment of AI in feedback systems offers problems and ethical considerations. Chen covers numerous topics such as potential biases in AI algorithms, the importance of data protection, and the need for transparency in system operations [20]. These problems underline the need for careful design, continual monitoring, and regular upgrades to guarantee AI feedback systems are egalitarian, fair, and respectful of student privacy. Additionally, they underline the significance of human oversight to complement AI feedback, ensuring the educational process stays sympathetic and student-focused.

5. Efficiency and scalability in AI-enabled grading

5.1 Enhanced resource management in education

5.1.1 The role of AI in optimizing educational resources

Effective resource management is vital in education, directly impacting the quality and efficacy of teaching and learning. AI technology plays a vital role in this field. Alqahtani analyzes the impact of AI on grading systems, stressing not only the rapidity of the grading process but also a considerable improvement in the efficiency of resource allocation [21].

AI-enabled systems can efficiently process massive volumes of student work, a task that normally demands significant time and effort from educators. This quick processing ability frees up significant time for teachers, allowing them to focus more on direct educational activities such as lesson planning, individual student interactions, and pedagogical research. Furthermore, these tools assist detect specific areas where students require additional support, enabling teachers to adapt their education and interventions with greater precision.

Beyond the classroom, AI increasingly influences resource management at the administrative level. Schools and educational institutions can gain large cost savings by decreasing the need for extra grading staff or outsourced grading services. This efficient deployment of human and financial resources can be channeled toward strengthening many educational facets, including student support services, technological improvements, and professional development opportunities for educators.

5.2 AI's role in expanding educational horizons

5.2.1 Adapting AI systems for large classes and online learning

The scalability and adaptability of AI systems are especially significant in big classes and online education, where the student-to-teacher ratio might be disproportionately high. Bauer highlights the inherent obstacles in these situations, such as ensuring equitable and uniform evaluation for a diverse and extensive student body and how AI effectively mitigates these issues [22].

In big classroom environments, traditional assessment procedures typically become unworkable due to the excessive volume of assignments needing examination. AI systems, unencumbered by the limits imposed by human graders, may adeptly manage thousands of assignments concurrently, assuring rapid and uniform grading. This scalability is critical for sustaining academic standards and insuring that students receive fast feedback, essential for their academic growth and progression.

In online education, AI systems demonstrate extraordinary versatility. They can adapt to numerous course styles, academic fields, and diverse levels of student academic readiness. This flexibility is crucial in online platforms that usually serve an international and culturally heterogeneous student community. AI grading systems can interpret and evaluate responses across different languages and dialects and can be adjusted to adapt to specific curriculum standards and cultural situations.

Moreover, in online education environments, AI systems can offer a level of customization that is impossible to attain in traditional classrooms. By evaluating individual student data, these systems can provide individualized feedback and learning ideas, enhancing the educational experience for each learner.

In summary, the efficiency and scalability of AI-enabled grading systems constitute a significant leap in educational technology. They provide solid solutions for managing resources in education, successfully addressing the particular issues provided by big classroom environments and online educational platforms. As these systems improve, they hold the potential to significantly revolutionize the delivery and assessment of education, increasing learning experiences worldwide.

6. Beyond traditional testing: AI in creative and subjective assessment

6.1 Evaluating creative work through AI

6.1.1 Advanced techniques for assessing creativity

The application of AI in grading has predominantly been linked to objective assessments with clear right-or-wrong answers. However, recent advancements have extended AI's capabilities to include the evaluation of creative work. This shift involves intricate processes wherein AI algorithms are trained to discern and assess elements such as creativity, originality, and artistic expression.

Johnson and Lee have investigated the methods through which AI can appraise creative assignments [19]. These methods often utilize advanced machine learning techniques, including neural networks and deep learning, capable of analyzing patterns, styles, and techniques in various creative works. For instance, in evaluating written creative work, AI systems can analyze aspects like narrative structure, language use, and originality. In visual arts, they can examine color usage, composition, and artistic techniques.

The subjective nature of creativity presents a unique challenge. Creative works, unlike objective tests, are open to various interpretations. Therefore, AI systems in this field aim to provide a structured evaluation, focusing on technical aspects and widely recognized standards of creativity while avoiding subjective judgment. This approach requires a delicate balance, ensuring that AI assessments both nurture and support creative expression, avoiding the imposition of overly rigid standards.

6.2 AI's flexibility in meeting diverse educational needs

6.2.1 Tailoring learning to individual preferences

The ability of AI to adapt to diverse educational needs is crucial to its effectiveness in contemporary learning environments. This flexibility is particularly evident in AI's capacity to personalize learning experiences for students with varying learning styles, abilities, and interests.

Alier discusses how AI systems can customize educational content and assessments based on individual student profiles [23]. These profiles are constructed from data such as previous academic achievements, learning speeds, preferred learning methods (visual, auditory, and kinesthetic), and even emotional reactions to specific types of content.

Personalization in AI goes beyond merely adapting to different learning styles. It encompasses accommodating students with special educational needs, like those with dyslexia or autism, by offering tailored resources and assessment approaches that align with their distinct learning requirements. For example, AI can provide more visually oriented content and interactive modules for learners who struggle with traditional text-based education.

Additionally, the adaptability of AI is vital in serving a culturally diverse student body. It can supply learning materials that are culturally sensitive and create assessments that avoid cultural biases. This inclusivity ensures that students from varied backgrounds receive equitable learning opportunities, promoting an inclusive educational atmosphere.

This chapter outlines the advancements and challenges in employing AI for creative and subjective assessments and its flexibility in addressing diverse educational requirements.

7. Fairness and bias in AI grading

7.1 Tackling inherent biases in AI systems

7.1.1 Identifying and reducing biases

The challenge of inherent biases in AI grading systems is a key issue in educational technology. AI systems inherently reflect the biases existing in their data and algorithmic basis. Zhang investigates the ways biases might penetrate AI systems, such as through prejudiced training data, the subjective opinions of developers, or the simplifying of complicated human actions [24]. Identifying these biases involves a complete investigation of the AI system's training data, algorithmic structure, and outputs. This process demands a multidisciplinary approach, incorporating not only technologists but also educators, sociologists, and students. Regular audits and updates are crucial to ensure in fairness and accuracy. Mitigating these prejudices is a continual undertaking. It needs broadening the training data to span a broad spectrum of student work across varied ethnicities, backgrounds, and learning styles.

Additionally, adding feedback mechanisms where educators and students can review the AI system's performance and fairness is crucial. This inclusive strategy assures the AI system is continually learning and changing, gradually decreasing biases.

7.2 Establishing fair and inclusive grading practices

7.2.1 Guidelines for equitable AI grading

Creating solutions for fair and inclusive AI grading is vital to guarantee these systems serve all students equitably. Qian recommends various best practices for educational institutions to employ in pursuit of this purpose [25].

A crucial strategy involves maintaining the transparency of AI systems. Such transparency allows educators and students to comprehend the workings of the AI system, its grading standards, and the logic behind its ratings. This clarity fosters trust and acceptability in AI systems.

Another essential method is the constant enrichment of AI systems with diverse and representative data. This technique enables the AI system to more precisely and fairly assess work from students of varied cultural backgrounds, learning styles, and capabilities.

Human oversight remains vital in this situation. AI grading systems are designed to complement, not displace, educators. Human educators add crucial context, empathy, and insight to the grading process, traits that AI systems cannot perfectly mimic.

Lastly, it is crucial to create continual collaboration and dialog among technologists, educators, and politicians. This partnership assures that AI grading systems are not only technically strong but also in harmony with educational objectives and ethical values.

8. Ethical and privacy considerations

8.1 Data privacy in AI systems

8.1.1 Protecting sensitive student data

The application of AI in education presents substantial concerns surrounding data privacy, especially considering the sensitivity of student data. These data comprise not only academic records but also personal information that could be subject to misuse if not securely protected. As stated by Kumar, preserving data privacy in AI systems is not simply a technological issue but also a legal and ethical one [26].

Protecting student data in AI systems involves several key strategies:

1. *Encryption and security measures:* Implementing effective encryption and cybersecurity measures is vital to safeguard data from unauthorized access and breaches.

2. *Data anonymization:* When AI systems use student data for learning and improvement, it is vital to anonymize these data. This involves eliminating any personally identifiable information to guarantee that individual students cannot be traced.

3. *Compliance with regulations:* Educational institutions must comply with data protection regulations such as the General Data Protection Regulation (GDPR) in Europe or the Family Educational Rights and Privacy Act (FERPA) in the United States. These regulations offer frameworks for how student data should be handled and protected.

4. *Transparency and consent:* Transparency regarding what data are gathered, how they are used, and obtaining approval from students or their guardians (in the case of minors) is vital.

5. Educational institutions must work closely with technology vendors to verify that the AI systems they utilize adhere to these privacy standards and legislation.

8.2 Ethical implications in automated grading

8.2.1 Ensuring fairness, transparency, and consent

The ethical issues of automated grading systems extend beyond only the accuracy and efficiency of grading. As Thomas notes, there are greater ethical questions that need to be addressed to ensure that these systems are fair and just [27].

1. *Fairness:* AI systems must be built to grade impartially, without prejudices based on race, gender, or socioeconomic background. This entails regular checks for biases and modifications in the AI algorithms.

2. *Transparency:* There should be information about how the AI system operates, the criteria it employs for grading, and how choices are reached. This transparency is vital for gaining the trust of students and educators.

3. *Consent and choice:* Students and educators should have a vote in whether or how AI grading is employed in their educational experience. Additionally, there should always be a possibility for human review of AI grading choices.

4. *Accountability:* There should be clear accountability for the judgments made by AI systems. In circumstances where a grading decision significantly effects a student's academic career, the process for appealing or revisiting the decision should be straightforward and fair.

5. *Ethical development and deployment:* The development and deployment of AI grading systems should adhere to ethical norms, ensuring they are utilized to enhance education rather than replace the human factors that are crucial to learning.

Addressing these ethical and privacy problems is vital for the proper use of AI in education. It not only assists in developing systems that are fair and just but also ensures that these systems are accepted and trusted by the educational community.

9. The future of AI in grading

9.1 Emerging trends and technologies

9.1.1 NLP advancements and integration with learning management systems (LMS)

The future of AI in grading is directly connected to continuous improvements in natural language processing (NLP) and the integration of AI technologies with learning management systems (LMS). Harry describes the rapid progress of NLP, with new models and algorithms being developed that can more properly read and process human language [28]. This progress is vital for grading systems, particularly in judging more subjective and complex student responses such as essays and open-ended questions.

One major trend is the development of AI models that can not only evaluate grammar and syntax but also examine the coherence, inventiveness, and depth of student responses. These models are being trained on enormous and diverse datasets, enabling them to grasp distinct writing styles and nuances in multiple languages.

Another noteworthy trend is the seamless integration of AI grading systems with LMS platforms. This integration is redefining how instructors and students interact with educational content and assessments. LMS platforms are rapidly adopting AI capabilities that provide real-time feedback on student assignments, suggest resources for improvement, and adapt learning content depending on individual student performance. This integration provides a more tailored and responsive learning experience, making education more flexible to individual learner needs.

9.2 Preparing for advanced AI integration in education

9.2.1 Infrastructure, ethical use, and training

Preparing for the advanced integration of AI in education demands a diverse approach, focused on infrastructure development, ethical application, and extensive training. As Khairi points out, securing the successful application of AI systems in education extends beyond just technological improvements [29].

1. *Infrastructure development:* Educational institutions must invest in the essential digital infrastructure to support advanced AI systems. This involves powerful internet connectivity, suitable hardware, and secure cloud storage solutions. Additionally, there needs to be a focus on designing interoperable systems that can combine diverse educational resources and platforms effortlessly.

2. *Ethical use:* The ethical application of AI in education is crucial. Institutions need to develop explicit standards and policies that regulate the usage of AI systems. This includes guaranteeing data protection, eliminating potential biases in AI systems, and assuring openness in AI-driven judgments. Ethical use also requires ensuring that AI does not replace the key human parts of teaching but rather complements and enriches them.

3. *Training and professional development:* Educators and administrative personnel need to be trained not only on how to use AI systems efficiently but also on understanding their limitations and potential biases. Continuous professional development programs should be created to keep educators current of the newest AI advances and best practices in integrating AI technologies into their teaching.

The future of AI in grading appears optimistic, with technology growing more sophisticated and integrated into educational procedures. However, this future also involves careful planning and deliberation, ensuring that AI tools are used responsibly and efficiently to enhance the educational experience.

10. Conclusion

10.1 Recap and future prospects

10.1.1 Summarizing key themes and looking ahead

As we conclude our exploration of AI in grading, it is necessary to recall the key aspects and discuss future opportunities. This book has traveled the landscape of AI in education, from its earliest implementation in grading systems to the ethical, privacy, and technical difficulties it raises. We have seen how AI can boost the efficiency and fairness of grading, provide tailored feedback, and adapt to varied educational demands.

The incorporation of advanced AI technologies like NLP and machine learning into educational procedures, particularly in grading, signifies a fundamental shift in how student achievement is judged. The opportunities of AI in this field are immense. As Pedró and Subosa imply, we are likely to see more advanced AI systems that not only evaluate with high accuracy but also provide nuanced feedback that helps lead students in their learning path [30].

Moreover, the incorporation of AI in grading is projected to expand in line with emerging educational technology. This includes the possible convergence of AI with augmented reality (AR) and virtual reality (VR) for more immersive learning experiences, as well as the usage of big data to get deeper insights into learning patterns and outcomes.

However, as AI systems become more incorporated into educational settings, attention about ethical considerations and data protection becomes increasingly crucial. The future of AI in education will be influenced not just by technology breakthroughs but also by the laws and frameworks designed to guide its ethical and fair use.

10.2 Embracing change and innovation

10.2.1 The broader impact of AI in education

The broader significance of AI in education extends beyond grading. As Rahman and Kodikal emphasize, AI has the ability to alter the entire educational scene [31]. It can lead to more personalized and adaptable learning settings, where instruction is adapted to individual student requirements, talents, and learning styles. This customization could help bridge gaps in education, delivering support where it needs most and challenging people to realize their full potential.

AI's significance in education also necessitates a reevaluation of established teaching methods and curriculum frameworks. The data-driven insights offered by AI can guide more effective teaching practices and curriculum development, leading to a more responsive and dynamic educational system.

However, embracing this transformation demands a coordinated effort from educators, lawmakers, technologists, and students. It includes not only accepting new technologies but also adapting to the cultural and educational shifts they bring. The role of educators, in particular, will shift, with a greater focus on guiding and mentoring students through a technology-enhanced learning journey.

In conclusion, the future of AI in education is not just about technological innovation; it is about utilizing this technology to produce more equal, effective, and inspiring educational experiences. As we welcome these developments, it is vital to handle them intelligently, ensuring that AI is used as a force for good in defining the future of education.

10.2.2 AI-driven tools for grading

Specific AI-driven solutions currently utilized for grading include automated essay scoring systems (like Turnitin's Gradescope), multiple-choice test graders, and platforms that assess coding assignments (such as CodeSignal). These technologies utilize natural language processing, machine learning algorithms, and pattern recognition to analyze student inputs against a set of criteria or answer keys.

10.2.3 Addressing bias and fairness

The difficulties of prejudice and fairness in AI-aided grading can be addressed by the following:

- Continually training AI models on varied datasets that represent a wide range of language, cultural, and educational backgrounds.
- Implementing transparent algorithms that allow instructors to see and understand the foundation for a grade.
- Incorporating human monitoring to validate AI grading conclusions, ensuring that the final grades reflect a fair assessment.

10.2.4 Ethical considerations and mitigation

The ethical considerations of utilizing AI in grading focus around data privacy, the potential for systematic bias, and the impact on student motivation. Mitigation strategies include the following:

- Ensuring rigorous compliance with data protection requirements to preserve student information.
- Engaging multidisciplinary teams in AI development to eliminate prejudice and assure ethical use.
- Communicating clearly with students about how AI is utilized in their evaluations to ensure confidence and integrity in the educational process.

10.2.5 Potential negative impacts

AI-aided grading could potentially:

- Diminish the qualitative feedback that is vital for student learning and progress.
- Increase dependence on technology, resulting to a loss in essential human evaluative abilities among instructors.

- Exacerbate educational inequities if access to advanced AI techniques is inconsistent across different institutions.

10.2.6 Future implication

The incorporation of AI in educational assessment could lead to significant upheavals in the job market, notably for educators. While AI can automate certain portions of grading, the role of educators could expand to focus more on curriculum development, personalized instruction, and mentoring. Additionally, there is potential for new occupations based around AI maintenance, development, and ethical monitoring within educational institutions.

Potential drawbacks of using AI for grading include the following:

- The possibility of mistakes in grading, especially for subjective projects like essays, where nuance and inventiveness could be disregarded.
- The potential for technology to produce errors due to biased training data or algorithmic restrictions.
- The requirement for significant upfront investment in technology and training for schools to effectively employ AI grading tools.

Acknowledgements

The author acknowledges the use of ChatGPT and Grammarly for language polishing of the manuscript.

Author details

Johnbenetic Gnanaprakasam* and Ravi Lourdusamy
Department of Computer Science, Sacred Heart College (Autonomous), Affiliated to Thiruvalluvar University, Tamil Nadu, India

*Address all correspondence to: johnbenetic543@gmail.com

References

[1] Dilon C. Grading discrepancy in global education. PEOPLE: International Journal of Social Sciences. 2018;**4**:1611-1624. DOI: 10.20319/pijss.2018.42.16111624

[2] O'Brien CL, Sanguino SM, Thomas JX, Green MM. Feasibility and outcomes of implementing a portfolio assessment system alongside a traditional grading system. Academic Medicine. 2016;**91**(11):1554-1560. DOI: 10.1097/ACM.0000000000001168

[3] Li H. Application of subjectivity education theory in sports teaching in university. Advanced Materials Research. 2011;**271-273**:1386-1391. DOI: 10.4028/www.scientific.net/AMR.271-273.1386

[4] Cain J, Medina M, Romanelli F, Persky A. Deficiencies of traditional grading systems and recommendations for the future. American Journal of Pharmaceutical Education. 2022;**86**(7):8850. DOI: 10.5688/ajpe8850 Epub 2021 November 23

[5] Kim HJ, Choi J, Lee S. Teacher experience of integrating tablets in one-to-one environments: Implications for orchestrating learning. Education Sciences. 2019;**9**(2):87. DOI: 10.3390/educsci9020087

[6] Owoc M, Sawicka A, Weichbroth P. Artificial Intelligence Technologies in Education: Benefits, Challenges and Strategies of Implementation. Switzerland: Springer; 2021. DOI: 10.1007/978-3-030-85001-2_4

[7] Regan K, Berkeley S, Hughes M, Kirby S. Effects of computer-assisted instruction for struggling elementary readers with disabilities. The Journal of Special Education. 2014;**48**(2):106-119. DOI: 10.1177/0022466913497261

[8] Garcia GE, Pearson PD. Literacy Assessment in a Diverse Society. Center for the Study of Reading Technical Report. 1991. p. 525

[9] Baidoo-anu D, Owusu AL. Education in the era of generative artificial intelligence (AI): Understanding the potential benefits of ChatGPT in promoting teaching and learning. Journal of AI. 2023;**7**(1):52-62. DOI: 10.61969/jai.1337500

[10] Kim J, Lee H, Cho YH. Learning design to support student-AI collaboration: Perspectives of leading teachers for AI in education. Education and Information Technologies. 2022;**27**:6069-6104. DOI: 10.1007/s10639-021-10831-6

[11] Darvishi A, Khosravi H, Sadiq S, Gašević D, Siemens G. Impact of AI assistance on student agency. Computers & Education. 2024;**210**:104967. ISSN 0360-1315. DOI: 10.1016/j.compedu.2023.104967

[12] Crompton H, Jones MV, Burke D. Affordances and challenges of artificial intelligence in K-12 education: A systematic review. Journal of Research on Technology in Education. 2024;**56**(3):248-268. DOI: 10.1080/15391523.2022.2121344

[13] Kompa B, Hakim JB, Palepu A, et al. Artificial intelligence based on machine learning in pharmacovigilance: A scoping review. Drug Safety. 2022;**45**:477-491. DOI: 10.1007/s40264-022-01176-1

[14] Patel JB, Christopher Bridge BS, Chang DK. Basic artificial intelligence

techniques, evaluation of artificial intelligence performance. Radiologic Clinics of North America. 2021;**59**(6): P941-P954. DOI: 10.1016/j.rcl.2021.06.005

[15] Goel PK, Singhal A, Bhadoria SS, Saraswat BK, Patel A. AI and machine learning in smart education: Enhancing learning experiences through intelligent technologies. In: Infrastructure Possibilities and Human-Centered Approaches with Industry 5.0. USA: IGI Global; 2024. p. 20. DOI: 10.4018/979-8-3693-0782-3.ch003

[16] Chiu TKF, Meng H, Chai C-S, King I, Wong S, Yam Y. Creation and evaluation of a pretertiary artificial intelligence (AI) curriculum. IEEE Transactions on Education. 2022;**65**(1):30-39. DOI: 10.1109/TE.2021.3085878

[17] Igbokwe IC. Application of artificial intelligence (AI) in educational management. International Journal of Scientific and Research Publications (IJSRP). 2023;**13**(03):201-214. ISSN: 2250-3153. DOI: 10.292/IJSRP.13.03.2023.p13536

[18] Al K'b A. Proposed artificial intelligence algorithm and deep learning techniques for development of higher education. International Journal of Intelligent Networks. 2023;**4**:68-73. ISSN 2666-6030. DOI: 10.1016/j.ijin.2023.03.002

[19] George B, Wooden O. Managing the strategic transformation of higher education through artificial intelligence. Administrative Sciences. 2023;**13**(9):196. DOI: 10.3390/admsci13090196

[20] Chen L, Chen P, Lin Z. Artificial intelligence in education: A review. IEEE Access. 2020;**8**:75264-75278. DOI: 10.1109/ACCESS.2020.2988510

[21] Alqahtani T, Badreldin HA, et al. The emergent role of artificial intelligence, natural learning processing, and large language models in higher education and research. Research in Social and Administrative Pharmacy. 2023;**19**(8):1236-1242. ISSN 1551-7411. DOI: 10.1016/j.sapharm.2023.05.016

[22] Bauer E, Greisel M, Kuznetsov I, Berndt M, Kollar I, Dresel M, et al. Using natural language processing to support peer-feedback in the age of artificial intelligence: A cross-disciplinary framework and a research agenda. British Journal of Educational Technology. Advance online publication. 2023. DOI: 10.1111/bjet.13336

[23] Alier M, García-Peñalvo FJ, Camba JD. Generative artificial intelligence in education: From deceptive to disruptive. International journal of interactive multimedia and artificial intelligence. Special Issue on Generative Artificial Intelligence in Education. 2024;**8**(5):5-14. DOI: 10.9781/ijimai.2024.02.011

[24] Zhang W, Cai M, Lee HJ, et al. AI in medical education: Global situation, effects and challenges. Education and Information Technologies. 2024;**29**:4611-4633. DOI: 10.1007/s10639-023-12009-8

[25] Qian X, Jingying H, Xian S, Yuqing Z, Lili W, Baorui C, et al. The effectiveness of artificial intelligence-based automated grading and training system in education of manual detection of diabetic retinopathy. Frontiers in Public Health. 2022;**10**:1025271. DOI: 10.3389/fpubh.2022.1025271

[26] Kumar R. Faculty members' use of artificial intelligence to grade student papers: A case of implications. International Journal for Educational Integrity. 2023;**19**:9. DOI: 10.1007/s40979-023-00130-7

[27] Thomas RL, Halim S, Gurudas S, Sivaprasad S, Owens DR. IDF diabetes

atlas: A review of studies utilising retinal photography on the global prevalence of diabetes related retinopathy between 2015 and 2018. Diabetes Research and Clinical Practice. 2019;**157**:107840. ISSN 0168-8227. DOI: 10.1016/j.diabres.2019.107840

[28] Harry A. Role of AI in education. Interdiciplinary Journal and Hummanity (INJURITY). 2023;**2**:260-268. DOI: 10.58631/injurity.v2i3.52

[29] Khairi F, Yhya Alhafidh F. Leveraging AI for Enhanced Teaching Efficiency and Effectiveness. ResearchGate (Self-Published); 2024. DOI: 10.13140/RG.2.2.27462.08007

[30] Pedró F, Subosa M, Rivas A, Valverde P. Artificial Intelligence in Education: Challenges and Opportunities for Sustainable Development (Document Code: ED-2019/WS/8). France: UNESCO; 2019 (Collation: 46 pages)

[31] Rahman H, Kodikal R. Revolutionizing education: Artificial intelligence empowered learning in higher education. Cogent Education. 2023;**11**:134-145. DOI: 10.1080/2331186X.2023.2293431

Chapter 8

Integrating AI into VET: Insights from AIM@VET's First Training Activity

Žiga Emeršič, Peter Peer, Gregor Hrastnik, Nataša Meh Peer, José María Bey, María Meizoso-García, António Pedro Silva, Cláudia Domingues, Carla Abreu, António Costa, Dalila Durães, Paulo Novais, Cristina Renda and Abraham Prieto

Abstract

This work presents the objectives, methodologies, and preliminary outcomes of the first training activity (TA1) within the AIM@VET project, an EU initiative aimed at integrating artificial intelligence (AI) into vocational education and training (VET) to align with labor market demands. Addressing the noticeable gap in AI education across various educational levels, AIM@VET, involving six partners from Spain, Portugal, and Slovenia, focuses on developing teacher-centered learning modules in key AI application areas: computer vision, robotics, and ambient intelligence. The project's methodology involves universities in content preparation and VET teachers in content delivery to students, with an iterative feedback loop enhancing the curriculum's relevance and effectiveness. TA1 demonstrated a practical approach to applying AI concepts through a mix of theoretical lessons and hands-on tasks, significantly improving students' technical AI skills and readiness for the digital workforce. The activity underscored the importance of standardizing lesson creation protocols to produce a unified curriculum, thereby facilitating improved coordination among partners. This chapter will detail the project's framework, its execution, and an analysis of the results obtained in the project's first steps.

Keywords: AI Literacy, artificial intelligence education, artificial intelligence modules (AIM), Erasmus+ program, project-based learning, digital transformation, vocational training, robotics, ambient intelligence, computer vision

1. Introduction

The technical revolution behind artificial intelligence (AI) impacts the society in a global way. Consequently, it is mandatory to educate the whole population in the fundamentals of this digital technology, not only to increase their job opportunities but as a fundamental right to be aware of the decisions that are taken by AI-based systems [1]. In this scope, the development of AI literacy is a key goal for most educational administrations and organizations around the world [2, 3]. It implies the definition of the skills and competencies that learners should acquire in this topic, as well as the procedures to achieve them, mainly through specific curricula and resources [4].

Many approaches to AI curricula development have arisen in the last 5 years, mainly focused on pre-university levels (secondary school). They come from official educational administrations [5], private initiatives [6], or directly from universities and schools [7]. All of them are still under development and testing because the creation of a formal and reliable curriculum in a new discipline takes time, reflection, and improvement. Most of the existing approaches are based on the guidelines of the AI4K12 initiative [8] and the five big ideas they propose (Perception, Representation and Reasoning, Learning, Natural Interaction, and Societal Impact). These ideas make up wide areas where educators should focus on training secondary school students in the fundamentals of AI. Following these guidelines, some remarkable educational resources in topics like perception, representation, and, mainly, machine learning can be found [9]. The majority of them have been developed for compulsory levels, with a general educational perspective. The technical background required to follow the lessons is low, as there is no warranty about the students' digital knowledge and skills in programming, statistics, or logics to be homogeneous in different countries and regions.

The current work is focused on Vocational Education and Training (VET), where the previous considerations must be reformulated [10]. These students have a clear technical focus, mainly practical, and their training must be aligned with the labor market requirements. Therefore, the existing curricula about AI for pre-university levels must be adapted to this scope, or new ones must be created. In addition, the curriculum organization in VET education is not the same as in secondary school, and here the typical teaching units are organized in "learning modules", which are more specific and shorter. Hence, there are many specific features to consider when developing an AI curriculum for VET. In this realm, some initiatives have arisen, although they are in the early stages of development. From all of them, the TaccleAI project must be highlighted, which has been analyzing the particularities of AI education in VET for the last few years. The main conclusions of their final report have been taken here as a guideline for future development [11]. A more detailed description of the AIM@VET project and its objectives can be found in [12].

The authors of this paper were granted an Erasmus+ project in 2022, called AIM@VET, with the aim of developing AI learning modules adapted to VET education in three areas: autonomous robotics, computer vision, and ambient intelligence. The project encompasses one VET school and one university from Spain, Portugal, and Slovenia, leading to a team of six partners. They have been organized into 'work islands'—a term referring to specialized working groups, each based in a different country and focused on one of the three distinct AI areas mentioned above. These work islands operate collaboratively yet independently, allowing each group to develop specialized teaching units (TUs) and then share insights and outcomes with other groups. This organizational structure promotes specialization and efficiency in the development of educational content tailored to VET education while facilitating broad collaboration among partners across different countries. Each university team creates specific teaching units (TUs) in their particular area, which are revised by the teachers at the corresponding VET school and tested with their students. From these results, the university team obtains feedback to improve the TUs. At the project's end, the TUs will be organized into three independent learning modules, and they will be published on the project website to make them available to the entire teaching community [13].

A key step in this project was the first Training Activity (TA1), held in Slovenia, as it tested the TUs developed by the different 'work islands' in a novel context. TA1 facilitated a deeper level of evaluation by involving students from various 'work islands', thus enabling the TUs to be tested across a more diverse student body. Specifically, the initial two teaching units, TU1 and TU2, which were developed as part of the project, were shared with and tested by students from mixed nationalities and backgrounds in collaborative groups. This approach is not only aimed at assessing the TUs' effectiveness but also at fostering an environment of cross-cultural learning and exchange.

The methodology and outcomes of TA1 are presented here, with a focus on analyzing the application of the developed TUs and their impact on the students. Specifically, the TU described here introduces students to the application of Q-learning in robotics for navigation and obstacle avoidance, basic principles and techniques of image processing and object detection in computer vision, and the integration and control of sensors and actuators in the context of ambient intelligence.

2. Methodology

This section details the first Training Activity (TA1) within the AIM@VET project, focusing on its dual role: assessing the effectiveness of teaching units from each work island and enhancing collaborative learning among diverse AI specialization students. TA1 was fundamental for evaluating lesson functionality, sharing challenges and insights, and introducing lessons to new students. It also proved very useful for refining practical aspects such as coordination, standardization of lesson format, content, and pedagogical philosophy.

2.1 Description of the activity

TA1 focused on the core themes of computer vision, robotics, and ambient intelligence, each managed by one of the three work islands. Each island was responsible for creating scenarios and tasks within their respective domain. The activity encompassed 18 hours of teaching for each thematic area, culminating in a 1:30-hour joint session for presentations and feedback exchange among students. This final session allowed students from different work islands to share their results and insights.

2.2 Participants

Participants in TA1 included students from each of the work islands, encompassing a range of educational backgrounds and AI expertise levels. This diversity was intentional to foster cross-disciplinary learning and interaction. In line with this approach, the participants were strategically mixed across different thematic areas, with each work island delivering lessons to students from the other islands. This structure, detailed in **Table 1**, was designed to encourage cross-disciplinary learning and foster a collaborative educational environment.

Teaching Island	Lesson Focus	Student Groups
Spain	Robotics	6 students (3 Slovenia, 3 Portugal)
Slovenia	Computer Vision	6 students (3 Spain, 3 Portugal)
Portugal	Ambient Intelligence	6 students (3 Spain, 3 Slovenia)

Table 1.
Distribution of mixed teaching approach in TA1.

2.3 Target age group and prerequisites

This activity is tailored for VET students aged 15–20. To accommodate this broad age range effectively, clear prerequisites are established to ensure students possess the foundational knowledge necessary for the proposed challenges. These prerequisites are:

- *Programming:* A basic level of programming, including an understanding of conditionals, loops, variables, and functions.
- *Python:* Basic familiarity with the Python programming language, suitable for engaging with simple programming tasks.
- *Mathematics:* Knowledge of Cartesian coordinates, angles, and time measurements is important for the mathematical aspects of the challenges.

These prerequisites aim to equip students with the essential skills required to engage with and benefit from the learning modules, ensuring a productive and enriching educational experience.

2.4 Feedback mechanisms

In TA1, several feedback mechanisms were employed to evaluate and enhance the effectiveness of the training sessions.

1. *General Lesson Evaluation Templates for Students:* These templates were used by teachers to assess student performance across the entire training activity unit. The evaluation focused on various aspects of the task, with teachers assigning scores based on a detailed rubric. This approach ensured a consistent and reliable method of evaluating student progress across different curriculum units. The criteria used in these templates included adequate selection of information, time management, design and construction of the solution, creativity, testing and debugging, programming and code, and teamwork.

2. *Specific Evaluation - Questionnaires:* Following the completion of each unit, students were required to answer questionnaires that focused on specific aspects of the lesson, including theory and implementation. These questions aimed to gauge the students' understanding and application of the material covered in the TU.

3. *Teacher Feedback:* Teachers filled out templates evaluating each section of the lesson, with a focus on potential improvements. They provided assessments on the

duration of each task, the theoretical content provided, and evaluation indicators for student performance. Teachers were also encouraged to propose new or modified evaluation indicators, contributing to the ongoing refinement of the teaching units.

These feedback mechanisms were integral to the continuous improvement of the teaching units and overall training activity, ensuring that both theoretical and practical components of the curriculum were effectively delivered and assimilated by the students. The evaluation templates presented here can be found in the modules section of the project website, accessible at [13].

3. Description of the teaching units

A cooperative Project-Based Learning (cPBL) approach has been used to implement the TUs that make the AIM@VET modules. Students are organized in teams to face a challenge that must be solved by means of an AI system they program in Python language. According to the PBL methodology, each TU establishes a set of design specifications, including a video displaying the final response the AI system must show, and provides a tentative organization in steps. The approach is based on a "hands-on" learning perspective in which the theoretical concepts are reinforced as they are required to solve a practical issue (learning by doing), and it is key for VET education.

All the materials associated with these TUs are available at [14], including a main file with the teacher's guide in pdf, the demonstrative video of the expected solution, the code with the Python programs, and other supportive materials.

3.1 Robotics

The current resource corresponds to the second TU of the robotics module of the AIM@VET project, focused on mobile robotics. The ultimate challenge of this TU is to enable a mobile robot to move autonomously within an industrial-like scenario, avoiding the risks of obstacles and floor level changes, and aiming to solve a patrolling task in an autonomous way. In the following subsections, more details regarding this resource and its scope in AI teaching for VET education are provided.

3.1.1 Temporal organization and required resources

The robotics unit is structured to be completed over 15 hours and tailored for students who have met the specified prerequisites. The curriculum is segmented into four main tasks, with the following time allocation: 5 hours for Task 1, 3 hours for Task 2, 4 hours for Task 3, and 3 hours for Task 4. This structure is designed to provide an in-depth exploration of each topic.

- In Task 1, students engage in learning how to detect and avoid potential hazards in the environment, with a significant emphasis on practical application and understanding.

- Task 2 introduces the basics of Q-learning, allocating sufficient time for both theoretical discussions and practical exercises.

- Tasks 3 and 4 continue to explore advanced concepts and applications of Q-learning, with time allotted for comprehensive engagement with the content, hands-on activities, and problem-solving.

Each task commences with a theoretical introduction by the teacher, detailing the scope and objectives. Students are provided with specific time frames to work on the challenges, encouraging active engagement and the development of problem-solving skills. Upon completion, collective discussions are held to review possible solutions, facilitating reflective learning and consolidation of knowledge.

The following resources are necessary to facilitate this TU:

1. Each student needs access to a laptop or computer capable of programming with Python.

2. The RoboboSim simulator should be downloaded from the Robobo wiki [15] and installed on each student's computer.

Robobo [16] is an educational robot comprised of a mobile base connected to a smartphone via Bluetooth, forming a single robotic platform with high technical specifications, low cost, and long durability. As depicted on the left side of **Figure 1**, Robobo is equipped with a high-resolution camera, microphone, tactile screen, speaker, WI-FI, Bluetooth, a powerful CPU, and other relevant features for teaching AI at all levels. It supports programming in Scratch and Python and is compatible with ROS [15], offering libraries for computer vision, sound processing, speech production, and recognition, among others.

Furthermore, different simulation tools suitable for various educational levels and skills are available [17]. One of these tools is RoboboSim [15], a 3D realistic simulator used in this TU to introduce students to reinforcement learning. Developed using Unity technology, RoboboSim provides a computationally light and engaging learning environment with video game-like usability and aesthetics. It supports most of the sensors and actuators of the real robot, especially those required for the challenges presented in this activity. The right image of **Figure 1** shows a snapshot of one RoboboSim simulation world, demonstrating its capabilities and interface.

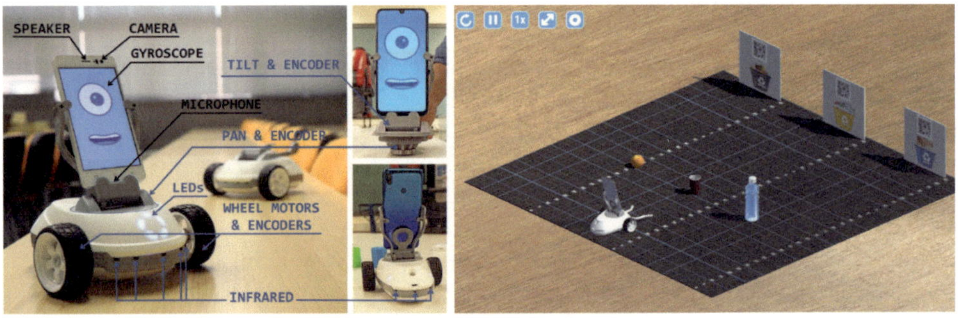

Figure 1.
Left: Robobo robot. Right: RoboboSim interface.

3.1.2 AI concepts addressed

Throughout this TU, students will learn the fundamentals of reinforcement learning by means of implementing a specific algorithm: Q-learning. They will use the Q-table to guide the robot's decision-making process. By updating the Q-values using the Bellman equation [18], the robot can learn from its experiences and improve its actions over time. In addition, students will explore how the robot actuators enable physical movements and actions, while the sensors provide important information about the environment. This knowledge about the interaction of the robot with its environment is essential for the effective integration of sensor feedback into the robot's decision-making. The robot learns a control program through a reward-based approach utilizing Q-learning. Specifically, through iterative experimentation, Robobo employs the Q-learning algorithm to optimize its actions and autonomously navigate toward illuminated areas. But it is important to note that other control programs required to support Q-learning have been previously developed by students, and they remain predefined during the reinforcement learning process. These control programs correspond to obstacle avoidance and fall prevention. They are programmed by the designer to ensure the safety of both the robot and its environment, a key aspect of autonomous robotics for VET education in terms of the legal, ethical, and responsibility considerations that these students must learn to face real-world jobs.

3.1.3 Challenge description

The primary goal of this Teaching Unit (TU) is to equip Robobo with the capability to autonomously navigate toward illuminated areas within unknown environments, identifying obstacles and avoiding falls. This requires students to program Robobo, blending predefined safety controls with the Q-learning algorithm for navigation. The simulation environment for this challenge, including obstacles and light sources, is depicted in **Figure 2**.

Students are tasked with implementing a Python program that integrates control programs with the Q-learning algorithm, enabling Robobo to move autonomously and safely. This involves using Robobo's sensors to measure light levels from various directions and determine the robot's next move based on the light intensity changes, rewarding actions through the Bellman equation, and updating the Q-table accordingly. This setup illustrates the balance between AI adaptability and safety, a crucial aspect of this TU. To achieve these objectives, the TU is structured into focused subtasks, each building upon the last to gradually enhance students' understanding and skills in autonomous navigation and AI adaptation:

1. *Autonomous Patrolling in an Industrial-like Setup:* Students start by familiarizing themselves with Robobo's sensors and actuators – wheels for movement, pans for smartphone rotation, IR sensors for distance measurement, and the light sensor for ambient light level detection. A theoretical background in reinforcement learning and the Q-learning algorithm is provided, encouraging concept acquisition through challenge-solving.

2. *Autonomous Safety: Avoiding Obstacles and Falls with IR Sensors:* The initial task is to program Robobo to autonomously navigate the environment shown in **Figure 2**, avoiding obstacles and falls. This involves dynamic adaptation to sensory inputs,

with a Python template (shown in **Figure 3**) guiding the implementation of fall and collision avoidance behaviors.

3. *Evaluating the State of the Robot: Using the Light Sensor to Define Orientation Toward the Light:* Students develop a function to determine Robobo's state based on light direction, introducing them to the concept of "state" in reinforcement learning. This task is essential for setting up the logic behind autonomous navigation decisions.

4. *Autonomous Navigation: Executing Actions Using the Q-Table:* This task involves filling in a Q-table (illustrated in **Table 2**) to link states with actions, ensuring

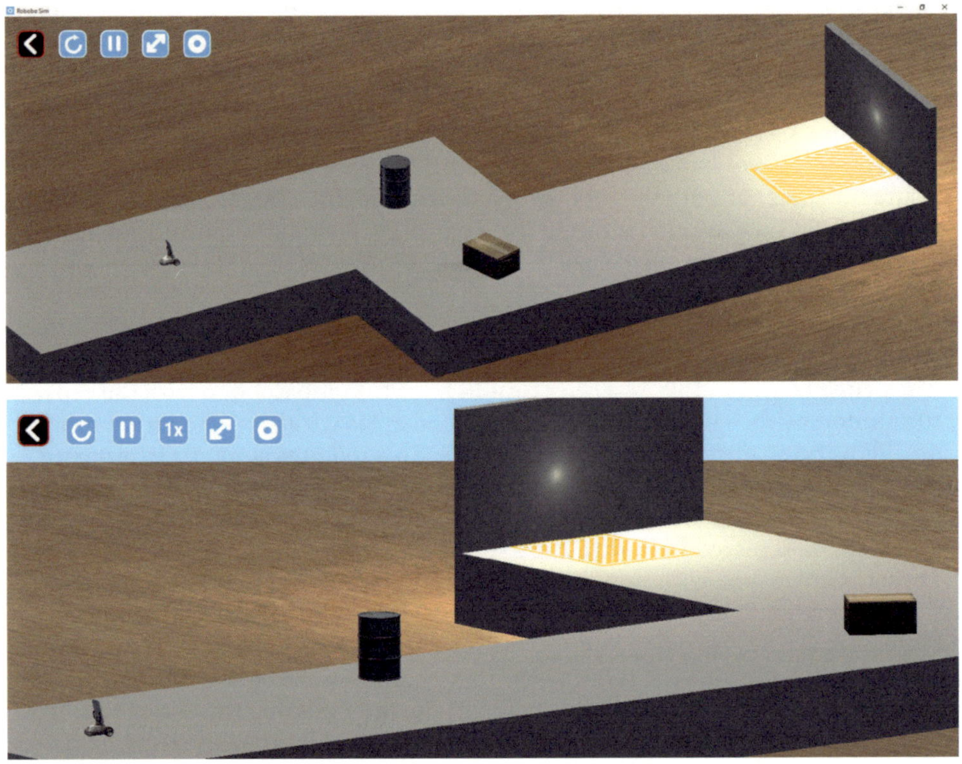

Figure 2.
Simulation environment used in the teaching unit.

```python
from robobopy.Robobo import Robobo  # import Robobo Library
from robobopy.utils.IR import IR # import IR module

rob = Robobo("localhost") # create a instance of Robobo class
rob.connect()  # connect robobo

speed = 15

rob.moveWheels(speed, speed)
rob.wait(0.5)

while True:
    # Avoid falls
    if (rob.readIRSensor(… :

    # Avoid collisions
    if ( …

    rob.wait(0.5)
```

```python
    # Avoid falls
    if (rob.readIRSensor(IR.FrontL) < minIR) or (rob.readIRSensor(IR.FrontR) < minIR):
        rob.stopMotors()
        rob.moveWheelsByTime(-speed, -speed, 3)
        if (rob.readIRSensor(IR.FrontL) < minIR): # Fall Left
            rob.moveWheelsByTime(speed, -speed, 1)
        else: # Fall Right
            rob.moveWheelsByTime(-speed, speed, 1)
        rob.wait(0.5)
        rob.moveWheels(speed, speed)

    # Avoid collisions
    if (rob.readIRSensor(IR.FrontC) > closeIRValue) or (rob.readIRSensor(IR.FrontRR) >
closeIRValue) or (rob.readIRSensor(IR.FrontLL) > closeIRValue):
        rob.stopMotors()
        rob.moveWheelsByTime(-speed, -speed, 2)
        rob.moveWheelsByTime(-speed, speed, 2)
        rob.wait(0.5)
        rob.moveWheels(speed, speed)
```

Figure 3.
Python template to face task 1 (left) and a possible solution for the teacher (right).

Q (S,A)	A0 (Turn right)	A1 (Turn left)	A2 (Go straight)
S0 (Right)	1	0	0
S1 (Left)	0	1	0
S2 (Front)	0	0	1

Table 2.
Q-table manually filled by students.

Q (S,A)	A0 (Turn right)	A1 (Turn left)	A2 (Go straight)
S0 (Right)	1.02	0.04	0.04
S1 (Left)	0.55	1.3	0.34
S2 (Front)	1.17	1.04	1.83

Table 3.
Final values of the Q-table obtained from the execution.

Robobo navigates effectively without encountering obstacles or falls. Students learn to execute actions based on the highest Q-value for the current state, demonstrating the practical application of Q-learning.

5. *Autonomous Learning: Reward Calculation and Q-Table Update:* The final task focuses on autonomous learning, where students calculate rewards and update the Q-table post-action (as seen in **Table 3**). This includes developing a function for reward calculation based on light intensity changes and updating the Q-table using the Bellman equation, enhancing Robobo's navigation strategies.

By sequentially addressing these subtasks, the TU offers a comprehensive learning experience, from sensor-based obstacle avoidance to sophisticated decision-making via Q-learning, ensuring students gain a thorough understanding of autonomous navigation in robotics. For a more detailed account of this implementation, refer to [19].

3.1.4 Outcomes

The initial implementation of this teaching unit occurred between April and May 2023 at CIFP Rodolfo Ucha Piñeiro in Spain, involving six vocational education and training (VET) students with prior RoboboSim experience. During this phase, all students completed task 1, four managed task 3 by identifying states and executing actions from the Q-table, and one student successfully applied the Bellman equation (see **Figure 4**) [18]. Time constraints were noted as a significant challenge, but the engagement and conceptual understanding of the students were evident.

Feedback highlighted the teaching unit's effectiveness, with students finding the balance of theory and practical tasks engaging. Despite some struggles with code errors, likely due to limited programming backgrounds, the overall response was positive. Four students gained new insights into AI, underscoring the importance and interest in the subject matter, which was further supported by teacher observations.

An analysis based on a general evaluation template and student questionnaire revealed that the theoretical content's complexity was seen as challenging but manageable, with students' perceptions evenly distributed across the complexity spectrum (see **Figure 5**).

Following an in-depth feedback review, we noticed several key issues with the original teaching unit (TU), necessitating a comprehensive review. Key challenges identified included the necessity for clearer, simpler explanations for advanced topics; sufficient time allocation for exploring complex subjects; a solid foundational understanding of mathematical concepts, notably matrices essential for Q-learning; and the need for explicit guidance in applying theoretical concepts, particularly the Bellman equation [18].

To address these challenges, we introduced significant enhancements in the TU:

- Enhanced Clarity and Accessibility: We refined the TU's language for greater accessibility and introduced explicit examples to explain complex ideas such as reward systems and state transitions within Q-learning.

- Optimized Time Management: We extended the time allotted to basic concepts, acknowledging the complexity of Q-learning, while making supplementary tasks optional.

- Mathematical Foundations Support: We provided students with essential coding tools, including ready-made Q-tables and the Bellman equation [18], facilitating a deeper comprehension of the mathematical concepts.

- Comprehensive Theoretical Application Guidance: The updated TU provides a detailed guide through critical tasks, with clear explanations of key operations such as reward calculation and Q-table updates and thorough clarification of function parameters and expected results, minimizing ambiguity.

At the training activity, a second implementation of the teaching unit took place with three students from Portugal and three students from Slovenia. Among the six students, five were male and one was female, with ages ranging between 15 and 18 years. Their digital competencies were slightly weaker than those of the Spanish group. Although they were familiar with programming fundamentals, their assimilation of these concepts was not as strong. The materials and questionnaires used in this second activity are accessible at [14].

On the first two days, a total of nine hours were dedicated to introducing RoboboSim and its fundamentals. The third day, with seven hours of class, was dedicated to the specific tasks of this work. During it, students were able to successfully identify states, create their own Q-tables, and execute the corresponding actions. They achieved better results in a significantly shorter time compared to the Spanish group, as displayed in **Figure 4**. This outcome, coupled with positive feedback from teachers, confirmed that feedback from the initial implementation of the first TUs at CIFP Rodolfo Ucha Piñeiro served to improve the preliminary versions, leading to a refined teaching approach.

After completing the robotics teaching unit with six new students who were new to robotics, our experience provided valuable insights into both learning and teaching within this field. Despite the students' varied educational backgrounds, they showed notable engagement and adaptability, especially when allowed sufficient time to delve into complex topics.

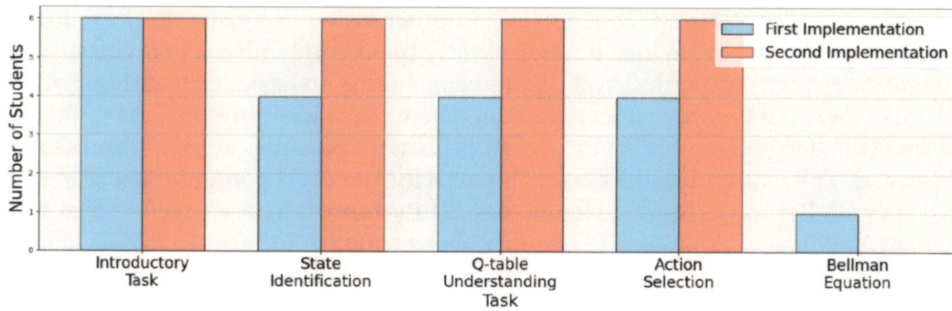

Figure 4.
Level of fulfillment of the tasks by the students among implementations.

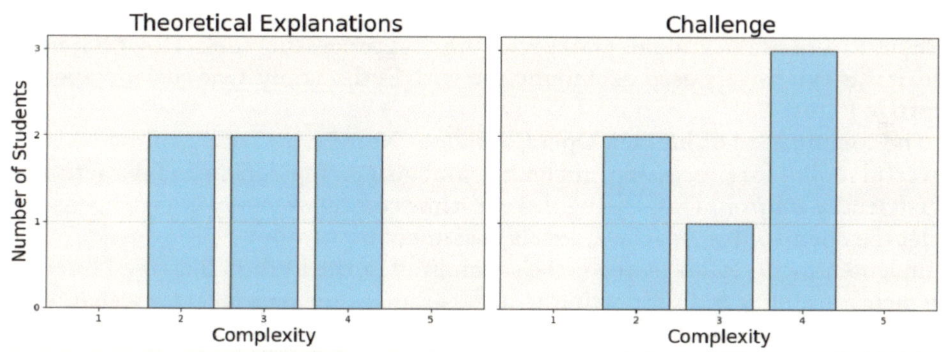

Figure 5.
Complexity rating of the TU contents perceived by students.

Teachers' feedback indicated the positive practical, hands-on approach of the project, though it was also mentioned that a stronger base in programming would be beneficial for future learning. The materials and questionnaires used in this second activity are accessible at [14].

3.2 Computer vision

This work package aims to develop computer vision teaching units for VET students, supporting the project's main goals. Computer vision, a key AI component, processes images to derive decision-making information. Grasping its concepts and applications is essential for future use in everyday life. Computer vision can be found in diverse areas like industry, smart homes, and autonomous vehicles. Teaching units therefore focus on critical topics such as unbiased data curation and object recognition, blending classical and deep learning approaches [20, 21]. Practical, real-world application examples are emphasized.

3.2.1 Temporal organization and required resources

The development of the TU takes approximately 10 hours if students have the necessary prior knowledge. It has been organized into 8 tasks, with an approximate load of 1 hour for the first 6 tasks and 2 hours for the last two tasks. In the first task, students get familiar with the concepts of computer vision in general. In the following tasks, they learn how to load images, modify them, apply filters, perform simple segmentation, continue with detection, and conclude with basics of tracking. For each task, there is a theoretical introduction where the teacher presents the topic, and small challenges are provided to the students with possible solutions discussed collectively. The required resources for this activity are: (1) A computer per student to program with Python in Jupyter Notebooks. (2) Python with necessary libraries such as OpenCV, Pillow, and Numpy. (3) Regular web camera is either available on laptop or plugged into the computer.

Jupyter Notebooks are an open-source web application that allows you to create and share documents containing live code, equations, visualizations, and narrative text. They are incredibly useful for data analysis, machine learning, scientific research, and educational purposes because they allow for interactive computing and data visualization. Jupyter supports over 40 programming languages, including Python, R, and Scala, and integrates with big data processing tools. However, in our case, it was extensively used in conjunction with Python only. One such notebook is shown in **Figure 6**.

The combination of Jupyter, OpenCV, Pillow, NumPy, and others provides a powerful toolkit for processing, analyzing, and visualizing data in an interactive and user-friendly environment. OpenCV, a real-time computer vision library licensed under the open-source Apache 2, excels in elementary computer vision tasks. Complementing it, Pillow extends the capabilities of the Python Imaging Library by enhancing usability and supporting an array of image formats ideal for web integration and batch processing. NumPy ties these tools together with its fundamental array and matrix operations, underpinning scientific packages like SciPy and Pandas, and proving indispensable in fields such as data analysis, machine learning, and engineering. Nevertheless, other libraries are also used to support teaching materials.

3.2.2 AI concepts addressed

Throughout their studies, students will delve into the complexities of computer vision, honing in on the algorithms that enable machines to interpret and understand visual data. By examining the nuances of simple biometric recognition systems, such as detecting face in live webcamera feeds, learners will grapple with the practicalities and challenges of implementing such systems. In the first TA, critical part of their exploration will be the scrutiny of bias in data sets, an issue of crucial importance in not only computer vision but in the wider field of AI. They will learn to identify and mitigate bias, understanding its implications for both the accuracy and fairness of decision-making. As they develop and test biometric algorithms, students will be encouraged to consider the ethical dimensions of AI, such as the potential for discrimination, thereby preparing them for the responsible development of technology in the future.

3.2.3 Challenge description

In this TU, the main goal was to get familiar with basics of common computer vision tasks. This includes loading images, understanding the matrix nature of images (e.g. order of channels), modifying them, performing filtering tasks, doing simple segmentation tasks, trying to detect important parts of images and deriving features out of them, and covering the basics of motion analysis. This establishes

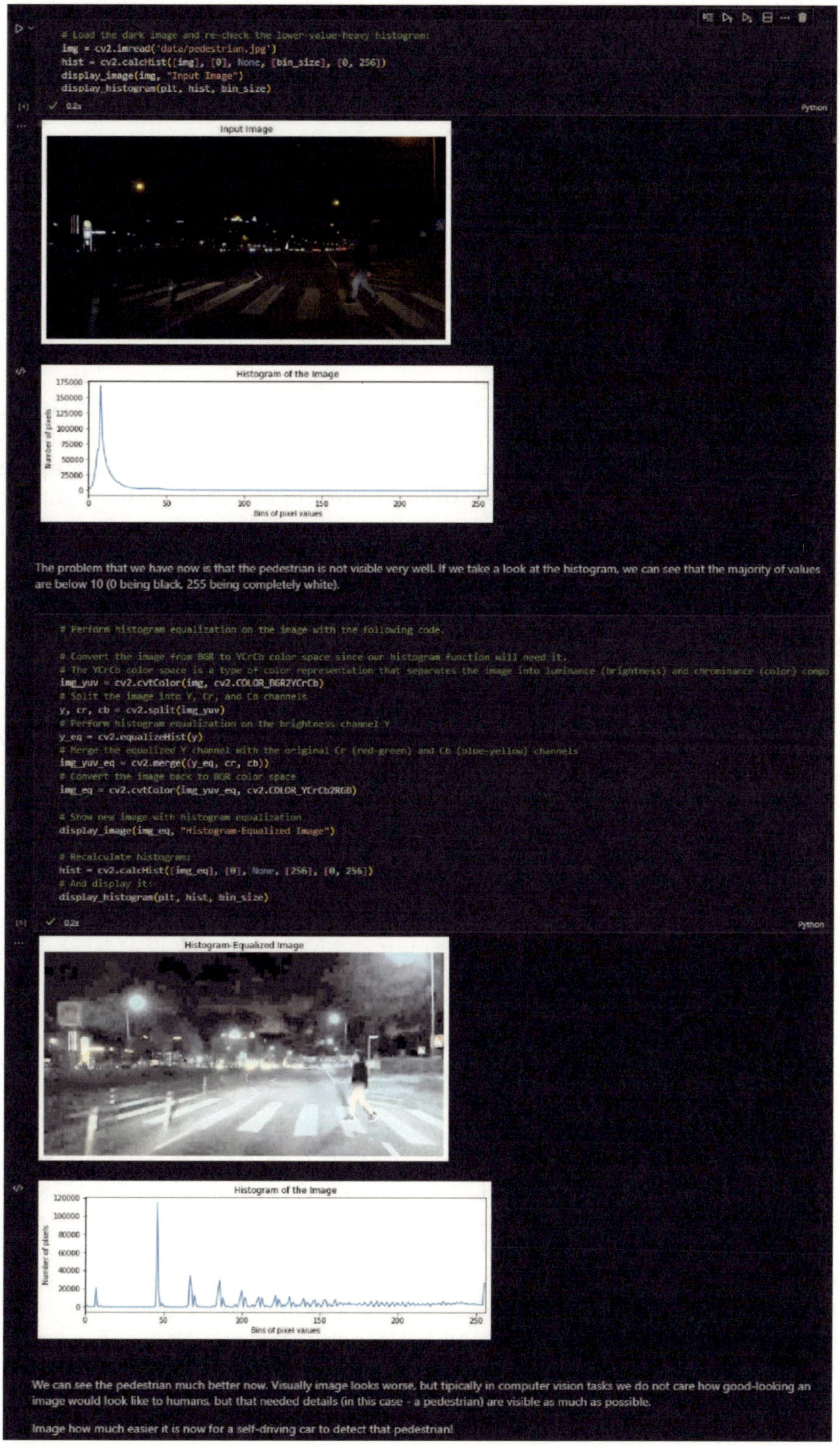

Figure 6.
An example of a Jupyter notebook with the code for students to play with.

a foundational understanding for students in the domain of computer vision and prepares them to understand more advanced topics, such as person recognition [22], the problems of image generation [23, 24], etc. Each topic covered by the tasks is described in more detail below.

Basics of Image Processing. Image processing forms the foundation of computer vision. It involves techniques to enhance raw images for further processing or to extract valuable information. Basic image processing tasks include loading, displaying, and storing images. Understanding how to manipulate the fundamental attributes of an image, such as color channels and intensity levels, is crucial. Grayscale conversion, thresholding, and basic filtering are typical tasks. The importance of these tasks lies in their ability to preprocess images for advanced analysis, improving the performance of complex algorithms in object detection, facial recognition, and more.

Modifying Images. Modifying images involves resizing, cropping, altering brightness and contrast, and geometric transformations. These techniques are critical in preparing images for specific applications. For instance, resizing and cropping can focus on regions of interest, enhancing the efficiency of object detection algorithms. Adjusting brightness and contrast makes features more discernible, which is crucial in medical imaging and surveillance. Understanding these transformations is key to customizing image data for varied applications, from automated inspection systems to augmented reality.

Advanced Image Processing. Advanced processing encompasses techniques like histogram equalization and convolution operations. Histogram equalization is a method for contrast adjustment using the image's histogram. This is particularly beneficial in improving the visibility of features in images, which is essential in areas like satellite imaging and night-vision processing. Convolutions, involving filters like blurring and sharpening, are fundamental in edge detection and noise reduction, paving the way for higher-level tasks such as feature extraction and image classification.

Image Segmentation. Image segmentation divides an image into meaningful regions, facilitating easier analysis. Techniques like thresholding, region-based segmentation, and watershed algorithms play a significant role. In medical imaging, for example, segmentation aids in identifying and isolating pathological regions. In autonomous vehicles, it helps in understanding the driving environment. The significance of segmentation lies in its ability to simplify complex images into analyzable segments, making it a cornerstone in both automated systems and interpretative analysis.

Feature Detection and Description. This process involves identifying key points in images and describing them in a way that facilitates matching and recognition. Algorithms like SIFT (Scale-Invariant Feature Transform) and ORB (Oriented FAST and Rotated BRIEF) are instrumental in this. These features are pivotal in tasks like image matching, 3D reconstruction, and motion tracking. They provide robust ways to represent and compare images, crucial in applications ranging from augmented reality to navigation in robotics.

Object Detection. Object detection is about identifying and locating objects within images. Techniques like template matching, Haar cascades, and HOG (Histogram of Oriented Gradients) are used for this purpose. Object detection is vital in various domains, such as security surveillance for unauthorized intrusion detection, in retail for inventory management, and in automotive for pedestrian and obstacle detection. The ability to accurately detect objects in diverse conditions is a testament to the evolution and importance of this field.

Optical Flow and Motion Analysis. Optical flow techniques, like the Lucas–Kanade method and dense optical flow, estimate the motion of objects between frames in a video sequence. These methods are crucial in understanding the dynamics of scenes and useful in video surveillance for anomaly detection, in sports analytics for player movement analysis, and in video editing for creating special effects. Optical flow provides a deeper understanding of motion patterns, making it invaluable in temporal analysis of videos.

3.2.4 Outcomes

Computer vision content presented a valuable learning experience for students, highlighting the practical applications and basic theoretical underpinnings of computer vision. Students engaged in a series of tasks that, hopefully, not only solidified their understanding of computer vision fundamentals but also prepared them for real-world applications or at least gave them motivation for it.

As described above, during the course, students demonstrated proficiency in basic image processing techniques, including image loading, manipulation, and the application of filters. They were introduced to more complex topics such as image segmentation, feature detection, and object detection, culminating in an understanding of motion analysis through optical flow methods. This progression from foundational skills to more advanced concepts allowed students to build a comprehensive knowledge base in computer vision.

One of the key outcomes was the students' ability to modify and apply code made available to them to actual scenarios. For instance, they successfully used and customized algorithms for detecting objects in images, a skill crucial for many applications in industries such as security, automotive, and healthcare. The hands-on approach, facilitated by the use of Jupyter Notebooks and Python libraries such as OpenCV, Pillow, and NumPy, was instrumental in bridging the gap between theory and practice.

Feedback from students and teachers alike underscored the relevance and engagement of the teaching units. Students appreciated the practical challenges, which enhanced their learning experience. They found the tasks engaging and relevant (albeit on the difficult side), reflecting the growing importance of computer vision in various technological domains.

However, similar to other subject areas, challenges were encountered. Time constraints were noted as a significant hurdle, with some students requiring more time to fully grasp complex concepts or not being able to grasp certain concepts at all. This feedback has been invaluable in planning future iterations of the teaching unit, with considerations for more flexible scheduling and possibly more in-depth preparatory sessions on programming fundamentals for students new to Python.

The successful completion of tasks by the students, coupled with their ability to understand and apply computer vision concepts, indicates the effectiveness of the teaching unit in achieving its educational objectives. It has laid a solid foundation for students, equipping them with the skills and knowledge to further explore the field of computer vision and its applications.

Moving forward, the curriculum will be simplified and refined based on the feedback received. This includes enhancing the clarity of theoretical explanations, adjusting the pacing of the course to better match student learning speeds, and incorporating more examples of real-world applications to further illustrate the impact of computer vision technologies. The ultimate goal remains to provide students with a robust understanding of computer vision, preparing them for future careers in this dynamic and rapidly evolving field.

3.3 Ambient intelligence

The current resource corresponds to the first and second TUs of the Ambient intelligence module of the AIM@VET project, which focuses on sensors and actuators. The ultimate aim of these TUs is for students to learn how sensors and actuators work, as well as how to control them with a microcontroller Arduino Uno. This is the basic principle of sensorization: to develop an entire intelligent ambient with Internet of Things (IoT) environment that can be controlled automatically and using AI. So, the idea is to understand how data is collected and how actuators work in an environment. The next step is to prepare the intelligent agent to act and react to this environment using AI techniques.

3.3.1 Temporal organization and required resources

The development of Modules 1 and 2, a comprehensive teaching unit on sensorization, requires approximately 24 hours to complete without students having the necessary prior knowledge. However, due to limitations in applying all planned activities during the training session, we have selected five key activities crucial for learning sensorization. Consequently, each of these five activities will require two hours to complete. Moreover, each activity may consist of one or more tasks. The temporal organization of the sensorization teaching unit is thoughtfully structured to facilitate a comprehensive and effective learning experience, focusing the entire curriculum into a total of 10 hours. This approach ensures a dedicated and intensive exploration of the subject matter.

The first activity introduces the foundational concepts within the "Required Knowledge" section, covering Ambient Intelligence (AmI), the Internet of Things (IoT), and the Internet of People (IoP). This introductory phase, lasting about 30 minutes, sets the stage for the hands-on work to follow, ensuring that students have a solid conceptual framework before diving into the more technical aspects. Following the introduction, the activity progresses into more practical tasks. Students start with understanding and setting up the Arduino Uno, a process that takes about 30 minutes. This is a crucial step, as it involves familiarizing themselves with the hardware that will be central to their subsequent activities. Next, they move on to preparing the Arduino Uno for integration with Python. This task, also estimated at 45 minutes, introduces students to the software aspect of their projects, bridging the gap between the physical hardware and the programming environment. A significant portion of the activity is dedicated to hands-on activities involving various sensors. Students spend time learning to use a breadboard, which is essential for assembling electronic circuits. This activity is designed to be brief yet informative, taking about 15 minutes, which finishes the first activity.

The second activity then delves deeper into the application of Python programming with Arduino. Students explore the use of PyFirmata and Pymata libraries, engaging in tasks that require them to build circuits and write code. Each of these tasks is allocated 60 minutes, allowing ample time for experimentation and learning.

The third activity has a unique aspect, which is the inclusion of virtual sensors. Students are introduced to the concept and application of virtual sensors, such as those used for climatological and air quality data. They learn how to access and use data from APIs like Open Weather Map and Open Air Quality, which involve writing Python scripts to fetch and process this data. These activities are given substantial time, with 45 minutes dedicated to learning the required knowledge and 75 minutes for practical implementation.

Throughout the fourth and fifth activity, students are progressively building their skills and knowledge, culminating in a final project or challenge. This project involves the integration of all the skills they have acquired, requiring them to develop a program for collecting data from sensors using Arduino and Python. The project is not only a test of their technical abilities but also of their capacity to integrate various components into a cohesive system.

The sensorization teaching unit necessitates a range of resources that are integral to the successful delivery and completion of the course. These resources span across various categories, including hardware, software, and additional support materials, ensuring that students have everything they need to effectively engage with and complete the unit.

At the forefront of the required hardware is a reliable WI-FI network with internet connectivity. This is crucial not only for accessing online resources and documentation but also potentially for some of the project tasks that might involve internet-based applications or data retrieval. Additionally, each student needs access to a laptop or computer. These are essential for coding in Python, interfacing with the Arduino, and potentially for other software-related tasks.

Smartphones, specifically Android-based, are also listed among the required hardware. The use of smartphones could be linked to interfacing with the Arduino projects or for testing and running applications developed during the course. Furthermore, the unit requires one Arduino Uno and a set of sensors for every two students. Arduino Uno serves as the foundational hardware for teaching sensor integration and data collection, and the sensors are crucial for hands-on experimentation and understanding real-world data acquisition.

In terms of software, Python is a primary requirement, as it is used extensively for programming the Arduino and data processing tasks. Additional software might include the Arduino Integrated Development Environment (IDE) and other related tools or libraries such as PyFirmata, which are used for interfacing Python with the Arduino.

An often-overlooked but equally important resource is a projector, recommended for displaying teaching unit materials and multimedia resources to all students. This facilitates a more interactive and engaging learning environment, allowing instructors to visually demonstrate concepts and procedures.

In addition to these, there might be requirements for basic electronic components such as breadboards, jumper wires, and power supplies, which are standard in any electronics lab setting. These components are necessary for building and testing circuits, and for interfacing the various sensors with the Arduino Uno.

To support the theoretical learning, there may also be a need for textbooks or online resources. These materials would provide additional reading and reference material to students, helping them to deepen their understanding of the concepts taught in the unit.

3.3.2 AI concepts addressed

The teaching unit begins with an introduction to the Arduino Uno, covering both its hardware and software components. Students will then be guided on how to prepare the Arduino for Python programming using PyFirmata, along with an explanation of the Firmata protocol. This foundational knowledge will be further enhanced by teaching them about breadboards and how to prepare them for creating electronic circuits. Students will gain practical experience by building simple circuits, programming them in Python, and exploring different approaches to problem-solving. A significant part of this phase involves the introduction and application of an ultrasonic sensor, which students will program and test to understand its functionality.

Virtual sensors represent another critical area of learning, where students will learn to apply these sensors through API calls and data interpretation. They will engage in tasks that involve accessing climatological and air quality data using APIs like Open Weather Map and Open Air Quality, enabling them to work with real-world data in a controlled environment.

Further into the materials, students will explore serial data communication with Arduino, learning to use Digital Read Serial with Python. This will include practical exercises like programming push buttons and LEDs, providing a deeper understanding of digital signal processing. Additionally, the course will cover the control of motor speed using Pulse Width Modulation (PWM), where students will experiment with DC motors and servo motors, learning to adjust their speed through programming in Python.

This structured approach will not only equip students with the necessary technical skills but also foster a deeper understanding of AI concepts through interactive and engaging activities.

3.3.3 Challenge description

The challenge presented in the sensorization teaching unit is a comprehensive and integrative project that encapsulates the skills and knowledge acquired throughout the course. The final objective of this challenge is to develop a program that allows for the collection of data from sensors using Arduino and Python, effectively combining hardware interfacing with software programming in order to, in future, use captured data from sensors, treat it with AI algorithms, and after that actuate on the environment.

Throughout the unit, students are introduced to various concepts and practical skills, starting with an understanding of Ambient Intelligence, the IoT, and the Internet of People. They learn the basics of Arduino Uno, setting it up, and preparing it for integration with Python. An essential part of this is learning to use a breadboard for circuit assembly, which is crucial for any hands-on electronics project.

As students progress, they delve into using specific libraries like PyFirmata and Pymata, which are instrumental in controlling the Arduino through Python. The challenge involves working with various sensors – such as ultrasonic, temperature, and humidity sensors – and understanding how to implement them in practical applications. This not only includes the physical assembly and programming of these sensors but also the interpretation of the data they collect.

A significant part of the challenge is working with virtual sensors. Students are expected to access climatological and air quality data using APIs such as Open Weather Map and Open Air Quality. They develop Python scripts to fetch and process this data, integrating it with the information collected from physical sensors.

The culminating project requires students to bring together all these elements. They need to design and program a system that can collect data from both the physical sensors attached to the Arduino and the virtual sensors accessed via APIs. This task tests their technical skills in programming and electronics, as well as their ability to process and interpret data. The challenge is not just about building a functional system but also understanding the practical applications of this technology in real-world scenarios.

The project is evaluated on various fronts, including technical skill in programming and circuit assembly, the effectiveness of data collection and processing, and the overall integration of different technologies and concepts. This challenge serves as a practical demonstration of the students' learning and proficiency in sensor technology and its applications in the digital world.

3.3.4 Outcomes

In completing the challenge of the sensorization teaching units, students are expected to achieve several significant outcomes that demonstrate their understanding and application of the concepts and skills learned throughout the course. Throughout the sensorization teaching unit, we have employed a methodologically rich and diverse approach to learning, which has effectively blended theoretical knowledge with practical skills. This multifaceted methodology has been pivotal in ensuring a comprehensive understanding of sensor technologies, Arduino programming, and Python integration. Here's a reflection on what we have learned so far methodologically and how this approach will continue to evolve and be applied as the course progresses.

Firstly, they will have developed a comprehensive program that successfully integrates Arduino and Python to collect data from various sensors. This involves not only the technical aspects of programming and circuit design but also a nuanced understanding of how different sensors operate and interact. By working with both physical sensors (like ultrasonic, temperature, and humidity sensors) and virtual sensors accessed via APIs, students will showcase their ability to handle diverse data sources. **Figure 7** is an example of a simple circuit.

So far, the course has emphasized a hands-on learning approach. This method is crucial in technical subjects, allowing students to directly interact with the tools and technologies they are studying. By working with Arduino and various sensors, students have gained practical skills that are often difficult to fully grasp through theoretical study alone. This hands-on approach not only aids in solidifying the students'

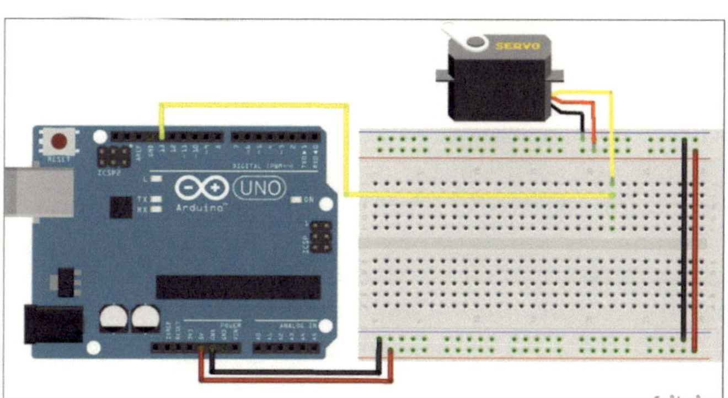

Figure 7.
An example of simple activity.

understanding of sensor technologies but also enhances their problem-solving and critical-thinking abilities.

An important outcome of this challenge is the practical application of the collected data. Students must demonstrate how to process, interpret, and potentially visualize this data in a meaningful way. This might involve identifying patterns, making predictions, or simply understanding the environmental conditions measured by the sensors. The ability to translate raw sensor data into usable information is a crucial skill in many fields, including IoT applications, environmental monitoring, and robotics.

The project also tests the students' problem-solving and critical-thinking skills. They need to integrate different technological components and concepts into a cohesive system. This includes troubleshooting issues that arise during the development process, whether related to hardware, software, or the interfacing between the two.

Additionally, students will have gained hands-on experience in programming with Python and working with Arduino, which are valuable skills in the fields of computer science, engineering, and technology. Their exposure to working with APIs to access virtual sensors like Open Weather Map and Open Air Quality broadens their understanding of how large-scale data can be accessed and used. The incremental learning model adopted in the course has been another key aspect of our methodological approach. We started with basic concepts and gradually moved to more complex topics. This step-by-step progression has ensured that students build a strong foundational knowledge before tackling more advanced aspects of sensor technology and programming. Such a structured approach helps in maintaining clarity and focus, preventing students from feeling overwhelmed by the complexities of the subject matter.

Integration of theory and practice has been a hallmark of our methodology. Theoretical concepts about Ambient Intelligence (AmI), the Internet of Things (IoT), and the role of sensors in digital transformation have been seamlessly tied to practical tasks and projects. This has enabled students to see the real-world applications of their learning, making the educational experience more relevant and engaging.

Lastly, the challenge offers an opportunity for creativity and innovation. Students can explore various applications of the technology, suggest improvements, or even propose new ways of using the sensor data. This fosters a mindset of exploration and innovation, which is crucial in the rapidly evolving field of technology. Also, collaborative learning has also been a focus. While the course has encouraged individual skill development, it has also fostered an environment where students can share knowledge, work together on projects, and learn from each other. This blend of individual and collaborative learning enriches the educational experience and prepares students for the teamwork-oriented nature of the professional world.

Looking forward, the course will continue to build upon these established methodological foundations. We will introduce more complex and challenging projects, encouraging students to apply their learned skills in new and innovative ways. These projects will not only reinforce the students' existing knowledge but will also push them to explore the limits of their creativity and problem-solving skills.

We will also continue to emphasize the real-world application of the skills and knowledge gained. Students will be encouraged to think about how the technologies and methods they are learning can be applied to solve real-world problems. This focus on practical application is intended to bridge the gap between academic learning and professional skills.

4. Results and discussion

This last section will go over feedback and learning outcomes from Teaching Activity 1 (TA1), pulling together what we've learned from areas like robotics, Ambient Intelligence, and computer vision. We'll look at what worked well, what challenges we ran into, and how we can make our teaching methods better for this AI curriculum.

One of the elements that proved to be essential to take good care of was the engagement of students and their progress in learning. Feedback from these sessions emphasized the critical need to allow students enough time to thoroughly explore and understand complex topics. Given adequate time, students were not only able to comprehend the foundational concepts but also found the tasks to be highly engaging. This underscores an important lesson: the pace at which new information is introduced must be thoughtfully aligned with the content's complexity, ensuring students remain fully engaged and able to grasp the material comprehensively.

There were, however, some important challenges to highlight, particularly when some students struggled with basic mathematical concepts necessary for progressing in more sophisticated tasks. This difficulty revealed a gap in the foundational knowledge of certain participants, highlighting the importance of implementing educational strategies that equip all students with a robust understanding of essential mathematical principles before moving forward with more complex subjects.

Therefore, important aspects noted based on comprehensive feedback focus on making complex subjects more accessible and engaging for all students. This included

Question	Rating
Do you think you had enough prior knowledge to tackle this lesson?	3.83
How would you rate the theoretical explanations? (1 star very easy, 5 stars very difficult)	3.28
How would you rate the level of difficulty of the challenges?	3.06
Do you think there has been a good balance between theory and practice? (1 star: too much theory; 5 stars: too much practice)	2.89
How interesting do you think it is to work with a partner compared to working alone?	3.50
What do you think about the number of days dedicated to workshops (3 days)? (1 star: too short; 5 stars: too long)	2.72
What do you think about the number of hours per day dedicated to workshops? (1 star: too short; 5 stars: too long)	4.00
What do you think about the number of hours per day dedicated to activities? (1 star: too short; 5 stars: too long)	2.78
Do you feel that this lesson has increased your understanding of AI?	3.89
Please rate your overall experience of the first training activity	4.17

Table 4.
Quantitative results from students' questionnaires.

Question	Rating
Do you think the students had sufficient prior knowledge to tackle this lesson?	2.50
Do you believe the students have acquired relevant knowledge?	3.33
How many difficulties have they encountered due to the language barrier? (1 star: few difficulties; 5 stars: many difficulties)	1.83
How positive has it been to work in international pairs?	4.33
How positive has it been to change partners in the middle of the activity?	4.33
How many students have actively participated and engaged in the activity?	4.50
What do you think about the number of days dedicated to workshops (3 days)? (1 star: too short; 5 stars: too long)	2.67
What do you think about the number of hours per day dedicated to workshops? (1 star: too short; 3 stars: ideal duration; 5 stars: too long)	4.17
What do you think about the number of hours per day dedicated to cultural and social activities? (1 star: too short; 3 stars: ideal duration; 5 stars: too long)	2.83

Table 5.
Quantitative results from teachers' questionnaires.

simplifying the instructional language to ensure clarity, adjusting the curriculum to allow more time for foundational concepts, providing practical tools like code templates for bridging gaps in mathematical understanding, and offering detailed walkthroughs for the theoretical application of concepts.

At the end of the TA1, a questionnaire was administered to the students and the teachers regarding various aspects of the workshop. **Tables 4** and **5** show the quantitative results provided.

The student feedback from the AI training session revealed a positive overall experience with a notable increase in AI understanding. However, while students felt somewhat prepared and engaged with the material, they also indicated that the number of workshop hours per day was higher than ideal. Despite this, the practical elements and partner collaboration were well-received, suggesting a successful integration of AI learning into the curriculum with room for adjustment in scheduling.

Teachers' feedback from the questionnaire indicates a mixed assessment of the AI training program. They observed that students came with a moderate understanding of the subject matter. The acquisition of relevant knowledge during the course was satisfactory. Language barriers did not significantly impede learning. Collaborative aspects of the program, such as working in international pairs and rotating partners, were highly regarded. Student engagement was notably high. However, teachers felt that the workshop days were slightly longer than ideal, and the hours allocated to workshops each day were deemed excessive. Conversely, they considered the time for cultural and social activities to be marginally insufficient.

The synthesis of feedback from Teaching Activity 1 (TA1) highlights a generally positive experience from both students and teachers, marking the project's success in integrating AI into vocational training. Students valued the practical application of AI, noting an effective balance between theory and practice, and showed keen interest in exploring more AI tools like ChatGPT. On the other hand, teachers pointed out the students' need for stronger foundational knowledge in AI and suggested shorter workshop hours for future sessions. Overall, the positive reception of TA1 underscores its contribution to enhancing AI education in vocational settings, reflecting its alignment with the evolving requirements of the job market.

5. Conclusions

The first Training Activity (TA1) within the AIM@VET project showcased a practical approach to integrating Artificial Intelligence (AI) into Vocational Education and Training (VET) curricula. Focusing on autonomous robotics, computer vision, and ambient intelligence, TA1 provided VET students with direct experience in applying AI concepts and techniques. This activity not only improved their technical skills in AI but also equipped them with essential knowledge of the impact of AI technologies, preparing them for the digital workforce.

The structured methodology adopted for the first Training Activity (TA1), involving a mix of theoretical lessons and practical tasks, has proven effective among students. The collaborative and cross-cultural learning environment also contributed to the educational experience, promoting a broader perspective on the impact of AI technologies.

Feedback from both students and teachers underscores the success of the project in achieving its educational objectives. Students have gained not only a theoretical understanding of AI principles but also hands-on experience in applying these concepts through projects and challenges. The positive reception of the teaching units, as evidenced by the high level of student engagement and interest, indicates a strong appreciation for the relevance and practicality of the curriculum.

Teachers have highlighted the importance of a balanced approach between theory and practice, and the feedback received has been instrumental in refining the teaching units for future implementations. The adjustments made based on this feedback, such as simplifying language, allowing more time for complex topics, and providing practical tools for mathematical concepts, have enhanced the accessibility and effectiveness of the curriculum.

The initial Training Activity (TA1) within the AIM@VET project was also key in refining our protocols for lesson creation and standardization. This refinement led to a more unified and standardized curriculum, improving coordination among project partners. The protocols addressed the creation of templates and set criteria for content creation, student assessment, and the standardization of complexity levels, lesson durations, and prerequisite knowledge.

The move toward standardization was necessary for collaborative curriculum development across different research groups. It brought together the work of various teams, resulting in a coordinated approach to educational content. Additionally, it allowed for the recognition of each partner's work and conceptual modules, finding a common base that will aid in future curriculum development with multiple working groups.

As these protocols are further refined and utilized, they also establish a framework for collaborative and coherent curriculum design, not only in AI education within VET programs but in general in any project-based learning curriculum.

Therefore, the AIM@VET project will serve as a model for future initiatives aimed at integrating AI education into VET programs. The insights gained from TA1 offer valuable guidance for educators and policymakers on how to design and implement AI curricula that are both engaging and informative. The project's focus on practical skills, collaboration, and cross-cultural exchange aligns with the evolving needs of the workforce, preparing students for the challenges and opportunities of the digital age.

Acknowledgements

The authors wish to acknowledge the Erasmus+ Programme of the European Union (grant 2022-1-ES01-KA220-VET-000089813).

Author details

Žiga Emeršič[1], Peter Peer[1], Gregor Hrastnik[2], Nataša Meh Peer[2], José María Bey[3], María Meizoso-García[3], António Pedro Silva[4], Cláudia Domingues[4], Carla Abreu[4], António Costa[5], Dalila Durães[5], Paulo Novais[5], Cristina Renda[6] and Abraham Prieto[6]*

1 University of Ljubljana, Ljubljana, Slovenia

2 School Center Velenje, Velenje, Slovenia

3 CIFP Rodolfo Ucha Piñeiro, Ferrol, Spain

4 Escola Secundária de Caldas das Taipas, Guimarães, Portugal

5 University of Minho, Guimarães, Portugal

6 Universidade da Coruña, Coruña, Spain

*Address all correspondence to: abraham.prieto@udc.es

References

[1] Dignum V, Penagos M, Pigmans K, Vosloo S. Policy Guidance on AI for Children. New York City, USA: UNICEF; 2021 [Online]. Available from: https://www.unicef.org/globalinsight/reports/policy-guidance-ai-children [Accessed: July 24, 2023]

[2] EC. Digital Education Action Plan 2021-2027. Brussels, Belgium: European Commission; 2021 [Online]. Available from: https://education.ec.europa.eu/focus-topics/digital-education/digital-education-action-plan [Accessed: July 24, 2023]

[3] Miao F, Holmes W, Huang R, Zhang H. AI and Education: Guidance for Policy-makers. Paris, France: UNESCO; 2021. [Online]. Available from: https://unesdoc.unesco.org/ark:/48223/pf0000376709 [Accessed: July 1, 2024]

[4] Ng DTK, Leung JKL, Chu SKW, Qiao MS. Conceptualizing AI literacy: An exploratory review. Computers and Education: Artificial Intelligence. 2021;**2**:100041

[5] UNESCO. K-12 AI curricula: A mapping of government-endorsed AI curricula. 2022 [Online]. Available from: https://unesdoc.unesco.org/ark:/48223/pf0000380602.locale=en [Accessed: July 24, 2023]

[6] M. Raise. MIT raise. [Online]. Available from: https://www.dayofai.org/curriculum [Accessed: July 24, 2023]

[7] Bellas F, Guerreiro-Santalla S, Naya M, Duro R. AI curriculum for European high schools: An embedded intelligence approach. International Journal of Artificial Intelligence in Education. 2023;**33**(2):399-426

[8] AI4K12. AI4K12. [Online]. Available from: https://ai4k12.org [Accessed: July 24, 2023]

[9] Rodríguez-García JD, Moreno-León J, Román-González M, Robles G. Introducing artificial intelligence fundamentals with LearningML: Artificial intelligence made easy. Eighth International Conference on Technological Ecosystems for Enhancing Multiculturality. New York City, USA: ACM; 2020. pp. 18-20

[10] Rott K, Lao L, Petridou E, Schmidt-Hertha B. Needs and requirements for an additional AI qualification during dual vocational training: Results from studies of apprentices and teachers. Computers and Education: Artificial Intelligence. 2022;**3**:100102

[11] Attwell G, Bekiaridis G, Deitmer L, Perini M, Roppertz S, Stieglitz D, et al. Artificial intelligence and vocational education and training: How to shape the future [Online]. Available from: https://taccleai.eu/artificial-intelligence-vocational-education-and-training-policy-recommendations/ [Accessed: July 24, 2023]

[12] Prieto A, Guerreiro S, Bellas F. Aim@vet: Tackling equality on employment opportunities through a formal and open curriculum about AI. In: Computational Intelligence in Security for Information Systems Conference. Berlin, Germany: Springer; 2023. pp. 228-237

[13] AIM@VET. AIM@VET web [Online]. Available from: https://aim4vet.udc.es [Accessed: July 24, 2023]

[14] A. Repositories. Teaching unit 2 and training activity 1 [Online].

Available from: https://bit.ly/3Or7StX; https://bit.ly/3DtuFyM [Accessed: July 24, 2023]

[15] Page RW. Robobo wiki page [Online]. Available from: https://github.com/mintforpeople/robobo-programming/wiki [Accessed: July 24, 2023]

[16] Llamas LF, Paz-Lopez A, Prieto A, Orjales F, Bellas F. Artificial intelligence teaching through embedded systems: A smartphone-based robot approach. In: Robot 2019, Ser. Advances in Intelligent Systems and Computing. Vol. 1092. Springer; 2020

[17] Naya-Varela M, Guerreiro-Santalla S, Baamonde T, Bellas F. Robobo smartcity: An autonomous driving model for computational intelligence learning through educational robotics. IEEE Transactions on Learning Technologies (early access). 2022;**16**(4):543-559

[18] Wikipedia. Bellman equation [Online]. Available from: https://en.wikipedia.org/wiki/Bellman_equation [Accessed: July 24, 2023]

[19] Renda C, García AP, Bellas F. Teaching reinforcement learning fundamentals in vocational education and training with robobosim. In: ROBOT2023: Sixth Iberian Robotics Conference. Berlin, Germany: Springer; 2023 [Online]. Available from: https://easychair.org/smart-program/ROBOT2023/bytalk-2023-11-24.html

[20] Emeršič Ž, Hrastnik G, Meh Peer N, Peer P. Adapting vet education to labor market needs with focus on artificial intelligence and computer vision. In: ROSUS. Maribor, Slovenia: University Press of the University of Maribor; 2023. pp. 77-86

[21] Kirn VL, Emeršič Ž, Hrastnik G, Meh Peer N, Peer P. Introductory computer vision teaching materials for vet education. In: ROSUS. Maribor, Slovenia: University of Maribor, University Press; 2024

[22] Emeršič Ž, Štepec D, Štruc V, Peer P. George A, Ahmad A, et al. The unconstrained ear recognition challenge. In: 2017 IEEE International Joint Conference on Biometrics (IJCB), Piscataway, New Jersey, USA: IEEE. Colorado, USA: Denver; 2017. pp. 715-724

[23] Meden B, Gonzalez-Hernandez M, Peer P, Štruc V. Face deidentification with controllable privacy protection. Image and Vision Computing. 2023;**134**:104678

[24] Markićević L, Peer P, Emeršič Ž. Improving ear recognition with super-resolution. In: 2023 30th International Conference on Systems, Signals and Image Processing (IWSSIP). Ohrid, North Macedonia: IEEE; 2023. pp. 1-5

Chapter 9

Teacher Educator Professionalism in the Age of AI: Navigating the New Landscape of Quality Education

Olivia Rütti-Joy, Georg Winder and Horst Biedermann

Abstract

This conceptual chapter discusses how requirements for teacher educator professionalism may be impacted by the integration of Artificial Intelligence (AI) in teacher education. With the aim to continuously facilitate high-quality teacher education, teacher education institutions must evolve in alignment with the rapidly changing landscape of AI and the respective shifting educational needs. Amidst this evolution, we argue that profound AI Literacy and AI-related ethical knowledge constitute two additional and inextricably intertwined knowledge facets of teacher educator professionalism essential for an ethical and effective integration of AI into teaching practices – and thus crucial for high quality teacher education. The paper explores avenues through which these facets of teacher professional competence and quality education can be fostered on the micro, meso and macro levels of institutional education. By consolidating the specific requirements in a framework for teacher educator professionalism in the age of AI, we highlight the necessity for continuous adaptation of teacher education institutions, ongoing multidisciplinary collaboration, and the provision of periodic professional development of educators. Finally, the chapter presents a concrete practical example and future research directions in AI and education with the aim to contribute to the advancement of quality education in the AI era.

Keywords: artificial intelligence in education (AIEd), quality education, quality teacher education, teacher educators' professional competence, teacher professionalism, professional development, AI literacy, AI ethics, organisational development, institutional adaptation

1. Introduction

The rapid proliferation of artificial intelligence (AI[1])-powered technologies [1, 2] carries the potential to reshape the foundations of education [3] and to significantly change the demands of the teaching profession [4]. With education's

[1] While the term AI is highly debated [5, 6] and there is no universally valid definition of AI [7], we adopt Southworth et al.'s [1] broad interpretation of AI to account for its interdisciplinary nature of AI in education, where "AI refers broadly to include related expertise and disciplines when used with AI (e.g., such as statistics and data science)" (p. 2).

IntechOpen

chief performance mandate to enable future generations for responsible participation in societal, political, cultural and economic processes [8], educational institutions are urged to adapt to the changing requirements of the educational and vocational domains to safeguard high-quality, relevant and fit-for-purpose teaching and research practices [1, 9]. To ensure education continues to optimally prepare forthcoming generations to succeed in their complex futures, curricula need to be aligned with the relevant skills and knowledge that will be relevant in the future [10]. The plethora of frameworks and guidelines that attempt to predict these skills [11, 12] – while not without their challenges [12, 13] – reinforce the importance of inclusive quality education and lifelong learning opportunities to contribute to a progressive and healthy society [14]. In the age of AI, such a society can harness the affordances of AI productively and mitigate its risks and challenges responsibly [15]. To ensure quality education, enhanced students' learning experience and improved learning outcomes [1], theories and applications of AI thus need to be constructively integrated as essential elements into education [1, 16]. As teachers and teacher educators bridge schools' and universities' AI policies and learners' needs, they play a critical role in fulfilling this task [16]. Accordingly, teacher educators need to acquire the professional competences needed to convey the relevant knowledge related to understanding, using, applying and teaching AI; validating knowledge and information and knowing about fairness, accountability, transparency and ethics related to AI [17–19].

Part of the responsibility of high-quality teacher education thus involves institutional adaptations and professional development initiatives to enable teacher educators (1) to use AI effectively and ethically in their teaching and research practice (as well as in their everyday lives) and (2) to teach AI skills and knowledge – including AI-related ethical knowledge – to their pre-service teachers to promote the cascading effects of passing on that knowledge to future generations.

This conceptual chapter addresses the evolving demands on teacher educator professionalism and the role of teacher education institutions in the age of AI. Synthesising the conceptual and empirical literature on Artificial Intelligence in Education (AIEd) and teacher professionalism, we propose that AI knowledge and skills and AI-related ethical knowledge, subsumed in the concept of AI Literacy, constitute additional and inextricably intertwined facets of knowledge of (teacher) educator professional competence that are crucial for enabling an ethical and effective integration of AI into education [20]. Based on these elaborations, we propose a framework of ten areas of development at the micro (learners, pre-service and in-service teachers, teacher educators), meso (schools, teacher education institutions, professional development providers, the teaching profession as a system) and macro level of teacher education (educational policy and politics at state, federal, and global levels) that can be seen as driving forces to facilitate avenues for up-to-date teacher professionalisation and high-quality teacher education amidst the AI transformation. Accordingly, we argue that AI Literacy needs to be incorporated in teacher education institutions, teacher education curricula and the conceptualisation of teacher professionalism to ensure equitable and inclusive AIEd practices [20, 21]. We outline the theoretical background, rationale and method for integrating AI Literacy into the conceptualisation of teacher educator professional competence [22], discuss the deduction of each area of development of the framework and present a case study of implementing AI in teacher education based on the developed framework before concluding with considering future empirical research trajectories.

DOI: http://dx.doi.org/10.5772/intechopen.1005030

2. Rethinking (teacher) educator professionalism in the age of AI

For future generations to be able to navigate AI-permeated societies responsibly, they will require profound AI-related knowledge [18, 23, 24]. To ensure educational institutions convey this knowledge in their curricula, educators need to become "fluent" in AIEd [21, 25, 26] and develop the respective professional competence to meet the changing demands of their vocation [27]. "Professional competence" is a core constituent of "professionalisation" and "professionalism" [27]. In the context of teacher education, professionalisation is generally understood as the development of professionalism, that is, developing professional competence and "becoming professional" [28]. Professionalism, on the other hand, refers to meeting the requirements of the profession [29]. Professionalisation and professionalism are understood to be both individual (micro level) and collective (micro and meso level) phenomena that are inextricably intertwined [30]. As an individual phenomenon, professionalism describes the extent to which educators' professional competence is developed to cope with the demands of their profession [31, 32]. As a collective phenomenon, professionalisation refers to the aim of an evidence-based modernisation of the teaching profession through clear task descriptions, the identification of educational goals and standardised professional development obligations [32].

Due to the multiplicity of theoretical models [31] and methodological approaches, the discourse on both concepts is multifaceted [30]. Building on Weinert's definition of competence [33], the competence-theoretical approach as one of the most prominent approaches in the German-speaking research tradition defines areas of competence and knowledge dimensions that are constitutive for mastering the tasks and challenges of the teaching profession [34]. Subsumed under the term "professional competence," these competences are conceptualised as profession-related abilities [31] that encompass the latent potential, the process that leads to the decision to act and the performance as the action itself [30]. The present chapter uses Baumert and Kunter's generic COACTIV model of teachers' professional competence as its theoretical foundation [22]. According to this model, professional competence is composed of the four aspects: (1) professional knowledge (i.e., knowledge and skills, including its distinct domains of knowledge and respective facets); (2) professional values, beliefs and goals; (3) motivational orientations and (4) professional self-regulation skills. While all aspects of competence are categorically separated, they interact with one another, and it is through their interaction that professionally competent behaviour arises [22]. The following sections of this chapter address professional competence and teacher educator professionalism in the context of AI in teacher education with a particular focus on teacher educators. Based on a review of the affordances of AI that may contribute to, and challenges that may hinder high-quality teacher education during rapid transitions, the subsequent sections investigate whether any of the above aspects of competence require revisiting and adaptation in alignment with the rapidly changing educational needs in an AI-mediated society and whether AI-related knowledge and skills can and should be conceptualised as additional domains or facets of knowledge of any of the model's aspects of competence [35].

3. AI in education: essentials and perspectives on benefits and challenges

AI is considered a powerful instrument to facilitate new opportunities for instructional design, independent and self-directed learning, educational innovation and educational research [4]. Successfully integrating, applying and teaching AI in education, however, raises profound questions and poses a range of challenges, such as revisiting and realigning research and teaching practices to ensure high-quality education that remains focused on the human learner [36]. This endeavour demands profound evidence-based knowledge about the background, affordances, challenges and prevailing questions related to AIEd. Among others, Bond et al.'s [20] meta-systematic review (i.e., a review of reviews) of AI in higher education provides a solid evidence base. By rigorously analysing a corpus of 66 reviews, they synthesised 12 categories of benefits and 17 categories of challenges of AIEd from a sub-corpus of 31 reviews. Among the 12 identified categories of the former, the top 6 benefits encompass: (1) personalised learning (e.g., customising educational materials to fit individual learners' needs), (2) greater insight into student understanding (e.g., using machine learning and analytics to classify patterns, model student profiles, identify learning issues or provide customised guidance or adaptive feedback), (3) positive influence on learning outcomes (despite very little empirical evidence of impact), (4) reduced planning and administration time for educators (e.g., streamlining administrative workflows, using AI to facilitate lesson planning or handle student inquiries), (5) greater equity in education (e.g., AI enhancing accessibility to education and expert systems) and (6) precise assessment and feedback (e.g., enhancing the likelihood for timely, objective and error-free grading or monitoring student progress).

Among the 17 identified categories of challenges, the 5 most imminent concerns of AIEd across all 31 reviews include: (1) a lack of ethical consideration, (2) issues related to curriculum development (e.g., disconnection between AI technology and educational systems), (3) infrastructure (e.g., technical, literacy and financial barriers, or access to stable high-speed internet), (4) a lack of educator knowledge (e.g., widespread misconceptions and unawareness of AI, lack of technological skills and knowledge, lack of pedagogical knowledge and pedagogical content knowledge to apply AI [35], or limited time resources to effectively integrate AI into the curriculum) and (5) shifting authority (e.g., misconceptions of AI's potentials and challenges that could lead to a transfer of authority from professionals to AI systems). While all the above categories may significantly shape how educational institutions appropriate AI [36], we argue that two categories are particularly central and demand more emphasis:

1. AI Literacy: The AI Literacy construct refers to AI-related knowledge and skills. By outlining their importance, the construct addresses questions such as what educators, learners and the public need to know about AI; how entire sectors of society can be trained in these competences; what pedagogy is most suitable to achieve this goal and who carries the responsibilities to initiate, fund and facilitate these processes.

2. Ethical Considerations: AI-related ethical knowledge and skills include, among others, the ability to address risks such as data privacy and unreflective use of information, understanding and uncovering biases in and ignorance regarding AI algorithms or overcoming the digital divide. Prioritising these concerns is necessary to ensure AI's integration into education is efficient and equitable and that it safeguards individual rights and societal standards.

If a lack of AI Literacy and an absence of AI-related ethical knowledge compound, problematic use and negative consequences of AIEd could propel and limit or reduce the quality of educational programmes. In alignment with Bond et al. [20], both categories are thus pivotal to ensure constructive collaboration with AI. As ethical knowledge about AI is inextricably intertwined with AI Literacy [23], it can be considered a subdimension of the broader AI Literacy construct.

4. AI literacy in education

AI Literacy is generally understood as "a set of competencies that enables [nonexpert] individuals to critically evaluate AI technologies, communicate and collaborate effectively with AI, and use AI as a tool online, at home, and in the workplace" [23]. While the publications on AI Literacy have increased significantly since the early 2020s [19, 25, 37], empirical research in this area is still in its infancy [25] and construct definitions vary [18, 23, 24]. For the purpose of this paper, we use Long and Magerko's definition of AI Literacy [38]. Their framework outlines 17 competences that make up AI Literacy, spanning from the ability to distinguish between tools that use and do not use AI to identify problems that AI can or cannot solve well, to recognise how computers reason and make decisions, to critically interpret data, or to understand ethical concerns related to AI. With education's responsibility to prepare individuals for personal and professional success [36], teacher education institutions are, among others, responsible for mitigating the benefits and concerns of education's AI-transformation [21]. Meeting this performance mandate crucially depends on teacher educators' readiness [16] and knowledge of AI, including the implications of applying and integrating AI to pedagogy, teaching and assessment [35]. In other words, in order to be able to convey the necessary AI-knowledge to their learners, educators need to become AI-literate [18, 19]. This prerequisite has come to be so prominent that AI Literacy has even been predicted to become as important as literacy in its original sense [18]—the ability to think (at least) in the first language and mathematical language, to use these abilities effectively to cope with professional and private tasks and the ability to read and write [39].

5. AI-related ethical knowledge and skills in education

Ethical concerns related to AI in education are among the most mentioned across the AIEd literature [35, 40–43] and are inextricably linked with AI Literacy. Among the 17 competences outlined in Long and Magerko's AI Literacy framework [38], competency 16 relates to ethics. Accordingly, AI-related ethical knowledge and skills refers to the ability to "identify and describe different perspectives on the key ethical issues surrounding AI (i.e., privacy, employment, misinformation, the singularity, ethical decision making, diversity, bias, transparency, accountability)" (p. 7). Further ethical concerns include apprehensions relating to fairness [41], issues relating to data sources and data sourcing (e.g., corporatisation and commodification of data, unethical treatment of staff who source the data), (data) ownership (e.g., increased potential to leak or misuse user data when AIEd systems are developed by for-profit organisations [44]), authorship and copyright (e.g., can AI tools be considered authors and own copyright [45]?), and authority and responsibility (e.g., can AI tools be held accountable and "responsible" for the content they produce [45]?). While ethical codes and policy on the use of AI-generated data [42, 46] and the implementation of AI-powered tools in professional practice [47] have been discussed in disciplines such as law, engineering and social science, the discussion (and scientific research) thereof is largely absent in education [35, 46]. This absence including a lack of a regulated AI ethics policy in education render ethical considerations in AI-permeated education especially pressing [47]. Aside from the need for a regulated AI ethics policy, enabling responsible and ethical use of AI in education will require a move away from closed, corporate-owned and guarded datasets to the generation of large, open datasets so that models appropriate and suitable for education can be developed, trained and validated [36]. Such a move necessitates the need for creating or improving ethical frameworks [48], alongside a deeper understanding of the social implications of AI more broadly. This development involves educating learners and educators about ethical principles, their own ethical behaviour and the ethical use of AI [41]; equipping them with necessary skills and knowledge required to navigate and shape a desirable future [3] and empowering educators to develop and uphold their professional ethics (i.e., professional ethos).

6. Connecting the dots: AI literacy and AI-related ethical knowledge as knowledge facets of professional competence

Despite AI's pervasive impact, a general lack of understanding about AI, its decision-making algorithms and the role humans play in AI interactions prevails [16, 23, 40, 49]. This lack is coupled with widespread confusion and misunderstanding surrounding AI [50], indicating a gap in teacher educators' AI-related professional competence that is paramount to be addressed. Indeed, carefully embedding AI in education and fostering teacher educators' AI Literacy [21] are crucial to empower not only learners but also themselves as professional educators and individuals to leverage AI for creating opportunities rather than incurring opportunity costs [18, 19]. To embed AI Literacy development - with a particular focus on the development of its sub-dimension "AI-related ethical knowledge" - effectively into education, an appropriate competence model is necessary. As mentioned above, the present chapter uses Baumert and Kunter's COACTIV model of teacher professional competence [22] as its theoretical foundation. Following this model, we propose reconsidering teacher professional competence by locating the AI Literacy construct with its subdimension "AI-related ethical knowledge" within the knowledge domain of "professional knowledge". We thereby conceptualise AI Literacy with its 17 competences as outlined by Long and Magerko [38] as facets of content knowledge (CK [51]) and, combined with its effective application to pedagogy, as facets of pedagogical content knowledge (PCK [51]). As professional values and beliefs are central functions of professional ethics, that is, teacher professional ethos [52], we situate AI Literacy's subdimension "AI-related ethical knowledge and skills" additionally as a domain in the professional competence aspect "professional values, beliefs and goals" [22].

In line with the competence-theoretical approach to teacher professionalism [34], these proposed conceptualisations of AI Literacy and AI-related ethical knowledge as knowledge facets include both teacher educators' latent potential, the process that leads to their decision to act and their performance visible in their actions itself (e.g., selecting suitable AI-tools based on ethical principles, using AI responsibly when planning and conducting lessons, incorporating AI equitably and effectively in pedagogy, using AI for enhanced student engagement and outcome, using AI for assessment and evaluation etc.). This proposal rests on mere theoretical reasoning. To determine whether AI Literacy and AI-related ethical knowledge indeed constitute facets of professional knowledge, and to verify whether they can indeed be situated in Baumert and Kunter's COACTIV model [22], empirical research and validation is indispensable. For now, we use the above theoretical elaborations as a foundation for reflecting on how these competence facets can be incorporated in institutional teacher education to contribute to a successful integration of AI—and consequently to contribute to safeguarding high-quality teacher education.

7. Establishing a framework for teacher educator professionalisation in the AI era

With teacher professionalism being both an individual (micro level) and collective (micro and meso level) phenomenon, questions regarding the development of teacher professional competence need to also be approached on the level of institutionalised teacher education including its structures, routines and cultures (meso and macro level) [30]. The elaborations on AI Literacy and AI-related ethical knowledge indicate that part of teacher education institutions' responsibility is to proactively and continuously address both facets by means of a coherent, strategic, holistic and adaptable approach—an approach that constructively intertwines the interrelated individual, micro level (professional development), the collective, organisational meso level (organisational development) and the overarching macro level (politics and policy). In the following sections, we propose a framework that outlines areas of development deducted from the literature to contribute to the development of AI Literacy and AI-related ethical knowledge. The aim of the framework is to thus suggest avenues for enabling a more aligned professionalisation process of teacher educators and teacher education from the perspective of AIEd (see **Figure 1**).

Underlying the framework is the need for a continuously growing degree of agility and flexibility in teacher education institutions—organisations known for their systemic resistance to adapt to change [36]—to sustain the fitness for purpose and continued high quality of their study programmes.

7.1 Micro level: teacher educators, in- and pre-service teachers and learners

In the process of teacher professionalisation, individuals develop the knowledge and skills necessary to cope with the demands of their profession [32]. In the age of

Figure 1.
Framework for teacher educator professionalisation in the AI era.

AI, the demands on teacher professional competence continue to change rapidly. Thus, it is not just individuals but also the teaching profession as a system that need to adapt continuously [32]. This section proposes two areas of development that may contribute to an up-to-date professionalisation process both at the individual as well as the systemic level. We thereby conceptualise the individual level to correspond with the micro level of institutional teacher education and the systemic level to bridge the micro and meso level of institutional education.

7.1.1 Periodic training and professional development

A prerequisite for effective professionalisation for AI is for educators to learn to understand, use, monitor and critically reflect AI [19, 23] and building thereupon to facilitate their teaching and develop their students' AI Literacy [26]. Both require knowing and using suitable AI technologies for learning (e.g., adaptive learning systems, intelligent learning environments, data analytics or automated scoring and feedback systems) to understand students' learning progress and needs in an AI-mediated education, to promote equitable personalised learning and the development of evaluative judgement and to foster students' ethically responsible, critical and constructive engagement with AI [19]. Teacher education institutions thus need to provide regular comprehensive training and professional development opportunities for developing educators' AI Literacy [21].

Both the development and support of teacher educators in their professionalisation as well as the integration of new technologies into their teaching are an ongoing and long-term tasks. Employing suitable and empirically sound educational models and didactic concepts (e.g., digitality-related pedagogical and content knowledge DPaCK [21, 53] or the Frankfurt triangle model for education in the digitally connected world[2] [17]) is crucial for achieving these tasks.

7.1.2 Strengthening the professional teacher-community

As a multifaceted and complex process, the professionalisation of educators transcends the individual and constitutes a collective endeavour. As such, strengthening the teaching community as teacher educators further develop their individual and collective professional competence is highly important. This becomes particularly salient in the context of AIEd. In this rapidly evolving domain, the facilitation of effective practices and discourse concerning diverse AI solutions is pivotal for successfully integrating AI into education. Consequently, the establishment of a supportive and collaborative environment is paramount. One way of supporting the collective teacher community is through initiatives such as platforms (online or analogue) to facilitate the knowledge transfer and encouraging sharing experiences as educators navigate the complexities of integrating AI into their teaching practices. This approach not only fosters an environment of continuous learning but also ensures that the educational community remains at the forefront of technological and methodological advancements.

[2] The "Frankfurt-Dreieck" is also known as and often referred to as "Dagstuhl-Dreieck" (cf. Brinda et al., [17]; Gesellschaft für Informatik e.V. [54]).

7.2 Meso level: educational institutions

The meso level encompasses schools, teacher education institutions, professional development providers, decision-makers in educational institutions, (educational) companies and the teaching profession as a system. This section presents avenues for promoting the teacher educator professionalisation process both at the level of organisational development as well as the level of teacher education as a system.

7.2.1 Agility as an organisational mindset

For individuals, organisations and policy makers to be able to establish favourable conditions for the development of AI-related professional competence and a successful appropriation of AI in education, stakeholders at all levels need to become increasingly more agile and flexible. This is relevant because the AI-environment is fast-paced, dynamic and unpredictable. As teacher education institutions are particularly known for systemic resistance to change [36], this area of development may present a considerable challenge for institutions to overcome. However, overcoming this resistance to change is considered pivotal for successful integration AIEd [55].

7.2.2 Institutional AI strategy and ethical principles

A unified institutional AI strategy including ethical principles on AI use is essential. The development of such strategy should involve a multidisciplinary team of all stakeholders and provide a clear framework for an ethically responsible use of AI in education [56]. At the same time, the strategy needs to be flexible enough to allow for adapting to rapid changes in the AI and educational landscape. Higher education institutions across a variety of countries have been developing such principles over the past years, and such developments have proliferated since the launch of Open-AI's Chatbot ChatGPT-3 in November 2022 (e.g., [57]). As institutional strategies may significantly influence the professionalisation of teacher educators [32], it is important that they align with initiatives to aid teacher educator professionalisation and teacher education institution development.

7.2.3 State-of-the-art ICT infrastructure

Maintaining up-to-date ICT infrastructure is vital for the effective use of AI in educational settings [58]. Without the respective infrastructure, AI cannot effectively be appropriated to education and its affordances cannot be efficiently harnessed. This maintenance is highly dependent on and intertwined with stakeholders of the macro level of institutional education, such as educational policy, state funding or political decisions.

7.2.4 Access to high-quality AI tools

Like the need for state-of-the-art ICT infrastructure, it is crucial to ensure equitable access to AI tools across all stakeholders in institutionalised teacher education. This aims to prevent a disparity in learning opportunities and seeks to avoid reproducing inequalities among students from different economic backgrounds.

7.2.5 Provision of resources for learning new paradigms

Institutions should provide resources that enable educators and students to explore and learn knowledge paradigms that may emerge with the AI-transformation of education and society as a whole [58]. In addition to making relevant resources available (e.g., time, targeted further training opportunities, highly qualified support staff and curricular freedom), it is important that an atmosphere of open discourse, innovative spirit and mutual support prevails across the institution.

7.3 Macro level: politics and policy

The macro level of institutional education spans dimensions such as educational policy and politics at state, federal and global levels. This section suggests foundational aspects that are relevant on this level to contribute to the development of professional competence and its cascading effects.

7.3.1 Interdisciplinary collaborative policy development and vision

Institutions must collaborate to update existing policies and develop a clear vision for the (ethical) use of AI in education [20]. Amidst the rapid technological advancements, teacher education institutions need to establish consensus among trans- and interdisciplinary teams (including students) that enables adopting and responding to technological AI-development to provide guidance to all its stakeholders. Furthermore, it is crucial to fund and conduct extensive research into the affordances, limitations, perils and impact on student and teacher outcomes and learning in (teacher) education [56]. This includes the establishment of full-time research positions at higher education institutions to foster and strengthen international collaboration on AI research [56]. Co-developing and sharing "good practices" and experiences across institutions will be crucial for establishing guiding principles for AI Literacy and AI-related ethical knowledge development across all levels of the educational sector.

7.3.2 Policies, ethical guidelines and data sovereignty as foundational documents

Establishing comprehensive ethical guidelines is paramount to a successful appropriation of AI in education. These guidelines should serve as the foundation for all AI-related activities in institutionalised education, ensuring that the selection and application of AI technologies aligns with ethical standards and values.

7.3.3 Focus on equity and inclusion

To support the creation of an ethical foundation for achieving a "Good AI Society" [15, 16], a clear ethical policy and integrity guidelines at the macro level are pivotal [59]. A prerequisite for meeting this requirement is to ensure that the adoption of new technologies does not widen existing disparities in education. Cascading from the macro level to the meso level, equitable access to AI tools and services needs to be provided to all students, pre-service and in-service teachers as well as teacher educators alike.

8. Implementing AI at the St. Gallen University of Teacher Education

In connection to the areas of development outlined in the present framework, this section discusses the St. Gallen University of Teacher Education's (PHSG) approach to appropriating AI in teacher education as a case study. With the goal to gradually incorporate AI into all its domains, the PHSG has taken a holistic approach by recently developing an institutional AI-Policy (meso level). The policy outlines five key initiatives that encompass the institution's commitment to (1) continuous assessment of stakeholders' needs and experiences with AI; (2) the provision of professional AI Literacy-development opportunities; (3) fair and reliable educational assessment including support on aspects such as academic integrity, ethics and legal implications of AI use; (4) the collaborative development of strategies for AI use in academic research and student written assignments and (5) a collaborative approach to appropriating AI in education that complies with guidelines, ensures knowledge transfer and facilitates faculty training.

Targeting the micro level, the PHSG is developing various learning modules on the continuing education platform www.aprendo.ch designed to enhance educators' digital and AI-related competences [60]. The platform aims to offer around 100 learning modules in the future that facilitate synchronous and asynchronous learning opportunities and target the development of six competence dimensions in relation to digitality and AI [60].

Another development project of the PHSG constitutes www.zitbox.ch [61], an online platform designed to promote discussions among educators, grow the (teacher) educator network and strengthen the teaching community. It facilitates discussions related to the digitalisation and AI-transformation of education beyond individual school units. First, user feedback indicates that the platform is effective in facilitating active communication and networking among educators. With regard to AI, this interaction focuses on discussing and sharing insights about the possibilities and limitations of AI in schools.

Further, to enhance educators' AI Literacy and empower them to effectively integrate AI into their teaching, the PHSG has developed a multitiered curriculum for its continuing education programmes (meso and micro level). Designed based on the Frankfurt triangle model [17], the curriculum constitutes a theory-based blended-learning approach to AI integration in higher education. It is designed to accommodate a diverse audience, including participants with no or little prior AI experience. The curriculum covers a wide range of topics, including machine processing of natural language, the impact of AI tools on didactic processes, experimentation with AI tools across various disciplines, the influence of AI on assessment and the application of AI in academic writing. The programme starts with a kickoff event to introduce participants to independent learning modules that focus on foundational AI concepts. Subsequent face-to-face sessions prepare participants for incorporating AI tools into their teaching. Participants' experiences thereof are then discussed and reflected in a follow-up session to reinforce learned concepts. An array of supplementary resources, including webinars, lectures and regular AI roundtable discussions, is designed to enhance participants' practical skills and knowledge, particularly in the performance dimension of AI-related professional competence. The programme aims to ensure participants not only understand AI's fundamental concepts and educational applications but also develop expertise in areas like prompt engineering and the critical assessment of AI's benefits and challenges in university teaching.

9. Future directions: research needs and development trajectories

Even though the use of AI and AI research in education has increased significantly since 2020, empirical literature on the application of AI to and impact of AI on teaching and learning [35, 62, 63], relationships between AI technologies and learning outcomes for students and teachers [35]—also beyond immediate concrete applications [41]—and theoretical concepts of AI in education [2, 25] are still scarce. Thus, the needs for further research in this area are manifold [20]. Most crucially, future research is required to further ethical considerations and attention within AIEd, both as a topic of research as well as a pivotal issue in conducting empirical research [20]. Research into ethical issues connected to bias, storage and use of teaching and learning data, fairness and accountability of AI use in education and AI Literacy development among learners and educators is urgently needed [20]. Closely intertwined with such endeavours are research required efforts into examining challenges associated with data sources, ownership and authorship as well as authority [36]. Furthermore, considerable efforts are required in the development of policy and institutional guidelines to increase the responsiveness of education systems to rapid changes driven by AI [36].

A specific call for more research relates to the current lack of educational frameworks underpinning the development and incorporation of AI tools in educational settings [20]. More research efforts are required in investigating the appropriateness, suitability and effectiveness of AI tools for pedagogical purposes—as well as their impact on learning, cognition, affect and overall competence development [20].

To ensure that the needs of minorities and marginalised people are reflected in AI development [35] and to prevent the potential of proliferating inequity and inequality caused by algorithmic bias, future research needs to involve a broader range of stakeholders across a wider range of disciplines [20].

Finally, to determine the appropriateness and validity of situating AI Literacy and AI-related ethical knowledge as facets of knowledge and aspects of professional competence in Baumert and Kunter's COACTIV model [22], extensive empirical research needs to be conducted. Similarly, to determine the effectiveness of initiatives to develop AI Literacy and to appropriate AI in education such as the PHSG-approach presented above, empirical investigations are necessary. Both strands would require intervention studies of pre-post designs with experimental and control groups.

10. Conclusion

In this chapter, we have argued that teacher professionalism in the age of AI requires the development of AI Literacy and AI-related ethical knowledge as new knowledge facets of professional competence. Fostering their development in teacher education not only aims to enable educators to become responsible navigators, ethical stewards and competent teachers of AI technology but also seeks to ensure high-quality teacher education that is up-to-date and fit-for-purpose as the AI-mediated landscape evolves. Continuously meeting the performance mandate of education requires a multifaceted and holistic approach that interlaces the micro, meso and macro levels of institutional education. As agents of change, teacher education institutions will need to harmonise the technological with the pedagogical and the ethical with the practical. Based on the above elaborations and identified needs, we proposed a framework for teacher educator professionalisation in the AI era that attempts to

DOI: http://dx.doi.org/10.5772/intechopen.1005030

support such endeavours. Based on the therein contained areas of development, we advocate for embedding the development of AI Literacy and AI-related ethical knowledge into teacher education's core curriculum and ongoing professional development programmes. We further suggest establishing structures that promote strengthening the teaching community as teacher educators further develop their individual and collective professional competence. Furthermore, we maintain that promoting the development of ethical frameworks on institutional and educational policy level to sustain an equitable and inclusive use of AI in educational settings, investing in state-of-the-art ICT infrastructure with access to cutting-edge AI tools and encouraging interdisciplinary collaboration constitute central components of teacher education professionalism in the age of AI. Finally, with AI Literacy and AIEd research still being in their infancies, extensive empirical research will be required to investigate the suitability and effects of the above recommendations on quality teacher education. As we move forward, embracing the transformative potential of AI in education with foresight and appropriate knowledge will be crucial for ensuring that the educators of today and tomorrow are empowered to harness the potential of AI.

Conflict of interest

The authors declare no conflict of interest.

Author details

Olivia Rütti-Joy*, Georg Winder and Horst Biedermann
St. Gallen University of Teacher Education, Switzerland

*Address all correspondence to: olivia.ruetti@phsg.ch

References

[1] Southworth J, Migliaccio K, Glover J, Glover J, Reed D, McCarty C, et al. Developing a model for AI across the curriculum: Transforming the higher education landscape via innovation in AI literacy. Computers and Education: Artificial Intelligence. 2023;**4**:100127

[2] Laupichler MC, Aster A, Raupach T. Delphi study for the development and preliminary validation of an item set for the assessment of non-experts' AI literacy. Computers and Education Artificial Intelligence. 2023;**4**:100126

[3] UNESCO. Futures of Education: Learning to Become [Internet]. Paris: United Nations Educational, Scientific and Cultural Organization; 2021. Available from: https://unesdoc.unesco.org/ark:/48223/pf0000370801

[4] Ouyang F, Jiao P. Artificial intelligence in education: The three paradigms. Computers and Education: Artificial Intelligence. 2021;**2**:100020

[5] Preston J. Artificial Intelligence in the Capitalist University: Academic Labour, Commodification, and Value [Internet]. New York: Taylor & Francis; 2022. Available from: https://library.oapen.org/handle/20.500.12657/51159 [Accessed: February 9, 2023]

[6] Zeide E. Artificial intelligence in higher education: Applications, promise and perils, and ethical questions. In: Educause Review Special Report: Artificial Intelligence: Where Are we Now? Educause; 2022. pp. 28-39. Available from: https://er.educause.edu/articles/2019/8/artificial-intelligence-in-higher-education-applications-promise-and-perils-and-ethical-questions [Accessed: March 28, 2024]

[7] BMK. Strategie der Bundesregierung für Künstliche Intelligenz [Internet]. 2018. Available from: https://www.bmk.gv.at/themen/innovation/publikationen/ikt/ai/strategie-bundesregierung.html [Accessed: October 13, 2023]

[8] Lilienthal J, Schroeder C. Reflexionsfähigkeit 4.0: Wie die Digitalisierung einen zentralen Bildungsuaftrag von Hochschulen stärkt. Empirische Begründung und Schlussfolgerungen. In: Deimann M, van Treeck T, editors. Digitalisierung der Hochschullehre: Aspekte und Perspektiven der Transformation. Berlin: DUZ Verlags- und Medienhaus GmbH; 2020. pp. 25-37. (DUZ Open)

[9] Strobel M, Welpe IM. Hochschule 4.0. Die Zukunft der Hochschule erfinden. Forschung & Lehre. 2017;**24**(4):316-318

[10] Adams C, Pente P, Lemermeyer G, Rockwell G. Ethical principles for artificial intelligence in K-12 education. Computers and Education: Artificial Intelligence. 2023;**4**:100131

[11] Ehlers UD. Future Skills: Lernen der Zukunft-Hochschule der Zukunft. Wiesbaden: Springer; 2020

[12] Samochowiec J. Future Skills: Four Scenarios for the World of Tomorrow. Rüschlikon: GDI Gottlieb Duttweiler Institute; 2020

[13] Kalz M. Zurück in die Zukunft? Eine literaturbasierte Kritik der Zukunftskompetenzen. MedienPädagogik: Zeitschrift für Theorie und Praxis der Medienbildung; 2023. pp. 332-352

[14] UNESCO. UNESCO Moving Forward the 2030 Agenda for Sustainable

Development [Internet]. Paris: United Nations Educational, Scientific and Cultural Organization; 2017. Available from: https://en.unesco.org/creativity/sites/creativity/files/247785en.pdf

[15] Floridi L, Cowls J, Beltrametti M. AI4People—An ethical framework for a good AI society: Opportunities, risks, principles, and recommendations. Minds and Machines. 2018;**28**:689-707

[16] Wang X, Li L, Tan SC, Yang L, Lei J. Preparing for AI-enhanced education: Conceptualizing and empirically examining teachers' AI readiness. Computers in Human Behavior. 2023;**146**:1-11

[17] Brinda T, Brüggen N, Diethelm I, Knaus T, Kommer S, Kopf C, et al. Frankfurt-Dreieck zur Bildung in der digital vernetzten Welt: Ein interdisziplinäres Modell. In: Knaus T, Merz O, editors. Interfaces and Interfaces. Digital Change in Educational Institutions. kopaed; 2019. pp. 157-167. Available from: https://dagstuhl.gi.de/frankfurt-dreieck

[18] Kandlhofer M, Steinbauer G, Hirschmugl-Gaisch S, Huber P. Artificial intelligence and computer science in education: From kindergarten to university. In: 2016 IEEE Frontiers in Education Conference (FIE), Erie, PA, USA. IEEE; 2016. pp. 1-9

[19] Ng DTK, Leung JKL, Chu KWS, Qiao MS. Conceptualizing AI literacy: An exploratory review. Computers and Education: Artificial Intelligence. 2021;**2**:100041

[20] Bond M, Khosravi H, Laat M, Bergdahl N, Negrea V, Oxley E, et al. A meta systematic review of artificial intelligence in higher education: A call for increased ethics, collaboration, and rigour. International Journal of Educational Technology in Higher Education. 2023;**21**:4

[21] Rütti-Joy O, Winder G, Biedermann H. Building AI literacy for sustainable teacher education. Zeitschrift für Hochschulentwicklung. 2023;**18**(4):175-189

[22] Baumert J, Kunter M. The COACTIV model of teachers' professional competence. In: Kunter M, Baumert J, Blum W, Klusmann U, Krauss S, Neubrand M, editors. Cognitive Activation in the Mathematics Classroom and Professional Competence of Teachers: Results from the COACTIV Project [Internet]. Boston, MA: Springer US; 2013. pp. 25-48. (Mathematics Teacher Education). DOI: 10.1007/978-1-4614-5149-5_2

[23] Long D, Magerko B. What is AI literacy? Competencies and design considerations. In: Proceedings of the 2020 CHI Conference on Human Factors in Computing Systems [Internet]. Honolulu HI USA: ACM; 2020. pp. 1-16. DOI: 10.1145/3313831.3376727

[24] Ng DTK, Leung JKL, Chu KWS, Qiao MS. AI literacy: Definition, teaching, evaluation and ethical issues. Proceedings of the Association for Information Science and Technology. 2021;**58**:504-509

[25] Laupichler MC, Aster A, Schirch J, Raupach T. Artificial intelligence literacy in higher and adult education: A scoping literature review. Computers and Education Artificial Intelligence. 2022;**3**:100101

[26] Ng DTK, Leung JKL, Su J, Ng RCW, Chu SKW. Teachers' AI digital competencies and twenty-first century skills in the post-pandemic world. Educational Technology Research and Development. 2023;**71**(1):137-161

[27] Werler T. Commodification of teacher professionalism. Policy Futures in Education. 2015;**14**(1):60-76

[28] Hericks U, Keller-Schneider M, Bonnet A. Professionalität von Lehrerinnen und Lehrern in berufsbiographischer Perspektive. In: Harring M, Rohlfs C, Gläser-Zikuda M, editors. Handbuch Schulpädagogik. 2nd ed. Waxmann: Münster & New York; 2022. pp. 647-658

[29] Cramer C, Rothland M. Pädagogische Professionelle in der Schule. In: Hascher T, Helsper W, Idel TS, editors. Handbuch Schulforschung. Wiesbaden: Springer VS; 2021. pp. 1-23

[30] Hinzke JH, Keller-Schneider M. Zur Pluralität des Diskurses um Professionalisierung und Professionalität im Kontext des Lehrer: innenberufs – Einleitung. In: Hinzke JH, Keller-Schneider M, editors. Professionalität und Professionalisierung von Lehrpersonen Perspektiven, theoretische Rahmungen und empirische Zugänge [Internet]. Bad Heilbrunn: Verlag Julius Klinkhardt; 2023. pp. 9-23. DOI: 10.35468/6043-01

[31] Blömeke S, Gustafsson JE, Shavelson RJ. Beyond dichotomies: Competence viewed as a continuum. Zeitschrift für Psychologie. 2015;**223**:3-13

[32] Kunter M. Professionalisierung von Lehrkräften im Dorsch Lexikon der Psychologie. 2022. Available from: https://dorsch.hogrefe.com/stichwort/professionalisierung-von-lehrkraeften [Accessed: December 22, 2023]

[33] Weinert FE. Concept of competence: A conceptual clarification. In: Rychen DS, Salganik LH, editors. Defining and Selecting Key Competencies. Göttingen: Hogrefe; 2001. pp. 45-66

[34] König J. Kompetenzorientierter Ansatz in der Lehrerinnen- und Lehrerbildung. In: Cramer C, König J, Rothland M, Blömeke S, editors. Handbuch Lehrerinnen- und Lehrerbildung. Bad Heilbrunn: Klinkhardt; 2020. pp. 163-171

[35] Chiu TKF, Xia Q, Zhou X, Chai CS, Cheng M. Systematic literature review on opportunities, challenges, and future research recommendations of artificial intelligence in education. Computers and Education: Artificial Intelligence. 2023;**4**:100118

[36] Gašević D, Siemens G, Sadiq S. Empowering learners for the age of artificial intelligence. Computers and Education: Artificial Intelligence. 2023;**100130**:1-4

[37] Crompton H, Burke D. Artificial intelligence in higher education: The state of the field. International Journal of Educational Technology in Higher Education. 2023;**20**(1):22

[38] Long D, Magerko B. What is AI literacy? Competencies and design considerations. In: CHI '20: Proceedings of the 2020 CHI Conference on Human Factors in Computing Systems. Honolulu, HI, USA: Association for Computing Machinery; 2020

[39] OECD. Literacy in the Information Age: Final Report of the International Adult Literacy Survey [Internet]. Paris: Organization for Economic Co-operation and Development; 2000. Available from: www.oecd.org/education/skills-beyond-school/41529765.pdf

[40] Educause. Artificial Intelligence: Where Are we Now? Educause; 2022. (Educause Review Special Report). Available from: https://er.educause.edu/-/media/files/articles/2022/6/er2225w.pdf [Accessed: March 28, 2024]

[41] Bearman M, Ryan J, Ajjawi R. Discourses of artificial intelligence in higher education: A critical literature review. Higher Education. 2022;**86**:369-385

[42] Brooks C. Artificial intelligence use in higher education. In: Educause Review Special Report: Artificial Intelligence: Where Are we Now? Educause; 2022. pp. 18-25. Available from: https://er.educause.edu/-/media/files/articles/2022/6/er2225w.pdf [Accessed: March 28, 2024]

[43] Fisher TL. Artificial intelligence and ethical accountability. In: Educause Review Special Report: Artificial Intelligence: Where Are we Now? Educause; 2022. pp. 12-17. Available from: https://er.educause.edu/-/media/files/articles/2022/6/er2225w.pdf [Accessed: March 28, 2024]

[44] Humble N, Mozelius P. The threat, hype, and promise of artificial intelligence in education. Discover Artificial Intelligence. 2022;**2**(1):1-13

[45] Kingsley D. Major Publishers Are Banning ChatGPT from Being Listed as an Academic Author. What's the Big Deal? [Internet]. Paris: The Conversation; 2023. Available from: https://theconversation.com/major-publishers-are-banning-chatgpt-from-being-listed-as-an-academic-author-whats-the-big-deal-198765?utm_source=twitter&utm_medium=bylinetwitterbutton [Accessed: March 10, 2023]

[46] Bozkurt BÜ, Erim A, Çelik-Demiray P. The effects of individual voice training on pre-service Turkish language teachers' speaking performance. Educational Sciences: Theory & Practice. 2018;**18**:151-171

[47] O'Brien J. Introduction: Moving beyond pockets of excellence in higher Ed AI. In: Educause Review Special Report: Artificial Intelligence: Where Are we Now? Educause; 2022. pp. 4-5. Available from: https://er.educause.edu/-/media/files/articles/2022/6/er2225w.pdf [Accessed: March 28, 2024]

[48] Zhai C, Wibowo S. A systematic review on artificial intelligence dialogue systems for enhancing English as foreign language students' interactional competence in the university. Computers and Education: Artificial Intelligence. 2023;**4**:100134

[49] Cetindamar D, Kitto K, Wu M, Zhang Y, Abedin B, Knight S. Explicating AI literacy of employees at digital workplaces. IEEE Transactions on Engineering Management. 2022;**71**:1-14

[50] Bewersdorff A, Zhai X, Roberts J, Nerdel C. Myths, mis- and preconceptions of artificial intelligence: A review of the literature. Computers and Education: Artificial Intelligence. 2023;**4**:100143

[51] Shulman L. Knowledge and teaching: Foundations of the new reform. Harvard Educational Review. 1987;**57**:1-22

[52] Zutavern M. Professionsethos - die Lehrer*innenbildung ist herausgefordert! Journal für LehrerInnenbildung. 2021;**3**:14-31

[53] Huwer J, Thomas I, Kuntze S, Schaal S, Thyssen C. From TPaCK to DPaCK – Digitalization in education requires more than technical knowledge. In: Mack S, Kiray A, editors. Education Research Highlights in Mathematics, Science and Technology. 2019. pp. 298-309

[54] Gesellschaft für Informatik e.V. Dagstuhl-Erklärung: Bildung in der digitalen vernetzten Welt. Eine gemeinsame Erklärung der

Teilnehmerinnen und Teilnehmer des Seminars auf Schloss Dagstuhl – Leibniz-Zentrum für Informatik GmbH [Internet]. Gesellschaft für Informatik e.V.; 2016. Available from: https://dagstuhl.gi.de/dagstuhl-erklaerung

[55] Sourani M. Artificial intelligence: A prospective or real option for education? Al Jinan الجنان. 2020;**11**(1):121-139

[56] Mah DK, Büching C. Künstliche Intelligenz in Studium und Lehre. In: Überblickstudie zu Professuren und Studiengängen der Künstlichen Intelligenz in Deutschland [Internet]. Berlin: Bundesministerium für Bildung und Forschung (BMBF); 2019. Available from: https://www.plattform-lernende-systeme.de/files/Downloads/Diverses/Studie_KI_in_Studium_und_Lehre.pdf [Accessed: March 3, 2023]

[57] Russell Group. Russell Group Principles on the Use of Generative AI Tools in Education [Internet]. The Russell Group of Universities; 2023. Available from: https://russellgroup.ac.uk/news/new-principles-on-use-of-ai-in-education/

[58] Dwivedi VJ, Joshi YC. Leadership pivotal to productivity enhancement for 21st-century Indian higher education system. International Journal of Higher Education. 2020;**9**(2):126

[59] McKnight L. Electric sheep? Humans, robots, artificial intelligence, and the future of writing. Changing English. 2021;**28**(4):442-455

[60] Winder G, Kern A, Zehetner K, Buchner J. Unlocking potential: Creating an innovative learning platform to Foster digital skills of educators. In: Singh S, editor. The 9th eLearning Excellence Awards an Anthology of Case Histories 2023 [Internet]. ACI; 2023. pp. 155-161. Available from: https://www.academic-bookshop.com/ourshop/prod_8077437-9th-eLearning-Excellenceawards-2023.html [Accessed: January 17, 2024]

[61] Pädagogische Hochschule St. Gallen, Kompetenzzentrum Digitalisierung & Bildung. zITBOx [Internet]. Vernetzungs-Plattform für die Volksschule, Mittelschulen und Berufsbildung. St. Gallen: zITBOx; 2023. Available from: https://zitbox.ch/ [Accessed: January 22, 2024]

[62] Bates T, Cobo C, Mariño O, Wheeler S. Can artificial intelligence transform higher education? International Journal of Educational Technology in Higher Education. 2020;**17**(1):1-12

[63] Bozkurt A, Karadeniz A, Baneres D, Guerrero-Roldán AE, Rodríguez ME. Artificial intelligence and reflections from educational landscape: A review of AI studies in half a century. Sustainability. 2021;**13**(2):800

Chapter 10

Perspective Chapter: Leveraging Artificial Intelligence in a Blotch Academic Environment

Ogunlade B. Olusola, Bahago S. Benedict and Shotayo E. Olusola

Abstract

We look into leveraging artificial intelligence (AI) to enhance the academic environment within our institution. We aim to streamline administrative tasks by integrating AI-powered tools, personalising student learning experiences and facilitating data-driven decision-making. This will improve efficiency, increase student engagement and improve resource allocation. AI can assist in automating routine administrative processes, providing data-driven insights for course improvement and offering personalised recommendations to students. By embracing AI, we can create a more dynamic and responsive academic ecosystem, ensuring that our institution remains at the forefront of educational innovation where learning problems can be thoroughly addressed in the learning environment.

Keywords: leveraging, artificial intelligence, blotch, academic, environment

1. Introduction

Artificial intelligence (AI) is the general term for developing computer systems that can do tasks that typically require human intelligence. These tasks include solving problems, learning, perceiving, understanding language and making decisions. AI seeks to create devices that can simulate or replicate human cognitive functions. This refers to AI systems developed and trained for a specific goal. Although they are less diverse and have a lower learning capacity than human intelligence, they are better in that sector. AI is used in education [1], including features like plagiarism detection and exam integrity [2]. Researchers in the field are constantly looking into new techniques, approaches and algorithms to enhance the capabilities of AI systems. Growing emphasis will be placed on ethical concerns, reducing bias and developing AI responsibly as technology advances. It is imperative to keep in mind that the sector is constantly evolving. Regretfully, however, organisations and scholars worldwide minimise the hazards associated with artificial intelligence while praising its negative impacts. Those unable to understand technology may soon be met with a world that increasingly resembles magic and makes them feel left behind [3].

Researchers worry that by 2030, the AI revolution will focus on enhancing benefits and societal control, but it will also raise contentious ethical questions. Regarding AI's positive effects on morality and life, opinions diverge significantly [4]. The advancement of AI also raises several concerns around morality, behaviour, privacy and trust, to name a few. There are various ethical concerns with the use of AI in education. Many scholars are exploring the field more thoroughly. Our classification of AI in education is based on three levels. First and foremost is the technology itself—its developer, manufacturer, etc. Third are the effects on the student or learner, then the teacher.

Artificial intelligence (AI) integration has become a revolutionary force in the constantly changing academic scene, providing hitherto unseen chances to improve and expedite administrative processes. AI-powered solutions offer a promising way to lessen the load of administrative duties as academic institutions need help with growing complexity and the requirement for efficiency.

Routine administrative work automation is a notable area where artificial intelligence can significantly impact. AI can analyse massive volumes of data quickly and accurately, saving human administrators much time and effort on tasks like monitoring student records and course registrations. A subset of artificial intelligence called machine learning algorithms can use data patterns to forecast future trends, allowing organisations to plan and allocate resources and student support services wisely [5].

Using chatbots with AI capabilities is an additional way to increase administrative effectiveness. These clever virtual assistants can respond quickly to often-asked queries and manage routine requests from staff, instructors and students. This improves the user experience while freeing human administrators to work on more challenging and worthwhile duties. Chatbots can be easily incorporated into various platforms, including mobile apps and websites, providing smooth and convenient communication for the academic community.

Beyond mere automation, AI's data processing capabilities provide insightful information for tactical decision-making. By predicting enrolment trends, predictive analytics can assist educational institutions in better allocating resources and optimising their course offerings. Furthermore, patterns in student performance data can be found using AI-driven analytics, allowing for prompt interventions to support students with academic difficulties.

AI integration can also be beneficial for educators' time-consuming grading procedures. AI-powered automated grading systems can quickly and accurately evaluate homework, tests and quizzes while giving students immediate feedback. This speeds up the grading process and frees teachers to work with students more individually, creating a more engaging learning environment.

AI can also help in the administration of research projects in academic settings. It can speed up the research process by assisting with literature reviews, recommending pertinent books and offering summaries. Furthermore, enormous volumes of research data may be analysed by AI algorithms, making it easier to find patterns and trends that may escape the notice of conventional techniques [6].

Even with all the benefits, ethical and privacy issues must be carefully considered when integrating AI into academic settings. Ensuring transparency in AI algorithms and protecting sensitive data is critical to preserving trust in the academic community.

To put it another way, the use of AI-powered tools in academic settings has the potential to transform administrative duties completely. Artificial intelligence (AI) has the potential to improve productivity, lessen burden and foster a more adaptable and responsive learning environment by automating repetitive tasks and delivering actionable insights. Institutions stand to gain from embracing these technological innovations regarding operational effectiveness and cultivating an innovative and adaptive culture.

2. Integration of AI-powered tools in university administrative tasks: perspectives of academics

Integrating artificial intelligence (AI) into university administrative tasks has sparked both academic agreement and disagreement. This debate concerns the potential benefits and challenges of adopting AI-powered tools to streamline various administrative processes.

2.1 Agreement

Proponents of AI integration in university administration highlight several advantages that could enhance efficiency and productivity. One key argument is the potential for AI to automate routine and time-consuming tasks, allowing administrative staff to focus on more complex and strategic aspects of their roles. This, they argue, could lead to a significant reduction in human error and increase overall operational efficiency [7].

Furthermore, AI-powered tools have the potential to analyse large datasets quickly, facilitating data-driven decision-making in areas such as admissions, enrolment management and resource allocation. This analytical capability is valuable for universities seeking to make informed, strategic decisions that align with their goals [8, 9].

In the context of student services, Duan et al. [10] stated that AI tools can provide personalised assistance, ranging from academic advising to student support services. Chatbots equipped with natural language processing capabilities can offer timely responses to student queries, enhancing student experience.

Financial considerations also come into play, as proponents argue that the initial investment in AI technology can lead to long-term cost savings through increased efficiency and reduced administrative workload. This, they assert, allows universities to allocate resources more effectively and invest in other critical areas, such as research and faculty development [11].

2.2 Disagreement

Despite the possible advantages, sceptical have doubts and worries regarding the extensive use of AI in university management. The fear of losing one's employment is a significant topic of dispute. Opponents contend that using AI to automate administrative activities could result in job redundancies, impacting administrative staff members' livelihoods [12].

Concerns about ethics and privacy are also significant points of contention. When AI is used, enormous volumes of sensitive data are handled, which raises concerns about data security, confidentiality and the possibility of bias in algorithms used to make decisions. Before AI is widely used, academics stress the significance of developing solid ethical standards and security safeguards to allay these worries [13].

Furthermore, there are concerns regarding AI technologies' adaptability and user-friendliness. According to sceptics, the learning curve of new technology might make it more challenging to incorporate AI seamlessly into current administrative procedures. Adoption of AI may encounter pushback from faculty and staff if changes upset established workflows [14].

Nonetheless, there is a complicated interaction of viewpoints in the discussion around integrating AI-powered tools into university administrative responsibilities. Supporters draw attention to the potential for increased productivity, data-driven decision-making and improved student services; detractors raise issues with job displacement, privacy and the flexibility of AI systems. Universities navigating the rapidly changing world of learning technology must balance utilising AI's benefits and attending to these justifiable concerns [13].

3. Conceptual model: the spectrum of academic perspectives on AI integration in university learning system

3.1 Agreement spectrum

Scholars who enthusiastically support and welcome the use of AI-powered tools in learning system tasks. They think AI has much to offer universities regarding possible advantages, efficiency gains and resource optimisation.

Academics who embrace AI integration pragmatically and approach it with a more measured and pragmatic attitude are those who recognise its benefits. They favour its implementation when AI proves beneficial and improves education without detaching the human element.

Though willing to investigate AI integration, open-minded explorers wait to make a firm endorsement until they have more data, study and real-world applications. Although they are open to the idea, they are somewhat sceptical once the benefits are demonstrated [15].

3.2 Disagreement spectrum

Academics who are sceptics regarding using AI in teaching and learning tasks likewise harbour misgivings and concerns. Concerns about potential biases in AI systems, ethical issues or job displacement might be on their minds. They think that until these issues are sufficiently resolved, we should continue to take a cautious approach.

Scholars vehemently disagree with the incorporation of AI and support the preservation of conventional techniques. They contend that when it comes to some parts of academic decision-making, human touch, experience, and judgement are irreplaceable. Academics emphasise issues with algorithmic unfairness, data privacy and the possible abuse of AI in administrative decision-making as the ethical foundation for their disagreement. Before any integration occurs, they need strict ethical guidelines and protections [16].

3.3 Connecting threads

Figure 1 illustrates how the degree of acceptance or rejection frequently depends on the availability of solid data and research demonstrating the benefits of integrating AI into academic administration and instruction. To find common ground, supporters

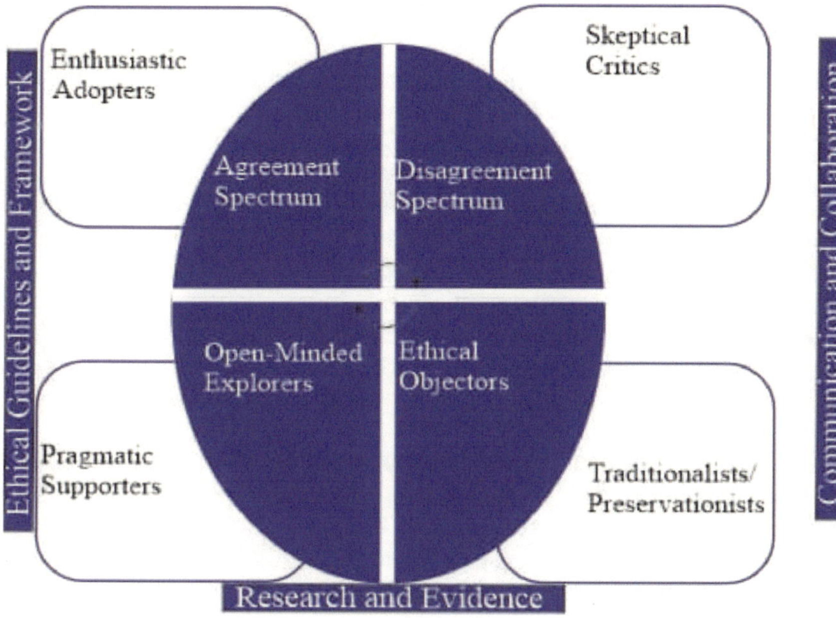

Figure 1.
Conceptual model of AI connecting threads.

and opponents must communicate effectively. A more balanced strategy can be formed *via* constructive communication, addressing issues and exchanging insights. To allay worries and guarantee responsible deployment, it is imperative to establish explicit ethical standards and frameworks for the application of AI.

This conceptual model acknowledges the range of viewpoints within academia concerning the integration of AI-powered tools for teaching and learning activities in academic settings.

4. Personalised student learning experiences through AI: enhancing education for the future

Artificial intelligence (AI) has been a transformational force in educational settings in recent years, with many benefits that reach deep into the core of academic experiences. The ability of AI to tailor student learning experiences to individual needs, strengths and learning styles is one of the most exciting developments in education. This method conforms to the changing expectations of academics on using AI in the classroom and creates a more dynamic and productive learning environment [16].

5. Advantages of tailored education for students

5.1 Tailored learning routes

AI-powered systems can generate customised learning routes by analysing individual students' performance data and preferences. This maximises comprehension and retention by ensuring students receive knowledge in a format that best fits their learning styles and is delivered promptly.

AI-driven tests have the ability to change their difficulty level and provide instant feedback in real time in response to a student's answer. This adaptive exam helps students grasp concepts more deeply and enables teachers to pinpoint and help each student's areas of difficulty [14].

Teachers can provide customised instruction to students of different skill levels in the same classroom because of personalisation. In order to satisfy each student's unique needs, AI technologies can suggest various materials, tasks and evaluations, promoting a more inclusive learning environment.

AI makes self-paced learning experiences easier, enabling students to move through the content at a speed that works for their schedules and skills. This adaptability fosters a sense of independence and accountability for one's education while accommodating a range of learning velocities.

AI can give prompt, helpful feedback on tasks, projects and evaluations. This quick feedback loop helps students learn more effectively by helping them recognise their errors, fix them and keep getting better. Through ongoing data analysis, AI can recognise early indicators of learning gaps or academic difficulties. This makes it possible for teachers to step in quickly, providing extra help to pupils with difficulties and halting the progression of any learning gaps [17].

6. Expectations of academics in AI usage in classroom activities

Academics anticipate that AI will simplify repetitive chores related to teaching and learning, freeing them up to concentrate more on individualised education, mentorship and developing creative and captivating learning opportunities. Time that might otherwise be spent on essential student interactions is freed up by automating administrative tasks such as attendance monitoring and grading. Teachers hope AI will enable them to get practical insights from data analysis. Teachers can improve the overall quality of education by using student performance data to influence decisions about curriculum design, instructional tactics and resource allocation [18].

Academics should consider using AI to create platforms and tools that simplify altering lesson plans and assignments. This personalisation helps educators create specialised learning environments that meet their pupils' varied needs and interests. AI tools are anticipated to make it easier for parents, students and instructors to collaborate and communicate. A more cohesive and encouraging learning community can be created with improved communication channels, online collaborative areas and AI-driven insights [19].

Academics emphasise the significance of moral AI practices in teaching. People anticipate transparent, bias-free and inclusively designed AI systems. For academics to embrace AI in the classroom, the algorithms must refrain from reinforcing or magnifying existing disparities. To stay current with the latest developments in artificial intelligence and educational technology, instructors should consider pursuing professional development opportunities and continuous training before implementing all these critical changes. Academics can maximise the benefits of AI technologies for student learning by integrating them into their teaching practices through continuous learning.

Furthermore, the individualised learning experiences that AI enables for students mark a paradigm change in the field of education, meeting the demands of scholars for a more effective, data-driven and student-centred method of instruction. For the benefit of students everywhere, the partnership between educators and AI promises to form an increasingly influential, flexible and inclusive educational environment as technology advances [17].

Teachers' perspectives on individualised learning with artificial intelligence (AI) in underdeveloped technical environments are varied, considering both the opportunities and difficulties of introducing such cutting-edge technologies in settings with inadequate infrastructure. The following are some viewpoints that teachers may have:

6.1 Educators' perspectives on personalised learning with AI in technology-disadvantaged areas: challenges and concerns

Teachers may voice worries about students' access to devices and internet connectivity in places with limited technology resources. Personalised learning frequently uses digital tools and online platforms, which can be problematic in places with differences in technology access. It is possible that educators should be more concerned with their own and their students' technological literacy. To effectively integrate AI-powered technologies into their teaching practices, educators may require additional training and a certain level of technological expertise [19].

It is possible to worry about making already-existing disparities worse. The educational divide could grow if confident kids can access AI-powered personalised learning tools while others do not. When using technology in underserved communities, it is imperative to ensure equity and inclusion.

6.2 Possible advantages

Teachers might know how AI might give kids with different learning needs individualised support. AI solutions can adjust to each learner's unique learning style and provide focused support to close educational gaps.

AI-powered personalised learning promises resource optimisation, freeing teachers to concentrate on areas that need more excellent care while technology takes care of regular teaching duties. This can be especially helpful in settings with limited resources. Students' acquisition of twenty-first century abilities, such as digital literacy, problem-solving and critical thinking, can be aided by AI integration. Teachers might value getting their kids ready for a world with more advanced technology.

6.3 Techniques for putting education front and centre in technologically disadvantaged areas

Steps should be taken to improve these conditions in locations with limited device access and dependable internet connectivity. In order to solve these infrastructural issues, public-private partnerships (PPPs), government programmes and community involvement can be significant.

6.4 Empowerment and community engagement

It is critical to include parents and the neighbourhood in the educational process. Even in places with limited access to technology, families can be empowered to assist their children's education through workshops, training sessions and awareness campaigns.

It is critical to implement tech solutions that can work in low-tech settings. In places with inadequate access to technology, personalised learning can be made possible by investigating mobile applications, offline resources and other creative strategies [20].

6.5 Teacher training and professional development

To improve teachers' technological literacy, it is crucial to give them continual training and assistance. Programmes for professional development can give educators the tools they need to incorporate AI-powered resources successfully.

6.6 Policies and assistance from the government

Policies that emphasise the use of technology in education can be put into place by governments and educational authorities, especially in underprivileged areas. The quality of education can be significantly impacted by funding and support for curriculum creation, teacher preparation and technology infrastructure.

6.7 Public-private partnerships

Working with tech companies *via* PPPs can help underserved areas of technology by providing resources, knowledge and creative solutions. These collaborations can help create and implement specialised educational technology solutions.

Stated differently, although educators in under-technologicalized regions might be apprehensive about integrating AI-powered personalised learning, noteworthy prospects exist for noteworthy benefits. Education can be promoted by strategies prioritising teacher preparation, community involvement, infrastructure development and supportive legislation. This will guarantee that all children, irrespective of their technological surroundings, have access to high-quality, customised learning experiences.

7. Empowering education through AI: facilitating data-driven decision-making in teaching and research

Data-driven decision-making has emerged as a potent tool for educators to improve teaching efficacy and make significant contributions to research in the constantly changing field of education. Through data-driven insights, educators can make well-informed decisions, customise instruction to meet the requirements of specific students and forward research goals. This essay explores how teachers and researchers may support data-driven decision-making.

7.1 Instructional environments

Real-time awareness of student comprehension can be gained by routinely gathering and evaluating formative assessment data. Teachers might modify their pedagogical approaches in light of this information to better meet their students' unique requirements and difficulties. Teachers can identify each student's strengths and weaknesses using personalised feedback from assessment data. Thanks to this tailored feedback, students can comprehend their success and areas for growth, which creates a more flexible learning environment.

These days, learning analytics-capable educational technology platforms provide helpful information on student performance, engagement and learning trends. Teachers can use these insights to modify the curriculum, change the pace and pinpoint children who are in danger and may require more assistance. Additionally, data are used by adaptive learning platforms to design customised learning pathways for students. Educators can guarantee that every student obtains a customised learning experience that aligns with their needs and skills by monitoring individual development and making content adaptations depending on performance.

7.2 Early intervention techniques

Students' patterns of learning gaps can be found through data analysis. Teachers can address these gaps early on and stop them from worsening by using targeted interventions such as extra resources, remedial sessions or customised instruction. Predictive analytics models may be used to identify kids at risk of falling behind. Teachers can take proactive measures to support kids experiencing academic or socio-emotional difficulties by examining a variety of indications [21].

7.3 Research settings

The development of research questions can benefit from data analysis. Educators can formulate hypotheses and study objectives using available data to discover trends, patterns or knowledge gaps. Educators can perform fundamental data analysis before undertaking large-scale research to determine their research questions' viability and possible impact. Thanks to this iterative approach, research activities are concentrated and aligned with educational objectives [7].

7.4 Design and execution of experiments

Experiment parameters can be refined with the help of data-driven insights. Educators can ensure that their study is rigorous and yields significant results by optimising their research design through the analysis of pilot data or small-scale experimentation. Educators can use data to pinpoint essential variables to measure their research. This guarantees that the research gathers complete and pertinent data, which advances our understanding of the educational phenomenon being studied [7].

7.5 Data-driven publication strategies

Teachers can choose which journals to submit their manuscripts to by looking at impact factors and publication trends. Educators may optimise sharing their study findings by carefully selecting publications with a large readership and impact. Research publications have a more significant impact when they use effective data visualisation. Educators can help readers understand and become more engaged by using data visualisation tools to communicate facts clearly and appealingly.

7.6 Cross-cutting techniques

Teachers can receive the necessary skills in data literacy through data literacy training. Effective data collection, analysis and interpretation are abilities that educators must possess. The data literacy of educators can be improved through workshops, courses and chances for continuous professional development.

However, working with experts or data scientists can increase the influence of data-driven decision-making. The knowledge of experts in data analysis can be beneficial to educators, providing them with more complex and nuanced views.

7.7 Moral aspects to take into account

Teachers need to put student privacy and research participant data protection first. Robust data protection protocols, informed permission and adherence to ethical standards are essential components of responsible data-driven decision-making. Transparency about the use of data is something that educators should uphold in both teaching and research contexts. Trust is developed among students, coworkers and the larger educational community when it is communicated how data influences decisions.

Data-driven decision-making has the potential to completely transform education by improving instructional strategies and generating influential research. Teachers can use data to discover areas for improvement, personalise learning and expand the body of knowledge in their subject. Teachers have a critical role in determining the direction of education by adopting data-driven practices and encouraging a culture of continuous improvement [22].

8. To create a more dynamic and responsive academic ecosystem that remains at the forefront of educational innovation and promotes inclusiveness, institutions can implement the following strategies

Dynamic and captivating learning environments can be produced through interactive learning platforms, virtual classrooms and online collaboration tools. To ensure that all students can take advantage of digital resources, give faculty members the assistance and training they need to incorporate technology into their teaching strategies successfully.

8.1 Encourage an innovative culture

Promote a culture that rewards creativity and innovation. Create avenues for staff, students and teachers to collaborate on creative projects. Establish innovation hubs or centres to facilitate research, experimentation and the creation of novel teaching strategies.

8.2 Put into practice adaptable curriculum framework

Provide a curriculum that is adaptable and flexible to different learning styles, inclinations and speeds. Permit students to select courses and learning pathways based on their interests and professional aspirations. Use multi-disciplinary methods to solve practical issues while encouraging teamwork and critical thinking.

8.3 Utilise data to make well-informed decisions

Use data analytics to monitor student satisfaction, engagement and performance. Utilise this information to pinpoint areas in need of development, tailor lessons to the needs of each student and offer early interventions to struggling students. Examine demographic information to ensure the programme is inclusive and to find and fix inequalities in academic performance.

8.4 Encourage cooperative education

Create a cooperative learning atmosphere that promotes group work and peer engagement. Teamwork, a sense of community and a diversity of ideas are all fostered *via* collaborative learning. Use technology-enabled collaboration solutions to promote inclusivity and break geographical barriers by facilitating communication and group work.

8.5 Set priorities development of faculty

Invest in your academic members' ongoing professional development and instruct staff on the newest teaching techniques, educational tools and classroom promotion tactics. Faculty members should be encouraged to publish their findings and research cutting-edge teaching techniques to enhance the institution's standing as an educational leader.

8.6 Make inclusive learning environments

Ensure all students, including those with impairments, have access to physical and virtual learning environments. Make sure that tools, online courses and classrooms are inclusively created. Provide a network of support for kids with a range of needs by providing resources, including counselling, tutoring and adjustments for accessibility.

8.7 Form collaborations and partnerships

Encourage collaborations with businesses, other academic institutions and neighbourhood associations. Collaborative projects expose students to various viewpoints and give education a real-world context. Engage in cooperative research initiatives, exchange schemes and internships to enhance the academic experience and equip students for the intricacies of the global labour market.

8.8 Encourage a student-centric perspective

Place the students at the centre of the learning process. Use focus groups, questionnaires and student involvement on committees to frequently collect feedback about your students' needs and preferences. Provide a system of support for students that addresses their emotional, social and academic needs to help them feel engaged and significant.

8.9 Ongoing assessment and enhancement

Build a robust AI-driven system to evaluate courses, instruction and institutional policies continuously. Make informed decisions and adapt to changing learning environments with the help of feedback and data. Encourage a culture of continuous development whereby the company searches for fresh, creative approaches to increase the calibre and relevance of its training.

By putting these concepts into reality, educational institutions may remain at the forefront of educational innovation while creating a more adaptable and dynamic academic ecosystem that actively supports diversity and meets various learning needs. This plan ensures that the school can keep up with the evolving needs of the students and prepares them for success in a world that is changing swiftly [17].

9. Addressing potential ethical issues that arises from integrating AI into education

While there are many chances to improve learning experiences when artificial intelligence (AI) is integrated into education, there are also ethical issues that must be carefully considered. Among the possible moral dilemmas are:

Fairness and bias: AI systems may inherit biases from the training data, leading to biased outcomes that affect pupils according to socioeconomic position, gender or race. This problem can be lessened by putting in place rigorous procedures for data validation and cleaning, using a variety of datasets and routinely checking AI systems for bias. Fairness and diversity should be given top priority in ethical norms.

Privacy issues: AI frequently requires gathering and examining enormous volumes of student data. Inadequate protection of sensitive information can result in misuse or unauthorised access, making privacy infringement dangerous. Obtain informed consent, anonymize data, create strong data protection policies and abide by privacy laws (such as family educational rights and privacy act (FERPA) and general data protection regulation (GDPR)). It is essential to communicate openly about data usage with parents, teachers and students.

Explainability and transparency: AI algorithms, particularly those used in machine learning, can be difficult to understand and intricate. Decisions that are difficult to grasp might become 'black box' problems due to a lack of openness and explainability. Enhancing trust and understanding can be achieved through emphasising transparency in algorithmic decision-making, giving explicit explanations for suggestions or assessments provided by AI and incorporating parents, teachers and students in the development process.

Equity in access: Inequalities in schooling can be made worse by unequal access to technology. Students who do not have as much access to AI-powered resources as their peers might be at a disadvantage. It is imperative to guarantee equitable access to technology, especially in impoverished communities. To close the digital divide, policymakers, educators and software developers must work together.

Decision-making accountability: When AI systems make choices that affect students' academic careers, accountability issues arise. It could be difficult to determine who is responsible for mistakes. The responsible use of AI in education can be ensured by putting in place explicit accountability structures, involving humans in crucial decision-making processes and creating appeal and redress channels.

Depersonalisation of education: If AI is used too much, it could result in a depersonalised learning environment where students require more individualised attention and important human interactions. It is crucial to strike a balance between preserving human ties and integrating AI. AI should not be seen as a replacement for human interaction in education, but rather as a helpful tool that allows for customised learning.

AI in education raises concerns about job displacement for teachers because it may allow automated systems to replace some teaching responsibilities. Reassuring people that AI is designed to support educators rather than take their place can allay fears.

Programmes for professional development can equip teachers to work efficiently using AI tools in collaboration.

Security risks: Artificial intelligence (AI) systems are susceptible to cyberattacks, which could result in student data breaches or interfere with instructional procedures. To protect against security concerns, it is crucial to implement strong cyber security measures, conduct frequent system audits and give administrators and instructors cyber security training [17].

Long-term effect on learning: It is important to evaluate how AI will affect students' ability to think critically, improve their cognitive abilities and be creative in the long run. This issue can be addressed by prioritising a balanced approach that incorporates both AI and conventional teaching approaches, integrating educators in the evaluation process, and doing continuing research on the educational implications of AI.

Ethical AI education: If ethical AI use is not taught in schools, students may misuse AI tools unknowingly or inadvertently. Students who comprehend the consequences of AI in education can be produced by addressing the integration of ethical AI education into school curricula, educating educators on responsible AI use and cultivating an ethically conscious culture [13].

In order to address these ethical issues, educators, legislators, technology developers and the general public must work together. For ethical AI integration into education to be implemented responsibly, it is imperative to establish transparent norms and give fairness and inclusivity first priority. The advantages of AI in education can be maximised while lowering any potential ethical hazards with the help of regular assessments, feedback loops and a dedication to continuing ethical discussions. OpenAI, 2024. Large-scale language model ChatGPT (3.5). OpenAI Chat https://chat

10. Some potential biases that faculty members should be aware of when using AI in the classroom

It is important for instructors to be aware of any potential biases in AI systems when they employ them in the classroom. These prejudices could unintentionally have an impact on pupils, grades and the general educational process. The following are some possible biases to be aware of:

Algorithmic bias: Preexisting biases in the training data may be inherited by AI algorithms. An AI system may reinforce and magnify biases if the historical data it was trained on is biased. It may lead to unfair results, giving some student groups less of an advantage due to things like socio-economic class, gender or colour. Verify and audit AI algorithms on a regular basis for bias. While training, make use of representative and heterogeneous datasets. Put in place fairness-aware algorithms that give equitable results priority.

Cultural bias: AI systems may reproduce cultural biases seen in the data, which could result in instructional content that is less sensitive to or relevant to other cultures. The information may not be as inclusive or relatable to some cultural groups, which could hinder learning and participation. Make sure that teams creating material are diverse. Verify the instructional content's sensitivity to and inclusion of different cultures. Include a range of viewpoints in datasets used for AI training.

Gender bias: Due to social assumptions or disparities in historical data, AI systems may display gender bias. Gender stereotypes may be reinforced if it leads to kids having different chances or expectations based on their gender. Examine comments and material carefully for gender neutrality. Ensure that all students, regardless of gender, receive fair treatment by routinely evaluating and modifying algorithms.

Socio-economic bias: AI systems may unintentionally maintain gaps in educational achievements if they are trained on data that reflects socioeconomic prejudices. There may be obstacles or prejudices that lower socio-economic backgrounds students must overcome in order to maximise their learning potential. Adopt strategies that are tailored to the requirements of pupils from various socio-economic backgrounds. To reduce socio-economic prejudices, algorithms should be regularly evaluated and adjusted.

Language bias: Due to biases in training data, AI systems may favour particular dialects or linguistic styles over others. AI-powered technologies may make it difficult for students from diverse linguistic backgrounds or skill levels to comprehend or express themselves. Make sure the training datasets are linguistically diverse. Use language models that are flexible and mindful of many dialects and linguistic styles.

Accessibility bias: Unintentionally favouring particular learning formats or styles, AI systems may exclude students who have particular accessibility needs. Participating in AI-driven activities or gaining access to educational content may provide challenges for students with disabilities. Give accessibility top priority while developing and deploying AI solutions. Audit accessibility to make sure AI systems can meet a range of learning requirements.

Historical data bias: AI systems that have been trained on historical data may reinforce biases that are reflected in that data, such as discriminatory or inequitable past actions. Inadvertent perpetuation of past prejudices could result in unequal chances or treatment for particular student populations. Update training data frequently to reflect prevailing social norms and values. Put procedures in place to recognise and address biases in historical data.

Involvement bias: Unintentionally favouring particular learning styles or behaviours might cause AI systems monitoring student involvement to produce biased evaluations of student participation. It is possible that students with diverse communication or learning styles will receive unjust evaluations. Put in place inclusive and diverse engagement metrics. In order to prevent favouring particular learning types, take into account a variety of engagement and comprehension markers.

In order to overcome prejudices, faculty members should actively assess and monitor AI systems, work with tech developers to eliminate biases and promote moral AI practices in educational environments. To guarantee that AI in the classroom fosters justice, inclusivity and equal opportunity for all students, awareness and proactive measures are crucial [13].

11. Use of generative AI

A kind of artificial intelligence called 'generative AI' uses patterns found in existing data to create new content, frequently text, graphics or other media. Applications for generative AI can be found in many different fields, demonstrating the technology's capacity for creativity, invention and problem-solving. Here are a few noteworthy applications of generative AI:

1. Creation of content:

   Generating AI is able to produce text, images and videos that are realistic and cohesive by identifying patterns from vast datasets.

Applications: Text generation: Composing poetry, essays or even conversation for computer games.

Producing lifelike representations of inexistent objects or scenes is known as picture synthesis.

2. Creative arts and design: Innovative compositions, artworks and designs are created in the creative arts through the application of generative AI.

   Applications: Creating original artwork, such as paintings, sculptures or digital art.

   Making original music or coming up with melodies is known as music composition.

3. Chatbots and virtual assistants: Conversational agents are powered by conversational AI, which comprehends and produces responses that resemble those of a human.

   Applications: Customer support: Offering instantaneous help and automated responses.

   Personal assistants: Using speech recognition to carry out tasks.

4. Video game design: Synopsis: Using generative AI, video game content may be made more dynamic and adaptable.

   Applications: Procedural content generation: Creating situations, game levels and characters.

   Creating branching tales according to player decisions is known as narrative generation.

5. Medical image synthesis: To create synthetic images for machine learning model training, medical imaging uses generative AI.

   Applications: Adding more medical photos to increase the variety of training datasets is known as data augmentation.

   Simulation: Creating plausible medical situations in order to provide instruction.

6. Language translation: Coherent and contextually correct translations are produced through the application of generative AI in language translation.

   Applications: Enabling smooth communication between speakers of several languages through the use of cross-language technology.

   Translation services: Improving language translation's accuracy and speed.

7. Code generation: Using high-level instructions as a guide, generative AI can help create code fragments or even whole programmes.

   Uses:

   Auto-completion: Provides code snippets to programmers to help them while they work.

   Code synthesis is the process of creating code from specifications in natural language.

8. Content summarisation: Long texts can be condensed and coherently summarised by generative AI.

   Uses:

   News articles: Condensing news stories for rapid reading.

   Document summarisation is the process of reducing long papers to make them easier to read.

9. Fashion design: To create original designs and forecast future trends, generative AI is applied in the fashion industry.

   Uses:

   Inventive and ground-breaking fashion designs serve as inspiration for designers.

   Customisation: Creating unique clothes designs according to personal tastes.

10. Scientific inquiry: By producing ideas, simulating experiments or analysing large, complicated datasets, generative AI supports scientific inquiry.

    Uses:

    Creating molecular structures for possible medications is known as drug discovery.

    Building models for scientific studies is known as simulation modelling.

The uses of generative AI are expected to grow as it develops further, opening up new avenues for anything from scientific research and healthcare to creative and entertainment industries. The ethical and appropriate application of generative AI is still vital to address potential biases and societal repercussions, even though it offers great benefits [10].

12. Integrating artificial intelligence (AI) into learning management systems (LMS)

There are numerous advantages of integrating artificial intelligence (AI) with learning management systems (LMS), which improve the quality of education for students and teachers alike. Here are a few main benefits:

Personalised learning: AI can customise content and activities to match each learner's preferences, learning styles and performance statistics, resulting in a personalised learning environment. Students receive individualised instruction that targets both their areas of strength and weakness, increasing comprehension and engagement.

Learning routes that are dynamically adjusted by AI algorithms in response to students' success guarantee that they are suitably challenged. This is known as adaptive learning paths. By moving at their own speed, students avoid becoming bored or frustrated and are encouraged to keep becoming better.

Predictive analytics: AI can evaluate data to forecast academic achievement and spot possible problems or areas in which help might be needed. Teachers are able to optimise their teaching tactics, proactively address learning gaps and offer timely help.

Automated grading and assessment: AI-driven solutions have the capacity to automatically grade assignments and tests, giving students immediate feedback. Lessens the administrative load on teachers, enabling them to concentrate on more intricate facets of instruction.

Natural language processing (NLP): With NLP in AI, chatbots and virtual assistants may converse in natural language and offer learners immediate assistance, improves responsiveness, accessibility and help availability, fostering a more encouraging learning environment.

Data-driven insights: AI is capable of analysing large amounts of data to produce useful insights about learning trends, student performance and engagement. Teachers get important knowledge that helps them make wise decisions, enhance their teaching methods and maximise the curriculum.

Content recommendation: Based on each student's interests and learning background, AI algorithms can recommend pertinent reading materials, websites and activities. It enhances learning, promotes experimentation and aids students in finding more material that is relevant to their interests.

Effective resource allocation: AI is capable of optimising the use of educational resources, such as budgetary allocation, classroom utilisation and faculty time. It guarantees a financially responsible learning environment, optimises resource utilisation and improves operational efficiency.

Routine administrative chores, including scheduling, enrolment and record-keeping, can be handled by AI for automated administrative chores. It lessens the administrative burden on teachers so they may concentrate on instruction and strategic planning [23].

13. Conclusion

AI has a significant impact on the education industry. Even though it helps with many academic and administrative activities and improves education, its worries regarding security, privacy, ethical considerations and the choice of learning resources should not be disregarded. Even in institutions with technology issues, incorporating artificial intelligence (AI) into teaching and learning can drastically change educational methods. While there are special issues when integrating AI capabilities into a university that is underperforming in its teaching programme, academics can effectively navigate this process by grasping the underlying concepts and using calculated strategies. When integrating AI into teaching and learning in a university that is lagging behind, academics should be aware of and take the following crucial insights and actions. Scholars ought to carry out a comprehensive evaluation of the university's existing IT setup. To ascertain whether AI integration is compatible, be aware of the hardware, software and internet connectivity that are available.

Acknowledge the unique technological limitations and difficulties faced by the university, such as restricted access to fast internet, obsolete technology or a dearth of technical assistance.

Clearly state the purposes and goals of AI integration as it relates to education. This can entail enhancing educational objectives, raising student involvement or simplifying administrative procedures. Determine the areas with the most potential for AI influence, taking into account resource limitations. This could entail emphasising data-driven decision-making in academic planning, automated assessment systems or adaptive learning platforms. Provide faculty members with thorough programmes in AI literacy. Make sure teachers are knowledgeable on the fundamentals of artificial intelligence (AI), how to use it in the classroom and how to incorporate AI technologies into their lesson plans.

Encourage academics, technologists and educational researchers to collaborate across disciplinary boundaries. The exchange of best practices and creative solutions can be encouraged by this cooperative approach. Seek AI solutions that can be tailored to the unique circumstances of the underperforming university. Seek for adaptable platforms that can take into account differences in educational objectives and technical infrastructure. Take into consideration integrating AI in stages or incrementally. Begin with pilot projects, evaluate their effects and then progressively expand them in light of positive results and lessons discovered.

Stress how crucial it is for AI algorithms and decision-making procedures to be transparent. Faculty members ought to be conscious of potential biases, ethical issues and the necessity of accountability and fairness in AI applications. Put strong data security procedures in place to protect the privacy of students. Make sure data usage adheres to ethical standards and that AI tools meet privacy requirements. Give inclusiveness in AI applications top priority. Take into account a variety of student demographics and make sure AI technologies accommodate varying learning preferences, aptitudes and cultural contexts.

Establish precise measurements to evaluate the effects of integrating AI. This can entail raising engagement levels, raising student achievement or increasing administrative process efficiency. Get input from educators, learners and other stakeholders on a regular basis. Utilise these comments to pinpoint problem areas, resolve issues and enhance AI deployments. Use an iterative process while integrating AI. Handle it as a continuous process of improvement, being willing to modify tactics in response to changing requirements and developments in technology.

Seek the leadership's assistance in promoting AI integration programmes at the university. A culture that supports experimentation and growth can be established by a top-down commitment to innovation. Engage the academic community in conversations around the integration of AI. Encourage knowledge and comprehension of the advantages and difficulties of using AI in the classroom. Encourage faculty members to participate in peer-to-peer learning opportunities. Promote the exchange of AI integration best practices, insights and experiences. Prepare for probable obstacles that may arise during the integration of AI and create backup strategies. This could entail fixing technical issues, offering more faculty assistance or modifying plans of action in response to changing conditions. Put strong cyber security measures in place to protect against such dangers. In AI applications, maintaining system security and data integrity is essential.

Demonstrate effective AI applications at the university. Honour the accomplishments of the faculty, the creative methods of instruction and the gains in student performance brought about by the incorporation of AI. Talk about your accomplishments with the larger educational community. Engage in publications, conferences and cooperative networks to increase awareness of the university's artificial intelligence programmes.

Investigate working together with AI-focused research universities and partners in the industry. Collaborate on projects together, pool resources and take advantage of outside knowledge to improve AI applications in education. Academics at a university that is falling behind can incorporate AI into their teaching and learning processes, resulting in better educational outcomes, increased innovation and increased inclusivity across the academic community by adopting these concepts and taking calculated action. It necessitates a deliberate, cooperative and flexible strategy that takes into account the particular opportunities and problems present by the university.

14. Recommendation

1. A certain amount of AI technology must be used in teaching and learning to preserve human cognition.

2. Training should be provided to educators and students before implementing AI technology.

3. The other issues with AI in education that need to be investigated can be researched.

4. Comparable research in different parts of the nation is possible.

Author details

Ogunlade B. Olusola[1*], Bahago S. Benedict[2] and Shotayo E. Olusola[3]

1 Bamidele Olumilua University of Education, Science and Technology, Ikere-Ekiti, Nigeria

2 Veritas University Abuja, Nigeria

3 Federal Ministry of Education, Abuja, Nigeria

*Address all correspondence to: ogunlade.bamidele@bouesti.edu.ng

References

[1] Karandish D. Benefits of AI in education. The Journal. 2021. Available from: https://thejourrnal.com/Aticle/2021/06/23/7-Benefits-of-AI-in-Education.aspx

[2] Ade-Ibijola A, Young K, Sivparsad N, Seforo M, Ally S, Olowolafe A. Teaching students about plagiarism using a serious game (plag warfare): Design and evaluation study. JMIR Serious Games. 2022;**10**(1):e33459. DOI: 10.2196/33459

[3] Maini V, Sabri S. Machine Learning for Humans. 2017. Available from: https://medium.com/machine-learning-for-humans

[4] Rainie L, Anderson J, Vogels EA. Experts Doubt Ethical AI Design Will be Broadly Adopted as the Norm within the Next Decade. Haley Nolan: Pew Research Center; 2021

[5] Shneiderman B. Responsible AI: Bridging from ethics to practice. Communications of the ACM. 2021;**64**(8):32-35. DOI: 10.1145/3445973

[6] Nemorin S, Vlachidis A, Ayerakwa HM, Andriotis P. AI hyped? A horizon scan of discourse on artificial intelligence in education (AIED) and development. Learning, Media and Technology. 2022:1-14. DOI: 10.1080/17439884.2022.2095568

[7] Schiff D. Out of the laboratory and into the classroom: The future of artificial intelligence in education. AI & Society. 2021. DOI: 10.1007/s00146-020-01033-8

[8] Ayling J, Chapman A. Putting AI ethics to work: Are the tools fit for purpose? AI and Ethics. 2022;**2**:405-429. DOI: 10.1007/s43681-021-00084-x

[9] Baron NS. Even Kids are Worried ChatGPT will Make Them Lazy Plagiarists, Says a Linguist Who Studies Tech's Effect on Reading, Writing and Thinking. Fortune; 2023. Available from: https://fortune.com/2023/01/19/what-is-chatgpt-ai-effect-cheating-plagiarism-laziness-education-kids-students/

[10] Duan Y, Edwards JS, Dwivedi YK. Artificial intelligence for decision making in the era of Big Data—Evolution, challenges and research agenda. International Journal of Information Management. 2019;**48**:63-71. DOI: 10.1016/j.ijinfomgt.2019.01.021

[11] Di Vaio A, Palladino R, Hassan R, Escobar O. Artificial intelligence and business models in the sustainable development goals perspective: A systematic literature review. Journal of Business Research. 2020;**121**:283-314. DOI: 10.1016/j.jbusres.2020.08.019

[12] Sutton S, Arnold V, Holt M. How much automation is too much? Keeping the human relevant in knowledge work. Journal of Emerging Technologies in Accounting. 2018;**15**(2). DOI: 10.2308/jeta-52311

[13] Venema L. Defining a role for AI ethics in national security. Nature Machine Intelligence. 2021;**3**:370-371. DOI: 10.1038/s42256-021-00344-9

[14] Suh W, Ahn S. Development and validation of a scale measuring student attitudes toward artificial intelligence. SAGE Open. 2022:1-12. DOI: 10.1177/21582440221100463

[15] Xie H, Hwang G-J, Wong T-L. From conventional AI to modern AI in education: Re-examining AI and analytic

techniques for teaching and learning. Educational Technology & Society. 2021;**24**(3):85-88

[16] Owoc ML, Sawicka A, Weichbroth P. Artificial intelligence technologies in education: Benefits, challenges and strategies of implementation. In: Artificial Intelligence for Knowledge Management: 7th IFIP WG 12.6 International Workshop, AI4KM 2019, Held at IJCAI 2019, Macao, China, August 11, 2019, Revised Selected Papers. Cham: Springer International Publishing; 2021. pp. 37-58

[17] Ross J Does the Rise of AI Spell the End of Education? 2021. Available from: https://www.timeshighereducation.com/features/does-rise-ai-spell-end-education

[18] Ogunlade BO. Creating a virtual community of practice through blended learning in teacher education in Southwest Nigerian Universities. Journal of Research Administration. 2024;**6**(1):397-414

[19] Weyerer J, Langer P. Garbage in, garbage out: The vicious cycle of ai-based discrimination in the public sector. In: 20th Annual International Conference on Digital Government Research. ACM Digital Library; 2019. pp. 509-511

[20] Ogunlade BO, Bahago SB, Ogunlade JO. Influence of mobile and blended learning on E-counselling teachers education in Nigeria Universities during Covid-19. Journal of Data Acquisition and Processing. 2023;**38**(4):765-777

[21] Vincent S, van R. Trustworthy Artificial Intelligence (AI) in Education. Organization for Economic Co-operation and Development. 2022. pp. 1-17

[22] Samtani S, Kantarcioglu M, Chen H. A multi-disciplinary perspective for conducting artificial intelligence-enabled privacy analytics: Connecting data, algorithms, and systems. ACM Transactions on Management Information Systems. 2021;**12**:1-18. DOI: 10.1145/3447507

[23] Huls A. Artificial intelligence and machine learning play a role in endpoint security. New York Times. 2021. Available from: https//www.nastel.com/artificial-intelligence-and-machine-learning-play-a-role-in-endpoint-security/

Chapter 11

Teachers in the Age of Artificial Intelligence: Preparation and Response to Challenges

Llaudett Natividad Escalona-Márquez, Stefanía Johanna Cedeño-Tapia, Luis Alberto Camputaro and Christian Oscar Orlando Aparicio-Escalante

Abstract

In the field of higher education, artificial intelligence (AI) stands as a transformative axis, presenting challenges and opportunities for both teachers and students. This chapter profiles the innovative teacher, whose responsibilities expand toward guaranteeing quality education that is adaptive to current technological demands. Students' digital competence is critically examined, identifying the gap that exists when entering higher education and focusing on the ethical and practical challenges inherent in the use of AI. The importance of continuous teacher training and strategies that promote active involvement in AI is emphasized, to cultivate a deep understanding and effective application of these technologies in the educational process. It is recognized that AI can provide personalized and self-regulated learning, but it does not replace the essence of human mentoring, essential for its capacity for empathy and creativity.

Keywords: automated learning, educational innovation, twenty-first century skills, teacher professional development, technological integration

1. Introduction

The question of the ability of machines to think was first raised in 1950 by Alan Turing [1]. Based on this, the theoretical foundations were established to develop a technology that was capable of imitating human thinking. With this background, the term Artificial Intelligence (AI) was first raised at the Dartmouth Founding Conference in 1956 by John McCarthy [2]. AI is a technology that is made up of algorithms that run continuously to predict and complete complex tasks in seconds. This sequencing algorithm takes back information learned through the machine learning (ML) process [3]. Machine learning is defined as "a subfield of artificial intelligence that includes software capable of recognizing patterns, making predictions, and applying the newly discovered patterns to situations that were not included or covered by its initial design" [4]. In education, main elements of AI are machine learning, educational data mining, and learning analytics [5]. UNESCO defines

Artificial Intelligence as "systems capable of processing data and information in a way that resembles intelligent behavior, and generally encompasses aspects of reasoning, learning, perception, prediction, planning or control" [6].

The focus on AI as an element of interest for society accelerated in 2023 with the emergence of an update to the ChatGPT application from the company OpenAi [7–9]. This application made it possible to integrate with other applications and digital technologies, which led to greater accessibility to AI for the population. Several economic and social sectors have begun their adapt to the use of this advanced digital technology as a basic and daily element of the activities demanded by their sector [10].

The educational field has not been the exception in the adoption of AI, since it provides the facilities for personalized training and self-regulated learning. A study from the United States of America reported that 38% of students use some AI at least once a month [11]. A global survey identified that 50% of university students, globally, have used AI to study and typically enter one or two questions a day [12]. As for teachers, other surveys identified the importance of training them on AI topics. The most notable results show that 65% should be trained to integrate it into their teaching methods, 70% think that teachers should teach students how to properly use AI and another important element is related to the ethical implications of the learning process, since 71% of people agree that teachers should receive training to identify the originality of educational tasks, such as essays, among other activities [13]. These results give indications that we are beginning a new educational era, that of Artificial Intelligence.

Due to the multiple challenges faced by the integration of AI in higher education, in this chapter, we want to reflect on the contributions of Artificial Intelligence in Education (AIEd) and the challenges they pose for teachers. Additionally, we want to talk about the ethical dilemma that this technology presents in the educational field.

2. Artificial Intelligence in education

Although we are facing a new era, interest in integrating Artificial Intelligence for educational uses has developed since the end of the last century. In 1993, to help with the advancement of knowledge and the promotion of research in the field of Artificial Intelligence in education, The International Society of Artificial Intelligence in Education (IAIED) [14] was created, which in turn time publishes a magazine related to the subject [15].

A meta-systematic review identified six main benefits of the integration of AIEd: personalisade learning; greater insight into student understanding; positive influence on learning outcomes; reduced planning and administration time for teachers; greater equity in education and precise assessment & feedback [16].

Likewise, four AIEd typologies have been identified: profiling and prediction, intelligent tutoring systems (ITS), assessment and evaluation, and adaptive systems and personalization [15]. In turn, these have given guidance on the classification of the application of AI in education (**Figure 1**) [16].

Profiling and prediction in the use of data collected by AI to provide information that helps in forecasting the possible academic trajectory of students. Likewise, it provides a diagnosis that will guide the decision-making of the senior teacher to design effective teaching strategies that improve academic performance. It provides the administrative body of the universities with the elements to identify the student retention route and avoid abandonment. Another of its virtues is that it offers a vision of the areas where the student will need support [15, 16].

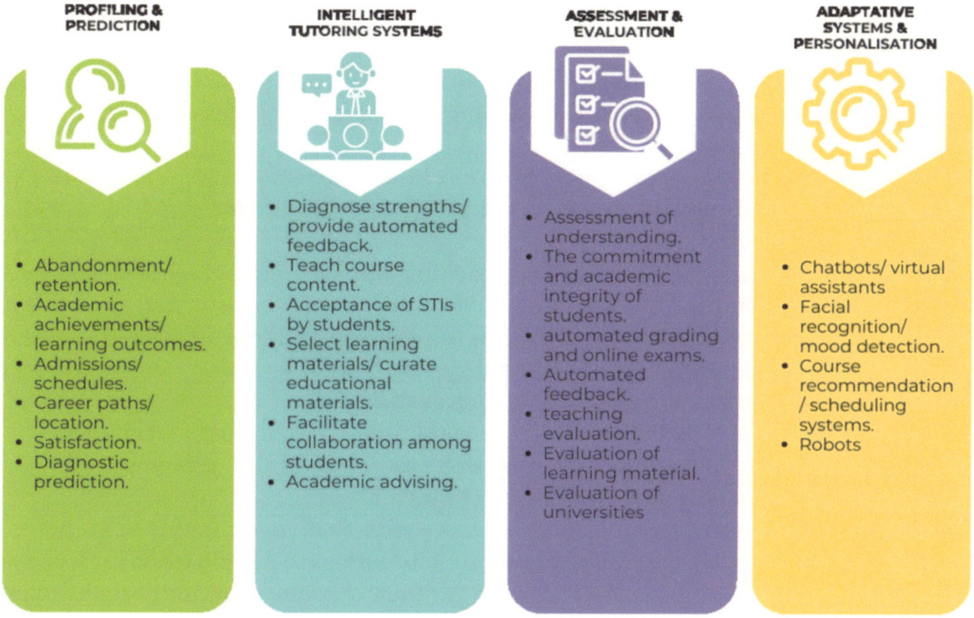

Figure 1.
AIEd applications. Source: Based on Bond et al. 2014 [16].

Intelligent tutoring systems (ITS) help students with self-regulated and personalized learning, giving them tools that allow them to enrich their learning experiences. ITS provide them with immediate, automated feedback. Among the benefits of these systems is that they help recommend teaching materials that fit the needs of the student. In addition, it makes it easier for them to collaborate with their peers. They provide teachers with the information necessary to diagnose students' learning needs and strengths [15].

Assessment and evaluation help educational institutions evaluate students, their understanding of content, and acquisition of skills and provides automated qualifications and online exams, among others. It also allows evaluating the quality of teaching and evaluates teachers and the effectiveness of the learning methodologies used [15, 16].

Adaptive systems and personalization contribute to personalized student learning. They guide them on different learning routes that fit their needs. They contribute to self-regulated learning and facilitate a deeper understanding of the knowledge they wish to acquire. They help teachers improve their teaching process through recommendations for the design of a pedagogical route, and they also help them monitor students [15, 16].

For example, chatbots can provide study guides tailored to learning needs and personalized educational resources [17]. Research showed that ChatGPT support for students who had self-regulated learning experiences performed well in activities related to the planning and facilitation of educational materials; however, it was not shown to be sufficiently appropriate for an evaluation that provides effective feedback [9].

The literature on the subject shows that human tutoring is more effective than that assisted by AI since it offers human qualities such as emotional support, intuition, compassion, empathy, and creativity [18, 19]. This makes it easier for students to obtain highly informative feedback that allows them to improve and acquire the expected learning results [20], which, when effective, helps improve motivation, self-regulation, self-efficacy, and metacognition [21]. Some studies have identified that students positively value the soft or socio-emotional skills of their teachers. These emotions provide elements in favor of not replacing the university professor with AI [22].

3. Challenges in the application of AI in higher education

Despite the imminent impact of AI on teaching and learning in higher education, a UNESCO survey revealed that less than 10% of universities have formal guidance on the use of AI [23]. This shows that, although AI offers great opportunities to enhance learning, its integration into the educational system is still beginning. This represents various challenges for pedagogical agents such as teachers.

A systematic review identified the advantages that AI can bring to teachers in the educational field. To do this, it takes up three moments of the pedagogical act: Planning, Implementation, and Evaluation (**Figure 2**) [24].

However, these advantages it offers also come with challenges. In this section, we highlight the limitations in teachers' preparation to integrate AI into teaching in ways that foster meaningful and situated learning for their students. Teachers are the main educational actors for improving educational quality [25]. For this reason, the training and preparation of university teachers in IAEd requires proposals that provide solid foundations on how to integrate it into the pedagogical act. In the literature, we identified some AI training proposals that consist of three stages: understanding AI, using AI tools, and experiencing AI programs [26].

In the first stage, teachers are intended to understand concepts related to AI, such as, for example, automated learning and how it works. Once this theoretical knowledge is obtained, it is expected that they will be more receptive to applying it for educational purposes. This will help them increase their confidence and overcome the

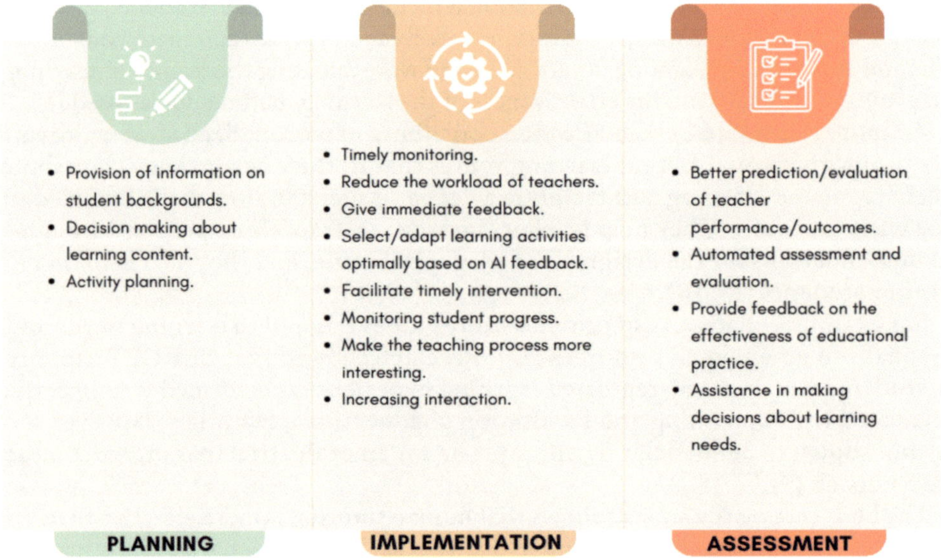

Figure 2.
AI advantages for university teaching. Source: Based on Celik et al. 2022 [24].

algorithm aversion effect, characterized by the loss of credibility in automated recommendation systems after making an error, and instead tolerating the same error when it is made by other people [27]. Overcoming this bias will make them understand that AI has limited capacity for some activities but still provides tools that can be leveraged to apply in teaching.

Furthermore, we consider that at this time, some myths related to educational technology must also be analyzed and overcome, such as pedagogical materialism, which defends that digital technologies are the only educational materials available on the Internet added to teaching [28].

The second stage is core; this seeks to encourage the teacher to explore and experiment with different AI tools to learn about their functions and usefulness for teaching. The importance of this stage is related to overcoming the three elements of the Technology Acceptance Model (TAM), which are perceived ease of use, perceived usefulness, and attitude toward use [29, 30].

A study identified that teachers' perception of risk [31] and feelings of anxiety [29] negatively influence the adoption of AI in education. That is why, at this stage, teachers are encouraged to develop a positive attitude about the use of AI through experimentation with different applications and tools; this will help them increase their confidence. Practice will help them reduce their anxiety and reduce the feeling of risk.

In the third stage, experimentation with different AI programs is interesting. These types of experiences will reaffirm knowledge about the uses and limitations of this technology in various tasks or activities. At this stage, it is considered necessary for teachers to return to the three elements of reflective practice: (1) collect data, (2) data analysis and evaluation, and (3) triangulation of information [32]. The integration of these three elements will help them with key moments of the pedagogical act, such as planning, implementing, and evaluating the educational task efficiently.

Data collection is important because it guides the teacher about what is happening in student learning, for example, learning styles, the impact of the strategies used, time use. For the second and third elements, teachers must have the basic knowledge to interpret AI-based learning analytics [27]; this skill will help them make more effective decisions for teaching.

4. Ethical and practical challenges of AI in education

To obtain data that feeds AI machine learning (ML), the applications must have confidential information from students and teachers, which represents an ethical dilemma for their use [15].

The lack of security and critical thinking about information shared through AI poses great risks for students, teachers, and educational institutions. Access to digital environments and advanced technologies does not guarantee that current and future students are competent in the use of technology [28, 33]. Despite being immersed in it since its birth, its effectiveness in safely using different technological tools has been questioned. It has been found that they are more likely to share personal and confidential information in different applications without investigating the possible consequences on their security and privacy [34].

These ethical dilemmas regarding the use of confidential and private data of students and teachers for educational purposes are not recent [35, 36]. For this reason, it is an issue that is on different political agendas to impose measures that guarantee the security and privacy of information [5]. For this reason, guidelines have emerged that allow the implementation of a safe, fair, responsible, transparent AI, guaranteeing the privacy and data protection [37, 38]. In the educational field, this concern has been shared by organizations such as UNESCO, which demands the creation of robust policies to ensure that advances in AI do not compromise privacy or exacerbate existing inequalities [39].

Researchers and educators must work together to ensure the responsible use of AI. This includes the development of regulations and policies that oversee its use, in line with the laws of each country.

5. Teaching competencies to integrate AI

Faced with the above, we ask ourselves: What are the key competencies that university teachers must develop in order to integrate AI efficiently into the teaching and learning process? Within the whole range of response possibilities, we consider digital teaching skills and socio-emotional skills to be fundamental. Digital teaching skills due to its relationship with the pedagogical use of digital technologies; and socio-emotional skills due to the need for humanization and management of emotions in the use of AI in education.

Teachers' digital competencies are the set of knowledge, skills, and abilities of teachers to provide solutions to the different challenges presented in their educational actions from an ethical, safe, and responsible point of view in synergy with the use of digital technologies and tools in the teaching and learning process.

It has been shown that one aspect that benefits teachers' self-efficacy is emotional intelligence. This is an individual's ability to understand and regulate their emotions [40, 41]. Accordingly, social-emotional skills are the "... a set of skills that allow us to manage our own feelings and emotions and those of others, with the intention of guiding thought and actions towards satisfactory performance, dealing in the best way with different emotional states as a way of motivation and meaning in life..." [41].

Considering that emotional intelligence and emotional skills are fundamental for teaching efficiency, there is a need to put them into practice in educational settings. This is why the need arises for the development of socio-emotional competencies, considered "Emotional awareness"; acquiring the ability to use and regulate them in different social contexts and educational environments [42].

Due to the predominance of digital technologies in education, the construct of digital emotional competence emerged, which is a fusion of digital competencies with emotional intelligence [43, 44].

Therefore, any educational activity in which AI is used must have a strong human component that supports the regulation and management of the emotions of the educational actors involved.

In addition, it is considered pertinent for higher education institutions to manage the incorporation of the aforementioned competencies for teacher education and training activities.

6. Conclusion

Exposure to technology does not guarantee the development of digital competencies; oftentimes, we possess intelligent technological resources, and their full potential is not exploited. It is necessary to take a critical and comprehensive position on these new generations, based on designing educational proposals that promote and evaluate the development of these skills.

Promoting critical thinking is crucial in a landscape where technology infiltrates all aspects of teaching. AI, although it is a powerful tool, must be the subject of detailed analysis for its effective incorporation into the educational environment. Educators must cultivate this quality, encouraging students to critically analyze and question both the information and the technological solutions they encounter.

For these methods to be effective, educators must be willing to acquire new competencies and keep up to date with technological and pedagogical developments. This includes participating in professional development programs and collaborating interdisciplinary to integrate AI into the classroom to enrich the learning experience. The creation of communities of practice and professional support networks will be essential for the exchange of ideas and strategies that facilitate the effective adoption of AI in teaching.

Understanding teacher professional development as a dynamic and adaptable process, rather than as a static requirement, is vital. By prioritizing continuous training, teachers can be prepared to lead the era of AI in education.

7. Key questions and reflection

Given the above, we ask ourselves: What are the key competencies that university teachers must develop to integrate AI efficiently into the teaching and learning process?

Conflict of interests

The authors declare that they have no conflicts of interest.

Notes/thanks/other declarations

We would like to thank the staff of the Ministry of Education of Science and Technology of El Salvador for their constant effort and commitment from their different workplaces. Thank you for all your support in the various programs and strategies for the implementation of digital innovation and for maintaining the quality of education at all levels. We also recognize and thank each teacher for their commitment to the digital transformation of the educational community.

Appendices and nomenclature

AI	Artificial Intelligence
AIEd	Artificial Intelligence in Education

Author details

Llaudett Natividad Escalona-Márquez[1], Stefanía Johanna Cedeño-Tapia[2], Luis Alberto Camputaro[3*] and Christian Oscar Orlando Aparicio-Escalante[4]

1 Higher Education, University of Concepción, Concepción, Chile

2 Higher Education Quality Accreditation Commission, San Salvador, El Salvador

3 Specialized Institute "Hospital El Salvador", San Salvador, El Salvador

4 Higher Education, Ministry of Education of El Salvador, San Salvador, El Salvador

*Address all correspondence to: luis.camputaro@salud.gob.sv

References

[1] Turing AM. Maquinaria de Computación e Inteligencia. In: Epstein R, Roberts G, Beber G, editors. Análisis de la prueba de Turing. Dordrecht: Springer; 2009. DOI: 10.1007/978-1-4020-6710-5_3

[2] Qu J, Zhao Y, Xie Y. Artificial intelligence leads the reform of education models. Systems Research and Behavioral Science. 2022;**39**(3):581-588. DOI: 10.1002/sres.2864

[3] Gaskins N. Interrogando el sesgo algorítmico: de la ficción especulativa al diseño liberador. Technology Trends. 2023;**67**:417-425. DOI: 10.1007/s11528-022-00783-0

[4] Popenici SAD, Kerr S. Explorando el impacto de la inteligencia artificial en la enseñanza y el aprendizaje en la educación superior. RPTEL. 2017;**12**:22. DOI: 10.1186/s41039-017-0062-8

[5] Maphosa V, Maphosa M. Inteligencia artificial en la educación superior: un análisis bibliométrico y un enfoque de modelado de temas. Applied Artificial Intelligence. 2023;**37**(1):2338-2358. e2261730. DOI: 10.1080/08839514.2023.2261730

[6] UNESCO. Recomendación sobre la ética de la inteligencia artificial. UNESCO; 2022. p. 43. Documento SHS/BIO/PI/2021/1. Disponible en: https://unesdoc.unesco.org/ark:/48223/pf0000381137_spa

[7] Mhlanga D. The Value of Open AI and Chat GPT for the Current Learning Environments and the Potential Future Uses. SSRN; 2023. DOI: 10.2139/ssrn.4439267

[8] Hadi MA, Abdulredha MN, Hasan E. Introducción a ChatGPT: una nueva revolución de la inteligencia artificial con algoritmos de aprendizaje automático y ciberseguridad. Arch Cient. 2023;**4**(4):276-285. DOI: 10.47587/SA.2023.4406

[9] Hartley K, Hayak M, Ko UH. Inteligencia artificial que respalda el aprendizaje independiente de los estudiantes: un estudio de caso evaluativo de ChatGPT y aprender a codificar. Ciencias de la Educación. 2024;**14**(2):120. DOI: 10.3390/educsci14020120

[10] Al Naqbi H, Bahroun Z, Ahmed V. Mejora de la productividad laboral mediante inteligencia artificial generativa: una revisión exhaustiva de la literatura. Sostenibilidad. 2024;**16**(3):1166. DOI: 10.3390/su16031166

[11] Edition US. AI in Higher Ed: Hype, Harm, or Help [Internet]. Anthology.com; 2024. Disponible en: https://www.anthology.com/sites/default/files/2023-11/White%20Paper-USA-AI%20in%20Higher%20Ed-Hype%20Harm%20or%20Help-v1_11-23.pdf

[12] Chegg. Global Student Survey 2023. [Internet]. Chegg.org; 2023. Available from:https://8dfb1bf9-2f43-45af-abce-2877b9157e2c.usrfiles.com/ugd/8dfb1b_e9bad0aef091478397e6a9ff96651f6d.pdf

[13] Ipsos [Internet] Ipsos.com. 2023. Disponible en: https://www.ipsos.com/sites/default/files/ct/news/documents/2023-09/Ipsos%20Monitor%20Global%20de%20Educaci%C3%B3n.pdf

[14] International AIED Society. [Internet] Iaied.org. 2023 . Disponible en: https://iaied.org/about

[15] Zawacki-Richter O, Marín VI, Bond M, Gouverneur F. Systematic

review of research on artificial intelligence applications in higher education—Where are the educators? International Journal of Educational Technology in Higher Education. 2019;**16**(1):1-27. DOI: 10.1186/s41239-019-0171-0

[16] Bond M, Khosravi H, De Laat M, Bergdahl N, Negrea V, Oxley E, et al. A meta systematic review of artificial intelligence in higher education: A call for increased ethics, collaboration, and rigor. International Journal of Educational Technology in Higher Education. 2024;**21**(4):1-41. DOI: 10.1186/s41239-023-00436-z

[17] Cotton DRE, Cotton PA, Shipway JR. Chatear y hacer trampa: garantizar la integridad académica en la era de ChatGPT. Innovations in Education and Teaching International. 2023;**61**(2):228-239. DOI: 10.1080/14703297.2023.2190148

[18] Sikström P, Valentini C, Sivunen A, Kärkkäinen T. How pedagogical agents communicate with students: A two-phase systematic review. Computers in Education. 2022;**188**:104564. DOI: 10.1016/j.compedu.2022.104564

[19] Cedeño, Tapia SJ. La inteligencia artificial como herramienta complementaria en la investigación y educación: responsabilidad ética y humana. RUSXXI. 2023, 8;**3**:6-10. DOI: 10.57246/rusxxi.v3i8.47

[20] Gombert S, Fink A, Giorgashvili T, et al. De la evaluación automatizada del contenido de los ensayos de los estudiantes a la retroalimentación altamente informativa: un estudio de caso. International Journal of Artificial Intelligence in Education. 2024. DOI: 10.1007/s40593-023-00387-6

[21] Huang X, Dong L, Chandru Vignesh C, Praveen KD. Aprendizaje autorregulado e investigación científica utilizando inteligencia artificial para sistemas de educación superior. International Journal of Technology and Human Interaction (IJTHI). 2022;**18**(7):1-15. DOI: 10.4018/IJTHI.306226

[22] Artyukhov A, Skowron L, ArtyukhovaN, DluhopolskyiO, CwynarW. Sustainability of higher education: Study of student opinions about the possibility of replacing teachers with AI technologies. Sustainability. 2024;**16**(1):55. DOI: 10.3390/su16010055

[23] UNESCO. Una encuesta de la UNESCO revela que menos del 10% de las escuelas y universidades disponen de orientaciones formales sobre IA [Internet]. Unesco.org; 2023 [citado el 9 de febrero de 2024]. Disponible en: https://www.unesco.org/es/articles/una-encuesta-de-la-unesco-revela-que-menos-del-10-de-las-escuelas-y-universidades-disponen-de

[24] Celik I, Dindar M, Muukkonen H, Järvelä S. The promises and challenges of artificial intelligence for teachers: A systematic review of research. TechTrends. 2022;**66**:616-630. DOI: 10.1007/s11528-022-00715-y

[25] Escalona Márquez LN, Cedeño-Tapia SJ, Virgili-Lillo MA. Competencia docente en el contexto de la evaluación universitaria en México. Educación Superior y Sociedad. 2022;**34**(2):376-398. DOI: 10.54674/ess.v34i2.653

[26] Ryu MY, Han SG. Educational perception of elementary teachers on artificial intelligence. Korean Journal. 2018;**22**(3):317-324. DOI: 10.14352/jkaie.2018.22.3.317

[27] Nazaretsky T, Ariely M, Cukurova M, Alexandron G. Teachers'

trust in AI-powered educational technology and a professional development program to improve it. British Journal of Educational Technology. 2022;**53**:914-931. DOI: 10.1111/bjet.13232

[28] Suárez-Guerrero C, Rivera-Vargas P, Raffaghelli J. Mitos EdTech: hacia una agenda educativa digital crítica. Technology, Pedagogy and Education. 2023;**32**(5):605-620. DOI: 10.1080/1475939X.2023.2240332

[29] Zhang C, Schießl J, Plößl L, Hofmann F. Acceptance of artificial intelligence among pre-service teachers: A multigroup analysis. International Journal of Educational Technology in Higher Education. 2023;**20**(1):1-22. DOI: 10.1186/s41239-023-00420-7

[30] Nja CO, Idiege KJ, Uwe UE, Meremikwu AN, Ekon EE, Erim CM, et al. Adoption of artificial intelligence in science teaching: From the vantage point of the African science teachers. Smart Learning Environments. 2023;**10**(1):1-19. DOI: 10.1186/s40561-023-00261-x

[31] Rahiman HU, Kodikal R. Revolutionizing education: Artificial intelligence empowered learning in higher education. Cogent Education. 2024;**11**(1):1-24. DOI: 10.1080/2331186X.2023.2293431

[32] Phillips TM, Saleh A, Ozogul G. Un conjunto de herramientas de inteligencia artificial para apoyar la reflexión de los docentes. International Journal of Artificial Intelligence in Education. 2023;**33**:635-658. DOI: 10.1007/s40593-022-00295-1

[33] Muñoz-Rodríguez JM, Dacosta A, Martín-Lucas J. ¿Nativos digitales o náufragos digitales? Procesos de construcción y reconstrucción de la identidad digital de los jóvenes y sus implicaciones educativas. In: Muñoz-Rodríguez JM, editor. Identidad en una sociedad hiperconectada. Cham: Springer; 2021. DOI: 10.1007/978-3-030-85788-2_2

[34] Hernández-Orellana MP, Pérez-Garcías A, Roco-Videla ÁG. Identidad digital y conectividad: conocimientos y actitudes de estudiantes universitarios chilenos. Formacion Universitaria. 2021;**14**(1):147-156. DOI: 10.4067/S0718-50062021000100147

[35] Kousa P, Niemi H. Ética y aprendizaje de la IA: desafíos y soluciones de las empresas de tecnología educativa. Interactive Learning Environments. 2023;**31**(10):6735-6746. DOI: 10.1080/10494820.2022.2043908

[36] Zobeida S, Yang Y. Artificial intelligence applications in Latin American higher education: A systematic review. International Journal of Educational Technology in Higher Education. 2022;**19**(1):1-20. DOI: 10.1186/s41239-022-00326-w

[37] European Commission. Ethics guidelines for trustworthy AI | shaping Europe's digital future [Internet]. Digital-strategy.ec.europa.eu. [Internet]. 2019. Available from: https://digital-strategy.ec.europa.eu/en/library/ethics-guidelines-trustworthy-ai

[38] de Leon P, Ferreira de Carvalho AC. Inteligência Artificial: riscos, benefícios e uso responsável. ESTUDOS AVANÇADOS / Inteligência Artificial. 2021;**35**(101):21-35. DOI: 10.1590/s0103-4014.2021.35101.003

[39] Unesco. La inteligencia artificial en la educación [Internet]. 2023 . Disponible en: https://www.unesco.org/es/digital-education/artificial-intelligence

[40] Lu Q, Ishak NA. Teacher's emotional intelligence and employee brand-based

equity: Mediating role of teaching performance and teacher's self-efficacy. Frontiers in Psychology. 2022;**13**:901019. DOI: 10.3389/fpsyg.2022.901019

[41] Llorent VJ, Zych I, Varo-Millán JC. Competencias socioemocionales autopercibidas en el profesorado universitario en España. Educación XX1. 2020;**23**(1):297-318. DOI: 10.5944/educxx1.23687

[42] Garner PW. Emotional competence and its influences on teaching and learning. Educational Psychology Review. 2010;**22**:297-321. DOI: 10.1007/s10648-010-9129-4

[43] Audrin C, Audrin B. Más que solo inteligencia emocional en línea: introducción de la "inteligencia emocional digital". Frontiers in Psychology. 2023;**14**:1154355. DOI: 10.3389/fpsyg.2023.1154355

[44] Fteiha M, Awwad N. Inteligencia emocional y su relación con el estilo de afrontamiento del estrés. Psicología de la Salud Abierta. 2020;7(2):1-9. DOI: 10.1177/2055102920970416

Chapter 12

The Influence of Affect in Help-Seeking Behaviors and Performance in a Math Intelligent Tutoring System

Ana Paula S. Loures-Elias and Matthew L. Bernacki

Abstract

This study investigates the association between help-seeking behaviors (hints, hints per step, hints with steps requests, and hint to error), affect (boredom, confusion, frustration, happiness, and engagement), and performance in seventh and eighth-grade students using the Cognitive Tutor Bridge to Algebra as a self-regulated learning environment. Analyses focused on correlations between students' help-seeking behaviors and their affect in units 4 and 14. Affect was also used to predict help-seeking behaviors in the next units (e.g., 5 and 15). Moreover, we examined how associations between help-seeking behaviors and performance differed as a function of affect. The results showed that a pattern emerged in which students showed more executive help-seeking behaviors rather than instrumental ones. Students feeling bored, confused, and frustrated tended to use more hints, and they were less likely to switch to external help-seeking sources. Also, those feeling happy or engaged were less likely to use hints.

Keywords: help-seeking, affect, academic emotions, math intelligent tutoring system, performance

1. Introduction

Help-seeking is a self-regulatory strategy in self-regulated learning (SRL) in which learners use help-seeking to sustain cognition, behavior, affect, and motivational factors in order to achieve goals in a domain-specific activity [1, 2]. According to Aleven [1], help-seeking is defined "[...] as episodes in which a learner, in the context of a specific learning activity [...] takes the initiative to seek assistance from a source within or outside of the learning environment, as opposed to persisting at trying to make progress independently" (p. 312). Help-seeking can be an instrumental (adaptive) strategy, in which learners seek to learn with understanding, or executive (maladaptive), when learners seek prompted answers without developing understanding [1, 3].

Previous research shows that help-seeking has been studied within the math domain [4–11]. Du and colleagues [12] showed that help-seeking was positively

IntechOpen

associated with math homework interest at the individual and class levels in eighth graders. The role of peers is a significant predictor of help-seeking in middle and high school students [11, 13–16]. Perfectionism and feedback were also positively related to help-seeking behaviors [4, 10, 17, 18]. Adaptative help-seeking behaviors significantly predicted mastery-approach goals in middle and high school students [13, 19]. In addition, perceived parent achievement goals predicted students' help-seeking and help-avoidance behaviors in middle school students' achievement goals [6]. Regarding gender, female students are more likely to demonstrate adaptative help-seeking behaviors, whereas boys are more likely to exhibit executive help-seeking behaviors [8, 15, 18, 20].

Help is a typical feature in intelligent tutoring systems (ITSs), and it is usually provided at students' request through the form of hints, internal help-seeking sources, or switches, and external help-seeking sources such as glossaries or hyperlinks. Previous research showed that high school students in a math help forum adopted executive help-seeking behaviors [9]. Best and colleagues [21] demonstrated that social support was increased and stigma was reduced by adolescent males seeking help online. Roll and colleagues [7] showed that high school students asking for help in online Geometry Cognitive Tutor had more productive learning, whereas those who overused it had poor learning. Contrary to previous research, avoiding help was associated with better learning. Otieno and colleagues [5] showed that hints and glossaries in the Cognitive Tutor Geometry differed from the answers in the questionnaire they gave to high school students, yet with a high prediction of learning. Roll and colleagues [4] showed that high school students were able to transfer their learning from a Geometry Cognitive Tutor to a new domain-level content.

The decision to seek help is influenced by metacognitive, affective, and social competencies. Affect can be understood "[...] as a generic term that includes emotions, mood, feelings, attitudes, etc." (p. 65) [22]. Students are likely to experience boredom, confusion, frustration, happiness, and engagement in intelligent learning environments (ILE) because they are affectively charged experiences [23]. Previous research shows that affect has been studied within the math domain. Positive academic emotions such as engagement, enjoyment, joy, contentment, and hope have been associated with adaptative behavior and lower math difficulties. However, negative academic emotions such as anxiety, confusion, frustration, shame, and hopelessness have been associated with executive behaviors and higher math difficulties [24–30]. Academic emotions and learning motivation were also studied in which students were more motivated when they exhibited positive academic emotions such as hope, pride, and enjoyment [31–33]. In addition, students exhibiting higher enjoyment and lower boredom had higher subsequent achievement, but pride was negatively correlated with achievement [29, 34].

Affect influences learning outcomes in face-to-face and mediated learning environments. Previous research has studied the effect of ITS on affect [33, 35–43]. In addition, an ITS provided transition from negative to positive affective states in high school students, and it improved their frustration and confidence by using online learning companions [24]. San Pedro and colleagues [25] showed that high school students who felt more engaged made more careless mistakes and those who felt confused or bored made fewer careless mistakes. Pre-school children demonstrated engagement and valence behaviors to nonverbal behaviors from the social robot tutor [44]. Padrón-Rivera and colleagues [28] showed in a math action units study that high school students experiencing confusion had better learning than those experiencing frustration.

However, to the best of our knowledge, no research was found about the influence of affect in students' help-seeking behaviors in math ITSs until the present date. Thus, more research is needed on how students' affect and motivation may influence learners' decisions to seek or not to seek help in ITSs [45]. According to D'Mello [23], there is still the need to understand "[...] how affective states arise, morph, decay, and impact learning outcomes [...]" (p. 2).

The purpose of this study was to examine affect and its implications for help-seeking as an SRL process. According to D'Mello [23], "Affective impacts of technologically infused learning environments are not very well understood" (p. 2). In addition, Karabenick and Gonida [3] acknowledged the need for additional research to better understand help-seeking as an SRL process. We aimed to investigate how students' academic emotions correlated with help-seeking within one task and if affect was a significant predictor of help-seeking on a subsequent task. Thereafter, we examined the effects of help-seeking on performance and the ways affect moderates this relationship. We attempted to answer the following research questions: (1) does students' affect after learning correlate with their help-seeking behavior within a unit of math problem-solving? (2) Is students' affect reported in one unit significant predictor of help-seeking in the next? (3) How do students' help-seeking behaviors and affect relate to their performance in a consecutive unit of math problem-solving tasks? We also hypothesized that, first, students presenting academic emotions such as happiness and engagement are less likely to switch to external help-seeking sources and make errors, and they are more likely to use hints as an internal help-seeking source and respond correctly to the questions. Secondly, students presenting academic emotions such as boredom, confusion, and frustration are less likely to use hints and respond correctly to the questions, and they are more likely to switch to external help-seeking sources and make errors.

2. Method

2.1 Participants

One hundred and ten students from math courses at a suburban, public middle school participated in the study. Students were 50% male and 50% female, and 96% were Caucasian. 71.8% reported low socioeconomic status (SES), and they were eligible for a free or reduced lunch from the school. Students were primarily enrolled in the seventh grade, but there were students from the eighth grade. Participation occurred in two math periods per week (45–50 minutes each) across the whole academic year as part of regular classroom activities through online questionnaires about math exercises, their academic emotions, and motivation using the Cognitive Tutor Bridge to Algebra [46].

2.2 Materials

The Cognitive Tutor Bridge to Algebra [47] collected fine-grained traced data about students' help-seeking behaviors in pre-algebra exercises in 54 units being units 4, 5, 14, and 15 used for this study, and it presented online questionnaires with a Likert scale ranging from 1 to 5 for students to report their academic emotions starting with boredom, confusion, frustration, happiness, and engagement, respectively [2, 48]. The unit topics comprised "Lowest Common Multiple" in unit 4,

"Greatest Common Factor" in unit 5, "Converting Fractions to Decimals" in unit 14, and "Decimal Operations" in unit 15. Units 4, 5, 14, and 15 were chosen because they represented the most challenging ones for students in which the difficulty was operationalized by time per unit and time per problem.

2.3 Procedures

This study focused on unit and problem level assessments employing a microgenetic and longitudinal approach. In the microgenetic approach, transactions were examined at unit and problem levels to understand the relationship between help-seeking, affect, and performance [2]. Researching at the transaction level allows to correlate help-seeking tools to affect and how the latter may predict the former in consecutive units. With longitudinal and microgenetic data, it is possible to determine if there is a correlation between help-seeking and academic emotions and how this correlation changes when it is investigated at unit and problem levels [2, 48].

In order to investigate the unit and problem level assessments, the Cognitive Tutor Bridge to Algebra uses the model of desired help-seeking behavior developed by Aleven and colleagues [49]. Briefly, students should start by thinking about the problem. If the problem is familiar and they know what to do, students make an attempt. If the problem is familiar but they do not know what to do, students switch to use the glossary to learn more about the problem. If they know what to do after consulting the glossary, students make an attempt. If not, they may either use the glossary again or ask for a hint. If the problem is not familiar, students ask for a hint, spend time reading it, and decide if the hint is helpful. If yes, they make an attempt. If not, students ask for another hint. If students make a correct attempt, the problem is complete. If not, students may require another hint and/or make a new attempt. In this study, hints reflect the total number of hints requested in raw terms. Hints per step normalize these requests to the amount of problem steps students had to complete. Steps with hints normalize the hint requests to account for steps in which students ask for one or more hints. Hint to error ratio reflects the proportion of times students opted to use a hint, rather than to make an attempt at a step that ultimately yielded an incorrect answer. Switch reflects the total number of glossary requests in raw terms [45, 49].

3. Results

Analyses focused on correlations between students' help-seeking behaviors and their affect in units 4 and 14. Affect was also used to predict help-seeking behaviors in the next units (e.g., 5 and 15). Moreover, we examined how associations between help-seeking behaviors and performance differed as a function of affect. The alpha level was set at .05 for simple correlations and multiple regressions.

Missing data were observed on 28 of the 34 variables examined. In unit 4, 20.9% of the data were missing on the boredom variable, 20% of data were missing on confusion, 19.1% of data were missing on frustration, happiness, engagement, hints, hints per step, steps with hint requests, and hint to error. In unit 5, 19.1% of data were missing on performance, hints, hints per step, steps with hint requests, and hint to error. In unit 14, 18.2% of data were missing on boredom, confusion, and frustration, 20% of data were missing on happiness and engagement, 12.7% of data were missing on hints, hints per step, steps with hint requests, and hint to error. In unit 15, 11.8% of data were missing on performance, hints, hints per step, steps with hint requests, and hint to error.

Rubin [50] proposed a taxonomy of missing data mechanisms to explain why the choice of missing data method impacts model parameters. The missing data were found to be consistent with a missing completely at random (MCAR) mechanism as determined by Little's [51] omnibus MCAR test, d2(199) = 213.907, p = .223. Therefore, listwise deletion (N = 97) was used to handle missingness for the present research problems because it is in accordance with the MCAR mechanism and it does not bias the parameter estimates generated by the correlations and multiple regression analyses. **Table 1** shows the descriptive statistics for affect, help-seeking behaviors, and performance in units 4, 5, 14, and 15. Happiness and engagement had the highest coefficients in units 4 and 14. Hints also had the highest coefficients with the exception of unit 5 in which switch exceeded hints. This may be due to the highest level of difficulty of the unit. Moreover, it demonstrated that students favored the usage of hints as an adaptative help-seeking behavior.

Table 2 answers the first research question if students' affect after learning correlates with their help-seeking behavior within a unit of math problem-solving. It shows correlations between affect components and help-seeking behaviors in units 4 and 14. In contrast to hypotheses one and two and the assumptions by Karabenick [52], a positive moderate correlation of .213 between hints and boredom was found in unit 4. Students feeling boredom were more likely to use hints in unit 4 as an instrumental help-seeking behavior. However, this correlation was nonsignificant in unit 14. Negative moderate correlations of −.218 and − .209 between happiness and hints per step and between happiness and steps with hints requests, respectively, were found in unit 14. They are also in contrast with hypotheses one and two. Students feeling happy were less likely to use hints per step and steps with hints requested in unit 14. It might indicate an executive type of help-seeking behavior. No other correlations were significant in units 4 and 14.

	Unit 4		Unit 5		Unit 14		Unit 15	
	M	SD	M	SD	M	SD	M	SD
Affect								
Boredom	4.70	1.84	—	—	4.94	1.76	—	—
Confusion	4.08	1.97	—	—	4.80	1.97	—	—
Frustration	3.81	2.19	—	—	4.54	2.01	—	—
Happiness	5.83	1.19	—	—	5.84	1.25	—	—
Engagement	5.43	1.48	—	—	5.45	1.47	—	—
help-seeking behaviors								
Switch	0.54	1.94	19.25	14.35	3.72	5.69	0.88	1.96
Hints	6.34	11.45	6.34	11.45	39.12	77.94	11.03	13.02
Hints per step	0.22	0.04	0.22	0.04	0.13	0.14	0.03	0.03
Steps with hints requested	274.82	115.77	274.82	115.77	227.8	110.8	332.62	113.11
Hint to error	0.18	0.24	0.18	0.24	0.77	0.77	0.26	0.23
Performance			0.94	0.03			0.93	0.03

Note. Listwise N = 110.

Table 1.
Descriptive statistics for help-seeking behaviors and affect in units.

Variable	Switch	Hints	Hints per step	Steps with hints requested	Hint to error
Unit 4					
Boredom	0.097	.213*	0.189	0.143	0.094
Confusion	0.034	0.186	0.164	0.102	0.017
Frustration	0.107	0.065	0.109	0.021	−0.051
Happiness	−0.182	−0.064	−0.089	0.012	−0.038
Engagement	−0.110	−0.022	−0.008	−0.039	−0.029
Unit 14					
Boredom	−0.062	−0.025	−0.001	−0.031	−0.046
Confusion	−0.044	−0.043	0.000	0.027	−0.042
Frustration	−0.001	0.033	0.072	0.09	0.053
Happiness	−0.146	−0.197	−.218*	−.209*	−0.186
Engagement	−0.039	−0.108	−0.137	−0.09	−0.128

*Note. Listwise N = 87. *p = 0.05.*

Table 2.
Correlations between affect components and help-seeking behaviors in units.

Multiple linear regression analyses with ordinary least squares estimations were conducted to answer the second research question. It was investigated if students' affect (boredom, confusion, frustration, happiness, and engagement) reported in units 4 and 14 were significant predictors of help-seeking behaviors (hints, hints per step, steps with hints requested, hint to error ratio, and switch) in the next units 5 and 15, respectively (**Table 3**).

The first significant regression equation was found when predicting hints in unit 5 from affect components in unit 4 ($F(5, 80) = 2.768$, $p = .023$, $r2 = .148$). Boredom reported in unit 4 was a statistically significant predictor of the total number of hints requested in unit 5 ($b = 3.344$, $SE = 1.139$, $p = .004$, 95% CI = 1.077, 5.610; $\beta = .406$), such that, for each unit increase in boredom, total hint requests were predicted to increase by 3.344. Frustration reported in unit 4 was also found to be a statistically significant predictor of hints in unit 5 ($b = -2.134$, $SE = 1.050$, $p = .045$, 95% CI = −4.224, −.044; $\beta = -.304$). For each unit increase in frustration, students were predicted to request 2.134 fewer hints. Predicting hints per step in unit 5 from boredom in unit 4 yielded another significant regression equation ($F(1,85) = 5.101$, $p = .026$, $r2 = .057$). Boredom reported in unit 4 was a statistically significant predictor of hints per step in unit 5 ($b = .002$, $SE = .001$, $p = .026$, 95% CI = .000, .005; $\beta = .238$), such that, for each unit increase in boredom, hints per step were predicted to increase by .002 points. Models regressing steps with hints, hint to error ratio, and switch on reported affect components were not statistically significant, $F(5,80) = 1.301$, $p = .272$, $r2 = .075$. Collectively, a pattern emerged where boredom increased tendency toward help-seeking, while frustration led to help avoidance. It demonstrates that frustration was the only significant predictor that corroborates with hypothesis two but not with hypothesis one, in which students feeling frustrated predicted to be less likely to use hints and respond correctly to questions in the subsequent unit. We next examined the stability of relations in units 14 and 15.

Variable	*F*	Df	p	R^2	b	SE	β	p
Unit 5 from unit 4								
Switch	1.301	5,80	0.272	0.075				
From boredom					1.194	0.952	0.181	0.214
From confusion					1.425	0.970	0.228	0.146
From frustration					−1.462	0.878	−0.260	0.100
From happiness					0.318	1.238	0.031	0.798
From engagement					−1.086	1.012	−0.132	0.287
Hints	2.768	5,80	0.023	0.148				
From boredom					3.344	1.139	0.406	0.004*
From confusion					0.598	1.161	0.077	0.608
From frustration					−2.134,	1.050	−0.304	0.045*
From happiness					1.622	1.48	0.127	0.276
From engagement					−2.253	1.211	−0.22	0.067
Hints per step	5.101	1.85	0.026	0.057				
From boredom					0.002	0.001	0.238	0.026*
From confusion					−0.001	0.001	−0.077	0.618
From frustration					0.000	0.001	−0.053	0.730
From happiness					0.002	0.002	0.096	0.425
From engagement					−0.002	0.002	−0.174	0.157
Steps with hints requested	0.955	5,80	0.450	0.056				
From boredom					11.984	24.336	0.072	0.624
From confusion					36.745	24.803	0.232	0.142
From frustration					−40.521	22.436	−0.285	0.075
From happiness					−11.455	31.631	−0.044	0.718
From engagement					−14.937	25.879	−0.072	0.565
Hints to error	1.274	5,80	0.283	0.074				
From boredom					0.039	0.019	0.300	0.041
From confusion					−0.021	0.019	−0.171	0.273
From frustration					0.006	0.017	0.052	0.738
From happiness					0.010	0.024	0.052	0.670
From engagement					−0.024	0.020	−0.150	0.227
Unit 15 from unit 14								
Switch	0.087	5,81	0.994	0.005				
From boredom					−0.016	0.205	−0.013	0.937
From confusion					−0.125	0.252	−0.116	0.621
From frustration					0.103	0.239	0.095	0.667
From happiness					0.072	0.239	0.042	0.764
From engagement					−0.032	0.206	−0.022	0.877

Variable	*F*	Df	p	R^2	b	SE	β	p
Hints	2.639	5,81	0.029	0.140				
From boredom					−1.735	1.133	−0.239	0.13
From confusion					0.135	1.392	0.021	0.23
From frustration					2.679	1.324	0.414	0.046*
From happiness					−1.326	1.321	−0.13	0.319
From engagement					−0.282	1.139	−0.033	0.805
Hints per step	3.018	5,81	0.015	0.157				
From boredom					−0.006	0.003	−0.32	0.041*
From confusion					0.001	0.003	0.078	0.720
From frustration					0.007	0.003	0.430	0.037*
From happiness					−0.002	0.003	−0.098	0.447
From engagement					−0.001	0.003	−0.039	0.764
Steps with hints requested	0.855	5,81	0.515	0.05				
From boredom					−5.017	10.788	−0.076	0.643
From confusion					3.996	13.255	0.069	0.764
From frustration					9.503	12.612	0.162	0.453
From happiness					−11.066	12.581	−0.12	0.382
From engagement					1.459	10.844	0.019	0.893
Hints to error	4.483	5,81	0.001	0.217				
From boredom					−0.075	0.019	−0.576	<.001*
From confusion					0.013	0.024	0.11	0.598
From frustration					0.059	0.023	0.506	0.011*
From happiness					−0.001	0.023	−0.005	0.965
From engagement					−0.004	0.02	−0.026	0.836

**p < or = .05.*

Table 3.
Multiple linear regression predicting help-seeking behaviors from affect.

When predicting hints in unit 15 from frustration in unit 14, a significant regression equation was found (F(5,81) = 2.639, p = .029, r2 = .140). Frustration reported in unit 14 was a statistically significant predictor of hints in unit 15 (b = 2.679, SE = 1.324, p = .046, 95% CI = .044, 5.314; β = .414), such that, for each unit increase in frustration, hints were predicted to increase by 2.679 points. Another significant regression equation was found (F(5,81) = 3.018, p = .015, r2 = .157) when predicting hints per step in unit 15 from affect components in unit 14. Boredom reported in unit 14 was a statistically significant predictor of hints per step in unit 15 (b = −.006, SE = .003, p = .041, 95% CI = −.011, .000; β = −.320). For each unit increase in boredom, hints per step were predicted to decrease by .006 points. Frustration reported in unit 14 was also found to be a statistically significant predictor of hints per step in unit 15 (b = .007, SE = .003, p = .037, 95% CI = .000, .013; β = .430), such that, for each unit increase in frustration, hints per step were predicted to increase by .007 points. The final significant regression equation was found ($F(5,81) = 4.483$, $p = .001$, $r2 = .217$) when predicting hint to error in unit 15 from affect components in unit 14. Boredom reported in unit 14 was a statistically significant predictor of hint to error in unit 15 ($b = -.075$, $SE = .019$, $p < .001$, 95% CI = −.114, −.037; $\beta = -.576$), such that, for each unit increase in boredom, hint to error were predicted to decrease by .075 points. Frustration reported in unit 14 was also found to be a statistically significant predictor of hint to error in unit 15 ($b = .059$, $SE = .023$, $p = .011$, 95% CI = .014, .104; $\beta = .506$), such that, for each unit increase in frustration, students' use of hints rather than to make an attempt at a step that resulted in an error was predicted to increase by .059 points. The pattern that emerged was flipped in units 14 and 15 in which frustration increased the tendency toward help-seeking, while boredom led to help avoidance. Contradicting hypotheses one and two, students feeling frustrated were predicted to be more likely to use hints and respond correctly to questions in the subsequent unit. On the contrary and in accordance with hypothesis two, students feeling bored were slightly less likely to use hints and respond correctly to questions in the subsequent unit. Models regressing steps with hints and switch on reported affect components were not statistically significant ($F(5,81) = .855$, $p = .515$, $r2 = .05$). All models appear in **Table 3**.

We next examined the third research question on whether the association between help-seeking behaviors and academic performance differs as a function of the affect. Prior to conducting substantive analyses, help-seeking behaviors were centered, and the help-seeking behaviors-by-affect components product term was generated via multiplication of centered help-seeking behaviors with the affect components. The continuous predictors, as well as their product terms, were entered into simultaneous regression models. Significant and nonsignificant regression equations were found, but only significant ones are presented in **Table 4**. There were no significant associations between switch as a help-seeking behavior and academic performance differing as a function of affect. It suggests that the interaction between the switch and affect was not a significant predictor of performance in subsequent units.

Significant interaction effects were found in multiple units. In unit 4, a hint × frustration interaction effect indicates that every unit increase in frustration led to a slight but statistically significant additional effect where hint use conducted while frustrated predicted an even poorer subsequent performance (e.g., as measured by the percent of first attempts at problem steps) over and above the negative effect of hint use as a help-seeking method. A hints per step × frustration interaction effect indicates that every unit increase in frustration led to a slight but statistically significant additional effect, where hints per step predicted an increase of .064 units on subsequent performance after help-seeking.

In unit 14, a hints per step × boredom interaction effect indicates that every unit increase in boredom led to a slight but statistically significant additional effect, where hints per step predicted an increase of .036 units on subsequent performance. A hints per step × confusion interaction effect indicates that every unit increase in confusion led to a slight but statistically significant additional effect, where hints per step predicted an increase of .030 units on subsequent performance. A hints per step × frustration interaction effect indicates that every unit increase in frustration led to a slight but statistically significant additional effect, where hints per step predicted an increase of .039 units on subsequent performance. A hint to error × boredom interaction effect indicates that every unit increase in boredom led to a slight but statistically significant

Variable	b	SE	β	t	p	*F*	df	p	R^2
Unit 4 predicting performance in unit 5									
Constant	.946	.006	–	170.610	<.001	3.862	(3,84)	.012	121
Hints	–.001	.000	–.560	–2.783	.007				
Frustration	–.002	.001	–.181	–1.766	.081				
Hints × frustration	.000	.000	.420	2.086	.040				
Constant	.945	.006	–	167.640	<.001	2.969	(3,84)	.036	096
Hints per step	–.403	.172	–.553	–2.349	.021				
Frustration	–.002	.001	–.178	–1.709	.091				
Hints per step X frustration	.064	.031	.484	2.056	.043				
Unit 14 predicting performance in unit 15									
Constant	.947	.009	–	109.715	<.001	5.652	(3,86)	.001	165
Hints per step	–.225	.073	–1.076	–3.080	.003				
Confusion	–.003	.002	–.156	–1.549	.125				
Hints per step X boredom	.036	.014	.869	2.485	.015				
Constant	.945	.008	–	124.354	<.001	5.961	(3,86)	.001	172
Hints per step	–.186	.055	–.887	–3.380	.001				
Confusion	–.002	.001	–.150	–1.489	.140				
Hints per step X confusion	.030	.011	.693	2.633	.010				
Constant	.946	.007	–	136.648	<.001	7.732	(3,86)	<.001	212
Hints per step	–.223	.059	–1.073	–3.772	<.001				
Frustration	–.003	.001	–.181	–1.838	.070				
Hints per step X frustration	.039	.012	.906	3.189	.002				
Constant	.950	.009	–	110.651	<.001	5.242	(3,86)	.002	155
Hints to error	–.041	.016	–1.113	–2.569	.012				
Boredom	–.003	.002	–.189	–1.882	.030				
Hints to error X boredom	.007	.004	.875	2.016	.047				
Constant	.946	.007	–	135.838	<.001	7.213	(3,86)	<.001	201
Hints to error	–.046	.014	–1.270	–3.385	.001				
Frustration	–.003	.001	–.192	–1.933	.056				
Hints to error X frustration	.009	.003	1.074	2.866	.005				

Table 4.
Interaction effect between help-seeking behaviors and affect in subsequent performance unit.

additional effect, where greater use of hints compared to making errors predicted an increase of .007 units on subsequent performance. A hint to error × frustration interaction effect indicates that every unit increase in frustration led to a slight but statistically significant additional effect, where hint to error predicted an increase of .009 units on

Variable	Unit 4	Unit 5	Unit 14	Unit 15	Total
Switch users	17	88	52	28	185
Non-switch users	93	22	58	82	255

Table 5.
Use of switch as a help-seeking behavior in units 4, 5, 14, and 15.

subsequent performance. These patterns contradicted hypotheses one and two in which students feeling bored, confused, and frustrated combined with the usage of hints predicted additional performance in subsequent units.

Subsequently, we examined how many students used switch as a help-seeking behavior in units 4, 5, 14, and 15. **Table 5** shows that students used more switch in unit 4. However, there were more students who overall preferred not to use switch across the four units analyzed in this study. It corroborates with descriptive findings from **Table 1** in which hints had the highest coefficients yielding students' preference.

Eventually, a series of independent t-tests were conducted in order to compare affect components including boredom, confusion, frustration, happiness, and engagement between students who used switch and those who did not use it as a help-seeking behavior. There were no systematic differences between cases with switch users and nonswitch users, according to **Table 6**. The group of students using switch did not have a significantly higher number of affect scores than the group of nonswitch users. These results suggest that switch does not have an effect on affect components such as boredom, confusion, frustration, happiness, and engagement. It contradicts both hypotheses stated in this article. Students presenting academic emotions such as happiness and engagement are less likely to switch to external help-seeking sources, whereas students presenting academic emotions such as boredom, confusion, and frustration are more likely to switch to external help-seeking sources.

Variables	N	M	SD	*t*	p
Unit 4					
Boredom				$t(85)= -1.192$	.236
Switch users	17	5.18	2.128		
Non-switch users	70	4.59	1.757		
Confusion				$t(87) = -.637$	.526
Switch users	17	4.35	2.290		
Non-switch users	72	4.01	1.895		
Frustration				$t(86) = -.773$	.442
Switch users	17	4.18	2.128		
Non-switch users	71	3.72	2.212		
Happiness				$t(87) = .256$	.799
Switch users	17	5.76	1.147		
Non-switch users	72	5.85	1.206		

Variables	N	M	SD	*t*	p
Engagement				$t(87) = -.134$	.894
Switch users	17	5.47	1.663		
Non-switch users	72	5.42	1.451		
Unit 14					
Boredom				$t(88) = .046$	.963
Switch users	47	4.94	1.686		
Non-switch users	43	4.95	1.851		
Confusion				$t(88) = .384$	.702
Switch users	47	4.72	1.942		
Non-switch users	43	4.88	2.014		
Frustration				$t(88) = .537$	.593
Switch users	48	4.44	1.945		
Non-switch users	42	4.67	2.103		
Happiness				$t(86) = .600$	.550
Switch users	47	5.77	1.255		
Non-switch users	41	5.93	1.253		
Engagement				$t(86) = .131$	.896
Switch users	46	5.43	1.241		
Non-switch users	42	5.48	1.700		

$^{*}p$ < or = .05.
This statistically significant result means that no effect was observed.

Table 6.
T-tests comparing switch as a help-seeking behavior with affect in units 4 and 14.

4. Discussion

This study was a response to Karabenick and Gonida's [3] call for additional research on help-seeking as a self-regulatory process, and it also examined the ways affect influenced their help-seeking intentions, behaviors, and performances within an ITS context. Students' affect—boredom, confusion, frustration, happiness, and engagement—related to students' concurrent problem-solving and help-seeking behaviors, predicted future help-seeking and moderated performance.

Students were more likely to use hints in unit 4 when they felt bored. This may indicate a trace of disengagement from the task and executive help-seeking behavior [52]. It contradicted hypothesis two that students feeling bored are less likely to use hints. On the other hand, students who reported they felt more positive were less likely to select multiple hints per step and rely on hints across more steps in unit 14. It does not corroborate with hypothesis one in which students feeling happy were thought to be more likely to use hints and respond correctly to the questions. According to Karabenick's [52] description, this may instead be construed to reflect engagement in the task and trace executive help-seeking behaviors.

When predicting future help-seeking behaviors from students' affect reported in previous units, results showed that students feeling bored in unit 4 were more likely to use hints and hints per step in unit 5 contradicting hypothesis two. On the other hand, students feeling frustrated were less likely to use hints in the subsequent unit which corroborates with hypothesis two. In unit 14, this relationship flipped and those who felt frustrated were more likely to use hints (overall, per step, and compared to making errors) in unit 15. Those who felt bored were less likely to use hints in the subsequent unit also corroborating with hypothesis two. This may indicate executive help-seeking behaviors because students were trying to keep engaged in the task by seeking help through hints while feeling bored or frustrated which contradicted the prior hypothesis. There is a need for future research to investigate the reasons why this relationship flipped from one unit to the other. Researchers should investigate more units and more students across time which was one of the limitations of this study. We hypothesized that it may be due to the level of difficulty of the unit in which Decimal Operations in unit 14 may have led to a greater frustration than Greatest Common Factor in unit 4.

Aleven and colleagues [45] acknowledged the need to investigate whether help-seeking as an SRL component could lead to improved students' performance in later units on ITSs. Thus, we examined whether the association between help-seeking behaviors and academic performance differs as a function of affect. In unit 4, the usage of hints when students felt greater frustration predicted an even poorer subsequent performance in unit 5 than hint usage alone. This may indicate an instrumental help-seeking behavior in unit 4 which slightly undermined students' performance in unit 5 due to the effect of frustration. On the contrary, the interaction between hints per step and frustration led to a slight increase in unit 5 performance. This may indicate an executive help-seeking behavior in unit 4 which slightly increased students' performance in unit 5 due to the effect of frustration. In unit 14, the association between hints per step and affect (boredom, confusion, and frustration) and between hint to error and affect (boredom and confusion) led to a slight increase in students' performance in unit 15. These results contradicted hypotheses one and two in which students feeling happy or engaged should be more likely to use hints, but no significant interactions were found. Additionally, students feeling bored, confused, and frustrated were more likely to use hints, and it led to subsequent performance. This may also indicate an executive help-seeking behavior in which students are trying to keep engaged in the task perhaps without developing much understanding in the present unit so they can progress to the next one.

We also examined how many students used switch as a help-seeking behavior in units 4, 5, 14, and 15, and we investigated if switch had an effect on affect components such as boredom, confusion, frustration, happiness, and engagement. The results showed that students used more switch only in unit 4, Greatest Common Factor. It also demonstrated that switch did not yield a significant effect on affect components. We hypothesized that it may be because students may consult other sources even if they know the answer to reach full certainty. It could also be because they were not aware of how helpful this resource could be in an ITS. There is also the need to investigate switches, as an external help-seeking source, in relation to other internal help-seeking sources and affect components.

5. Conclusion

These findings contribute to the growing body of SRL literature because it demonstrates new ways students are regulating their actions and seeking for help in ITSs in relation to affect. The results from the present article showed that a pattern emerged in which students showed more executive help-seeking behaviors rather than instrumental ones. Students feeling bored, confused, and frustrated tended to use more hints, and they were less likely to switch to external help-seeking sources. Also, those feeling happy or engaged were less likely to use hints. It may be because students were feeling disengaged from the task or because of the level of difficulty of the task in the ITS. There is a need to research more units across time and investigate if these patterns hold.

Future research may assess students' affective states which could shed light on the affective conditions that lead students to use hints and to profit from hint use. These results suggest a complex relationship between affect and hint use where affect explains variance in the tendency to use hints, but the directions of affect sometimes contradict the prior theory [52].

Moreover, future research may also focus on students' help-seeking behaviors and affect in relation to their self-efficacy beliefs in ITSs. Efficacy beliefs influence peoples' thinking process, goals, effort, commitment, outcomes, resilience, and how much stress or depression they tend to feel when performing a task [53]. Students' usage of hints may be related to their sense of efficacy, and it may be a product of hint usage across the span of different units.

Acknowledgements

The authors gratefully acknowledge the support of the National Science Foundation (NSF) that funded this research. The collaboration of Steve Ritter and Tristan Nixon at Carnegie Learning is also especially appreciated regarding their assistance in adapting the Cognitive Tutor to include the instruments used in this study.

Declaration of interest

The authors declare no conflict of interest.

Author details

Ana Paula S. Loures-Elias[1*] and Matthew L. Bernacki[2]

1 The Loures-Elias Educational Psychology Lab, Lisbon, Portugal

2 University of North Carolina at Chapel Hill, School of Education, The United States of America

*Address all correspondence to: anapaula.loureselias@gmail.com

References

[1] Aleven V. Help seeking and intelligent tutoring systems: Theoretical perspectives and a step towards theoretical integration. In: Aleven V, Azevedo R, editors. International Handbook of Metacognition and Learning. New York: Springer; 2013. pp. 311-335

[2] Bernacki ML, Nokes-Malach TJ, Aleven V. Fine-grained assessment of motivation over long periods of learning with an intelligent tutoring system: Methodology. advantages, and preliminary results. In: Aleven V, Azevedo R, editors. International Handbook of Metacognition and Learning. New York: Springer; 2013. pp. 629-644

[3] Karabenick SA, Gonida EN. Academic help seeking as a self-regulated learning strategy. In: Schunk DH, Greene JA, editors. Handbook of Self-Regulation of Learning and Performance. New York: Routledge; 2018. pp. 421-433

[4] Roll I, Aleven V, McLaren BM, Koedinger KR. Improving students' help-seeking skills using metacognitive feedback in an intelligent tutoring system. Learning and Instruction. 2011;**21**(2):267-280. DOI: 10.1016/j.learninstruc.2010.07.004

[5] Otieno C, Schwonke R, Salden R, Renkl A. Can help seeking behavior in intelligent tutoring systems be used as online measure for goal orientation? Cognitive Science. 2013;**35**:1103-1108

[6] Gonida EN, Karabenick SA, Makara KA, Hatzikyriakou GA. Perceived parent goals and student goal orientations as predictors of seeking or not seeking help: Does age matter? Learning and Instruction. 2014;**33**:120-130. DOI: 10.1016/j.learninstruc.2014.04.007

[7] Roll I, Baker RSJD, Aleven V, Koedinger KR. On the benefits of seeking (and avoiding) help in online problem-solving environments. Journal of the Learning Sciences. 2014;**23**(4):537-560. DOI: 10.1080/10508406.2014.883977

[8] Schenke K, Lam AC, Conley AMM, Karabenick SA. Adolescents' help seeking in mathematics classrooms: Relations between achievement and perceived classroom environmental influences over one school year. Contemporary Educational Psychology. 2015;**41**:133-146. DOI: 10.1016/j.cedpsych.2015.01.003

[9] Puustinen M, Bernicot J, Volckaert-Legrier O, Baker M. Naturally occurring help-seeking exchanges on a homework help forum. Computers and Education. 2015;**81**:89-101. DOI: 10.1016/j.compedu.2014.09.010

[10] Shim SS, Rubenstein LDV, Drapeau CW. When perfectionism is coupled with low achievement: The effects on academic engagement and help seeking in middle school. Learning and Individual Differences. 2016;**45**:237-244. DOI: 10.1016/j.lindif.2015.12.016

[11] Zander L, Chen IC, Hannover B. Who asks whom for help in mathematics? A sociometric analysis of adolescents' help-seeking within and beyond clique boundaries. Learning and Individual Differences. 2019;**72**:49-58. DOI: 10.1016/j.lindif.2019.03.002

[12] Du J, Xu J, Fan X. Investigating factors that influence students' help seeking in math homework: A multilevel analysis. Learning and

Individual Differences. 2016;**48**:29-35. DOI: 10.1016/j.lindif.2016.03.002

[13] Roussel P, Elliot AJ, Feltman R. The influence of achievement goals and social goals on help-seeking from peers in an academic context. Learning and Instruction. 2011;**21**(3):394-402. DOI: 10.1016/j.learninstruc.2010.05.003

[14] Ryan AM, Shim SS. Changes in help seeking from peers during early adolescence: Associations with changes in achievement and perceptions of teachers. Journal of Educational Psychology. 2012;**104**(4):1122-1134. DOI: 10.1037/a0027696

[15] Kiefer SM, Shim SS. Academic help seeking from peers during adolescence: The role of social goals. Journal of Applied Developmental Psychology. 2016;**42**:80-88. DOI: 10.1016/j.appdev.2015.12.002

[16] Shin H. The role of friends in help-seeking tendencies during early adolescence: Do classroom goal structures moderate selection and influence of friends? Contemporary Educational Psychology. 2018;**53**:135-145. DOI: 10.1016/j.cedpsych.2018.03.002

[17] Fletcher AK. Help seeking: Agentic learners initiating feedback. Educational Review. 2018;**70**(4):389-408. DOI: 10.1080/00131911.2017.1340871

[18] Parker JS, Shum KZ, Suldo SM, Shaunessy-Dedrick EM, Ferron J, Dedrick RF. Predictors of adaptive help seeking across ninth-grade students enrolled in advanced placement and international baccalaureate courses. Psychology in the Schools. 2019;**56**(5):652-669. DOI: 10.1002/pits.22223

[19] Ryan AM, Shin H. Help-seeking tendencies during early adolescence: An examination of motivational correlates and consequences for achievement. Learning and Instruction. 2011;**21**(2):247-256. DOI: 10.1016/j.learninstruc.2010.07.003

[20] Kessels U, Steinmayr R. Macho-man in school: Toward the role of gender role self-concepts and help seeking in school performance. Learning and Individual Differences. 2013;**23**(1):234-240. DOI: 10.1016/j.lindif.2012.09.013

[21] Best P, Gil-Rodriguez E, Manktelow R, Taylor BJ. Seeking help from everyone and no-one: Conceptualizing the online help-seeking process among adolescent males. Qualitative Health Research. 2016;**26**(8):1067-1077. DOI: 10.1177/1049732316648128

[22] Efklides A, Schwartz BL, Brown V. Motivation and affect in self-regulated learning: Does metacognition play a role? In: Schunk DH, Greene JA, editors. Handbook of Self-Regulation of Learning and Performance. New York: Routledge; 2018. pp. 64-82

[23] D'Mello S. A selective metaanalysis on the relative incidence of discrete affective states during learning with technology. Journal of Educational Psychology. 2013;**105**(4):1082-1099

[24] Arroyo I, Woolf BP, Burelson W, Muldner K, Rai D, Tai M. A multimedia adaptive tutoring system for mathematics that addresses cognition, metacognition and affect. International Journal of Artificial Intelligence in Education. 2014;**24**(4):387-426. DOI: 10.1007/s40593-014-0023-y

[25] San Pedro MOZ, Baker RSJD, Rodrigo MMT. Carelessness and affect in an intelligent tutoring system for mathematics. International Journal of Artificial Intelligence in Education.

2014;**24**(2):189-210. DOI: 10.1007/s40593-014-0015-y

[26] Tornare E, Czajkowski NO, Pons F. Children's emotions in math problem solving situations: Contributions of self-concept, metacognitive experiences, and performance. Learning and Instruction. 2015;**39**:88-96. DOI: 10.1016/j.learninstruc.2015.05.011

[27] Ganotice FA, Datu JAD, King RB. Which emotional profiles exhibit the best learning outcomes? A person-centered analysis of students' academic emotions. School Psychology International. 2016;**37**(5):498-518. DOI: 10.1177/0143034316660147

[28] Padrón-Rivera G, Rebolledo-Mendez G, Parra PP, Huerta-Pacheco NS. Identification of action units related to affective states in a tutoring system for mathematics. Educational Technology and Society. 2016;**19**(2):77-86

[29] Villavicencio FT, Bernardo ABI. Beyond math anxiety: Positive emotions predict mathematics achievement, self-regulation, and self-efficacy. Asia-Pacific Education Researcher. 2016;**25**(3):415-422. DOI: 10.1007/s40299-015-0251-4

[30] Sainio PJ, Eklund KM, Ahonen TPS, Kiuru NH. The role of learning difficulties in adolescents' academic emotions and academic achievement. Journal of Learning Disabilities. 2019, 2019;**52**(4):287-298. DOI: 10.1177/0022219419841567

[31] Valinasab S, Zeinali A. The mediational pathway among academic emotions, self-regulated learning, and academic motivation with academic achievement of high school students. International Journal of Educational and Psychological Researches. 2018;**3**(4):255-261. DOI: 10.4103/jepr.jepr_3_17

[32] Westphal A, Kretschmann J, Gronostaj A, Vock M. More enjoyment, less anxiety and boredom: How achievement emotions relate to academic self-concept and teachers' diagnostic skills. Learning and Individual Differences. 2018;**62**:108-117. DOI: 10.1016/j.lindif.2018.01.016

[33] Yu FY, Wu WS, Huang HC. Promoting middle school students' learning motivation and academic emotions via student-created feedback for online student-created multiple-choice questions. Asia-Pacific Education Researcher. 2018;**27**(5):395-408. DOI: 10.1007/s40299-018-0398-x

[34] Putwain DW, Becker S, Symes W, Pekrun R. Reciprocal relations between students' academic enjoyment, boredom, and achievement over time. Learning and Instruction. 2018;**54**:73-81. DOI: 10.1016/j.learninstruc.2017.08.004

[35] Zakharov K, Mitrovic A, JohnstonL.Towardsemotionally-intelligent pedagogical agents. In: Woolf BP, Aïmeur E, Nkambou R, Lajoie S, editors. Proceedings of Intelligent Tutoring Systems, 9th International Conference, ITS 2008. Berlin, Germany: Springer; 2008. pp. 19-28. DOI: 10.1007/978-3-540-69132-7_7

[36] Chaffar S, Derbali L, Frasson C. Inducing positive emotional state in intelligent tutoring systems. In: Dimitrova V, Mizoguchi R, Du Boulay B, Graesser A, editors. Proceedings of the 2009 Conference on Artificial Intelligence in Education: Building Learning System that Care: From Knowledge Representation to Affective Modelling; Amsterdam. Netherlands: IOS Press; 2009. pp. 716-718. Available from: https://ebooks.iospress.nl/publication/5255

[37] Mao X, Li Z. Implementing emotion-based user-aware e-learning. In: Proceedings of the 27th International

Conference Extended Abstracts on Human Factors in Computing Systems. New York, NY: ACM Press; Apr 2009. pp. 3787-3792. Available from: https://dl.acm.org/doi/proceedings/10.1145/3334480

[38] Woolf B, Burleson W, Arroyo I, Dragon T, Cooper D, Picard R. Affect-aware tutors: Recognising and responding to student affect. International Journal of Learning Technology. 2009;**4**(3/4):129-164. DOI: 10.1504/IJLT.2009.028804

[39] Strain A, D'Mello S. Emotion regulation during learning. In: Biswas G, Bull S, Kay J, Mitrovic A, editors. Proceedings of Artificial Intelligence in Education, 15th International Conference, AIED 2011. Berlin, Germany: Springer; 2011. pp. 566-568. DOI: 10.1007/978-3-642-21869-9_103

[40] Rodrigo MMT, Baker RSJD, Agapito J, Nabos J, Repalam MC, Reyes SS, et al. The effects of an interactive software agent on student affective dynamics while using. An intelligent tutoring system. IEEE Transactions on Affective Computing. 2012;**3**(2):224-236. Available from https://ieeexplore.ieee.org/document/6122013

[41] D'Mello S, Graesser A. AutoTutor and affective AutoTutor: Learning by talking with cognitively and emotionally intelligent computers that talk back. ACM Transactions on Interactive Intelligent Systems. 2012;**2**(4):1-39. DOI: 10.1145/2395123.2395128

[42] Tian F, Gao P, Li L, Zhang W, Liang H, Qian Y, et al. Recognizing and regulating e-learners' emotions based on interactive Chinese texts in e-learning systems. Knowledge-Based Systems. 2014;**55**:148-164. DOI: 10.1016/j.knosys.2013.10.019

[43] Malekzadeh M, Mustafa MB, Lahsasna A. A review of emotion regulation in intelligent tutoring systems. Educational Technology and Society. 2015;**18**(4):435-445. Available from: http://www.jstor.org/stable/jeductechsoci.18.4.435

[44] Gordon G, Spaulding S, Korywestlund J, Lee JJ, Plummer L, MartinezM, et al. Affectivepersonalization of a social robot tutor for children's second language skills. In: Proceedings of the 30th AAAI Conference on Artificial Intelligence. Palo Alto, CA: MIT; 2016. pp. 3951-3957. DOI: 10.1609/aaai.v30i1.9914

[45] Aleven V, Roll I, McLaren BM, Koedinger KR. Help helps, but only so much: Research on help seeking with intelligent tutoring systems. International Journal of Artificial Intelligence in Education Society. 2016;**26**(1):205-223. DOI: 10.1007/s40593-015-0089-1

[46] Ritter S, Anderson JR, Koedinger KR, Corbett A. The cognitive tutor: Applied research in mathematics education. Psychonomics Bulletin & Review. 2007;**14**(2):249-255. DOI: 10.3758/BF03194060

[47] Bernacki ML, Nokes-Malach TJ, Aleven V. Examining self-efficacy during learning: Variability and relations to behavior, performance, and learning. Metacognition and Learning. 2015;**10**(1):99-117. DOI: 10.1007/s11409-014-9127-x

[48] Bernacki M, Ritter S. Hopewell 2011-2012 Algebra (Learning Orientation Survey). Dataset 718 in DataShop. 2013. Available from: https://pslcdatashop.web.cmu.edu/DatasetInfo?datasetId=718

[49] Aleven V, McLaren BM, Roll I, Koedinger KR. Toward meta-cognitive

tutoring: A model of help-seeking with a cognitive tutor. International Journal of Artificial Intelligence. 2006;**16**(2):101-130

[50] Rubin DB. Inference and missing data. Biometrika. 1976;**63**(3):581-592. DOI: 10.2307/2335739

[51] Little RJA. A test of missing completely at random for multivariate data with missing values. Journal of the American Statistical Association. 1988;**83**(404):1198-1202. DOI: 10.1080/01621459.1988.10478722

[52] Karabenick SA. Classroom and technology-supported help seeking: The need for converging research paradigms. Learning and Instruction. 2011;**21**(2):290-296. DOI: 10.1016/j.learninstruc.2010.07.007

[53] Bandura A. Exercise of human agency through collective efficacy. Current Directions in Psychological Science. 2000;**9**(3):75-78. Available from: http://www.jstor.org/stable/20182630

Chapter 13

Using ChatGPT to Create Constructively Aligned Assessment Tasks and Criteria in the Context of Higher Education Teaching

Niclas Schaper

Abstract

The chapter deals with the question, how ChatGPT can be used to support and facilitate the planning of competence-oriented written university examinations. Therefore, the possibilities and potentials of ChatGPT in the formulation of learning outcomes, the derivation of examination tasks from the outcomes and the formulation of assessment criteria are examplarily demonstrated and analyzed. First, it is explained what ChatGPT actually is and which aspects should be considered when using it. Then didactic concepts and requirements for designing higher education examinations (especially the constructive alignment concept) are presented, which ChatGPT can support. The focus is on the competence-oriented design of written examinations with open answer formats. Therefore, the use of ChatGPT in the generation of learning outcomes, examination tasks and assessment criteria as well as the critical review and optimization of the drafts will be presented using an example from the domain of work and organizational psychology.

Keywords: ChatGPT, constructive alignment, competence oriented examinations, higher education teaching, written exam tasks, assessment criteria, learning outcomes

1. Introduction

Several authors claim that the fastest impact of ChatGPT in university teaching was seen in the area of examinations after this generative AI was made freely available to the public in November 2022 [1, 2]. In particular, for unproctored written exams in the form of seminar papers, for example, students were already experimenting with how ChatGPT could take over writing tasks in December 2022 [3], and in January the first students were already caught having their papers written by ChatGPT [4]. In a survey of US-American students conducted at the same time, 30% of respondents stated that they were already using ChatGPT for assignments [5]. University teachers also tested relatively early on whether their exams could be solved by ChatGPT, which was confirmed for exams in Business Information Systems [6] and Astrophysics [7]. Naturally, such examples prompted alarmists to think about how to prevent the misuse of chatbots in higher education (cf. e.g., Fleck [8], Gamage et al. [2] or Sullivan et al. [9]). According to a recent survey by Solis [10] at German universities, the use of ChatGPT is therefore completely prohibited in a small number of universities (approx. 2%) and only permitted for certain purposes (e.g., research) in just over a fifth of the universities surveyed.

However, ChatGPT can also be put to good use, for example, as an effective learning aid for students [1, 11]. ChatGPT can, for example, help with reading academic texts (e.g., by asking the bot to summarize texts or explain sections of text that are difficult to understand in more detail) or help with learning (e.g., by asking ChatGPT to extract the most important key terms from a textbook chapter or ask questions about specific course content) or help with writing (e.g., by having ChatGPT suggest outlines or key questions on a topic or give feedback on text passages). University teachers can also use ChatGPT not only for writing academic texts, but also for teaching tasks. For example, texts can be generated from keyword lists or text summaries can be created (e.g., for exercises in seminars). To illustrate teaching content, ChatGPT can help to develop examples and counter-examples. ChatGPT can also be used to test comprehension and application skills in seminars or lectures by creating quiz questions or short tests (see, e.g., Patel [12]). ChatGPT therefore harbors both risks and potential for scientific activities and university teaching. Although, it is understandable to try to avoid or minimize the risks, also the potential and benefits of AI can and should be used for one's own work.

As someone who teaches and designs teaching myself, but also accompanies other teachers in workshops to deal with questions about the design of competence-oriented teaching and in particular corresponding examination formats, I am familiar with how laborious it often is to determine and formulate learning outcomes or competence-oriented learning objectives and to develop teaching processes and learning objective-oriented examination tasks and assessment criteria on this basis. This requires not only subject-specific but also didactic expertise at an advanced level, as well as time and cognitive resources, which are often scarce in the everyday life of an academic. For inexperienced university teachers in particular, such tasks are not only laborious but also associated with uncertainties, doubts and considerable motivational challenges.

This raises the question of whether the design of these didactic components—in particular the learning outcomes and the corresponding examination tasks and assessment criteria—can be meaningfully supported and made more efficient or less time-consuming with the help of AI technologies. If this can actually be realized and ChatGPT can be an effective support for teaching and examination tasks, it would make things much easier for many teachers and offer support for a qualitative improvement of teaching and examinations.

The aim of this chapter is therefore to test and analyze the extent to which the AI tool ChatGPT can be used sensibly to support and facilitate the conceptual tasks and requirements described, especially in the area of examinations. On the one hand, the possibilities and potentials of ChatGPT in the formulation of learning outcomes, the derivation of examination tasks from the outcomes and the formulation of assessment criteria and schemes for the learning goal-oriented evaluation of learning achievements are to be demonstrated and analyzed by way of example. Since generative language models are prone to errors and inaccuracies in their answers, we will also look at how AI answers can be concretized and improved with regard to incorrect, incomplete and too vague information.

2. What is ChatGPT and what should I bear in mind when using it? (hier weiterlesen)

ChatGPT is a ChatBot that not only answers (knowledge) questions, but can also write its own texts (e.g., term papers or speech manuscripts), summarize articles, program or solve mathematical problems. The core of ChatGPT is an AI-supported language model that has been trained to compose texts that correspond to a high degree to human conversation or text contributions with regard to the respective topics addressed. For this purpose, a neural network of the language model was fed with an enormous amount of texts in order to learn the characteristics and content of the linguistic representation. In simple terms, the program's language-generative mechanism works by using a highly complex statistical algorithm to generate one word after another so that they fit together in a meaningful way with a specific context or linguistic requirement.

The language model is also further refined using human feedback. Output generated by the model is manually evaluated by humans. However, this is usually not done by experts but by the so-called "clickworkers" in the form of students or unemployed academics (e.g., from India or Kenya), who evaluate the responses generated by the model with regard to content and linguistic characteristics (e.g., with regard to the use of offensive words) [13]. Clickworkers often only have 2–3 minutes for the content fact check to examine the truth of a statement with a short research. Wikipedia articles (which should also be treated with caution as a source of information) are often used as a benchmark to determine whether a statement made by the chatbot is true or false. In view of the precarious working conditions and the high time pressure under which the clickworkers carry out such checks, the fact checks can be characterized as rather inaccurate and unreliable. They only superficially conceal the susceptibility of the language model to errors. Against this background, the best open source language models achieve an average accuracy of 73% at best [13, 14]. Even ChatGPT 4 only achieves a hit rate of just over 80% for two topics, while the rest fluctuate between 60% and 80%. Overall, it can therefore be assumed that the quality of ChatGPT's responses varies greatly.

The procedure described above attempts to train the language models so that they answer as many questions as possible correctly and helpfully. In principle, however, the program itself cannot check and decide whether its answers are correct or incorrect or good or bad [13, 14]. This applies in particular to answers about complex issues or questions where there are several possible answers. Although a language model such as ChatGPT can calculate an answer, it cannot check how good or correct this answer is. This requirement is therefore left to the user of ChatGPT. In areas where a high degree of accuracy is required and even the smallest errors can cause major damage (e.g., in medicine), other expert systems are needed to provide reliable answers. However, if one is aware of these problems of language models, they can be used in areas where the accuracy of the solution is not that important, but rather the clarification of the basic solution path. In such cases, precision and quality can also be generated and guaranteed by human users themselves. This includes in particular the aspect that the quality of chatbot answers can be significantly improved by the way in which they are prompted (e.g., giving detailed instructions, providing examples or a step-by-step instruction) [15, 16].

A major benefit of ChatGPT, as well as other language models, is the program's ease of use. A simple text interface is used to formulate a question, to which ChatGPT responds directly (usually after a few seconds). However, as already mentioned, the quality and fit of the answers to the questions depends to a large extent on the way in which the questions or prompts are formulated. In general, it is recommended to develop an incremental, dialogic approach, that is, instead of a difficult and complex question in the form of a long prompt, start with simple questions and gradually increase the complexity of the questions [17]. The structure of the prompt should also be based on the ROMANE scheme [18]:

- R stands for role assigned to the AI (who or what is being simulated)
- O stands for the goal of the prompt
- M stands for instructions or assistance (what to do and in what order)
- A stands for application examples or content specifications (what exactly or specifically is it about)
- N stands for useful details (e.g., secondary conditions or restrictions; what should also be taken into account)
- E stands for the recipient of the text (for whom the text is written)

Furthermore, it has proven useful to develop an understanding of the effects of prompts by trying them out for yourself. It is usually also important to take into account that language models such as ChatGPT take into account the previous context or course of the conversation, so that it makes sense to generate answers to a complex task step by step. This means, among other things, that the chat should be restarted when changing or starting a new task. In this context, it can also be helpful to use additional plug-ins such as "Text Blaze" or "Prompt Box" to efficiently save text snippets or prompts and quickly insert them into the input fields of ChatGPT or to extract and summarize texts from videos with the help of "Chrome" [1, 19].

3. Didactic concepts and requirements for competence-oriented examination planning

A competence-oriented planning and design of examination arrangements that are appropriate in terms of the Bologna reform requires that this is based on learning outcomes which are formulated in a curriculum- and target group-oriented way [20, 21]. Corresponding learning outcomes for courses or course units are derived on the one hand from the overarching qualification objectives of the degree program and refer to the competences or sub-competences to be acquired during the course. On the other hand, they describe as specifically as possible the knowledge, skills, and abilities to be acquired in a course or associated teaching units in a defined subject domain, which should contribute to the development of the overarching competencies. On the basis of the learning outcomes, the teaching and learning activities as well as the associated content must then be planned in further steps, which not only aim to acquire knowledge, but also to initiate and practice skills, abilities and professional attitudes. The step of determining and developing the forms and tasks of the examinations is also carried out before or immediately after this. This must be done in a way that they are suitable for observing and recording the knowledge and skills or competencies to be acquired under conditions of action and performance. It also includes the identification and description of assessment criteria and schemes in order to analyze and evaluate the extent of competence acquisition during the examination tasks. These criteria can also be used to provide informative and learning-promoting feedback to the learners. The next step is to check whether these didactic elements are sufficiently related or coordinated, which is referred to as constructive alignment [22]. This concept is the central didactic principle of competence-oriented teaching and examination design. ChatGPT or other generative language models can be used to support the design of the above mentioned didactic elements. Before we come to this, it should be clarified which design requirements these core elements of competence-oriented teaching and examination design should fulfill.

3.1 Definition and formulation of learning outcomes

Learning outcomes describe the competence facets related to the relevant knowledge, skills, abilities and attitudes that have to be acquired in a course or course unit [20]. The outcomes should be formulated as specific as possible so that they can be operationalized for examination purposes. This is done in such a way that the description of objectives makes clear which behavior—especially in terms of cognitive performance—illustrates the application of knowledge or its mastery in certain situations [23]. On the one hand, reference must be made to the subject-specific content of the knowledge and application domain covered in the course, as well as to the tasks and requirements that must be mastered in this context in order to solve certain problems. However, the social-communicative, metacognitive and motivational-affective aspects that additionally motivate, control, and support professional action should not be neglected.

On the other hand, certain linguistic and conceptual principles must be taken into account when formulating learning outcomes in order to ensure that the learning outcomes provide a sufficient basis for further planning steps. The following aspects should be taken into account:

- Learning outcomes should describe actions to be learned, that is, they should refer to both a content component (What does the learning action relate to in terms of content?) and a behavioral component in the form of a verb (What is to be done? What are the requirements for dealing with the knowledge?)

- Learning outcomes should describe concretely observable modes of action so that they can be operationalized for teaching and assessment purposes; the criterion of concreteness and specificity refers both to the behavior to be demonstrated or the required cognitive activities and to the relevant knowledge concepts.

- It is also helpful if, in connection with the learning outcome, it is described which methods, instruments or procedures are required to implement the action to be learned (How is the learning outcome achieved?) and why the action to be learned is needed in a comprehensive sense.

- It is also advisable to use proven learning objective taxonomies to derive and formulate learning objectives in order to create a description of the learning content that is based on learning theory and comprehensible to other participants. The taxonomy by Anderson and Krathwohl [24], which provides categories for the systematization and derivation of cognitive learning objectives or learning outcomes, has become widespread and proven its usefulness in the higher education context. For the formulation, categorization and analysis of cognitive learning objectives, a distinction is made between two dimensions: (1) the process dimensions, with which different cognitive processes in dealing with knowledge (remembering, understanding, applying, analyzing, evaluating, synthesizing/creating) are characterized, and (2) the knowledge dimension, which differentiates between different types of knowledge (factual knowledge, conceptual/contextual knowledge, procedural knowledge, metacognitive knowledge).

3.2 Construction of examination tasks

The next step is to design the examination tasks on the basis of the learning outcomes and the learning content. The learning outcomes can be assessed in different formats, which can also be combined with each other. In principle, a choice or distinction can be made between written, oral and practical examination forms, which in turn can be implemented and methodically designed in a variety of ways [25, 26].

The further discussion is focused on written examinations, especially the more open formats. This is done because of space restrictions and in order to provide the most concrete advice possible for the use of ChatGPT in the generation of examination tasks.

Written examinations are the main examination format for many degree programs. In this context, test performance is achieved by answering or completing written tasks or requirements. With regard to the design of written examinations, a distinction is usually made between strongly pre-structured (e.g., in the form of multiple choice tasks) and more open forms (e.g., based on open questions on a topic) as well as different time periods that are made available (at a defined rather short timeslot in the form of written exams or tests vs. longer timeframes in the form of, e.g., term papers) [26]. The more structured formats are suitable for testing knowledge, whereas more open formats are also able to test more complex performance or competencies. In the more open forms of written examinations, a distinction is also made between restricted tasks (which can be answered with keywords or limited explanations and which relate more to requirements for knowledge reproduction and understanding or simple analysis and application requirements) and tasks that require a more detailed answer in the form of more complex statements, case analyses and solutions or requirements for more extensive conceptual drafts.

When formulating open written tasks, certain principles must also be considered [25, 27]. The tasks should be formulated as clearly, completely and structured as possible; that is, the examinees should understand exactly what is expected of them (e.g., with regard to the focus of the content, the scope and level of detail and the structure of the answers) as well as the criteria that the answers should meet. The questions or assignments should therefore clearly state what kind of knowledge is required to answer them (especially for less demanding learning outcomes) and what cognitive requirements are to be met when answering the tasks (especially for more demanding learning outcomes; e.g., describe, compare, justify, evaluate, develop).

Since the learning outcomes should relate to cognitive requirement levels based on learning taxonomies (e.g., the Anderson and Krathwohl taxonomy [24]), certain task forms and variants must also be taken into account when designing appropriate tasks with reference to a specific requirement level (Schaper [28]). For the "Understanding" level, it is therefore particularly important to construct tasks that require the explanation or clarification of a theory or tasks in which it is necessary to justify why a certain method is particularly suitable for the analysis of certain facts, while for the "Applying" level, tasks such as the transfer of a theory to a specific situation should be set. At the "Analyzing" level, for example, tasks should be constructed in which two or more variants of a concept or case are compared or errors in an application description are to be identified. In connection with requirements at the "Evaluating" level, tasks are to be constructed, for example, in which a solution approach for a specific case is to be evaluated on the basis of various criteria or a case or a topic is to be viewed and evaluated from different perspectives. Finally, for the "Creating" level, tasks are required in which, for example, a complex plan for solving a specific problem must be designed using certain principles or heuristics or a concept or technical solution for a specific task or function must be designed or constructed.

In addition, the tasks must be designed with regard to difficulty-determining aspects (e.g., the number of aspects to be considered, the scope of the text to be analyzed, the contextual relevance of the analysis and the request for additional justifications at the "Analyzing" level). It should also be considered to what extent the tasks should be better illustrated or made more difficult by providing additional documents (e.g., statistics, illustrations, calculation examples, descriptions).

3.3 Determining and designing assessment criteria and schemes

In a further design step, the focus is on the question of how examination performance can be evaluated and assessed. Basically, this step involves comparing the examination performance shown with the expectations that were formulated in advance of the examination with regard to certain requirements [26, 29]. According to the purpose of the examination, the type of performance expected and the examination format, an assessment scheme should be developed on which the assessment is based [30]. An assessment scheme requires the determination of assessment-relevant criteria that can be used for a transparent and objective assessment. In most cases, the individual tasks are presented in tabular form in the assessment scheme with response categories, their definition or characterizing description, anchor examples and the allocation of values (e.g., points) [31, 32]. For each response category, the characteristics that a response must fulfill in order to be assigned to the category are described. The descriptions therefore contain characteristics or indicators that allow different answers to be assigned without any doubt. In some cases, it is advisable to list all possible correct answers, in other cases, especially more complex questions, examiners should provide a generic description of characteristics for assessment purposes. The naming of anchor examples can also facilitate the assignment to a specific assessment category [32].

Furthermore, it is usually necessary to convert the assessments into points and grades; that is, the assessments of the individual aspects of an examination performance must be converted into a numerical value in the form of points. These points must then be aggregated and transformed into a grade using appropriate schemes. This requires that the assessment schemes should also specify how the individual aspects of the examination performance are to be scored and how the point totals are to be converted into grades. For more complex examination requirements (e.g., when assessing case analyses or more complex answer texts), where it makes little sense to refer only to individual aspects of the performance, but rather to assess the solution as a whole on the basis of certain criteria, it is advisable to use so-called rubrics. Rubrics are assessment schemes that consist of a list of competence-related assessment criteria (including a definition of the criterion) and, for each criterion, a description of competence levels that reflect different levels of competence in relation to the criterion (e.g., excellent performance, satisfactory but improvable performance, unsatisfactory performance) [33].

Once the central methodological elements of an examination arrangement have been designed, the final step is to check the fit of the learning outcome, examination task and assessment scheme and adjust them if necessary. This corresponds to the construction alignment concept [22], in which the fit of the teaching/learning activities with the learning outcomes and the examination tasks and requirements must also be checked and ensured. Methodologically, the corresponding fit aspects can be checked with the help of learning objective taxonomies. Anderson and Krathwohl's approach is particularly suitable for this (see Universität Zürich—Arbeitsstelle für Hochschuldidaktik [34]). Essentially, this involves analyzing to which knowledge categories and requirement levels the learning outcome(s), the examination tasks and the assessment criteria and schemes can be assigned. After this step, a comparison is made as to whether the learning outcome, task and assessment criteria can be assigned to the same categories and levels or not (e.g., procedural knowledge and application for both the learning outcome, task and assessment schemes). If this is the case, a successful constructive alignment can be assumed. However, if there are deviating assignments, adjustments or modifications to the three components are necessary in order to achieve or guarantee constructive alignment. In addition, content-related and representation-related aspects (e.g., whether the content area is sufficiently represented by the tasks) must also be taken into account when assessing the fit.

The conceptual and design aspects described provide an overview of the key conditions that need to be taken into account when designing examination arrangements, particularly for written but also other examination forms. These aspects are also decisive in the instruction of ChatGPT for the generation of corresponding examination elements as well as in the evaluation of the elements generated by ChatGPT. The following section presents how to proceed when instructing or prompting ChatGPT for the generation of learning outcomes, examination tasks and assessment schemes for specific learning objectives.

4. Generation of a competence-oriented examination arrangement with the help of ChatGPT

4.1 Generation of learning outcomes

Teachers at universities—be they research assistants, academic assistants, professors or lecturers—often have little training in didactics and instructional psychology—with a few exceptions. However, they usually have a high level of professional and academic expertise in the subject matter they are teaching. When planning courses and teaching units, including the examination arrangements required in this context, they often approach this task from a subject-systematic, content-oriented perspective or are guided by traditional concepts of teaching design. It is therefore often difficult for them to determine learning objectives and the learning outcomes to be achieved for their courses and to formulate them in a didactically appropriate way. This difficulty often results in learning objectives that only vaguely represent what is to be achieved. Or they are too knowledge-heavy in their objectives and do not adequately describe which cognitive and non-cognitive skills are to be acquired in the course. It is difficult to derive learning activities and examination requirements in an appropriate form on the basis of such deficient learning objectives respective learning outcomes.

In this context, a generative, AI-supported language model such as ChatGPT can provide helpful support in order to partially take on the task of formulating adequate learning outcomes for certain learning domains. In particular, ChatGPT is able to take into account the conceptual and linguistic aspects described in chapter 3. The more precisely and skillfully certain aspects of prompting are considered, the better the system is able to do this. Therefore, various aspects and levels should be addressed when generating learning outcomes for a course with the help of ChatGPT:

- The assignment for ChatGPT can be formulated as follows, for example: "Formulate learning outcomes for [specification of the overarching topic] with the learning content [keywords of the learning content and aspects to be covered]."

- Furthermore, the target group of the course should be stated (e.g., Bachelor's students of Computer Science in the 3rd semester) and how the course or learning unit is integrated into the curriculum (e.g., the course builds on the introductory lecture of informatics)

- Another crucial aspect of prompting is to specify the taxonomic system according to which the learning outcomes are to be developed (e.g., "Base your formulation on the taxonomy of Anderson and Krathwohl."). It is also advisable to specify the exact learning objective levels that you want to address in the course (e.g., understanding, applying, analyzing) or to specify each level individually using a separate prompt.

- It should also be specified what the overarching qualification objective of the course or teaching unit is (e.g., ability to carry out requirements analyses for specific issues) and also which format the course is based on (e.g., lecture, seminar, tutorium, project seminar) and which didactic principle or approach the course is based on (e.g., problem-oriented didactics).

- Didactic planning usually requires the consideration of different levels of planning, which refer to different levels of concretization of the instructional approach; at least two, if not three or more levels are to be distinguished (e.g., program, module, and course level). These different levels of planning also refer to different levels of goal formulation. For the prompting of learning outcomes, this means that these different levels should be addressed by specifying the respective curricular level.

- Last but not least, it may be useful to specify which additional aspects should be taken into account when formulating the learning outcomes (e.g., explain the why and wherefore of the learning outcomes).

Two examples of generating learning outcomes using ChatGPT are presented in **Tables 1** and 2, which take the aforementioned aspects of prompting into account in

The request for generating learning outcomes for this course unit was: "Formulate learning outcomes for a lecture with 10 sessions on the topic of personnel development based on personnel psychology with the following learning contents: definitions and objects of personnel development, methods for determining learning and training needs, approaches to promoting different areas of competence, evaluation of personnel development and approaches to enhance transfer. Participants in the lecture are Bachelor of Psychology students in their 4th semester. The overarching qualification objective of the lecture is to familiarize students with personnel psychology concepts and to enable them to develop solutions for practical problems in corporate personnel development on the basis of scientific concepts and findings. When formulating the learning outcomes, use Anderson and Krathwohl's learning objective taxonomy as a guide." ChatGPT provided me with the following answer (this is limited to a selection of learning outcomes for reasons of space): "After completing the teaching unit 'Psychology-based Personnel Development', students should be able to achieve the following learning outcomes based on Anderson and Krathwohl's learning objective taxonomy: 3. Applying: • Students will be able to apply appropriate methods to determine learning and training needs in specific organizational contexts. • Students can apply various approaches to promote different areas of competence in practice. 4. Analyzing: • Students can analyze the evaluation of personnel development and select suitable evaluation methods. • Students can analyze approaches to transfer enhancement and assess their effectiveness. 5. Evaluating: • Students can evaluate the effectiveness of personnel development measures and make suggestions for improvement. • Students can evaluate the effectiveness of transfer enhancement approaches and suggest alternative approaches. These learning outcomes should ensure that students gain a sound understanding of personnel development based on personnel psychology and are able to apply and evaluate this knowledge in practical situations."

Table 1.
Generation of learning outcomes for a lecture in personnel psychology by ChatGPT.

In a second trial the query was focused on a specific learning unit (approaches to enhance transfer in company training measures): Inquiry: "Formulate learning outcomes at the Understand, Apply, Analyze, Evaluate and Create learning objective levels of Anderson and Krathwohl's learning objective taxonomy for the teaching unit 'Approaches to transfer enhancement for in-company training courses'. Approaches, influencing factors and methods of transfer enhancement in occupational psychology research are discussed with examples of operational application." ChatGPT provided me with the following answer (presentation is limited to excerpts): • "Understanding: Participants will understand the basic concepts and theories of transfer enhancement. They will learn about the various approaches to promoting the transfer of training content into workplace practice and understand their importance to performance development and organizational success. • Apply: Participants will be able to apply the approaches and methods of transfer enhancement they have learned to specific in-company training programs. They will be able to identify suitable measures to promote transfer and integrate these into the design and implementation of training courses. • Analyze: Participants will be able to analyze the factors influencing the transfer of training content into operational practice. They will understand the individual, organizational and contextual factors that can influence the transfer process and will be able to evaluate these and derive appropriate measures to optimize transfer."

Table 2.
Generation of learning outcomes for a learning unit of the lecture by ChatGPT.

different ways. First, an example of generating learning outcomes with the help of ChatGPT for a lecture covering topics of "Psychology based personnel development" for psychology students in the 4th semester of their bachelor's degree program and second learning outcomes of a lecture unit concerning the transfer of learning respective training results is presented:

*Comments on the first example (**Table 1**):* The learning outcomes generated by ChatGPT are consistently based on the cognitive process dimension of Anderson and Krathwohl's taxonomy so that at least two learning outcomes are formulated for each of the learning objective levels. However, the knowledge dimension of the taxonomy is not taken into account. It is therefore not addressed which type of knowledge (declarative, conceptual, procedural or metacognitive) is to be acquired. It should also be noted that the learning objectives do not systematically refer to the various learning content areas addressed in the prompt. The formulated outcomes therefore only incompletely cover the breadth of the mentioned learning contents. However, the generated learning outcomes specify meaningful learning requirements of the HR development domain. They are therefore rather exemplary suggestions for relevant learning outcomes of this domain, although this is not made clear by the system. Overall, it should be noted that the specification of the taxonomy leads to a comprehensible addressing of cognitive learning requirements and relevant learning content in relation to the relevant knowledge and action domain.

*Comments on the second example (**Table 2**):* The first thing that is noticeable here with regard to the individual learning outcomes is that they are formulated more extensively than the learning outcomes for the overarching subject area. In addition to an introductory, rather general description of the learning outcome at the respective level, further sentences are used to specify what characterizes the cognitive activity or action at this learning objective level and with regard to which aspects of the action the learners should be enabled. Overall, it can be stated for this example that the three learning outcomes address important aspects of the respective learning objective level for the addressed content area. Also, the cognitive requirements of the content area are better and more completely covered than in the previous example. However, no reference is made to specific concepts of transfer enhancement. These may still need to be specified.

In both cases, there is a need to revise the learning outcomes generated by ChatGPT. The language model is therefore not able to provide readymade solutions for the formulation of learning outcomes for a course or teaching unit. However, the system provides useful suggestions as to the form in which the learning outcomes should be formulated and which content aspects should be addressed in the learning outcomes. It is also made clear how the learning objective levels differ in terms of specific learning requirements. In my opinion, this creates a good basis for further revision.

Therefore, in the next step you should revise the generated learning outcomes yourself (possibly in collaboration with colleagues or as part of an expert consultation). In a first step, this includes analyzing the learning outcomes generated for a specific case in terms of their strengths and deficits, in a similar way to that described in the comments. In further steps, depending on the initial situation, either minor additions or extensions (e.g., with regard to the mentioned concepts in the learning outcomes) should then be made. In some cases, however, it may also be necessary to formulate additional learning outcomes with regard to subject-specific content not yet covered or to significantly expand the description of the learning outcomes in order to describe more specifically which aspects of acting and thinking the learners should be enabled to do.

Furthermore, it can also be recommended that ChatGPT should be used to optimize the learning outcomes—for example, through more limited and more specific prompts. For example, it is advisable to limit yourself to only one learning objective level in order to achieve a greater breadth of content in the generated learning outcomes. It is also possible to provide ChatGPT with a best practice example of a learning outcome for the learning domain and ask the system to generate learning outcomes for other content areas or learning objective levels in a similar way.

4.2 Generating examination tasks on the basis of predefined learning outcomes

On the one hand, examination tasks should cover the relevant learning area broadly enough in terms of subject matter content in order to test a representative cross-section of knowledge and skills in a domain. On the other hand, the tasks should address the cognitive and non-cognitive skills to be learned in the course, for example, with regard to different requirement levels. While most teachers can still meet the first requirement with relative certainty due to their subject matter expertise, many teachers are often unsure with regard to the second condition. That means they are unsure whether they are addressing more demanding learning objective levels appropriately with their examination tasks. This uncertainty or, in some cases, missing knowledge of how certain cognitive learning requirements can be specifically addressed with the help of certain tasks leads teachers to focus only on the requirements that they can confidently implement and assess (often tasks that only require knowledge reproduction and comprehension). As a result, certain requirement areas are not sufficiently addressed in the examination. Or tasks are constructed that lead to students being under- or overchallenged or that pose demands that are irrelevant in relation to the skills to be learned (e.g., by disproportionately emphasizing the fulfillment of formal requirements).

ChatGPT can also provide support with regard to the design of examination tasks. By specifying and prompting the relevant learning outcomes, ChatGPT is able to construct examination tasks that represent the relevant cognitive requirements of the addressed learning objective levels in a valid and appropriate manner. However, when generating examination tasks for a specific teaching area with the help of ChatGPT, certain aspects should be taken into account with regard to effective prompting:

- The assignment for ChatGPT can be formulated for example as follows: Suggest to me [number, e.g., 2] examination tasks for a [name the desired task or examination format; e.g., written exam] on the following learning outcome: [original formulation of the learning outcome]

- Furthermore, the target group for the examination should be specified and the prerequisites that the candidates have for the examination (e.g., whether they already possess skills to apply certain concepts or in dealing with certain tasks through exercises).

- It is also advisable to provide information on the framework conditions for the examination (e.g., the average amount of time available for a task or additional material/devices available for working on the task—e.g., reference books or calculators).

- Further information that may be relevant for the design of examination tasks relates to the relevant contextual reference or the degree of realism to be taken into account. Or it should be specified in which way certain aspects should be considered concerning the difficulty of the task (e.g., requesting a justification).

Three examples of how ChatGPT generates examination tasks are presented in **Table 3** to illustrate the potential that the bot has for creating examination tasks.

*Comments on the examples in **Table 3**:* In my opinion, the suggestions generated by ChatGPT for examination tasks on the topic of transfer enhancement address the cognitive requirements relevant to the respective learning objective level adequately. The tasks could certainly be adopted in this form. Nevertheless, they are more likely to be assessed as draft samples, as they are formulated rather vaguely with regard to certain aspects. These aspects that are not sufficiently addressed in the presented task examples I will discuss below for each of the three tasks:

- In the task for the Understanding level, it would be useful to specify the approach for classifying or systematizing the basic concepts to which the examinees should refer here, as various systematization approaches exist for this.

- In my opinion, the task for the Applying level should not only describe the measures for enhancing transfer but also explain why these are relevant for the stated training topic for enhancing transfer. If necessary, the overarching objective of

1. Request: "Suggest examination tasks for a written exam on the following learning outcome (learning objective level Understanding): Participants will understand the basic concepts and theories of transfer enhancement in workplace training programs. They will become familiar with the various approaches for promoting the transfer of training content into operational practice and understand their significance for performance development and organizational success."

   Answer: "Examination task (exemplary): Explain the basic concepts of transfer enhancement in in-company training measures and give an example of their application in each case. Also describe how these concepts can contribute to performance development and organizational success."

2. Request: "Suggest examination tasks for a written exam on the following learning outcome (learning objective level Apply): Participants will be able to apply the learned approaches and methods of transfer enhancement in training measures to specific in-company training programs. They will be able to identify suitable measures to promote transfer and integrate these into the design and implementation of training courses."

   Answer: "Examination task (exemplary): You are working as a trainer for a company and are to carry out a training measure on the topic of 'Effective communication in a team'. Describe three specific measures that you would take to promote the transfer of what you have learned into the participants' everyday working life. Explain how these measures can be integrated into the design and delivery of the training."

3. Request: "Suggest examination tasks for a written exam on the following learning outcome (learning objective level Analyze): Participants will be able to analyze the factors influencing the transfer of training content into operational practice. They will understand the organizational, contextual and individual factors that can influence the transfer process and be able to evaluate these and derive appropriate measures to optimize transfer."

   Answer: "Examination task (exemplary): Consider the organizational context and contextual factors that can influence the transfer of training content into workplace practice. Describe three organizational or contextual factors and analyze how they can influence the transfer process. In addition, evaluate what measures could be taken to optimize transfer in relation to these factors."

Table 3.
Generation of examination tasks based on the learning outcomes by ChatGPT.

the training measure should also be stated so that it is clear what the communication skills are specifically needed for or are to be improved.

- In the task for the Analyzing level, the task is somewhat general and is aimed rather at requirements of the knowledge reproduction and comprehension level. I would therefore make the application context more concrete or contextualize it with a specific training scenario so that the examinees are required to perform analytical tasks to a greater extent. In my opinion, the additional instruction in this assignment ("Explain how these measures can be integrated into the design and delivery of the training.") also requires a correction: The examinee should first be instructed to create measures to adapt and optimize the context factors and then it should be asked for reasons why they lead to an improved transfer.

The comments make clear that there are several needs to optimize the task suggestions by ChatGPT. At best, the suggestions generated can be used as a basis for further elaboration. In my opinion, however, they are still useful as a source of ideas. They also substantially facilitate or support the conceptual work designing examination tasks. When generating examination tasks, ChatGPT draws relatively systematically on knowledge about suitable task formats for certain learning objective levels and contextualizes these in relation to the subject domains addressed. The bot thus provides valuable support in the construction of tasks tailored to learning objectives, as the consistent design of such examination tasks is often difficult to achieve on one's own or without appropriate assistance, especially for teachers with little didactic experience.

However, it must also be stated that further revision steps are required in order to achieve a usable result. This requires also an analysis of which aspects already fulfill the proposed tasks, but where there are still deficits and a need for changes or additions. This can relate to technical aspects to make the task more specific and to better focus on the intended cognitive requirements or performance. Additionally, information should be provided in the task instructions with regard to the criteria used to assess the answers to the task. This gives the examinees information on what they need to pay attention to when answering the question or task and improves the clarity of the examination task. Furthermore, ChatGPT can be used to further develop and optimize examination tasks. For this purpose, it is recommended to provide the bot with the previous version of the task and ask it to specify certain aspects (e.g., generation of a suitable application scenario). The bot can also be used to generate additional tasks for a specific learning outcome in order to increase the covered breadth of content.

4.3 Generation of assessment criteria and schemes for examination tasks

The assessment of examination performance should be as objective, reliable, valid, transparent and fair as possible. In addition to examination tasks that are geared toward learning objectives and the guarantee of standardized and fair implementation conditions, assessment criteria also essentially contribute to the examination being experienced as appropriate and fair and that the examination fulfills the above-mentioned quality criteria. Assessment criteria and assessment schemes support examiners in ensuring that the assessment of examination performance is not carried out arbitrarily or intuitively, but on the basis of comprehensible and content-appropriate standards, and that quality criteria of objectivity, reliability, validity and fairness are fulfilled to a sufficient extent. If this is not the case, the result of the examination is not only questionable and distorted by uncontrolled assessment processes, but also legally contestable. Assessment criteria for evaluating examination performance—not only for written examinations—are therefore essential. They also provide examiners with certainty when evaluating examination performance, increase transparency with regard to how the examination result is arrived at and are helpful in providing detailed feedback on the strengths and weaknesses of performance. Incidentally, this applies not only to inexperienced examiners but also to experienced and didactically skilled examiners. In my opinion, assessment criteria for the tasks used in an examination that are appropriate in terms of content and requirements are therefore an indispensable part of the examination arrangement. For various reasons, appropriate assessment criteria should also be developed during the construction of the examination tasks, that is, parallel to or immediately after the task construction. On the one hand, they are necessary for a solid task design in order to avoid errors or ambiguities in the formulation of the task (e.g., to check whether the task is sufficiently focused on the skills and knowledge to be tested and whether this is also clearly stated in the task instructions for the examinees). Since the assessment criteria should also refer to a large extent to the learning outcome and the corresponding learning objective level as well as the knowledge required for this, they also ensure that the examination arrangement fulfills the principle of constructive alignment. In addition, the assessment criteria can be used to transparently communicate to examinees what is important in the examination and what standards their examination performance should meet.

In my opinion, ChatGPT can also effectively support examiners in this step. Various application tests show that the system is able to develop meaningful suggestions for determining and formulating assessment criteria and schemes for certain examination tasks. The bot is oriented toward common assessment dimensions in examinations and is also able to take into account specific methodological concepts of assessment schemes (e.g., the rubrics concept). When generating assessment criteria for a specific examination task or even a class/group of tasks with the help of ChatGPT, certain aspects of prompting should be taken into account:

- The core prompt for ChatGPT can be formulated, for example, as follows: "How should the answers to the following examination task be scored: [state the original wording of the exam task]. Also, give me a detailed evaluation scheme for scoring the performance."

- In addition to stating the original task, it may be useful to state the learning outcome on which the task is based, which is also an important reference point for generating the assessment criteria.

- Additionally, it may be necessary to specify the type of assessment scheme required (e.g., whether this should take the form of a rubric). As described above, rubrics consist of an evaluation matrix in which the evaluation criteria relevant to a specific task or examination performance are named and defined on one axis. On the second axis, the type and number of quality levels for evaluating the individual criteria or dimensions are also specified (e.g., (1) good to very good, (2) satisfactory and (3) insufficient fulfillment of criteria). The cells of the matrix then describe notes and aspects for evaluating the respective quality level for each criterion. Another tested evaluation scheme refers to following aspects:

Evaluation category, definition/characterization of the level, anchor example, points. This scheme is particularly useful for tasks in which a distinction is made only between "correct" and "incorrect" or "partially correct." In this way, a well-founded decision or assignment can be made on the basis of the definitions of each level and the anchor examples given in each case.

In the following (see **Tables 4** and **5**), two examples are presented in which ChatGPT was asked to generate suitable assessment criteria for two specific examination tasks. These in turn relate to the teaching unit "Enhancing transfer in in-company training measures" and include examination tasks for the learning objective levels "Understanding" and "Evaluating."

*Comments on the first example (**Table 4**):* The assessment criteria for the "Understanding" level task refers to the various subtasks and proposes a scoring

Question: "How should the answers to the following examination task be assessed? Explain the basic concepts of transfer enhancement in in-company training measures and give an example of their application in each case. Also describe how these concepts can contribute to performance development and organizational success. Give me a detailed assessment scheme for scoring performance."
Answer: "The following criteria and a detailed scoring scheme could be used to score performance when evaluating responses to this assessment task:

1. Understanding of the basic concepts of transfer enhancement in workplace training programs (e.g., repetition, practical application, feedback, transfer climate):
   - Complete and accurate explanation of concepts: 3 points
   - Partially correct explanation of the concepts: 2 points
   - Superficial or incomplete explanation of concepts: 1 point
   - Missing or incorrect explanation of concepts: 0 points
2. Exemplary application of the concepts:
   - Clear and relevant examples that illustrate the application of the concepts: 3 points
   - Partially appropriate examples that partially illustrate the application of the concepts: 2 points
   - Unclear or inadequate examples that do not clearly illustrate the application of the concepts: 1 point
   - Missing or incorrect examples: 0 points
3. Description of the contribution of the concepts to performance development and organizational success:
   - Clear and comprehensive connection between the concepts and their contribution to performance development and organizational success: 3 points
   - Partially correct connection between the concepts and their contribution to performance development and organizational success: 2 points
   - Superficial or incomplete description of the link: 1 point
   - Missing or incorrect description of the connection: 0 points
4. The evaluation scheme could then add up the points for each category and determine an overall score for the task. For example, a possible scale could be as follows:
   - 9–10 points: Very good performance
   - 7–8 points: Good performance
   - 5–6 points: Adequate performance
   - 3–4 points: Poor performance
   - 0–2 points: Insufficient performance"

Table 4.
Generation of assessment criteria by ChatGPT for the task at the understanding level.

Question: "How should the answers to the following assessment task be evaluated: Critically analyze the effectiveness of a particular approach to transfer enhancement in workplace training (e.g., collegial mentoring) and discuss its advantages and disadvantages. Assess in which workplace contexts this approach might work best and what factors might influence its effectiveness. Provide recommendations on how the approach could be improved to facilitate the transfer of training content into practice. Give me a detailed assessment scheme for scoring performance."
Answer: "A possible scheme for scoring performance in response to this examination task could be as follows:

1. Analyze the effectiveness of the transfer enhancement approach (max. 10 points):
   - Critically evaluate the effectiveness of the approach (e.g., using studies or case studies)
   - Consideration of advantages and disadvantages of the approach
   - Understanding of the underlying principles and mechanisms of the approach
2. Discussion of the operational contexts and influencing factors (max. 10 points):
   - Identification of operational contexts in which the approach might work best
   - Analysis of factors that could influence the effectiveness of the approach (e.g., corporate culture, employee motivation)
3. Recommendations for improving the approach (max. 5 points):
   - Creative and realistic suggestions for optimizing the approach
   - Consideration of the challenges and potentials in the implementation of the recommendations
4. Structure and clarity of the argumentation (max. 5 points):
   - Logical and well-structured presentation of the arguments
   - Comprehensible and precise formulation of the statements
5. References and use of specialist terminology (max. 5 points):
   - Citation of relevant sources to support the statements
   - Use of specialist terminology appropriately and correctly
6. The maximum score is therefore 35 points. The assessment should be based on these criteria, with particular consideration given to the quality of the argument, the depth of the analysis and the originality of the recommendations."

Table 5.
Generation of assessment criteria by ChatGPT for the task at the evaluating level.

scheme for different levels of response quality for each subtask. However, the levels are characterized in relatively general terms (from "complete and precise" to "missing or incorrect explanation of concepts"), making it difficult to differentiate between the various levels or to assign a specific answer to a level. If this assessment scheme is to be used, it is therefore advisable to add more specific information to assign different answers and the delimitation of scoring levels. In particular, the insertion of anchor examples, that is, typical response examples for the respective level, could be helpful here. Also, the recommendation for the grading scheme is relatively generic and can or must also be adapted to specific application contexts (e.g., with regard to the point ranges selected for the respective grade).

*Comments on the second example (**Table 5**):* The assessment criteria for the task at the "Evaluate" level refer on the one hand to the various subtasks of the task and on the other to overarching aspects of the quality of the answer ("Structure and clarity of argumentation" and "Sources cited and use of specialist knowledge"). However, no levels of the response quality are described for the assessment of the individual subtasks, but rather specific criteria to which reference should be made in the response (e.g., "understanding of the underlying principles and mechanisms of the approach" for the first subtask). In my opinion, this achieves a more detailed specification of the assessment criteria than in the first case. However, with regard to a concrete application of this assessment scheme, an additional concretization of these specific criteria is recommended (e.g., through sample solutions or anchor examples) in order to obtain even clearer indications for the evaluation of individual answers. Also, the scoring scheme should be defined more clearly. The indication of the maximum possible point totals for each criteria is not very comprehensible and primarily illustrates a different weighting of the subtasks. Overall, ChatGPT provides useful information on structuring the assessment and determining relevant assessment criteria and aspects, which, however, require further elaboration as described.

My comments on the two examples have already illustrated the potential, but also the need for revision, of the suggestions generated by ChatGPT for the assessment of examination tasks. For this step, it can also be stated that ChatGPT is able to generate useful suggestions for the assessment of examination performance for defined examination tasks. In this case too, ChatGPT's suggestions have the status of useful ideas and hints rather than readymade solutions. Suggestions for the independent revision of such proposals have already been mentioned and relate primarily to the specification of the assessment criteria through anchor examples and sample solutions as well as the clearer definition of the aspects for which points are to be awarded and how the points can be transformed into grades. Chat GPT can also be used in this context for the further development of assessment schemes, in particular by specifying certain forms of assessment schemes (e.g., rubrics or true/false assessment schemes).

Finally, in a further test step, it is advisable to analyze and evaluate the constructive alignment of the various elements. On this basis, adjustments can then be made between the elements and the fit between learning outcome, examination tasks and assessment criteria as well as the associated teaching/learning activities can be improved. This coordination of the various elements is not only important for reasons of successful didactic coherence, but also has an effect on the quality of the learning process and has been proven to promote motivation to learn. Various studies have shown that learning arrangements that follow the constructive alignment principle result in or effectively promote the use of learning strategies for more in-depth learning and greater learning effort as well as the achievement of more challenging learning goals [35–37]. However, attempts to determine the extent to which ChatGPT is also able to support teachers in checking constructive alignment proved to be less successful. Unfortunately, the language model was not able to make appropriate assessments of the fit between learning outcome, examination task and assessment criteria in relation to this didactic principle.

5. Conclusion and outlook

In my opinion, the trial to use ChatGPT for the design of written examinations, in particular for the design of learning outcome and competence-oriented examination arrangements, has made clear that ChatGPT can indeed provide support for this task and the process in various ways. For example, the AI tool can be used to generate drafts and ideas for the appropriate formulation of learning outcomes, the learning outcome-oriented design of examination tasks and the construction of suitable assessment criteria and schemes. However, it also became clear that ChatGPT does not provide readymade solutions, but at best drafts and ideas that require further elaboration. It also became clear that ChatGPT creates errors and inaccuracies in its proposals, which need to be identified and corrected through careful and critical analysis. In this context, it should be pointed out once again that the generative language models are not trained to produce true and correct solutions, but rather probable word sequences. The fact that these word sequences make sense and are true is to a certain extent only a by-product [1]. Although between 60% and 80% of the answers are correct in this way, conversely this means that 20–40% of the answers can also be or are incorrect [13]. ChatGPT therefore repeatedly presents answers that are partially incorrect or even invented. The providers and developers of generative language models will certainly work on increasing this rate of correct answers, but the fundamental problem will remain permanently, as it is inherent in the design of the language model. For the application context under consideration, this means that ChatGPT's proposals for the formulation of learning outcomes, the design of examination tasks and assessment criteria/schemes must be reviewed and revised in any case—it cannot be repeated often enough.

Finally, it should be pointed out that ChatGPT can be used to generate not only open response formats for exams, but also closed task formats or examination tasks. Reference should be made here to the presentations by Patel [12] or Workera [38], which show that ChatGPT can also be used to generate useful suggestions for multiple choice questions, cloze texts, true/false questions, matching tasks or open short questions for all possible subject areas. As described above, information is required not only on the content to be tested, but also on the target group, the level of difficulty or the type of cognitive demand, the style of wording, etc. when prompting. These descriptions also address what needs to be done and considered with regard to the revision of the proposals (in particular [38]). ChatGPT can also be used to check answers to exam questions. Examples of this can be found in Herft [39], where it is particularly pointed out that these reviews can be used for individual feedback to students regarding their performance in formative (exercise) tasks. Potential for change through ChatGPT in the area of examinations also arises from the fact that this tool can be used specifically when writing unsupervised examination formats (such as seminar papers and term papers) and when writing academic theses. ChatGPT thus becomes a partner in the writing process (in the sense of hybrid writing) and is therefore not only a relief for certain writing requirements, but also enables further learning processes through its support. However, in order to allow such hybrid processes in written examinations, clear framework conditions are also required that specify the form in which the bot may be used for the creation of such texts and what must be considered (e.g., that all passages in which ChatGPT or another generative AI was used in writing must be marked).

Let me resume with a statement of an experienced colleague in this field [1]: "I assume that sooner rather than later ChatGPT will become a tool that we use as a matter of course - especially at universities. We should understand this tool (with all its strengths and weaknesses), use it confidently and reflectively and be vigilant about the risks and side effects."

Author details

Niclas Schaper
Faculty of Cultural Sciences, University of Paderborn, Paderborn, Germany

*Address all correspondence to: niclas.schaper@upb.de

References

[1] Fleischmann A. ChatGPT in der Hochschullehre. Wie künstliche Intelligenz uns unterstützen und herausfordern wird. In: Berendt B, Fleischmann A, Schaper N, Sczcyrba B, Wiemer M, Wildt J, editors. Neues Handbuch Hochschullehre A 1.30. Ausgabe 110 ed. Berlin: DUZ; 2023. pp. 13-56

[2] Gamage K, Dehideniya SCP, Xu Z, Zu XT. ChatGPT and higher education assessments: More opportunities than concerns? Journal of Applied Learning and Teaching. 2023;**6**(2):1-12

[3] Kalnitski M. Bachelorarbeit mit KI Schreiben? TiTok vom 21.12.2022. 2023. Verfügbar unter: https://www.tiktok.com/@botscaler/video/719688580804087046

[4] Tangermann V. College Student caught Submitting Paper Using ChatGPT. 2023. Available from: https://futurism.com/college-student-caught-writing-paper.chatgpt

[5] Pollfish Surveys. Nearly 1 in 3 College Students Have Used ChtaGPT on Written Assignments. 2023. Available from: https://www.inteligent.c0m/nearly-1-in-3-college-students-have-used-chatgpt-on-written-assignments

[6] Jouanne-Diedrich Hv. Professor Gegen KI! Kann ChatGPT Meine Klausur Lösen? Video. 2023. Verfügbar unter: https://owww.youtube.com/watch?v=SS24iTTOn7o

[7] Kipping D. How does ChatGPT Do on a College Level Astrophysics Exam? Video. 2023. Verfügbar unter: https://www.youtube.com/watch?v=KOcmmKPKlp4

[8] Fleck T. Prüfungsrechtliche Fragenzu ChatGPT. Erlangen: Handreichung der Stabsstelle IT-Recht der bayerischen Universitäten; 2023. Verfügbar unter: https://www.rz.uni-wuerzburg.de/fileadmin/42010000/2023/ChatGPT_und_Prüfungsrecht.pdf

[9] Sullivan M, Kelley A, McLaughlin P. ChatGPT in higher education: Considerations for academic integrity and student learning. Journal of Applied Learning and Teaching. 2023;**6**(1):1-10

[10] Solis T. Die ChatGPT-Richtlinien der 100 Größten Deutschen Universitäten. Amsterdam: Scribbr; 2023. Verfügbar unter: https://www.scribbr.de/ki-tools-nutzen/chatgpt-universitaere-richtlinien/

[11] Rasul T et al. The role of ChatGPT in higher education: Benefits, challenges, and future research directions. Journal of Applied Learning and Teaching. 2023;**6**(1):1-16

[12] Patel N. Using ChatGPT Prompts to Generate Assessment Questions. 2023. Available from: https://www.playpowerlabs.com/post/using-chatgpt-prompts-to-generate-effective-formative-assessment-questions

[13] Gieselmann H. Die 80-Prozent-Maschinen. Warum KI-Sprachmodelle weiterhin Fehler machen und was das für den produktiven Einsatz bedeutet. c't, Heft. 2023;**21**:17-19

[14] Caulfield J. Is ChatGPT Trustworthy? Accuracy Tested. 2023. Available from: https://www.scribbr.com/ai-tools/is-chatgpt-trustworthy/

[15] Ekin S. Pprompt enginnering for ChatGPT. A quick guide to techniques,

tips, and best practices. 2023. DOI: 10.36227/techrxiv.22683919.v1

[16] Gptbot.io. Mastering ChatGPT: How to Craft Effective Prompts. 2023. Available from: https://gptbot.io/master-chatgpt-prompting-techniques-guide/

[17] Bager J. Instruieren und verifizieren. Tipps und Tools, mit denen Sie Sprachmodelle produktiv nutzen. c't, Heft. 2023;**21**:26-19

[18] Coursera. How To Write ChatGPT Prompts: Your 2024 Guide. 2024. Available from: https://www.coursera.org/articles/how-to-write-chatgpt-prompts

[19] Crabtree M. 10 of The Best ChatGPT Plugins to Get The Most From AI in 2023. 2023. Available from: https://www.datacamp.com/blog/best-chat-gpt-plugins

[20] Schaper N, Soyka C. Kompetenzorientiertes Prüfen. Grundlagen, präsenz- und onlinegestützte Formate, Bewertung und Rückmeldung kompetenzorientierter Prüfungsleistungen. In: Neues Handbuch Hochschullehre, H 6.8. Ausgabe 100 ed. Berlin: DUZ Verlags- und Medienhaus; 2021. pp. 95-122

[21] Wesselink R, Biemans H, Gulikers J, Mulder M. Models and principles for designing competence-based curricula, teaching, learning and assessment. In: Mulder M, editor. Competence-Based Vocational and Professional Education. London: Springer; 2017. pp. 533-553

[22] Biggs J, Tang K. Teaching for Quality Learning at University. 4th ed. Maidenhead: Oxford University Press; 2011

[23] Pointz TS. Learning outcomes: What are they? Who defines them? When and where are they defined? Educational Assessment, Evaluation and Accountability. 2010;**22**:119-137

[24] Anderson LW, Krathwohl DR. A Taxonomy for Learning, Teaching and Assessing: A Revision of Bloom's Taxonomy of Educational Objectives: Complete Edition. New York: Longman; 2001

[25] Gamage KAA, Pradeep RGG, de Silva EK. Rethinking assessment: The future of examinations in higher education. Sustainability. 2022;**14**:3552. DOI: 10.3390/su14063552

[26] Schaper N. Prüfen in der Hochschullehre. In: Kordts-Freudinger R, Schaper N, Scholkmann A, Szczyrba B, editors. Handbuch Hochschuldidaktik. Bielefeld: wbv/utb; 2021. pp. 87-102

[27] Klapproth F. Von den Lehrzielen zur schriftlichen Prüfung. Ein Leitfaden für Lehrende der Psychologie. Göttingen: Hogrefe; 2023

[28] Schaper N. Konzeption schriftlicher Prüfungsaufgaben auf Basis der Lernzieltaxonomie von Anderson & Krathwohl. In: Schaper N, Vogelsang C, editors. Prüfen in der Hochschullehre—Konzeptionell Fundiert, Evidenzbasiert und Qualitätsorientiert. Köln: FIHB-Reihe, TH Köln; 2024. pp. 58-72

[29] Bloxham S, Boyd P. Developing Assessment in Higher Education: A Practical Guide. New York, NY: Open University Press, McGraw-Hill; 2007

[30] Metzger C, Nüesch C. Fair prüfen. Ein Qualitätsleitfaden für Prüfende an Hochschulen, Hochschuldidaktische Schriften. Vol. 6. StaGallen: Institut für Wirtschaftspädagogik, Universität St. Gallen; 2024

[31] Glion Institute of Higher Education. How to Write Asssessment Criteria? 2023.

Available from: https://library.glion.edu/celt/how-to-write-assessment-criteria/

[32] Schlomske-Bodenstein N, Strasser A, Schindler C, Schulz F. Handreichungen zum Kompetenzorientierten Prüfen. 2018. Verfügbar unter: https://www.lehren.tum.de/fileadmin/w00bmo/www/Downloads/Themen/Pruefungen/Handreichungen_kompetenzorientiert_Pruefen.pdf

[33] Stevens DD, Levi A. Introduction to Rubrics: An Assessment Tool to Save Grading Time, Convey Effective Feedback, and Promote Student Learning. Sterling, VA: Stylus Publishing LLC; 2012

[34] Universität Zürich—Arbeitsstelle für Hochschuldidaktik. Taxonomie-Matrix zur Analyse und Selbstevaluation von Hochschullehre (TAMAS). 2010. Verfügbar unter: https://www.weiterbildung.uzh.ch/dam/jcr:ffffffff-9a08-8cca-0000-000037b2e4ce/DU_Tamas_def.pdf

[35] Biggs J. Constructive alignment in university teaching. HERDSA Review of Higher Education. 2014;**1**:5-22. Available from: https://www.herdsa.org.au/herdsa-review-higher-education-vol-1/5-22

[36] Hailikari T, Virtanen V, Vesalainen M, Postareff L. Student perspectives on how different elements of constructive alignment support active learning. Active Learning in Higher Education. 2022;**23**(3):217-231. DOI: 10.1177/1469787421989160

[37] Raeburn P, Muldoon N, Bookallil C. Blended spaces, work-based learning and constructive alignment: Impacts on student engagement. In: Atkinson RJ, McBeath C, editors. Same Places, Different Spaces. Proceedings of the 26th Annual Ascilite International Conference. Auckland: Ascilite; 2009. pp. 820-831

[38] Workera. Making the Right Choices: How to Generate Outstanding Multiple-Choice Questions using ChatGPT. 2023. Verfügbar unter: https://workera.ai/blog/chatgpt-learning-assessments

[39] Herft A. Leitfaden für Lehrkräfte zur Nutzung von ChatGPT. Vortrag Gehalten am Pädagogischen Tag am Deutsch-Französischen Gymnasium in Saarbrücken am 13.02.2023. 2023. Verfügbar unter: https://material.rpi-virtuell.de/material/leitfaden-fuer-lehrkraefte-zur-nutzung-von-chatgpt/